Mathematical Methods for Physics

Mathematical Methods for Physics

Problems and Solutions

Farkhad G. Aliev

Antonio Lara

Thoroughly expanded and updated version of the book

Problemas resueltos de métodos matemáticos de la Física: Método de Fourier,

UAM Ediciones (2020)

JENNY STANFORD

PUBLISHING

Published by

Jenny Stanford Publishing Pte. Ltd.
101 Thomson Road
#06-01, United Square
Singapore 307591

Email: editorial@jennystanford.com
Web: www.jennystanford.com

British Library Cataloguing-in-Publication Data
A catalogue record for this book is available from the British Library.

Mathematical Methods for Physics: Problems and Solutions

ISBN 978-981-4968-71-3 (Hardcover)
ISBN 978-1-003-41088-1 (eBook)

Contents

Preface

Mathematical Methods for Physics is a course which describes mathematical tools that allow to solve partial differential equations (PDE), typical of physical problems. Despite a large number of textbooks on this topic, few explain in a detailed manner the process of solving the problems that typically arise in the context of physics. Another original feature of this book is the emphasis on the mathematical formulation of the problems, as well as the analysis of several alternative ways to solve them, and a graphical analysis of the results when appropriate.

The book uses teaching material from the mandatory course Mathematical Methods of Physics III at the Autonomous University of Madrid (Universidad Autonoma de Madrid - UAM). It presents substantially (about one-third) expanded and essentially updated version of the book previously published by the authors (in Spanish) in the "Collecion de estudios" series (Ediciones UAM, 2020). It can also be considered as a supplementary resource to another textbook (only in Spanish): *Métodos matemáticos de la Física: Método de Fourier*, by Arkadi P. Levanyuk and Andrés Cano. The current book contains the solution to several problems from this book.

The book starts with an introduction to the Fourier Method, analyzing simple problems as the vibrations of a harmonic oscillator. In this context, Green's functions are introduced in system of zero (oscillators) or one (strings) dimensions, since the search for these functions only represents a specialized method to find particular solutions to the non-homogeneous part of PDEs.

The rest of the chapters contain solved problems in Cartesian coordinates in one, two, and three dimensions, in polar, cylindrical, and spherical coordinates. Graphic illustrations and representations of the solutions (using MATLABs PDE Toolbox) are frequently introduced. Finally, the book shows the problems related to the Fourier transform operation. It includes several related theorems, such as the theorem of convolution, with practical applications. Furthermore, this book also shows the methods to apply the Fourier transform to the process of solving PDE, mostly in one dimension.

We would like to thank all the physics grade students of Universidad Autonoma de Madrid who have contributed to the book, especially those who helped finding typos in the draft of the Spanish version, particularly Cristina Viviente, Carolina Alvarez, Aitana Hurtado, Alejandro Blanco, Miguel Turad, Santiago Agui, Gonzalo Morras, and Javier Robledo. We would also like to express our gratitude to the comments and support shown by our colleagues of the Condensed Matter Physics Department at Universidad Autonoma de Madrid, José Vicente Álvarez, Guillermo Gómez Santos, Juan José Palacios, Arkadi Levanyuk, Sebastián Vieira and Raúl Villar during the preparation of the book.

We also acknowledge "UAM Ediciones" for granting permission to Jenny Stanford Publishing for using in this book, *Mathematical Methods for Physics: Problems and Solutions*, by Farkhad G. Aliev and Antonio Lara, the figures from the book *Problemas Resueltos de Métodos Matemáticos de la Física, Método de Fourier*, by Farkhad G. Aliev and Antonio Lara, edited by UAM Ediciones in 2019.

Finally, Farkhad Aliev acknowledges Tatiana Alieva for helping clarify several aspects of the Fourier transform and for her patience.

Farkhad G. Aliev and Antonio Lara

Chapter 1

Harmonic Oscillator and Green's Function

The Fourier method, in its simplest version, consists in expanding in harmonic functions (solutions of the harmonic oscillator) the solution of a differential equation. For this reason, we consider useful to begin this book with some simple examples of the solution of problems involving a harmonic oscillator. The application of initial conditions of some simple, as well as complicated, equations will be considered (as, for example, friction terms or non-homogeneous terms in the equation, such as external forces).

Using the solution of the oscillator problem, it will be useful to introduce the concept of Green's functions, solving some examples that explain the application of this concept, very useful in solving non-homogeneous problems. In the last chapter, about the application of the Fourier transform for solving partial differential equations, we will use some of the solutions obtained in this chapter, in particular those of an oscillator with friction.

In general, the obtained solutions for an oscillator will be frequently used as part of the solutions (in time and space) when dealing with problems up to three spatial dimensions, where the solutions will be expanded in sums of orthogonal functions, multiplied by coefficients or other non-orthogonal functions that satisfy the

Mathematical Methods for Physics: Problems and Solutions
Farkhad G. Aliev and Antonio Lara

ISBN 978-981-4968-71-3 (Hardcover), 978-1-003-41088-1 (eBook)
www.jennystanford.com

boundary conditions. We hope that these simple examples facilitate the understanding of the solutions of the problems in later chapters.

1.1 Damped Harmonic Oscillator

Consider a damped harmonic oscillator such that $\omega_0^2 < \eta/2m$ (consider also the case $\omega_0^2 > \eta/2m$), where ω_0 is the eigenfrequency of the oscillator, m is the mass and η is the friction constant of the oscillator moving in a viscous medium (take the friction force proportional to the oscillator velocity).

(a) Find the mass position as a function of time if at $t = 0$ it is at rest, but at a distance x_0 from its equilibrium position.
(b) Find the mass position as a function of time if at $t = 0$ it is at its equilibrium position, and is hit in such a way that it has a velocity v_0.
(c) Show that, for large enough dampings, these movements could have been deduced from the equation $\eta\dot{x} + kx = 0$. Explain why.

See also problem 1.4 from [1].

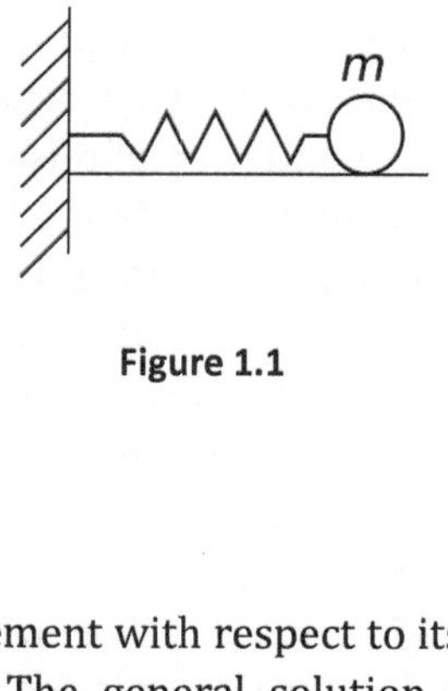

Figure 1.1

Solution:

(a) The mass displacement with respect to its equilibrium position will be called x. The general solution for $\omega_0^2 > \eta/2m$ is the (equation 1.24 from [1]):

$$x(t) = Ae^{-\gamma t}\sin(\Omega_0 t + \delta) \tag{1.1}$$

being: $\Omega_0 = \sqrt{\omega_0^2 - \gamma^2}$; $\omega_0^2 = \frac{k}{m}$; $\gamma^2 = \frac{\eta}{2m}$. From the zero initial velocity condition we have: $\left.\frac{\partial x}{\partial t}\right|_{t=0} = A(-\gamma)\sin(\delta) +$

$A\Omega_0 \cos(\delta) = 0$, where $\tan(\delta) = \frac{\Omega_0}{\gamma}$. From the condition of initial position we have: $x(0) = x_0 = A \sin\delta$, with $A = \frac{x_0}{\sin\delta}$. For the case $\omega_0^2 < \eta/2m$ we have equation (1.25) from [1]):

$$x(t) = C_1 e^{-\gamma_+ t} + C_2 e^{-\gamma_- t} \tag{1.2}$$

being $\gamma_\pm = \gamma \pm \sqrt{\gamma^2 - \omega_0^2}$. From the condition of zero initial velocity we have: $\left.\frac{\partial x}{\partial t}\right|_{t=0} = C_1(-\gamma_+) + C_2(-\gamma_-) = 0$, where $\frac{C_1}{C_2} = -(\frac{\gamma_-}{\gamma_+})$ From the condition of initial position: $x(0) = x_0 = C_1 + C_2$, where $C_1 = \frac{x_0}{\frac{\gamma_-}{\gamma_+} - 1}(\frac{\gamma_-}{\gamma_+})$; $C_2 = \frac{x_0}{1 - \frac{\gamma_-}{\gamma_+}}$

(b) In this case only the initial conditions change:
General solution for $\omega_0^2 > \eta/2m$ (equation 1.24 from [1]):

$$x(t) = Ae^{-\gamma t}\sin(\Omega_0 t + \delta) \tag{1.3}$$

From the zero initial displacement we have: $x(0) = 0 = A\sin(\delta)$. Then we have δ=0, and $x(t) = Ae^{-\gamma t}\sin(\Omega_0 t)$. From the condition of zero initial speed we have, using $\delta = 0$:

$$\left.\frac{\partial x}{\partial t}\right|_{t=0} = v_0 = A(-\gamma)\sin(0) + A\Omega_0\cos(0)$$

which yields $A = \frac{v_0}{\Omega_0}$, so that:

$$x(t) = \frac{v_0}{\Omega_0} e^{-\gamma t}\sin(\Omega_0 t) \tag{1.4}$$

The general solution for $\omega_0^2 < \eta/2m$ is (equation 1.25 from [1]):

$$x(t) = C_1 e^{-\gamma_+ t} + C_2 e^{-\gamma_- t} \tag{1.5}$$

From the condition of zero initial displacement: $x(0) = 0 = C_1 + C_2 \rightarrow C_1 = -C_2$

From the condition of zero initial velocity: $\left.\frac{dx}{dt}\right|_{t=0} = C_1(-\gamma_+) + C_2(-\gamma_-) = v_0$

With this we obtain the constants values: $C_2 = \frac{v_0}{\gamma_+ - \gamma_-}$; $C_1 = \frac{v_0}{\gamma_- - \gamma_+}$ and finally the position as a function of time:

$$x(t) = \frac{v_0}{\gamma_+ - \gamma_-}[e^{-\gamma_- t} - e^{-\gamma_+ t}] \tag{1.6}$$

(c) This is the case of a "relaxator" $\eta \rightarrow \infty$. In this limit, γ is much larger than ω_0 and the following simplifications are possible:

$$\gamma_+ \to 2\gamma$$

$$\gamma_- \to 0$$

With this, the solution of the equation of motion also simplifies:

$$x(t) = C_1 e^{-\gamma_+ t} + C_2 e^{-\gamma_- t} \to x(t) \simeq C_2 e^{-\gamma_- t} \tag{1.7}$$

The value of γ_- can be approximated with a power series: $\gamma_- = \gamma - \sqrt{\gamma^2 - \omega_0^2} \simeq \gamma\left(1 - 1 + \frac{\omega_0^2}{2\gamma^2}\right) = \frac{\omega_0^2}{2\gamma} = \frac{k}{m}\frac{1}{2\frac{\eta}{2m}} = \frac{k}{\eta}$, where $\sqrt{1 - \omega_0^2} \simeq 1 - \frac{\omega_0^2}{2}$ has been used.

On the other hand, treating the movement of the relaxator from the oscillator equation:

$$m\frac{d^2x}{dt^2} + \eta x_t + kx = 0 \to \eta\dot{x} + kx = 0 \tag{1.8}$$

Once again the decaying solution is obtained: $x(t) = e^{-at}$, where $a = \frac{k}{\eta}$.

1.2 Properties of the Ordinary Linear Differential Equation for a Forced Oscillator

This section discusses the linearity properties of the equation describing the movement of a forced oscillator, as well as methods for its resolution.

Solution:

(1) The equation of motion for a harmonic oscillator with mass m and spring constant k in a viscous medium with friction coefficient η and under the influence of an external force $F(t)$ is:

$$m\frac{d^2X}{dt^2} + \eta\frac{dX}{dt} + kX = F(t) \tag{1.9}$$

The same equation normalized by the mass is:

$$\frac{d^2X}{dt^2} + 2\beta\frac{dX}{dt} + \omega_0^2 X = f(t) \tag{1.10}$$

with $\beta = \frac{\eta}{2m}$, $\omega_0^2 = \frac{k}{m}$, $f(t) = \frac{F}{m}$. This is an linear, second order, non-homogeneous, ordinary differential equation (i.e. the variable $X(t)$ only depends on one parameter, t). The goal of this

exercise is to clarify the meaning of the linearity of differential equations using the oscillator problem as an example.

(2) If we rewrite the equation as:

$$D[X(t)] = f(t) \tag{1.11}$$

With the operator $D = \frac{d^2}{dt^2} + 2\beta\frac{d}{dt} + \omega_0^2$, the linearity property of the differential operator of the equation is formulated as:

$$D[aX_1(t) + bX_2(t)] = aD[X_1(t)] + bD[X_2(t)] \tag{1.12}$$

Then, if we separate the non-homogeneous term of the equation in two parts:

$$f(t)) = f_1(t) + f_2(t) \tag{1.13}$$

The solution will be divided in two parts as well:

$$X = X_1(t) + X_2(t) \tag{1.14}$$

corresponding to $D[X_i(t)] = f_i(t)$ because:

$$D[X_1(t)+X_2(t)] = D[X_1(t)]+D[X_2(t)] = f_1(t)+f_2(t) \tag{1.15}$$

(3) If we now propose that $f_1 = f(t)$ and $f_2 = 0$, we will separate the ordinary differential equation (or, as will be shown later, also the partial differential equation) in two, one non-homogeneous equation with a particular solution X_p and another homogeneous X_h:

$$D[X_p(t)] = f(t) \tag{1.16}$$

$$D[X_h(t)] = 0 \tag{1.17}$$

so that to find the final solution we will just need to apply the initial conditions to the total solution:

$$X(t) = X_p(t) + X_h(t) \tag{1.18}$$

(4) Finding a particular solution might be at times a matter of luck, and sometimes the Green's function method must be used. The property of linearity is used to introduce the method of Green's functions by separating the non-homogeneous part of the equation (for example, in this problem, the applied force) into infinitesimal parts (for example, here, Dirac's delta-like hits on the mass), which facilitate the search for each of the corresponding particular solutions.
Green's function is obtained by solving a specific problem for each situation, with its boundary conditions, etc.. For example, in the case of a frictionless oscillator, Green's function found in [1], which can be deduced from the equivalence of the problems with an instantaneous hit, resulting in a non-zero initial velocity, can be employed for any external force on a frictionless oscillator. To conclude we will mention that another method for solving homogeneous differential equations is the separation of the non-homogeneous function (and its solution) into Fourier series, consisting in orthogonal functions obtained from the solution of the corresponding Sturm–Liouville problem. In the case of infinite or semi-infinite systems, the Fourier transform method is used, solving the problem in reciprocal space and later applying the inverse Fourier's transform.

1.3 General Definition of Green's Functions

This exercise discusses the general aspects of Green's functions from a non-homogeneous differential equation in a general form in a unidimensional space.

Solution: We suppose a differential equation of the form:

$$Lu(x) = f \tag{1.19}$$

where L is a differential operator (with derivatives with respect to x) and f being a function of x. The Green's function $G(x, x_0)$ is

the solution that corresponds to an inhomogeneous equation with the same operator L than the original equation but with a non-homogeneous, Dirac's delta like term:

$$LG(x, x_0) = \delta(x - x_0) \tag{1.20}$$

Note: the L operator acts on a variable x, while x_0 is a parameter.

Next we consider the following equation:

$$u(x) = \int_a^b G(x, x_0) f(x_0) dx_0 \tag{1.21}$$

We check that this u function is a particular solution of the equation, applying the L operator to both sides of the previous equation:

$$\begin{aligned} Lu(x) &= L \int_a^b G(x, x_0) f(x_0) dx_0 = \int_a^b LG(x, x_0) f(x_0) dx_0 \\ &= \int_a^b \delta(x - x_0) f(x_0) dx_0 = f(x) \end{aligned} \tag{1.22}$$

Note 1: The integration range $[a, b]$ depends on the space where the problem is defined, finite or infinite.

Note 2: Once the problem for Green's function has been solved, we can get the solution for every non-homogeneous integrable part.

Note 3: Analogous expressions are obtained in higher dimensions (2D, 3D).

1.4 Expansion of Green's Function in a Series of Orthogonal Eigenfunctions

Show that Green's function of a particular problem can be expanded in a series of orthogonal eigenfunctions, which are the solutions of the corresponding Sturm–Liouville problem for those boundary conditions.

Solution: We will suppose that we need to solve the following differential equation with homogeneous boundary conditions:

$$Lu(x) = f(x) \tag{1.23}$$

Being L a homogeneous second order differential operator. We will suppose that we can solve the Sturm–Liouville problem by finding a set of orthogonal functions v_n which fulfill the following conditions.

$$Lv_n(x) = \lambda_n v_n \tag{1.24}$$

with boundary conditions. The solution will be expanded as a combination of the v_n functions and we will suppose that there is an infinite number of such eigenvalues and eigenfunctions (in principle, nothing forbids this assumption).

$$u(x) = \sum_{n=1}^{\infty} c_n v_n \tag{1.25}$$

Replacing this solution into the equation (1.23) we have:

$$Lu(x) = \sum_{n=1}^{\infty} c_n Lv_n = \sum_{n=1}^{\infty} c_n \lambda_n v_n = f(x) \tag{1.26}$$

Both parts are multiplied by the former expression by the orthogonal functions $v_m(x)$ with the index m independent from the index n and we will integrate in the interval where the $v_m(x)$ are defined.

$$\int_a^b f(x) v_m(x) dx = \sum_{n=1}^{\infty} c_n \lambda_n \int_a^b v_n v_m dx = c_n \lambda_n |v_n|^2 \tag{1.27}$$

Simplifying, for the case $|v_n|^2 = 1$, we have:

$$c_n = \frac{1}{\lambda_n} \int_a^b f(x') v_n(x') dx' \tag{1.28}$$

With this expression the solution can be rewritten as:

$$\begin{aligned} u(x) &= \sum_{n=1}^{\infty} \frac{1}{\lambda_n} \int_a^b f(x') v_n(x') dx' v_n(x) \\ &= \sum_{n=1}^{\infty} \frac{1}{\lambda_n} \int_a^b v_n(x') v_n(x) f(x') dx' \end{aligned} \tag{1.29}$$

Comparing this with the definition of the solution obtained by the Green's function method:

$$u(x) = \int_a^b G(x, x') f(x') dx' \tag{1.30}$$

The following is Green's function is obtained:

$$G(x, x') = \sum_{n=1}^{\infty} \frac{1}{\lambda_n} v_n(x') v_n(x) \tag{1.31}$$

1.5 Green's Function of an Oscillator with Friction

Find Green's function of an oscillator in a viscous medium in the limit of very low friction.

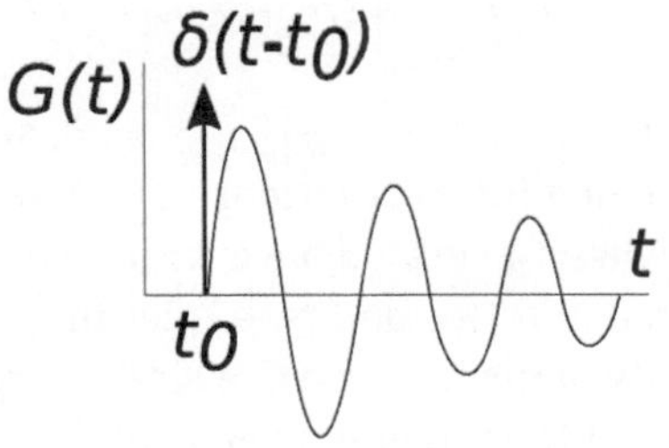

Figure 1.2

Solution: We can write the following equation of motion $u(t)$ (normalized by the mass m) of a harmonic oscillator with a spring constant k in a viscous medium with friction coefficient η, and under the action of an external force $F(t)$:

$$\frac{d^2u}{dt^2} + 2\beta\frac{du}{dt} + \omega_0^2 u = f(t) \tag{1.32}$$

with $\beta = \frac{\eta}{2m}$, $\omega_0^2 = \frac{k}{m}$, $f(t) = \frac{F}{m}$. If we solve this very equation for an instant hit (described by a delta function) the solution would be Green's function $G(t)$:

$$\frac{d^2G}{dt^2} + 2\beta\frac{dG}{dt} + \omega_0^2 G = \delta(t - t_0) \tag{1.33}$$

One of the methods to find the oscillator Green's function is replacing the hit (non-homogeneous equation) by the solution of a homogeneous equation with initial conditions: $G(t = t_0) = 0$; $\left.\frac{dG}{dt}\right|_{t=t_0} = C$ where $C = 1$ in the case $f(t) = \delta(t - t_0)$.

Using this method, in [1] the equation 1.24 is deduced. This solution represents the movement of an oscillator corresponding to the initial conditions, which is proportional to Green's function of a oscillator with friction (with proportionality constant A). For the case of a very low friction $\Omega_0 = \sqrt{\omega_0^2 - \gamma^2} \rightarrow \omega_0$:

$$G_p = Ae^{-\beta(t-t_0)} \sin[\omega_0(t - t_0)] \tag{1.34}$$

To find the A coefficient, equation (1.33) is integrated in an infinitesimal range, centered in the instant t_0:

$$\left.\frac{dG}{dt}\right|_{t_0-\epsilon}^{t_0+\epsilon} + \int_{t_0-\epsilon}^{t_0+\epsilon} \left[2\beta\frac{dG}{dt} + \omega_0^2 G\right] dt = 1 \tag{1.35}$$

As the initial velocity is null, $\left.\frac{dG}{dt}\right|_{t=t_0-\epsilon} = 0$, since the oscillator is at rest prior to being hit. Furthermore, the integral is equal to 0: the first term $\frac{dG}{dt}$, when integrated once with respect to time gives G. When G is evaluated in $t_0 + \epsilon$ and $t_0 - \epsilon$, letting $\epsilon \rightarrow 0$, $G|_{t_0-\epsilon}^{t_0+\epsilon} = 0$, due to the continuity of the function G. A similar argument is applied to the second term of the integral, which will be an order higher in t. Then, replacing the function 1.34 in the condition $\left.\frac{dG}{dt}\right|_{t_0+\epsilon} = 1$ the $A = \frac{1}{\omega_0}$ coefficient is obtained.

Green's function is:

$$G = \begin{Bmatrix} 0 & (t < t_0) \\ \frac{1}{\omega_0} e^{-\beta(t-t_0)} \sin[\omega_0(t - t_0)] & (t > t_0) \end{Bmatrix} \tag{1.36}$$

Note that for $\beta = 0$ the solution is converted into Green's function of a frictionless harmonic oscillator. Let us now consider a gaussian pulse at $t = 0$:

$$f(t) = e^{-t^2} \tag{1.37}$$

To find the particular solution the following integral must be solved first:

$$u_p(t) = \int_0^t \left[e^{-x^2} \frac{1}{\omega_0} e^{-\beta(t-x)} \sin[\omega_0(t - x)]\right] dx \tag{1.38}$$

This integral doesn't have an analytic solution.

Another possible problem which can be solved using Green's function: We can, for example, seek the particular solution of an oscillator with very low friction if the external force acts upon the oscillator from the instant $-a$ to the instant $+a$ and transfers a linear momentum I. To do this, the next steps should be followed:

- Evaluate the force density $f(t)$ from the transferred momentum and the mass m.

- Find the solution using the Green's function method and Green's function found in this problem.
- To solve the integral we will need to use Wolfram Alpha or some other computer algebra system.

1.6 Movement of an Oscillator under the Influence of a Constant Force, Solved by Two Methods

Find the solution for the movement of an oscillator with natural frequency ω_0, which is initially at rest ($t = 0$) if, at $t = 0$ a force f_0 starts acting upon it.

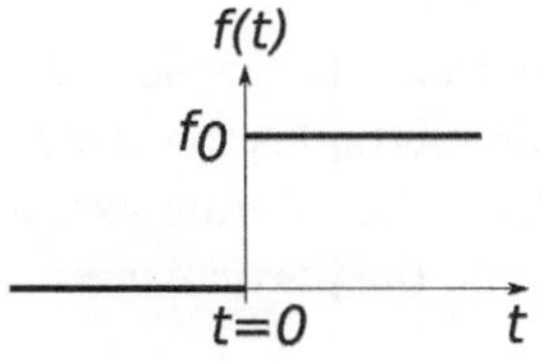

Figure 1.3

Solution: Problem to be solved:

$$u = \left\{ \begin{array}{l} \frac{d^2u}{dt^2} + \omega_0^2 u = f(t) \\ \\ u(0) = \frac{du}{dt}\Big|_{t=0} = 0 \end{array} \right\} \tag{1.39}$$

$$f = \left\{ \begin{array}{ll} f(t) = 0 & (t < 0) \\ f(t) = f_0 & (t \geq 0) \end{array} \right\} \tag{1.40}$$

We will compare two different methods for solving the problem, a) and b):

(a) The first method is based on the search for a particular solution. The total solution is the sum of the particular solution of the non-homogeneous equation and the solution of the homogeneous equation.

$$u(t) = C_1 \cos(\omega_0 t) + C_2 \sin(\omega_0 t) + \frac{f_0}{\omega_0^2} \tag{1.41}$$

Applying the initial conditions, we have $C_1 = -\frac{f_0}{\omega_0^2}$ and $C_2 = 0$. Then:

$$u(t) = \frac{f_0}{\omega_0^2}[1 - \cos(\omega_0 t)] \qquad (1.42)$$

(b) The second method for searching the solution is based on the application of the Green's function method, with the known function for a harmonic oscillator (equation 1.65 from [1]).

$$u_p(t) = \int_{-\infty}^{+\infty} f(t_0)G(t - t_0)dt_0 = \frac{f_0}{\omega_0}\int_0^t \sin[\omega_0(t - t_0)]dt_0 \qquad (1.43)$$

With the change of variable $\omega_0(t - t_0) = x \rightarrow -\omega_0 d(t_0) = dx$

$$u_p(t) = -\frac{f_0}{(\omega_0)^2}\int_0^t \sin(x)dx = \frac{f_0}{\omega_0^2}[1 - \cos(\omega_0 t)] \qquad (1.44)$$

Note. in the latter case, the particular solution obtained with the Green's function method is the same as the final solution. To show it, we search for the solution as the sum of the solution of the homogeneous and the particular equations:

$$u(t) = C_1 \cos(\omega_0 t) + C_2 \sin(\omega_0 t) + \frac{f_0}{(\omega_0)^2}[1 - \cos(\omega_0 t)] \qquad (1.45)$$

Applying the initial conditions, we would have: $C_1 = C_2 = 0$

1.7 Oscillator Forced by a Rectangular Hit, Solved with Green's Functions

An oscillator with mass $m = 1$ Kg and a natural frequency $\omega_0 = 1$ rad/s is initially at rest. Starting at $t = -1$ and until $t = +1$, the oscillator is subject to a force of value $f = 1$ N/kg. Find the particular solution using the Green's function method.

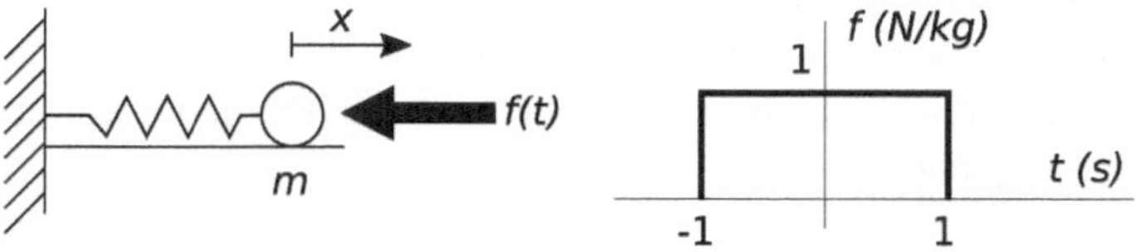

Figure 1.4

We will use the Green's function method, integrating it for the duration of the external force. Since this force acts for a limited time, there will be two parts: one for $t < 1$ and another for $t > 1$.

Mathematical formulation:

$$\left\{ \begin{array}{c} \frac{d^2X}{dt^2} + X(t) = f(t) \\ f = 1 \quad (-1 < t < 1) \end{array} \right\} \tag{1.46}$$

Solution:

$$X_{part}(t) = \int_{-\infty}^{+\infty} f(x)G(t,x)dx \tag{1.47}$$

Starting from the known form of Green's function for an oscillator of mass m and natural frequency $\omega_0 = 1$, for a given time t we have:

$$G(t,x) = \begin{pmatrix} \sin(t-x) & x < t \\ 0 & (x > t) \end{pmatrix} \tag{1.48}$$

Given the form of the force, the solution will be split into parts. Before applying the force, the mass is at rest:

$$X_{part}(t < -1) = 0 \tag{1.49}$$

Then for $t > -1$ the solution will be of the form:

$$X_{part}(t > -1) = \int_{-\infty}^{t} f(x)\sin(t-x)dx = \int_{-1}^{t} \sin(t-x)dx \tag{1.50}$$

The particular solution is different for $t < +1$ and $t > +1$. In $t < +1$ it will be:

$$\int_{-1}^{t} \sin(t-x)dx = -(-\cos(t-x)|_{-1}^{t})$$

$$= \cos(0) - \cos(t+1) = 1 - \cos(t+1) \tag{1.51}$$

And in the second, $t > +1$:

$$X_{part}(t) = \int_{-\infty}^{t} f(x)\sin(t-x)dx$$

$$= \int_{-1}^{1} \sin(t-x)dx + \int_{1}^{t} 0 * \sin(t-x)dx$$

$$= \cos(t-1) - \cos(t+1) \tag{1.52}$$

The final solution is presented graphically:

$$X_{part}(t) = \begin{cases} 0 & (t < -1) \\ 1 - \cos(t+1) & (-1 < t < 1) \\ \cos(t-1) - \cos(t+1) & (t > 1) \end{cases} \tag{1.53}$$

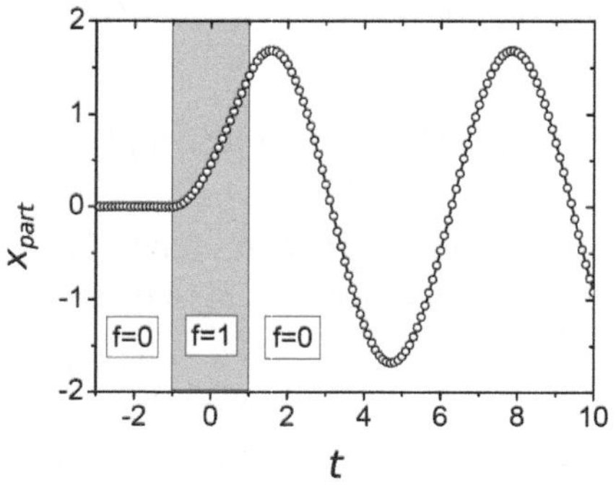

Figure 1.5

1.8 Movement of a Mass after an Instantaneous Exponential Hit

Using the Green's function method, find the response of a mass $m = 1$ to a hit with force density e^{-t}, which occurs at $t = 0$. The mass is free to move and there is no friction. Consider the case in which the mass is at rest prior to being hit.

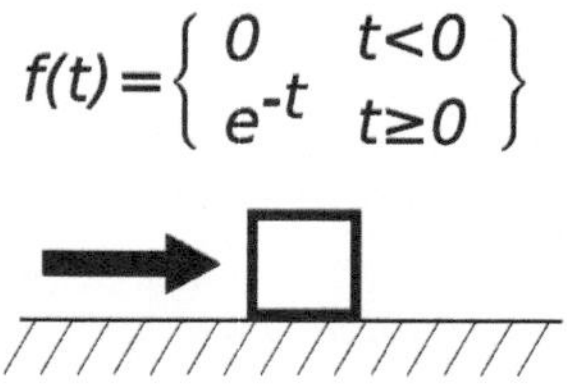

Figure 1.6

Mathematical formulation:

$$\left\{\begin{array}{l} m\frac{d^2u}{dt^2} = f(t) \\ u(0) = 0;\ \left.\dfrac{du}{dt}\right|_{t=0} = 0 \qquad (t < t_0) \end{array}\right\} \tag{1.54}$$

We are searching for Green's function $G(t, t_0)$ of the following differential equation, which describes the instantaneous hit (Dirac's delta function like) at $t = t_0$.

$$\left\{\begin{array}{l} \frac{d^2G(t,t_0)}{dt^2} = \delta(t - t_0) \\ G(0) = 0;\ \left.\dfrac{dG}{dt}\right|_{t=0} = 0 \qquad (t < t_0) \end{array}\right\} \tag{1.55}$$

We seek a solution separated into parts for the different time ranges where the equation is homogeneous. Later both solutions will be joined using the initial condition and integrating the equation in the environment of the anomalous point $t = 0$.

$$G(t, t_0) = \left\{\begin{array}{ll} A(t_0)t + B(t_0) & (t < t_0) \\ C(t_0)t + D(t_0) & (t > t_0) \end{array}\right\} \tag{1.56}$$

From the general definition of Green's function, we have (for time values a and b in the interval taken into account):

$$x = \int_a^b G(t, t_0) f(t_0) dt_0 \tag{1.57}$$

$$x = \int_{-\infty}^{t} (At_0 + B) \cdot 0 \cdot dt_0 = 0 \rightarrow x = 0 \quad (t < 0) \tag{1.58}$$

But if the equation of motion for G is integrated around the delta-like hit:

$$\frac{d^2G}{dt^2} = \delta(t - t_0) \rightarrow \int_{t_0-\epsilon}^{t_0+\epsilon} \frac{d^2G}{dt^2} dt = \int_{t_0-\epsilon}^{t_0+\epsilon} \delta(t - t_0) dt \tag{1.59}$$

The right hand side of this equation is just the integral of a delta function, which is equal to 1. The left hand side is the integral of a derivative, which means that the order of the derivative is decreased by 1:

$$\left.\frac{dG}{dt}\right|_{t_0+\epsilon} - \cancel{\left.\frac{dG}{dt}\right|_{t_0-\epsilon}} = 1 \tag{1.60}$$

From that the value of the constants after the hit is obtained: $\frac{dG}{dt}|_{t_0} = C = 1, G|_{t_0} = 0 = Ct_0 + D \to D = -t_0$,

$$G(t, t_0) = \begin{Bmatrix} 0 & (t < t_0) \\ t - t_0 & (t > t_0) \end{Bmatrix} \tag{1.61}$$

Solution for $t > 0$ (when the force is applied):

$$u(t) = \int_0^{+\infty} e^{-t_0} G(t, t_0) dt_0 = \int_0^t e^{-t_0} G(t, t_0) dt_0 + \int_t^{+\infty} e^{-t_0} G(t, t_0) dt_0 \tag{1.62}$$

$$u(t) = \int_0^t e^{-t_0} G(t, t_0) dt_0 \tag{1.63}$$

as $G = 0$ for $t < t_0$

$$u(t) = t - 1 + e^{-t} \tag{1.64}$$

Representing the result graphically:

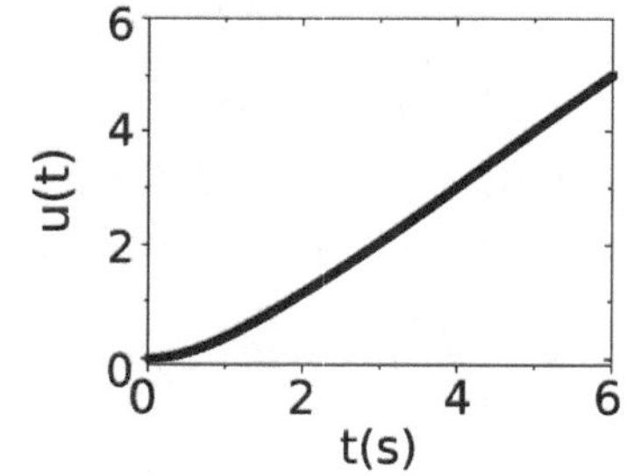

Figure 1.7

1.9 Shape of a String in Mechanical Equilibrium, Solved by the Green's Function Method

Solve the equation:

$$\frac{d^2u}{dx^2} + u = x \tag{1.65}$$

with boundary conditions $u(0, \pi/2) = 0$ using the Green's function method.

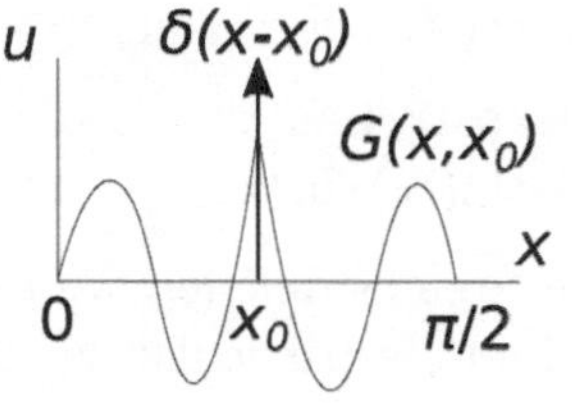

Figure 1.8

Solution: To find the Green's function the following differential equation must be solved:

$$\frac{d^2G(x, x_0)}{dx^2} + G(x, x_0) = \delta(x - x_0) \tag{1.66}$$

with the boundary conditions $G(x = 0) = G(x = \pi/2) = 0$

We apply an analogous method to the one already used to solve the problem considered in section 3.2.5 from [1], by dividing space in two region ($x < x_0$ and $x > x_0$) and solving first two independent homogeneous problems.

The resulting Green's function is defined by parts:

$$G(x, x_0) = \begin{Bmatrix} A(x_0)\sin(x) + B(x_0)\cos(x) & (x < x_0) \\ C(x_0)\sin(x) + D(x_0)\cos(x) & (x > x_0) \end{Bmatrix} \tag{1.67}$$

We apply boundary conditions and bind these functions to find the coefficients A, B, C and D. From the boundary conditions we arrive at:

$$G(x, x_0) = \begin{Bmatrix} A(x_0)\sin(x) & (x < x_0) \\ D(x_0)\cos(x) & (x > x_0) \end{Bmatrix} \tag{1.68}$$

The continuity equation of the solution is:

$$A(x_0)\sin(x_0) = D(x_0)\cos(x_0) \tag{1.69}$$

The difference between the derivatives to both sides is equal to 1 (it is obtained by integrating equation 1.66 around an infinitesimal surrounding of the delta function.)

$$-D(x_0)\sin(x_0) - A(x_0)\cos(x_0) = 1 \tag{1.70}$$

Solving the system of two equations with two unknown variables we obtain:

$$\begin{aligned} A(x_0) &= -\cos(x_0) \\ D(x_0) &= -\sin(x_0) \end{aligned} \tag{1.71}$$

Then the Green's function has a value:

$$G(x, x_0) = \begin{cases} -\cos(x_0)\sin(x) & (x < x_0) \\ -\sin(x_0)\cos(x) & (x > x_0) \end{cases} \quad (1.72)$$

Once obtained the Green's function we can use it to solve the original problem. To find the particular solution we need to solve the following integral:

$$u(x) = \int_0^{\pi/2} z \cdot G(x, z)dz = \int_0^{x} z \cdot G(x, z)dz + \int_x^{\pi/2} z \cdot G(x, z)dz \quad (1.73)$$

To solve the integral with the z variable it's important to distinguish the parts of the Green's function, corresponding to $z > x$ and $z < x$. The following figure can help clarify this:

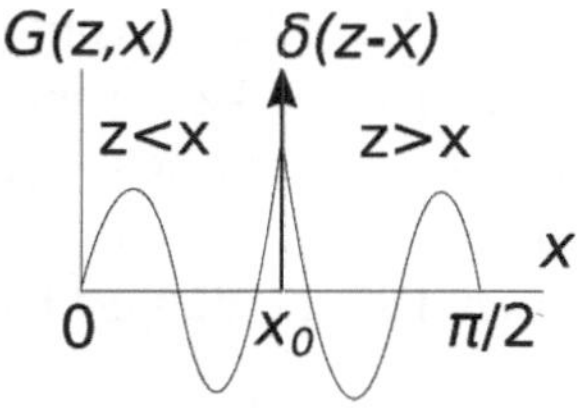

Figure 1.9

We finally arrive at the result:

$$\begin{aligned} u(x) &= \int_0^{\pi/2} z \cdot G(x, z)dz \\ &= -\cos(x)\int_0^{x} z \cdot \sin(z)dz - \sin(x)\int_x^{\pi/2} z \cdot \cos(z)dz \end{aligned} \quad (1.74)$$

Integrating by parts we arrive at:

$$u(x) = x - \pi/2\sin(x) \quad (1.75)$$

1.10 Case Study: Transversal Displacement of a Tense String Glued to an Elastic Plane

Determine the transversal displacement of a tense string of length L which is attached to an elastic plane and upon which a force acts with the following density:

$$f(x) = \left\{ \begin{array}{ll} f_0, & \frac{L}{4} < x < \frac{3L}{4} \\ 0, & x < \frac{L}{4},\ x > \frac{3L}{4} \end{array} \right\} \tag{1.76}$$

Consider the case in which the two ends of the string cannot move. Solve the problem with the Green's function method.

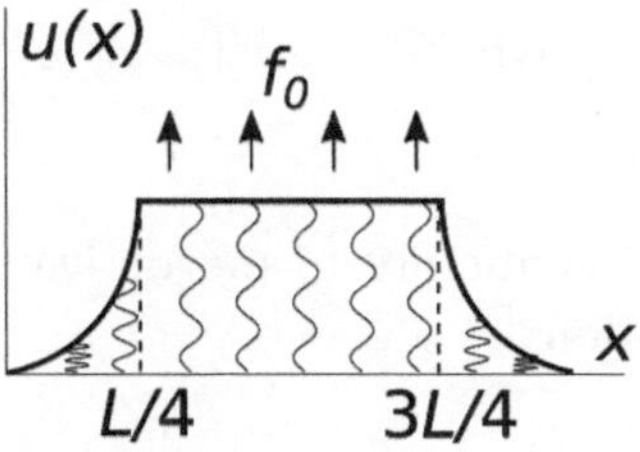

Figure 1.10

Mathematical formulation:

$$\left\{ \begin{array}{l} \frac{d^2u(x)}{dx^2} - a^2u(x) + \frac{f(x)}{T} = 0 \\ \\ u(0) = u(L) = 0; \quad a^2 = \frac{\beta}{T} \end{array} \right\} \tag{1.77}$$

Since we will solve the problem using Green's functions, we will first use a density of forces $f(x) = \delta(x - x_0)$.

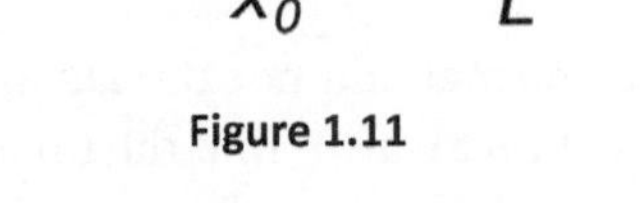

Figure 1.11

We have once again a solution with the form corresponding to an infinite plane:

$$u(x) = \left\{ \begin{array}{ll} u_1(x) = A_1e^{-ax} + B_1e^{ax} & (x < x_0) \\ u_2(x) = A_2e^{-ax} + B_2e^{ax} & (x > x_0) \end{array} \right\} \tag{1.78}$$

From the boundary conditions, we get:

$$\left\{\begin{array}{c} u(0) = A_1 + B_1 = 0 \rightarrow A_1 = -B_1 \\ u(L) = A_2 e^{-aL} + B_2 e^{aL} = 0 \rightarrow B_2 = -A_2 e^{-2aL} \end{array}\right\} \tag{1.79}$$

From the condition of continuity of the string, and replacing these values of B_1 and B_2:

$$u_1(x_0) = u_2(x_0) \rightarrow A_1[e^{-ax_0} - e^{ax_0}] = A_2[e^{-ax_0} - e^{-2aL}e^{ax_0}] \tag{1.80}$$

The condition of continuity of the derivatives is obtained by integrating the equation:

$$\frac{d^2u(x)}{dx^2} - a^2u(x) = -\frac{\delta(x - x_0)}{T} \tag{1.81}$$

around the point x_0 in an interval 2ε, where we will later let $\varepsilon \rightarrow 0$:

Doing as in previous problems we get:

$$\left.\frac{du_2(x)}{dx}\right|_{x_0+\varepsilon} - \left.\frac{du_1(x)}{dx}\right|_{x_0-\varepsilon} = -\frac{1}{T} \tag{1.82}$$

From here:

$$A_2[(-a)e^{-ax_0} - ae^{-2aL}e^{ax_0}] + aA_1e^{-ax_0} + aA_1e^{ax_0} = -\frac{1}{T} \tag{1.83}$$

Here we arrive at a system of two equations with two unknowns to solve $A_{1,2}$

$$\left\{\begin{array}{c} A_2[e^{-ax_0} + e^{-2aL}e^{ax_0}] - A_1[e^{-ax_0} + e^{ax_0}] = \frac{1}{aT} \\ A_1[e^{-ax_0} - e^{ax_0}] = A_2[e^{-ax_0} - e^{-2aL}e^{ax_0}] \end{array}\right\} \tag{1.84}$$

This method is rather formal and not the fastest one. We will now explore another more direct way to find the final solution.

Observing carefully the symmetry of the solution and the boundary conditions we can present the solution in a more transparent way, which already includes other boundary conditions (we also arrive at this form if we seek a solution as a sum of hyperbolic functions).

$$u(x) = \begin{cases} u_1(x) = A\sinh(ax) & (x < x_0) \\ u_2(x) = B\sinh(a(L-x)) & (x > x_0) \end{cases} \tag{1.85}$$

Again we apply the condition of continuity of the solutions:

$$u_1(x_0) = u_2(x_0) \rightarrow A\sinh(ax_0) = B\sinh(a(L-x_0)) \tag{1.86}$$

Replacing these values, we get:

$$\begin{cases} u_1(x) = A\sinh(ax) & (x < x_0) \\ u_2(x) = A\frac{\sinh(ax_0)}{\sinh(a(L-x_0))}\sinh(a(L-x)) & (x > x_0) \end{cases} \tag{1.87}$$

Applying the previous expression to the condition of continuity of the derivatives:

$$\left.\frac{du_2(x)}{dx}\right|_{x_0+\varepsilon} - \left.\frac{du_1(x)}{dx}\right|_{x_0-\varepsilon} = -\frac{1}{T} \tag{1.88}$$

we have:

$$(-a)A\frac{\sinh(ax_0)}{\sinh(a(L-x_0))}\cosh(a(L-x_0)) - aA\cosh(ax_0) = -\frac{1}{T} \rightarrow \tag{1.89}$$

$$A = \frac{1}{aT}\frac{\sinh(a(L-x_0))}{\sinh(ax_0)\cosh(a(L-x_0)) + \cosh(ax_0)\sinh(a(L-x_0))} \tag{1.90}$$

In this manner we get Green's function:

$$G(x, x_0) = \begin{cases} u_1(x) = A(x_0)\sinh(ax) & x < x_0 \\ u_2(x) = A(x_0)\frac{\sinh(ax_0)}{\sinh(a(L-x_0))}\sinh(a(L-x)) & x > x_0 \end{cases} \tag{1.91}$$

To find the solution of a non-homogeneous equation starting from a Green's function:

$$u(x) = \int_0^L f(x')G(x, x')dx' = f_0\int_{\frac{L}{4}}^{\frac{3}{4}L} G(x, x')dx' \tag{1.92}$$

where x' is the point of application of the point force. In Green's function previously obtained, $x' = x_0$.

Note. we can get to the solution of the problem without using Green's functions:

$$\left\{ \begin{array}{c} \frac{d^2u(x)}{dx^2} - a^2u(x) + \frac{f(x)}{T} = 0 \\ u(0) = u(L) = 0 \end{array} \right\} \tag{1.93}$$

The solution again is composed of three parts: two of them correspond to the ranges where the equation is homogeneous and the third one, to the central part (non-homogeneous equation). The first two parts are sought in a similar manner to how we did for finding Green's function:

$$u(x) = \left\{ \begin{array}{ll} u_1(x) = A\sinh(ax) & (x < L/4) \\ u_2(x) = B\sinh(a(L-x)) & (x > 3L/4) \\ \multicolumn{2}{c}{\text{By symmetry it must be: } A = B} \\ \multicolumn{2}{c}{u_3(x)?} \end{array} \right\} \tag{1.94}$$

We search for $u_3(x)$ as the sum of the solutions of the homogeneous equation $u_h(x)$ and the particular solution $u_p(x)$ of the non-homogeneous equation. The homogeneous equation has solution:

$$\frac{d^2u_h(x)}{dx^2} - a^2u_h(x) = 0 \rightarrow u_h(x) = C\cosh\left[a\left(x - \frac{L}{2}\right)\right] \tag{1.95}$$

The non-homogeneous equation has a particular solution:

$$\frac{d^2u_p(x)}{dx^2} - a^2u_p(x) = -\frac{f_0}{T} \tag{1.96}$$

A particular solution of this equation consists in taking $u_p(x)$ as a constant and replacing we obtain its value: $u_p(x) = \frac{f_0}{a^2T}$. Then:

$$u_3(x) = C\cosh\left[a\left(x - \frac{L}{2}\right)\right] + \frac{f_0}{a^2T} \tag{1.97}$$

We introduce the conditions of continuity of the functions and their derivatives at the point $3L/4$, integrating the differential inhomogeneous equation in an environment ϵ around this point.

$$u_3\left(\frac{3L}{4}\right) = u_2\left(\frac{3L}{4}\right) \tag{1.98}$$

$$\left.\frac{du_3}{dx}\right|_{x=\frac{3L}{4}} = \left.\frac{du_2}{dx}\right|_{x=\frac{3L}{4}} \tag{1.99}$$

We obtain a system of two equations with two unknowns, C and A.

$$\left\{\begin{array}{c} C\cosh\left(a\frac{L}{4}\right) + \frac{f_0}{a^2 T} = A\sinh\left(a\frac{L}{4}\right) \\ -C\sinh\left(a\frac{L}{4}\right) = A\cosh\left(a\frac{L}{4}\right) \end{array}\right\} \tag{1.100}$$

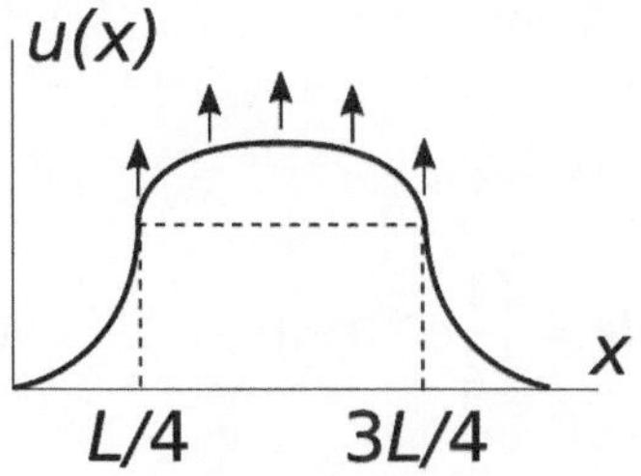

Figure 1.12

1.11 Forced Harmonic Oscillator, Solved with Green's Functions

An oscillator with mass $m = 1$ kg and an eigenfrequency $\omega_0 = 1$ Rad/s is initially at rest.

Starting at $t = 0$ a force starts acting on the oscillator with a value that exponentially decays with time e^{-t}. Find the particular solution for this case.

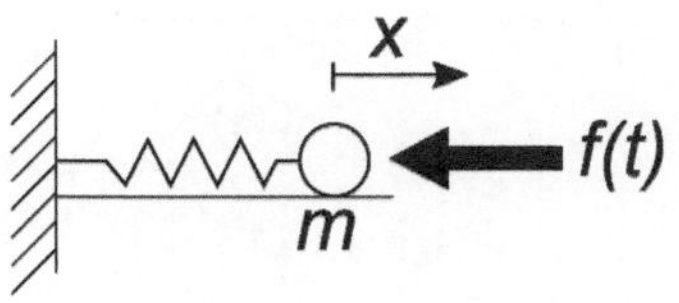

Figure 1.13

Mathematical formulation:

$$\left\{\begin{array}{l} \dfrac{d^2u(t)}{dt^2} + u(t) = \dfrac{f(t)}{m} \\ f = \left\{\begin{array}{ll} e^{-t} & (t > 0) \\ 0 & (t < 0) \end{array}\right\} \end{array}\right\} \tag{1.101}$$

For this oscillator with natural frequency $\omega_0 = 1$, Green's function is $\sin(t - x)$, which allows to write the particular solution as:

$$u_{part}(t) = \int_{-\infty}^{t} f(x)\sin(t - x)dx = \int_{0}^{t} e^{-x}\sin(t - x)dx \quad (1.102)$$

From integral tables we have:

$$\int_{0}^{t} e^{-x}\sin(t - x)dx = \left[\frac{1}{2}e^{-x}[\cos(t - x) - \sin(t - x)]\right]_{0}^{t} \quad (1.103)$$

$$u_{part}(t) = \frac{1}{2}[e^{-t} - \cos(t) + \sin(t)] \quad (1.104)$$

Finally the result is shown graphically, from which we can observe the oscillator's larger amplitude of motion at the beginning (and which decays due to the exponential decay).

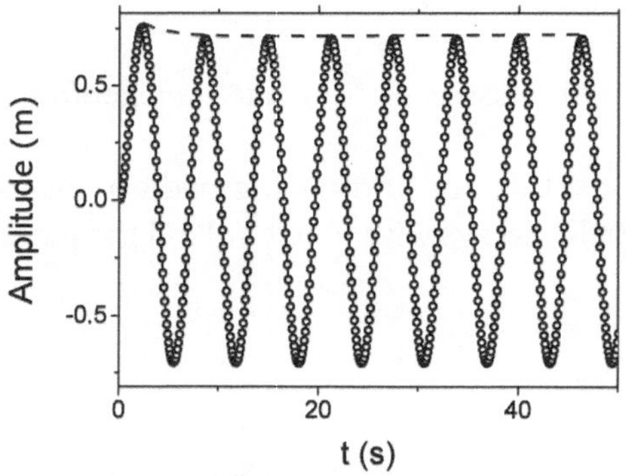

Figure 1.14

Chapter 2

Problems in One Dimension

In this chapter the method to seek solutions will be generalized and applied to confined systems in one dimension. We will analyze several physical processes, described by linear differential equations in partial derivatives in one spatial dimension. The solution will be found using the Fourier method, expanding it in series of harmonic functions.

Of course, in the cases of a static string the problem will consist in solving a second order differential equation with boundary conditions. Whereas in the case of transversal oscillations along a string or longitudinal ones in a solid rod, it is obvious that the solution needs to consist of harmonic functions, in the case of the diffusion equation the interpretation will be more abstract.

To summarize, the goal of this chapter is to start from some simple examples to show the basic methods to solve the wave and diffusion equation, and static Poisson problems. In the case of a string with forces applied to it, sometimes we will use the Green's function method introduced in the previous chapter.

The appendix contains a summary of the different equations of the physical processes described, using second order PDEs, as

Mathematical Methods for Physics: Problems and Solutions
Farkhad G. Aliev and Antonio Lara

ISBN 978-981-4968-71-3 (Hardcover), 978-1-003-41088-1 (eBook)
www.jennystanford.com

well as solutions of the Sturm–Liouville problems for a harmonic oscillator with different types of homogeneous boundary conditions.

2.1 Closed String

Consider a string (tension T, linear density of mass ρ, length L) with the shape of a circular ring. The string is free to move across its entire length (with no gravity).

Find the oscillations of the string if we know the displacement $u(x, 0) = f(x)$ and the velocity $u_t(x, 0) = \psi(x)$ at the instant $t = 0$ with respect to the plane of equilibrium (we can call x the variable that indicates the position along the string).

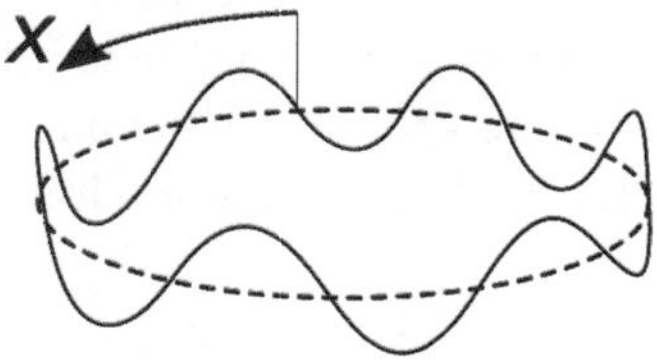

Figure 2.1

Mathematical formulation:

$$\left\{\begin{array}{ll} \dfrac{\partial^2 u}{\partial t^2} = c^2 \dfrac{\partial^2 u}{\partial x^2} & \left(0 < x < L;\ c^2 = \dfrac{T}{\rho}\right) \\ u(0, t) = u(L, t) & \\ \left.\dfrac{\partial u}{\partial x}\right|_{x=0} = \left.\dfrac{\partial u}{\partial x}\right|_{x=L} & \\ u(x, t = 0) = f(x) & \\ \left.\dfrac{\partial u}{\partial t}\right|_{t=0} = \psi(x) & \end{array}\right\} \quad (2.1)$$

Sturm–Liouville problem: We use separation of variables:

$$u(x, t) = T(t) \cdot X(x) \quad (2.2)$$

We arrive at the Sturm–Liouville problem for the spatial part $X(x)$:

$$\frac{d^2X}{dx^2} + \lambda X = 0 \qquad (2.3)$$

The general solution (having considered $\lambda > 0$) is: $X(x) = A\cos\sqrt{\lambda}x + B\sin\sqrt{\lambda}x$.

The eigenvalues are found with the boundary conditions, which are adapted to the spatial part of the solution:

$$\left\{\begin{array}{l} u(x=0,t) = u(x=L,t) \\ \left.\dfrac{\partial u}{\partial x}\right|_{x=0} = \left.\dfrac{\partial u}{\partial x}\right|_{x=L} \end{array}\right\} \rightarrow \left\{\begin{array}{l} X(x=0) = X(x=L) \\ \left.\dfrac{\partial X}{\partial x}\right|_{x=0} = \left.\dfrac{\partial X}{\partial x}\right|_{x=L} \end{array}\right\} \qquad (2.4)$$

We arrive at the following system of equations to find the eigenvalues:

$$\left\{\begin{array}{l} A = A\cos\sqrt{\lambda}L + B\sin\sqrt{\lambda}L \\ B\cancel{\sqrt{\lambda}} = -A\cancel{\sqrt{\lambda}}\sin\sqrt{\lambda}L + B\cancel{\sqrt{\lambda}}\cos\sqrt{\lambda}L \end{array}\right\} \qquad (2.5)$$

Equating DET=0:

$$\cos\sqrt{\lambda}L = \pm 1 \quad \rightarrow \sqrt{\lambda}L = \pi n \rightarrow \lambda_n = \left[\frac{\pi n}{L}\right]^2 \qquad (2.6)$$

General solution:

$$u(x,t) = \sum_{n=0}^{\infty}\left[A_n\cos\left(c\frac{\pi n}{L}t\right) + B_n\sin\left(c\frac{\pi n}{L}t\right)\right]\cos\left(\frac{\pi n}{L}x\right) + \sum_{n=1}^{\infty}\left[C_n\cos\left(c\frac{\pi n}{L}t\right) + D_n\sin\left(c\frac{\pi n}{L}t\right)\right]\sin\left(\frac{\pi n}{L}x\right) \qquad (2.7)$$

Note: if we try to isolate the values of A or B in the previous system (replacing the obtained condition for the eigenvalues) we arrive at expressions of the type $A = A$ and $B = B$. Precisely, due to the uncertainty in the origin of coordinates we cannot fix the phase of every eigenfunction (that is, we cannot fix the ratio A_n/B_n) contrary to the case of open strings. In the latter the position of the string ends fixes the phase of each eigenfunction. To illustrate this, consider the following figure:

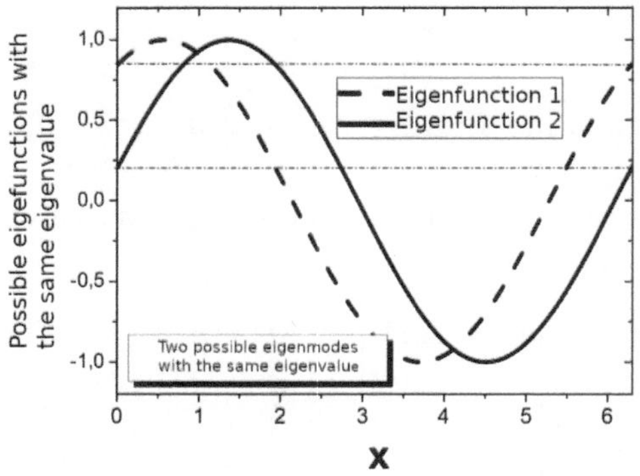

Figure 2.2

Final solution: The coefficients are obtained by imposing the initial conditions and using the properties of orthogonality (in this case to each eigenvalue $\lambda_n = \left[\frac{\pi n}{L}\right]^2$ correspond two orthogonal degenerated eigenfunctions $\cos\left(\frac{\pi n}{L}x\right)$ and $\sin\left(\frac{\pi n}{L}x\right)$):

$$A_n = \frac{2}{L}\int_0^L f(x)\cos\left(\frac{\pi n}{L}x\right)dx \qquad (n > 0) \tag{2.8}$$

$$A_0 = \frac{1}{L}\int_0^L f(x)dx \tag{2.9}$$

$$C_n = \frac{2}{L}\int_0^L f(x)\sin\left(\frac{\pi n}{L}x\right)dx \tag{2.10}$$

$$B_n = \frac{2}{\pi nc}\int_0^L \psi(x)\cos\left(\frac{\pi n}{L}x\right)dx \qquad (n > 0) \tag{2.11}$$

$$D_n = \frac{2}{\pi nc}\int_0^L \psi(x)\sin\left(\frac{\pi n}{L}x\right)dx \tag{2.12}$$

Notes: the B_0 term does not exist, since it enters the summation as $0 \cdot B_0$

2.2 Sturm–Liouville Problem with Boundary Conditions of the Second and Third Kind

Find the eigenvalues and eigenfunctions of the Sturm–Liouville problem in a string whose right end is free to move and its left end is connected to a spring.

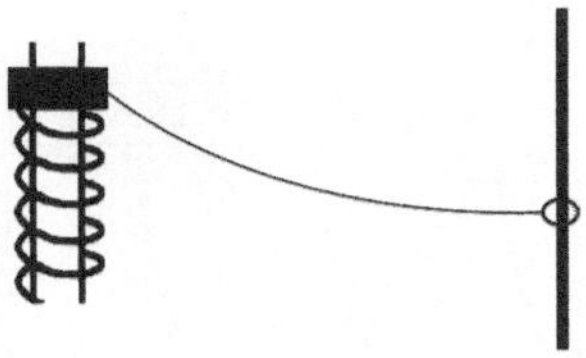

Figure 2.3

Mathematical formulation:

$$\begin{pmatrix} \dfrac{du^2}{dx^2} + \lambda u = 0 \\ \left.\dfrac{du}{dx}\right|_{x=0} - hu(0) = 0; \\ \left.\dfrac{du}{dx}\right|_{x=L} = 0 \end{pmatrix} \tag{2.13}$$

Here $h = \beta/T$ is the ratio between the spring constant and the tension (in the general case of a string).

The more formal method to solve this problem consists in finding a solution as the sum of two linearly independent solutions and to apply the boundary conditions.

$$u(x) = A\sin(\sqrt{\lambda}x) + B\cos(\sqrt{\lambda}x) \tag{2.14}$$

However, in this case we can take a shortcut noting that the only solution than can possibly satisfy the second boundary condition must have the form:

$$u(x) = B\cos(\sqrt{\lambda}[x - L]) \tag{2.15}$$

Applying the first boundary condition:

$$-B\sqrt{\lambda}\sin(\sqrt{\lambda}[-L]) - hB\cos(\sqrt{\lambda}[-L]) = 0 \tag{2.16}$$

$$\sqrt{\lambda}\sin(\sqrt{\lambda}L) - h\cos(\sqrt{\lambda}L) = 0 \tag{2.17}$$

The zeroes of the equation: $\tan(\sqrt{\lambda}L) = \frac{h}{\sqrt{\lambda}}$ give us the eigenvalues of the problem.

The following graphical representation shows how the eigenvalues of the problem are found:

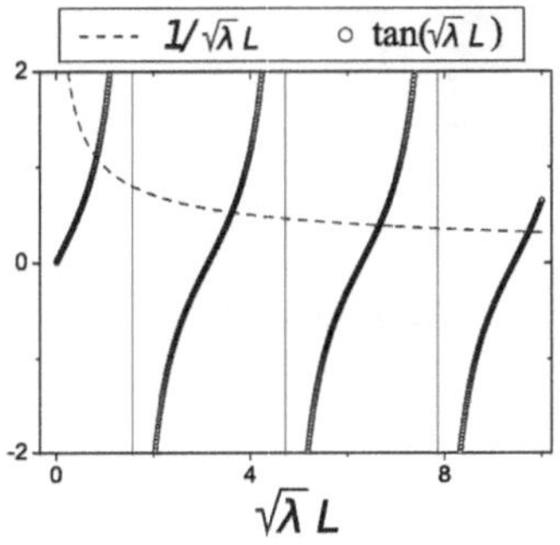

Figure 2.4

Note. These orthogonal functions could be used to seek the solution of, for example, the heat propagation in a 1D bar with its right end thermally insulated and the left end in contact with an outer medium, with which it exchanges heat according to Newton's law.

2.3 Stationary String in a Gravitational Field

A string of length L is stretched with a tension T and is fixed at $x = 0$. The other end (without a gravitational field) is kept at the same height as the first end. Furthermore, it is subject to an elastic force (it is connected to a spring with elastic constant β). The elastic force is zero when (in the absence of gravitational field) the right end is at the same height as the one at $x = 0$. Find the shape of the string when there is a gravitational field. Consider the linear density of mass ρ is uniform. What will the solution be in the limit $\beta \to \infty$?

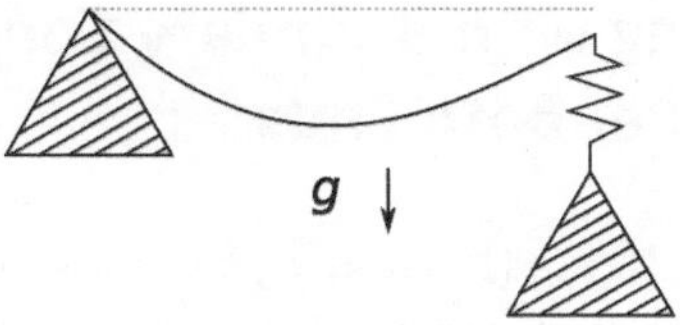

Figure 2.5

Mathematical formulation:

$$\left\{\begin{array}{l} T\dfrac{d^2u(x)}{dx^2} = \rho g \\ u(x=0) = 0 \\ T\left.\dfrac{du}{dx}\right|_{x=L} = -\beta u(L) \end{array}\right\} \quad (2.18)$$

Solution: We look for the general solution and apply the boundary conditions.

$$u(x) = \frac{\rho g}{2T}x^2 + Cx + D \quad (2.19)$$

From the first boundary condition $u(0) = 0$ we have $D = 0$. From the second condition:

$$T\left.\frac{du}{dx}\right|_{x=L} + \beta u(L) = 0 \to \rho g L + CT + \beta\left[\frac{\rho g L^2}{2T} + CL\right] = 0 \quad (2.20)$$

$$C = -\frac{\rho g L}{2T}\frac{2T + \beta L}{T + \beta L} \quad (2.21)$$

C must be negative to guarantee $u(L) < 0$.

The solution in the limit $\beta = \infty$ corresponds to both ends fixed and is (as can be checked by imposing $\beta = \infty$ in the result) obtained in [1]:

$$u(x) = \frac{\rho g}{2T}(x^2 - Lx) \quad (2.22)$$

Comment: in the case of also solving the vibrations of the string with the same contours, the shape of the solution could include the static shape of the string and vibrations of a string with homogeneous boundary conditions and no external forces could be summed to the previous to describe oscillations of the string with the new equilibrium position.

2.4 Static String with Boundary Conditions of the Third Kind at Both Ends

The ends of a tense string (tension T) can move transversally, since they are attached to two springs (of elastic constant β). Find the form of the string when, under the action of a constant point force (F) applied at x_0 it is in mechanical equilibrium. See also problem (3.4) from and section 3.2.2 from [1].

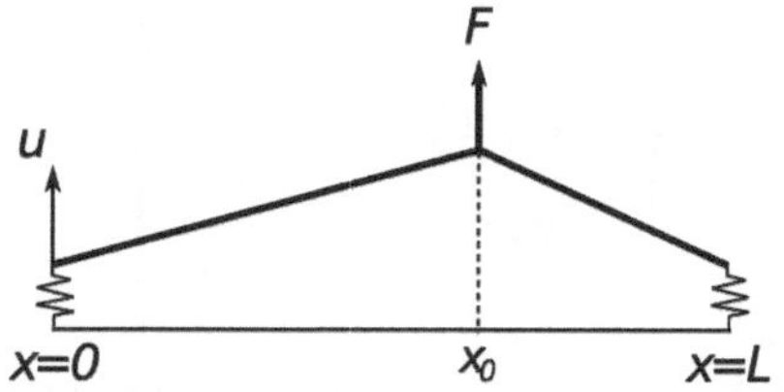

Figure 2.6

Mathematical formulation and general solution:

Gravity is neglected in this problem:

$$T\frac{d^2u(x)}{dx^2} = -f(x) = -F\delta(x - x_0) \tag{2.23}$$

The boundary conditions are:

$$T\left.\frac{\partial u_1}{\partial x}\right|_{x=0} = \beta u_1(0)$$
$$-T\left.\frac{\partial u_2}{\partial x}\right|_{x=L} = \beta u_2(L)$$

We seek a solution by parts:

$$u(x) = \begin{Bmatrix} u_1(x) & (0 < x < x_0) \\ u_2(x) & (x_0 < x < L) \end{Bmatrix} \tag{2.24}$$

Since in the indicated ranges the equation is homogeneous (there are no forces applied):

$$T\frac{d^2u(x)}{dx^2} = 0 \rightarrow \tag{2.25}$$

$$u(x) = \left\{ \begin{array}{ll} u_1(x) = A_1 x + B_1 & (0 < x < x_0) \\ u_2(x) = A_2 x + B_2 & (x_0 < x < L) \end{array} \right\} \quad (2.26)$$

Now the boundary conditions and the continuity of the functions $u_{1,2}(x)$. Also the parameter ε is used for the discontinuity at x_0.

First continuity condition: $u_1(x_0 - \varepsilon) = u_2(x_0 + \varepsilon)$ $(\lim \varepsilon \to 0)$

Second continuity condition: We integrate equation (2.23) in an interval 2ε centered at x_0 and take the limit $\varepsilon \to 0$:

$$T \cdot \lim_{\varepsilon \to 0} [u_{2x}(x_0 - \varepsilon) - u_{1x}(x_0 + \varepsilon)] = -F \quad (2.27)$$

Applying the four conditions to the solution (2.26) we arrive at a system of four equations with four unknowns $A_{1,2}$; $B_{1,2}$:

$$\left\{ \begin{array}{l} A_1 x_0 + B_1 = A_2 x_0 + B_2 \\ A_2 - A_1 = -\dfrac{F}{T} \\ T A_1 = \beta B_1 \\ -T A_2 = \beta(A_2 L + B_2) \end{array} \right\} \quad (2.28)$$

Final solution

From this system of equations, the coefficients A_1, A_2, B_1, B_2 are determined and we arrive at the solution $u(x)$. From the two last equations we can write the solution as:

$$u(x) = \left\{ \begin{array}{ll} A_1 \left(x + \dfrac{T}{\beta} \right) & (0 < x < x_0) \\ 8pt[A_2 \left(x - L - \dfrac{T}{\beta} \right) & (x_0 < x < L) \end{array} \right\} \quad (2.29)$$

From the two first equations from the system, we finish obtaining the value of the coefficients.

$$\left\{ \begin{array}{l} A_1 = \dfrac{F}{T} \left(1 - \dfrac{\beta x_0 + T}{2T + L\beta} \right) \\ B_1 = \dfrac{F}{\beta} \left(1 - \dfrac{\beta x_0 + T}{2T + L\beta} \right) \\ A_2 = -\dfrac{F}{T} \left(\dfrac{\beta x_0 + T}{2T + L\beta} \right) \\ B_2 = \dfrac{F}{T} \left(L + \dfrac{T}{\beta} \right) \left(\dfrac{\beta x_0 + T}{2T + L\beta} \right) \end{array} \right\} \quad (2.30)$$

We can check that the solution is converted in that which is explained in [1] for the case of fixed ends ($\beta \to \infty$).

Note 1: the function $\frac{u(x,x_0)}{F}$ represents the Green's function of the problem.

Note 2: Since it's a static problem this solution can be extended to problems described by an equation like the heat equation but without first order time derivative.

2.5 String with a Point Mass Hanging from One of Its Ends

Consider a string of length L, tension T and linear density of mass ρ. The string has the left end fixed and the right end can move freely in the transversal direction. The string is placed in the Earth's gravitational field (g). Determine the stationary form of the string after an infinite amount of time if there is a point mass m hanging from its right end ($x = L$).

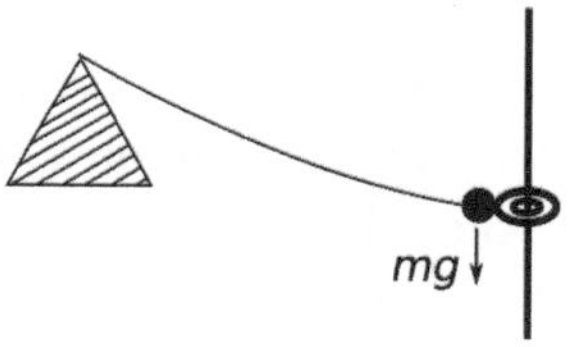

Figure 2.7

Mathematical formulation

We assume that the mass is at a distance $\varepsilon \to 0$ from the right end at $x = L$. First we will discuss how to find the density of forces. The total force applied to the string is directed in the negative direction:

$$f(x) = \frac{F}{L} = -\left[\rho g + mg\delta(x - L + \varepsilon)\right] \tag{2.31}$$

This form describes two contributions:

Without the point mass:

$$F_1 = \int_0^L g\rho dx = \rho g L \tag{2.32}$$

Force due to the point mass:

$$F_2 = \int_0^L mg\delta(x - L + \varepsilon)dx = mg \tag{2.33}$$

General solution Then the equation to be solved will be:

$$\rho\frac{\partial^2 u}{\partial t^2} - T\frac{\partial^2 u}{\partial x^2} = -[\rho g + mg\delta(x - L + \varepsilon)] \tag{2.34}$$

since $\frac{d^2u}{dt^2} \to 0$

$$-T\frac{d^2u}{dx^2} = -[\rho g + mg\delta(x - L + \varepsilon)] \tag{2.35}$$

And we have the boundary conditions:

$$u(0, t) = 0; \quad \left.\frac{\partial u}{\partial x}\right|_{x=L} = 0 \tag{2.36}$$

We will solve the problem by parts. The first part $u_1(x, t)$ will be comprised between the points $x = 0$ and $x = L - \varepsilon$. There,

$$\frac{d^2u_1}{dx^2} = \frac{\rho g}{T} \tag{2.37}$$

Whose function is, considering the left boundary condition:

$$u_1(x) = \frac{\rho g}{T}\frac{x^2}{2} + Bx \tag{2.38}$$

On the other hand, the equation for the solution $u_2(x)$, to the right of the point mass, when $\varepsilon \to 0$ will be:

$$-T\frac{d^2u_2}{dx^2} = 0 \tag{2.39}$$

from the right boundary condition:

$$\left.\frac{du_2}{dx}\right|_{x=L} = 0 \tag{2.40}$$

Final solution Integrating the wave equation in the surroundings of the point $L - \varepsilon$ we find the change of derivative in the point where the point mass is:

$$-\left.\frac{du_1}{dx}\right|_{x=L-\varepsilon} - \left[-\left.\frac{du_2}{dx}\right|_{x=L-\varepsilon}\right] = \frac{mg}{T} \tag{2.41}$$

since $\left.\frac{du_2}{dx}\right|_{x=L-\varepsilon} = 0$ due to the right boundary condition:

$$\left.\frac{du_1}{dx}\right|_{x=L-\varepsilon} = -\frac{mg}{T} \tag{2.42}$$

This derivative determines the angle of the string at the point where the mass is attached. Appling this last condition to $u_1(x)$ in the limit $\varepsilon \to 0$:

$$\left.\frac{du_1}{dx}\right|_{x=L-\varepsilon} = \frac{\rho g}{T}L + B = -\frac{mg}{T} \tag{2.43}$$

Then:

$$B = -\frac{\rho g}{T}L - \frac{mg}{T} \tag{2.44}$$

The form of the string is:

$$u_1(x) = \frac{\rho g}{T}\frac{x^2}{2} - \left(\frac{\rho g}{T}L + \frac{mg}{T}\right)x \tag{2.45}$$

Note. In principle, the possible alternative way would be to ignore the point mass in the equation (it would be $T\frac{\partial^2 u}{\partial x^2} = g\rho$, with $\rho = const$) and to include the point mass only as the effect that gravity would have on it through the boundary condition at the end $x = L$: $T\left.\frac{\partial u}{\partial x}\right|_{x=L} - mg = 0$. However, the advantage of the method used is that it first detaches the mass from the border so that it could be used to find the dynamics of the problem (formulate and resolve the Sturm–Liouville problem) and then to search for the dynamic solution asymptotes at the limit $\epsilon \to 0$.

2.6 String with a Point Mass in Its Center and Second and Third Type Boundary Conditions

A string with tension T and linear density of mass ρ has a point mass m in its center ($x = L/2$). The right end can move freely and the left one is connected to a spring with constant β. Determine the eigenfunctions and eigenvalues of the Sturm–Liouville problem and the static form of the string in the gravitational field.

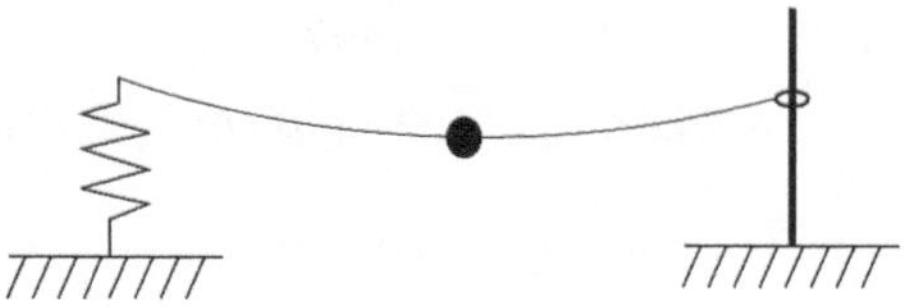

Figure 2.8

Mathematical formulation

$$\left\{ \begin{array}{l} \rho(x)\dfrac{\partial u^2}{\partial t^2} - T\dfrac{\partial u^2}{\partial x^2} = 0 \\ \left.\dfrac{\partial u}{\partial x}\right|_{x=0} - hu(0) = 0 \\ \left.\dfrac{\partial u}{\partial x}\right|_{x=L} = 0 \end{array} \right\} \tag{2.46}$$

Being $h = \frac{\beta}{T}$. The density of mass is $\rho(x) = \rho + m\delta(x - \frac{L}{2})$. The total mass of the string, for this density is the expected one: $\int_0^L \rho(x)dx = \rho L + m$. If the mass were delocalized we would need to formulate the density in different parts, but we take it to be point-like.

Sturm–Liouville problem Separating variables we arrive to the Sturm–Liouville problem:

$$\left\{ \begin{array}{l} \dfrac{d^2X}{dx^2} + \lambda\rho(x)X = 0 \\ \left.\dfrac{dX}{dx}\right|_{x=0} - hX(0) = 0 \\ \left.\dfrac{dX}{dx}\right|_{x=L} = 0 \end{array} \right\} \tag{2.47}$$

To find the eigenfunctions avoiding the anomalous point where the linear density of mass diverges, we need to separate the eigenfunctions in two parts.

$$X_1(x) = A\sin(\sqrt{\lambda\rho}x) + B\cos(\sqrt{\lambda\rho}x) \tag{2.48}$$

$$X_2(x) = C\cos(\sqrt{\lambda\rho}[x - L]) \tag{2.49}$$

(here the second boundary condition has already been applied). The first condition is applied for $X_1(x)$:

$$A\sqrt{\lambda\rho}\cos(\sqrt{\lambda\rho}0) \quad -B\sqrt{\lambda\rho}\sin(\sqrt{\lambda\rho}0)$$
$$-h[A\sin(\sqrt{\lambda\rho}0)+B\cos(\sqrt{\lambda\rho}0)]=0 \quad (2.50)$$

$$A\sqrt{\lambda\rho}-hB=0 \quad (2.51)$$

Then $B=A\frac{\sqrt{\lambda\rho}}{h}$

$$X_1(x)=A\left[\sin(\sqrt{\lambda\rho}x)+\frac{\sqrt{\lambda\rho}}{h}\cos(\sqrt{\lambda\rho}x)\right] \quad (2.52)$$

We impose the continuity condition for the eigenfunctions:

$$A\left[\sin\left(\sqrt{\lambda\rho}\frac{L}{2}\right)+\frac{\sqrt{\lambda\rho}}{h}\cos\left(\sqrt{\lambda\rho}\frac{L}{2}\right)\right]-C\cos\left(\sqrt{\lambda\rho}\frac{L}{2}\right)=0 \quad (2.53)$$

We now apply the discontinuity of the first derivatives of the eigenfunctions by integrating equation (2.47) around the point $x=L/2$.

$$\left.\frac{dX_1}{dx}\right|_{x=\frac{L}{2}}-\left.\frac{dX_2}{dx}\right|_{x=\frac{L}{2}}=\lambda m X_2\left(\frac{L}{2}\right)=C\lambda m\cos\left(\sqrt{\lambda\rho}\frac{L}{2}\right) \quad (2.54)$$

$$A\left[\sqrt{\lambda\rho}\cos\left(\sqrt{\lambda\rho}\frac{L}{2}\right)-\frac{\lambda\rho}{h}\sin\left(\sqrt{\lambda\rho}\frac{L}{2}\right)\right]$$
$$+C\left[\sqrt{\lambda\rho}\sin\left(\sqrt{\lambda\rho}\frac{L}{2}\right)-\lambda m\cos\left(\sqrt{\lambda\rho}\frac{L}{2}\right)\right]=0 \quad (2.55)$$

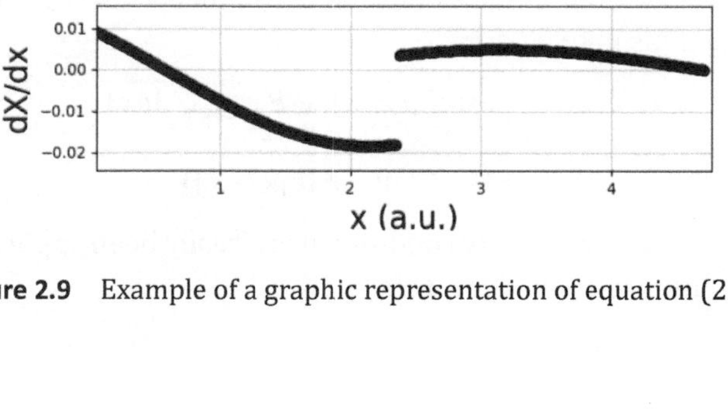

Figure 2.9 Example of a graphic representation of equation (2.54.

With the system of equations (2.53, 2.55) we can form a determinant which, when equated to zero, gives the eigenvalues λ_n. Furthermore, from equation (2.53), we have:

$$\frac{A}{C} = \frac{\cos(\sqrt{\lambda\rho}\frac{L}{2})}{\sin\left(\sqrt{\lambda\rho}\frac{L}{2}\right) + \frac{\sqrt{\lambda\rho}}{h}\cos\left(\sqrt{\lambda\rho}\frac{L}{2}\right)} \quad (2.56)$$

Then:

$$X_n(x) = A \times \left\{ \begin{array}{ll} \sin(\sqrt{\lambda_n\rho}x) + \frac{\sqrt{\lambda_n\rho}}{h}\cos(\sqrt{\lambda_n\rho}x) & \left(x < \frac{L}{2}\right) \\ \frac{\sin\left(\sqrt{\lambda_n\rho}\frac{L}{2}\right) + \frac{\sqrt{\lambda_n\rho}}{h}\cos\left(\sqrt{\lambda_n\rho}\frac{L}{2}\right)}{\cos\left(\sqrt{\lambda_n\rho}\frac{L}{2}\right)}\cos\left(\sqrt{\lambda_n\rho}[x-L]\right) & \left(x > \frac{L}{2}\right) \end{array} \right\} \quad (2.57)$$

Static form of the string in the presence of the gravitational field The general mathematic formulation (without specifying the position of the mass. x_0) is:

$$T\frac{d^2u(x)}{dx^2} = -f(x) = -[-mg\delta(x - x_0) - g\rho] \quad (2.58)$$

We look for a solution by parts:

$$u(x) = \left\{ \begin{array}{ll} u_1(x) & \left(0 < x < \frac{L}{2}\right) \\ u_2(x) & \left(\frac{L}{2} < x < L\right) \end{array} \right\} \quad (2.59)$$

Since in the indicated ranges the equation is non-homogeneous the following type (assuming a uniform density):

$$T\frac{d^2u(x)}{dx^2} = g\rho \quad (2.60)$$

$$\rightarrow u(x) = \left\{ \begin{array}{ll} \frac{g\rho}{2T}x^2 + A_1x + B_1 & \left(0 < x < \frac{L}{2}\right) \\ \frac{g\rho}{2T}x^2 + A_2x + B_2 & \left(\frac{L}{2} < x < L\right) \end{array} \right\} \quad (2.61)$$

We formulate the boundary and continuity conditions for the functions $u_{1,2}(x)$. We use the parameter ε, which measures the proximity to the point $\frac{L}{2}$.

Continuity of the solution: $u_1\left(\frac{L}{2}-\varepsilon\right)=u_2\left(\frac{L}{2}+\varepsilon\right)$ ($\lim \varepsilon\rightarrow 0$)

$$\frac{g\rho}{2T}\left[\frac{L}{2}\right]^2+A_1\left[\frac{L}{2}\right]+B_1=\frac{g\rho}{2T}\left[\frac{L}{2}\right]^2+A_2\left[\frac{L}{2}\right]+B_2 \qquad (2.62)$$

$$(A_2-A_1)\frac{L}{2}=B_1-B_2 \qquad (2.63)$$

Discontinuity of the derivatives: We integrate equation (2.58) in the proximity ε of the point $\frac{L}{2}$ and take the limit $\varepsilon\rightarrow 0$.

$$T\int_{\frac{L}{2}-\varepsilon}^{\frac{L}{2}+\varepsilon}\frac{d^2u(x)}{dx^2}=-\int_{\frac{L}{2}-\varepsilon}^{\frac{L}{2}+\varepsilon}f(x)dx=\{\varepsilon\rightarrow 0\}=mg\int_{\frac{L}{2}-\varepsilon}^{\frac{L}{2}+\varepsilon}\delta(x-x_0)dx=mg \qquad (2.64)$$

We arrive at:

$$T\left[\frac{g\rho}{T}\frac{L}{2}+A_2\right]-T\left[\frac{g\rho}{T}\frac{L}{2}+A_1\right]=mg \qquad (2.65)$$

$$A_2-A_1=\frac{mg}{T} \qquad (2.66)$$

Boundary condition for the right end: $\left.\frac{du_2}{dx}\right|_{x=L}=0$

$$A_2=-\frac{g\rho}{T}L \qquad (2.67)$$

Condition for the left end:

$$T\left.\frac{du}{dx}\right|_{x=0}-\beta u(0)=0 \qquad (2.68)$$

$$TA_1-\beta B_1=0\rightarrow B_1=\frac{TA_1}{\beta} \qquad (2.69)$$

Final solution We have a system of four equations with four unknowns $A_{1,2}$; $B_{1,2}$:

$$\left\{\begin{array}{l}(A_2-A_1)\frac{L}{2}=B_1-B_2\\ A_2-A_1=\frac{mg}{T}\\ A_2=-\frac{g\rho}{T}L\\ B_1=\frac{TA_1}{\beta}\end{array}\right\} \qquad (2.70)$$

$$\left\{\begin{aligned} A_1 &= -\left[\frac{\rho g L}{T} + \frac{mg}{T}\right] \\ B_1 &= -\frac{1}{\beta}[\rho g L + mg] \\ A_2 &= -\frac{g\rho}{T} L \\ B_2 &= -\frac{1}{\beta}[\rho g L + mg] - \frac{mgL}{2T} \end{aligned}\right\} \tag{2.71}$$

$$u(x) = \left\{\begin{aligned} &\frac{\rho g}{2T}[x^2 - 2Lx] - \frac{mg}{T}x - \frac{1}{\beta}(\rho g L + mg) \; \left(0 < x < \frac{L}{2}\right) \\ &\frac{\rho g}{2T}[x^2 - 2Lx] - \frac{mgL}{2T} - \frac{1}{\beta}(\rho g L + mg) \; \left(\frac{L}{2} < x < L\right) \end{aligned}\right\} \tag{2.72}$$

2.7 Static Form of a String with a Mass

In the middle point of a tense string a mass m is placed. The string has a fixed end and the other can move transversally. Determine the shape of the string when, under the action of gravity, is in mechanical equilibrium.

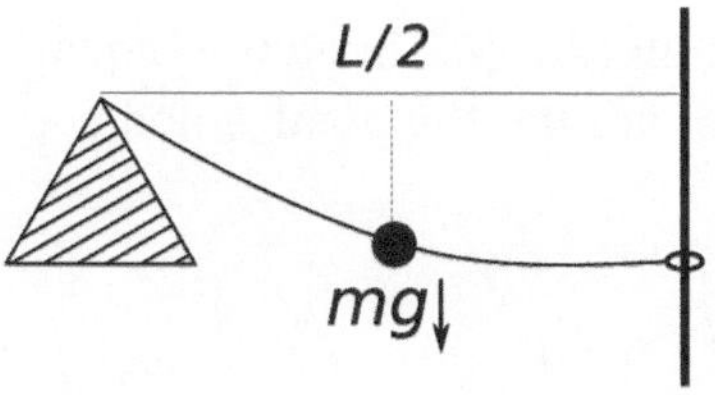

Figure 2.10

Mathematical formulation:

$$T\frac{d^2u(x)}{dx^2} = -f(x) = -[-mg\delta(x - x_0) - g\rho] \tag{2.73}$$

(Note that the forces are directed in the negative direction).

We will look for the solution in the form of two functions:

$$u(x) = \begin{Bmatrix} u_1(x)\ (0 < x < \frac{L}{2}) \\ u_2(x)\ (\frac{L}{2} < x < L) \end{Bmatrix} \tag{2.74}$$

In the indicated ranges the equation is non-homogeneous (uniform density approximation).

$$T\frac{d^2u(x)}{dx^2} = g\rho \tag{2.75}$$

$$\rightarrow u(x) = \begin{Bmatrix} \frac{g\rho}{2T}x^2 + A_1x + B_1 \left(0 < x < \frac{L}{2}\right) \\ \frac{g\rho}{2T}x^2 + A_2x + B_2 \left(\frac{L}{2} < x < L\right) \end{Bmatrix} \tag{2.76}$$

We formulate the boundary conditions and the continuity conditions for the functions $u_{1,2}(x)$, introducing the parameter ε, which measures the proximity to $\frac{L}{2}$.

Continuity condition 1:

$$u_1\left(\frac{L}{2} - \varepsilon\right) = u_2\left(\frac{L}{2} + \varepsilon\right) \qquad (\varepsilon \rightarrow 0) \tag{2.77}$$

$$\frac{g\rho}{2T}\left[\frac{L}{2}\right]^2 + A_1\left[\frac{L}{2}\right] + B_1 = \frac{g\rho}{2T}\left[\frac{L}{2}\right]^2 + A_2\left[\frac{L}{2}\right] + B_2 \tag{2.78}$$

Continuity condition 2: We integrate equation (2.73) in the proximity of size ε around the point $\frac{L}{2}$ and consider the limit: lim $\varepsilon \rightarrow 0$:

$$T\int_{\frac{L}{2}-\varepsilon}^{\frac{L}{2}+\varepsilon} \frac{d^2u}{dx^2} = -\int_{\frac{L}{2}-\varepsilon}^{\frac{L}{2}+\varepsilon} f(x)dx = \{\varepsilon \rightarrow 0\} = mg\int_{\frac{L}{2}-\varepsilon}^{\frac{L}{2}+\varepsilon} \delta(x - x_0)dx = mg \tag{2.79}$$

We arrive at the second continuity condition:

$$T \cdot \lim_{\varepsilon \rightarrow 0}\left[\left.\frac{du_2}{dx}\right|_{\frac{L}{2}+\varepsilon} - \left.\frac{du_1}{dx}\right|_{\frac{L}{2}-\varepsilon}\right] = mg \tag{2.80}$$

$$T\left[\frac{g\rho}{T}\left[\frac{L}{2}\right] + A_2\right] - T\left[\frac{g\rho}{T}\left[\frac{L}{2}\right] + A_1\right] = mg \tag{2.81}$$

$$A_2 - A_1 = \frac{mg}{T} \tag{2.82}$$

Now we consider the boundary conditions:

$$u_1(0) = 0 \rightarrow \ B_1 = 0 \tag{2.83}$$

$$\left.\frac{du_2}{dx}\right|_{x=L} = 0 \rightarrow \frac{g\rho}{T}L + A_2 = 0 \tag{2.84}$$

Applying these four conditions to the solution (2.76) we arrive at a four equation system, with four unknowns $A_{1,2}$, $B_{1,2}$.

$$\left\{\begin{array}{l} A_1\frac{L}{2} + B_1 = A_2\frac{L}{2} + B_2 \\ A_2 - A_1 = \frac{mg}{T} \\ B_1 = 0 \\ \frac{g\rho}{T}L + A_2 = 0 \end{array}\right\} \tag{2.85}$$

$$u(x) = \left\{\begin{array}{l} \frac{\rho g}{2T}[x^2 - 2Lx] - \frac{mg}{T}x \ \left(0 < x < \frac{L}{2}\right) \\ \frac{\rho g}{2T}[x^2 - 2Lx] - \frac{mgL}{2T} \ \left(\frac{L}{2} < x < L\right) \end{array}\right\} \tag{2.86}$$

Note 1: The correct result must imply continuity in the derivative at $x = L/2$ for $m = 0$.

Note 2: A similar problem can be formulated, consisting of the heat conduction in a thin bar or rod (one dimensional problem) which has a similar solution.

2.8 Heat Conduction through a Semi-Insulated Bar

Consider a bar of length L thermically insulated on the left end ($x = 0$) and in thermal contact with a reservoir at $T = 0$ on the right end ($x = L$). The material has a thermal diffusion coefficient equal to χ.

Find the temperature distribution as a function of time if at $t = 0$ both ends are at $T = 0$ and the temperature distribution has a triangular shape with maximum value $T = \Theta$ at the center of the bar.

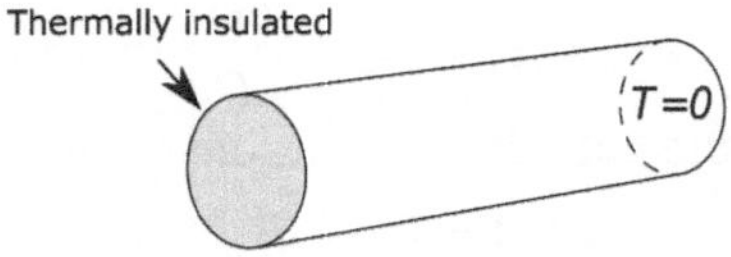

Figure 2.11

Solution:

Mathematical formulation

$$\frac{\partial u(x,t)}{\partial t} = \chi \frac{d^2u(x,t)}{dx^2} \tag{2.87}$$

$$u(x,t) = \begin{Bmatrix} \frac{2\Theta}{L}x & \left(x < \frac{L}{2}\right) \\ \frac{2\Theta}{L}(L-x) & \left(x > \frac{L}{2}\right) \end{Bmatrix} \tag{2.88}$$

$$\left.\frac{du}{dx}\right|_{x=0} = 0; \quad u(x=L) = 0 \tag{2.89}$$

Sturm–Liouville problem We replace a solution of the type $u(x,t) = A(t)X(x)$ in the heat equation and separate variables:

We have the next Sturm-Liouville problem for $X(x)$:

$$\frac{dX^2}{dx^2} + \lambda X = 0 \tag{2.90}$$

$$\left.\frac{dX}{dx}\right|_{x=0} = 0; \quad X(L) = 0; \tag{2.91}$$

The general solution for $X(x)$ is:

$$X(x) = C\sin(\sqrt{\lambda}x) + B\cos(\sqrt{\lambda}x)$$

from $\left.\frac{dX}{dx}\right|_{x=0} = 0$ we get $C = 0$ and from $X(L) = 0$, $\sqrt{\lambda}L = \frac{\pi}{2}(2n+1)$, with $n = 0, 1, 2, \ldots$

The eigenvalues are: $\lambda_n = \left[\frac{\pi}{2L}(2n+1)\right]^2$

The eigenfunction are: $X_n(x) = \cos\left[\frac{\pi}{2L}(2n+1)x\right]$

General solution We take into account now also the temporal part which resulted from the previous separation of variables:

$$\frac{dA_n}{dt} + \chi A_n = 0 \tag{2.92}$$

With which we get to the general solution:

$$u(x, t) = \sum A_n \exp(-\chi \lambda_n t) \cos\left[\frac{\pi}{2L}(2n+1)x\right] \tag{2.93}$$

Final solution Initial conditions:

$$u(x, 0) = \left\{ \begin{array}{ll} \dfrac{2\Theta}{L}x & \left(x < \dfrac{L}{2}\right) \\ \dfrac{2\Theta}{L}(L-x) & \left(x > \dfrac{L}{2}\right) \end{array} \right\} \tag{2.94}$$

Equating them to the solution at $t = 0$: $\sum A_n \cos[\frac{\pi}{2L}(2n+1)x]$

Using the orthogonality of $\cos\left[\frac{\pi}{2L}(2n+1)x\right]$ we find the A_n coefficients:

$$\begin{aligned} A_n &= \frac{2}{L}\left[\int_0^{L/2} \frac{2\Theta}{L} x \cos(\lambda_n x) dx + \int_{L/2}^{L} \frac{2\Theta}{L}(L-x)\cos(\lambda_n x) dx\right] \\ &= \frac{4\Theta}{L^2}\int_0^{L/2} x\cos(\lambda_n x)dx + \frac{4\Theta}{L}\int_{L/2}^{L}\cos(\lambda_n x)dx - \frac{4\Theta}{L^2}\int_{L/2}^{L} x\cos(\lambda_n x)dx \\ &= \frac{4\Theta}{L^2}\int_0^{L/2} x\cos(\lambda_n x)dx + \frac{4\Theta}{L}\int_{L/2}^{L}\cos(\lambda_n x)dx - \frac{4\Theta}{L^2}\int_{L/2}^{L} x\cos(\lambda_n x)dx \\ &= \frac{4\Theta}{L^2}\left\{\frac{x\sin(\lambda_n x)}{\lambda_n}\Bigg|_0^{L/2} - \frac{\cos(\lambda_n x)}{\lambda_n}\Bigg|_0^{L/2}\right\} \\ &\quad + \frac{4\Theta}{L}\left\{-\frac{\sin(\lambda_n x)}{\lambda_n}\Bigg|_{L/2}^{L}\right\} - \frac{4\Theta}{L^2}\left\{\frac{x\sin(\lambda_n x)}{\lambda_n}\Bigg|_{L/2}^{L} - \frac{\cos(\lambda_n x)}{\lambda_n}\Bigg|_{L/2}^{L}\right\} \end{aligned} \tag{2.95}$$

From where we get:

$$A_n = \frac{16\Theta}{\pi^2(2n+1)^2}\left[2\cos\left(\frac{\pi}{2}\frac{2n+1}{2}\right) - 1\right] \tag{2.96}$$

2.9 Variation of the Temperature of a Thin Rod as a Function of Time

Suppose a bar that is at a temperature T_0 until $t = 0$. At this moment both ends of the bar are brought into contact with two thermal reservoirs at temperatures T_1 and T_0. Find the distribution of temperature in the bar as a function of time.

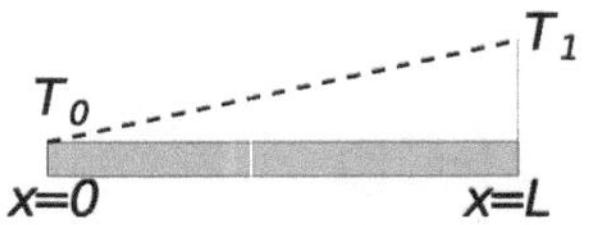

Figure 2.12

Mathematical formulation:

$$\left\{\begin{array}{l} \dfrac{\partial u(x,t)}{\partial t} = \chi \dfrac{d^2u(x,t)}{dx^2} \\ u(0,t) = T_0 \\ u(L,t) = T_1 \\ u(x,0) = T_0 \end{array}\right\} \tag{2.97}$$

We look for a solution as the sum of a function $u_0(x,t)$ (which satisfies the non-homogeneous boundary conditions) and another function $v(x,t)$ (which satisfies the homogeneous boundary conditions).

$$u(x,t) = u_0(x) + v(x,t) \tag{2.98}$$

Since the boundary conditions do not depend on time, we can choose the $u_0(x)$ function to be independent of time and, as we will see, correspond to the stationary solution.
Boundary conditions for $u_0(x)$.

$$v(0,t) = 0 \rightarrow u_0(0) = T_0$$

$$v(L,t) = 0 \rightarrow u_0(L) = T_1$$

Then the simplest solution is:

$$u_0(x) = T_0 + \frac{T_1 - T_0}{L}x \tag{2.99}$$

Mathematical formulation of the problem for $v(x, t)$

$$\begin{cases} \dfrac{\partial v}{\partial t} = \chi \dfrac{\partial^2 v}{\partial x^2} \\ v(0, t) = 0 \\ v(L, t) = 0 \\ v(x, 0) = \frac{T_0 - T_1}{L} x \end{cases} \tag{2.100}$$

We will use the method of separation of variables for the eigenfunctions in the x direction:

$$X_n(x) = A_n \sin\left[\frac{\pi n}{L} x\right] \tag{2.101}$$

$$\lambda_n = \left[\frac{\pi n}{L}\right]^2 \tag{2.102}$$

The solution for the temporal part is obtained by solving the equation:

$$\frac{dC_n}{dt} + \chi \lambda_n C_n = 0 \tag{2.103}$$

General solution The general solution is:

$$u(x, t) = \sum A_n e^{-\chi \lambda_n t} \sin\left[\frac{\pi n}{L} x\right] \tag{2.104}$$

From the initial conditions $v(x, 0) = \left\{\frac{T_0 - T_1}{L} x\right\} = \sum A_n \sin\left[\frac{\pi n}{L} x\right]$ using the orthogonality of the eigenfunctions $\sin[\frac{\pi n}{L} x]$ we find the coefficients of the expansion:

$$A_n = \frac{2}{L} \int_0^L \left(\frac{T_0 - T_1}{L} x\right) \sin\left[\frac{\pi n}{L} x\right] dx = 2 \frac{(T_0 - T_1)}{\pi n} (-1)^n \tag{2.105}$$

Final solution The final solution is:

$$u(x, t) = T_0 + \frac{T_1 - T_0}{L} x + v(x, t) \tag{2.106}$$

2.10 Thermal Conduction in a Bar with Insulated Ends

Find the variation of temperature as a function of time in a bar of length L with both ends thermally insulated if the initial distribution of temperature is $T(x, 0) = f(x)$.

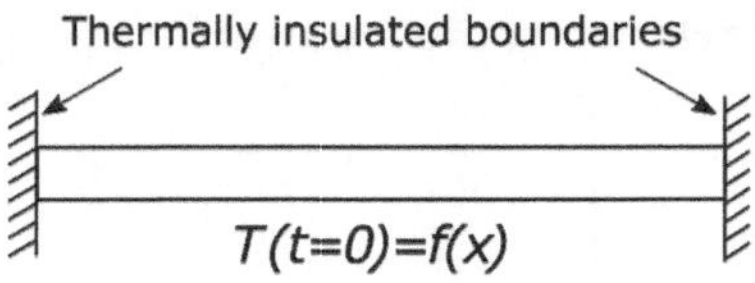

Figure 2.13

Mathematical formulation:

$$\begin{cases} \dfrac{\partial u}{\partial t} = c^2 \dfrac{\partial^2 u}{\partial x^2} \\ \left.\dfrac{\partial u}{\partial x}\right|_{x=0} = \left.\dfrac{\partial u}{\partial x}\right|_{x=L} \\ u(x, 0) = f(x) \end{cases} \tag{2.107}$$

Sturm–Liouville problem We use the separation of variables method:

$$u(x, t) = T(t)X(x) \tag{2.108}$$

To find the eigenfunctions $X_n(x)$ we have a problem analogous to a string, but free on both ends. The Sturm–Liouville problem for $X(x)$ is:

$$\begin{Bmatrix} \dfrac{dX}{dx} + \nu X = 0 \\ \left.\dfrac{dX}{dx}\right|_{x=0} = \left.\dfrac{dX}{dx}\right|_{x=L} = 0 \end{Bmatrix} \tag{2.109}$$

Eigenfunctions and eigenvalues:

$$X_n(x) = \cos\left(\sqrt{\nu_n} x\right) \tag{2.110}$$

$$\nu_n = \left(\frac{\pi n}{L}\right)^2 \qquad n = 0, 1, 2, 3, \ldots \tag{2.111}$$

General solution The general solution is $u(x, t) = \sum T_n(t)X_n(x)$

Final solution Replacing this solution in the heat equation and using the orthogonality properties of the eigenfunctions we get to the equation for the $T_n(t)$ coefficients, which determines the initial conditions.

$$T_n(t) = A_n e^{-\left(\frac{c\pi n}{L}\right)^2 t} \tag{2.112}$$

Solution:

$$u(x,t) = A_0 + \sum_{n=1}^{\infty} A_n e^{-\left(\frac{c\pi n}{L}\right)^2 t} \cos\left(\frac{\pi n}{L}x\right) \tag{2.113}$$

Using the initial conditions:

$$f(x) = A_0 + \sum_{n=1}^{\infty} A_n \cos\left(\frac{\pi n}{L}x\right) \tag{2.114}$$

and taking advantage of the orthogonality of the $X_n(x)$ eigenfunctions we determine the coefficients of the expansion:

$$A_n = \frac{2}{L}\int_0^L f(x)\cos\left(\frac{\pi n}{L}x\right) dx \tag{2.115}$$

2.11 Variation of the Temperature of a Bar as a Function of Time

A bar of length $L = 1$ has a thermal diffusivity coefficient equal to 1 (the ratio between the heat capacity times density and the thermal conductivity). At the initial instant ($t = 0$) the temperature of the bar has an exponential distribution $T(x, 0) = e^{-x}$. At $t = 0$, both ends are connected to a thermal reservoir with temperature equal to zero. Find the time variation of temperature of the bar.

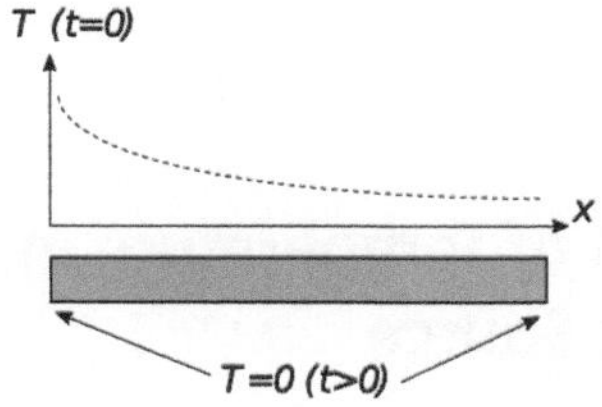

Figure 2.14

Mathematical formulation:

$$\frac{du(x,t)}{dt} = c^2 \frac{d^2u(x,t)}{dx^2}, \qquad c^2 = \frac{k}{c_p \rho} \tag{2.116}$$

First boundary condition: $u(0) = 0$

Second boundary condition: $u(L) = 0$

Initial condition: $u(x, 0) = e^{-x}(0 < x < L)$

$c =$ thermal diffusivity; k- thermal conductivity; $c_p =$ heat capacity; ρ- linear density of mass.

General solution Considering $L = c = 1$, when we use a solution obtained by separation of variables $u = T(t)X(x)$ we get the following general solution of the problem (with "fixed borders"), expanded in eigenfunctions for the given boundary conditions.

$$u(x, t) = \sum_{n=1}^{\infty} B_n e^{-(\pi n)^2 t} \sin(\pi n x) \tag{2.117}$$

Using the boundary conditions of the eigenfunctions and the initial condition we find the value of the coefficients:

$$\begin{aligned} B_n &= 2\int_0^1 e^{-x} \sin(\pi n x) dx \\ &= \frac{2e^{-x}}{[1+(\pi n)^2]}[-\sin(\pi n x) - \pi n \cos(\pi n x)]\Big|_0^1 \\ &= \frac{2\pi n}{[1+(\pi n)^2]}[1+(-1)^{n+1}e^{-1}] \end{aligned} \tag{2.118}$$

Final solution The final solution is:

$$u(x, t) = \sum_{n=1}^{\infty} \frac{2\pi n}{[1+(\pi n)^2]}[1+(-1)^{n+1}e^{-1}]e^{-(\pi n)^2 t} \sin(\pi n x) \tag{2.119}$$

2.12 Relaxation of Temperature in a Rod with a Local Heat Source

The stationary distribution of temperature of a very thin rod whose lateral surfaces are kept thermally insulated is initially determined by the presence of a local heat source with density $f(x) = Q\delta(x - x_0)$.

The left end of the bar is in contact with a thermal reservoir at zero temperature, while the right end exchanges heat according to Newton's law with the outer medium, which is at $T = 0$.

Find the temperature distribution along the bar when, starting at $t = 0$, the heat source is turned off.

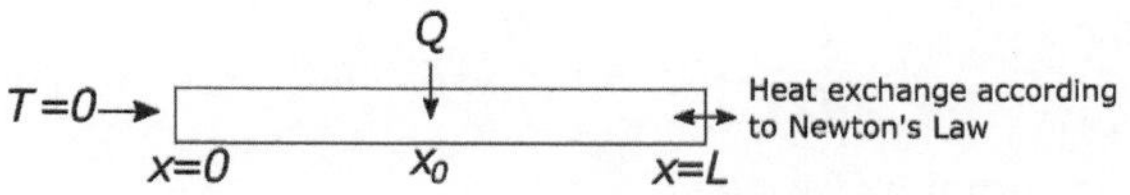

Figure 2.15

Mathematical formulation (for times prior to $t = 0$)

$$\left\{\begin{array}{c} C\rho\dfrac{du}{dt} - k\dfrac{d^2u}{dx^2} = f(x) \\ u(0) = 0 \\ \left.\dfrac{\partial u}{\partial x}\right|_{x=L} - hu(L) = 0 \end{array}\right\} \tag{2.120}$$

We will first find the stationary distribution at $t < 0$ in order to determine the initial condition.

$$\left\{\begin{array}{c} \dfrac{d^2u}{dx^2} = -\dfrac{q}{k}\delta(x - x_0) \\ u(0) = 0 \\ \left.\dfrac{\partial u}{\partial x}\right|_{x=L} - hu(L) = 0 \end{array}\right\} \tag{2.121}$$

We seek a solution in two parts:

$$u(x) = \left\{\begin{array}{lll} u_1(x) = & Ax + B & (x < x_0) \\ u_2(x) = & Cx + D & (x > x_0) \end{array}\right\} \tag{2.122}$$

From the boundary conditions:

$$\left\{\begin{array}{l} u(0) \to B = 0 \\ \left.\dfrac{\partial u}{\partial x}\right|_{x=L} - hu(L) = 0 \to C - hCL - hD = 0 \\ C = \dfrac{hD}{1 - hL} \qquad (1) \end{array}\right\} \tag{2.123}$$

From the condition of continuity:

$$\left\{\begin{array}{l} u_1(x_0) = u_2(x_0) \\ Ax_0 = Cx_0 + D \qquad (2) \end{array}\right\} \tag{2.124}$$

The condition of continuity of the derivatives is obtained by integrating the heat equation:

$$\frac{d^2}{dx^2}u(x) = -\frac{Q}{k}\delta(x - x_0) \tag{2.125}$$

In the surroundings of $x_0 \pm \varepsilon$

$$\left[\frac{du_2}{dx}\right]_{x_0+\varepsilon} - \left[\frac{du_1}{dx}\right]_{x_0-\varepsilon} = -\frac{Q}{k} \tag{2.126}$$

From here we get: $A - C = -\frac{Q}{k}$ (3)

We have a system of three equations (1-3) with three unknowns, A, B and C. Once we have found the form of the distribution of temperature at $t = 0$ (the $\psi(x)$ function) we can find its relaxation after the local heat source has been turned off ($f(x) = 0$). The

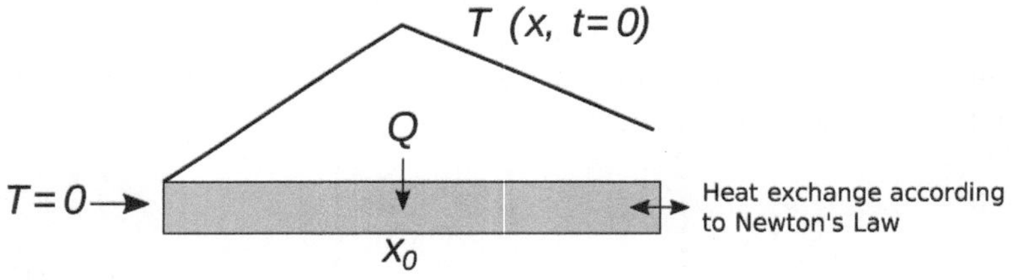

Figure 2.16

resulting time dependent problem to determine the time evolution of the temperature after turning off the heat source is:

$$\begin{cases} C\rho\frac{\partial u}{\partial t} - k\frac{\partial^2 u}{\partial x^2} = 0 \\ u(x = 0) = 0 \\ \left.\frac{\partial u}{\partial x}\right|_{x=L} - hu(x = L) = 0 \\ u(x.t = 0) = \psi(x) \end{cases} \tag{2.127}$$

Sturm–Liouville problem We seek the solution of the heat equation using the method of separation of variables and expand the solution in orthogonal eigenfunctions.

$$u(x, t) = \sum T(t)v(x) \tag{2.128}$$

being $v(x)$ the solutions of the Sturm–Liouville problem:

$$\left\{\begin{array}{l} \dfrac{d^2v}{dx^2} + \lambda v = 0 \\ v(0) = 0 \\ \left.\dfrac{\partial v}{\partial x}\right|_{x=L} - hv(x = L) = 0 \end{array}\right\} \tag{2.129}$$

Carefully observing the boundary conditions, we can present the solution in a more transparent form that contains the left boundary condition (we can also get to this form if we seek the solution as the sum of trigonometric functions).

$$v(x) = A\sin(\sqrt{\lambda}x) \tag{2.130}$$

Applying the condition of the right boundary:

$$\left.\frac{dv}{dx}\right|_{x=L} - hv(x = L) = 0 \tag{2.131}$$

$$\sqrt{\lambda}\cos(\sqrt{\lambda}L) - h\sin(\sqrt{\lambda}L) = 0 \tag{2.132}$$

We get to a transcendent equation for the λ_n eigenvalues:

$$\tan(\sqrt{\lambda}L) = \frac{\sqrt{\lambda}}{h} \tag{2.133}$$

General Solution Replacing $u(x) = \sum T(t)v(x)$ into the heat equation we arrive at the equation for the temporal part:

$$\frac{dT(t)}{dt} + \lambda_n T(t) = 0 \tag{2.134}$$

with the solution $T(t) = E_n e^{-\lambda_n t}$

The general solution is:

$$u(x, t) = \sum E_n e^{-\lambda_n t}\sin(\sqrt{\lambda_n}x)$$

Final solution From the initial condition, and using the orthogonality of the eigenfunctions we find the E_n coefficients:

$$\psi(x) = \sum_n E_n \sin(\sqrt{\lambda_n}x) \tag{2.135}$$

$$\int_0^L \psi(x)\sin(\sqrt{\lambda_n}x)dx = E_n\left|\sin(\sqrt{\lambda_n}x)\right|^2 \tag{2.136}$$

$$E_n = \frac{\int_0^L \psi(x)\sin(\sqrt{\lambda_n}x)dx}{\left|\sin(\sqrt{\lambda_n}x)\right|^2} \tag{2.137}$$

2.13 Heat Transfer in an Insulated Bar According to Newton's Law

Consider a one-dimensional bar of length L, insulated in its left end ($x = 0$). The right end ($x = L$) is in contact with the outer medium, which is at a temperature T_0 and they exchange heat according to Newton's law (with constant h). The thermal diffusivity coefficient is χ and the thermal conductivity coefficient is $k = 1$. Find the variations of the temperature as a function of time if at $t = 0$ the central part of the bar ($L/4 < x < 3L/4$) is at a temperature Θ, different from that of the outer medium.

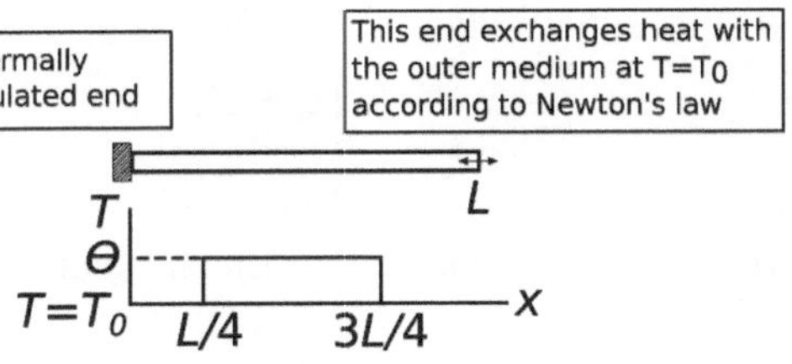

Figure 2.17

Mathematical formulation

$$\frac{\partial u(x,t)}{\partial t} = \chi \frac{d^2u(x,t)}{dx^2} \tag{2.138}$$

Initial conditions:

$$u(x,0) = \left\{ \begin{array}{ll} T_0 + \Theta & \left(\frac{L}{4} < x < \frac{3L}{4}\right) \\ T_0 & \left(x < \frac{L}{4}; x > \frac{3L}{4}\right) \end{array} \right\} \tag{2.139}$$

Boundary condition 1: $\left.\frac{du}{dx}\right|_{x=0} = 0$

Boundary condition 2: $\left.\frac{du}{dx}\right|_{x=L} = -h[u(L,t) - T_0]$

The sign imposed in the second boundary condition is determined by the form of the Fourier law and the direction of the vector normal to the boundary.

Sturm–Liouville problem We seek the solution by subtracting the thermal background T_0 so that we have homogeneous boundary conditions:

$$u(x, t) = v(x, t) + T_0 \tag{2.140}$$

The problem is solved by separation of variables: $v(x, t) = A(t)X(x)$. That is substituted in the heat equation. The spatial part is:

$$\frac{dX_n^2}{dx^2} + \lambda_n X = 0 \tag{2.141}$$

To obtain the boundary conditions for the Sturm–Liouville problem from the boundary conditions for the general solution:

$$v(x, t) = \sum A_n(t)X_n(x) \tag{2.142}$$

The boundary conditions must be imposed. For condition 1:

$$\left.\frac{dv}{dx}\right|_{x=0} = \sum A_n(t) \left.\frac{dX_n}{dx}\right|_{x=0} = 0 \tag{2.143}$$

Since $A_n(t)$ is independent from the position, $\left.\frac{dX_n}{dx}\right|_{x=0} = 0$

In a similar fashion we obtain for boundary condition 2:

$$\left.\frac{dX_n}{dx}\right|_{x=L} + hX(L) = 0 \tag{2.144}$$

The Sturm–Liouville problem we need to solve then is:

$$\left\{ \begin{array}{l} \dfrac{d^2X}{dx^2} + \lambda X = 0 \\ \left.\dfrac{dX}{dx}\right|_{x=0} = 0 \\ \left.\dfrac{dX}{dx}\right|_{x=L} + hX(L) = 0 \end{array} \right\} \tag{2.145}$$

General solution The general solution for $X(x)$ is:

$$X(x) = A\sin(\sqrt{\lambda}x) + B\cos(\sqrt{\lambda}x) \tag{2.146}$$

From the condition $\left.\frac{dX}{dx}\right|_{x=0}$ we have $A = 0$

And from $\left.\frac{dX}{dx}\right|_{x=L} + hX(L) = 0$ we get:

$$-\sqrt{\lambda}\sin(\sqrt{\lambda}L) = -h\cos(\sqrt{\lambda}L) \tag{2.147}$$

The eigenvalues λ_n are given by the equation

$$\tan(\sqrt{\lambda}L) = \frac{h}{\sqrt{\lambda}} \tag{2.148}$$

Then

$$X_n(x) = B_n \cos[\sqrt{\lambda_n}x] \tag{2.149}$$

The temporal part is:

$$\frac{dA_n(t)}{dt} + \chi\lambda_n A_n(t) = 0 \tag{2.150}$$

The general solution for the complete problem is:

$$v(x,t) = \sum_n A_n e^{-\chi\lambda_n t}\cos[\sqrt{\lambda_n}x] \tag{2.151}$$

Final solution We have the initial condition:

$$v(x,0) = \begin{Bmatrix} \Theta & \left(\frac{L}{4} < x < \frac{3L}{4}\right) \\ 0 & \left(x < \frac{L}{4}; x > \frac{3L}{4}\right) \end{Bmatrix}$$

Using the orthogonality of the eigenfunctions $\cos(\sqrt{\lambda_n}x)$ it is possible to find the coefficients A_n:

$$\sum_n A_n \cos[\sqrt{\lambda_n}x] = \begin{Bmatrix} \Theta & \left(\frac{L}{4} < x < \frac{3L}{4}\right) \\ 0 & \left(x < \frac{L}{4}; x > \frac{3L}{4}\right) \end{Bmatrix} \tag{2.152}$$

$$\begin{aligned} A_n &= \frac{\Theta}{\int_0^L \left|\cos[\sqrt{\lambda_n}x]\right|^2 dx}\int_{\frac{L}{4}}^{\frac{3L}{4}} \cos[\sqrt{\lambda_n}x]dx \\ &= \frac{\Theta\{\sin[\sqrt{\lambda_n}\frac{3L}{4}] - \{\sin[\sqrt{\lambda_n}\frac{L}{4}]\}}{\sqrt{\lambda_n}\int_0^L \left|\cos[\sqrt{\lambda_n}x]\right|^2 dx} \end{aligned} \tag{2.153}$$

2.14 Case Study: Heat Transfer in a Semi-Infinite 1D Bar: Periodically Varying Temperature

Consider a semi-infinite bar in 1D (that is, we do not consider the contour of the bar, only the left end, at $x = 0$ and the right end, at $x = \infty$). The temperature at $x = 0$ has been changing in time as $T = T_0 + T_1 \cos(\omega t)$ since $t = -\infty$. The thermal diffusivity coefficient is χ, the density of the material is ρ and the thermal conductivity coefficient is k. Find the stationary (although time dependent) distribution of temperature of the bar, as a function of distance to $x = 0$.

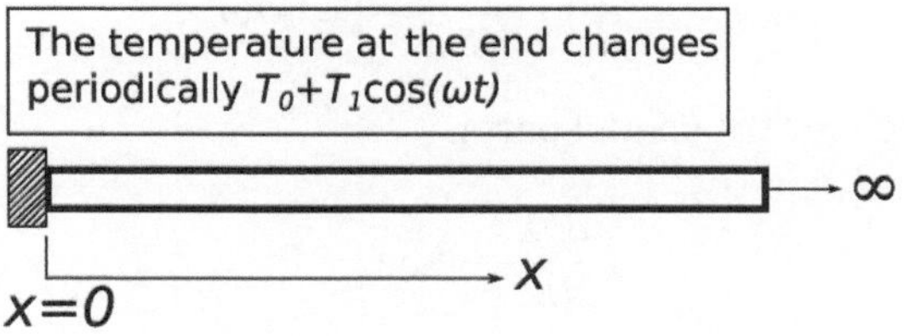

Figure 2.18

Mathematical formulation We first subtract the temperature T_0, to study the variations relative to it:

$$u(x, t) = v(x, t) - T_0 \tag{2.154}$$

The heat equation to be solved is:

$$\frac{\partial u(x, t)}{\partial t} = \chi \frac{\partial^2 u(x, t)}{\partial x^2} \tag{2.155}$$

where $\chi = \frac{k}{C\rho}$ is the thermal diffusivity coefficient.

The initial conditions are not specified.

Boundary condition 1: $u(0, t) = T_1 \cos(\omega t)$. We will present this condition as $u(0, t) = T_1 e^{-i\omega t}$, and once the problem is solved, only the real part of the solution will be considered.

Boundary condition 2: $u(\infty, t) = 0$

We seek the solution using the method of separation of variables $u(x, t) = T(t)X(x)$. Substituting this solution in the heat equation, and separating variables:

$$\frac{1}{T}\frac{dT}{dt} = \frac{\chi}{X}\frac{d^2X}{dx^2} = -\lambda = \text{const} \tag{2.156}$$

In the stationary case the temporal dependence will be the same as that of the external excitation (same frequency).

$$T(t) \sim e^{-i\omega t} \tag{2.157}$$

Then

$$\frac{1}{T}\frac{dT}{dt} = \frac{\chi}{X}\frac{d^2X}{dx^2} = -i\omega \tag{2.158}$$

The equation for the spatial part is:

$$\chi\frac{d^2X}{dx^2} + i\omega X = 0 \tag{2.159}$$

Note that this is not a Sturm–Liouville problem. We are looking for solutions of the form $X \propto e^{\beta x}$, and substituting it we arrive at:

$$\chi\beta^2 e^{\beta x} = -i\omega e^{\beta x} \to \beta = \pm\sqrt{\frac{-i\omega}{\chi}} \tag{2.160}$$

The following relations will help isolate $-i$:

$$\sqrt{-i} = \sqrt{e^{-i\frac{\pi}{2}}} = e^{-i\frac{\pi}{4}} = \frac{1-i}{\sqrt{2}} \tag{2.161}$$

General solution If we define $\beta_0 = \sqrt{\frac{\omega}{2\chi}}$, the general solution for $X(x)$ is:

$$X = De^{(1-i)\beta_0 x} + Ce^{-(1-i)\beta_0 x} \tag{2.162}$$

From the second boundary condition we have: $u(\infty, t) = 0 \to D = 0$. Then, the solution will be:

$$u(x,t) = T(t)X(x) = Ce^{-i\omega t}e^{-(1-i)\beta_0 x} \tag{2.163}$$

From the first boundary condition: $u(0,t) = T_1 e^{-i\omega t} \to C = T_1$

Then:

$$u(x,t) = T_1 e^{-i\omega t} e^{-(1-i)\beta_0 x} = T_1 e^{-\beta_0 x} e^{i(-\omega t+\beta_0 x)} \tag{2.164}$$

Final solution Considering the real part of the solution will give us the variations in temperature as a function of time and distance to the bar left end.

$$u(x,t) = T_1 e^{-\beta_0 x}\cos(-\omega t + \beta_0 x) \tag{2.165}$$

The thermal wave decay length is $\frac{1}{\beta_0} = \sqrt{\frac{2\chi}{\omega}}$

The wave half period is obtained from the phase of the temporal part $-\omega t + \beta_0 x$, taken for a given instant t_1, the spatial separation $x_0 = x_1 - x_2$ between the points x_1 and x_2, with phases 0 and π respectively. This separation is:

$$\beta_0 x_o = \pi \tag{2.166}$$

Along this distance (x_0), the maximum temperature of the oscillations decays by a factor: $e^{-\pi} = 0.043$.

Analyzing the solution one can see that at every instant the temperature in the zone $\beta_0 x > \frac{\pi}{2}$ has an opposite sign to the temperature in the surroundings of $x = 0$.

An analogous problem would be the calculation of the temperature as a function of time at a certain depth underground. We would see that the temperature would increase in winter with respect to its value in summer. This possibility is shown in the graph, analyzing the profile of temperature at the instants t_1 (summer) and t_2 (winter).

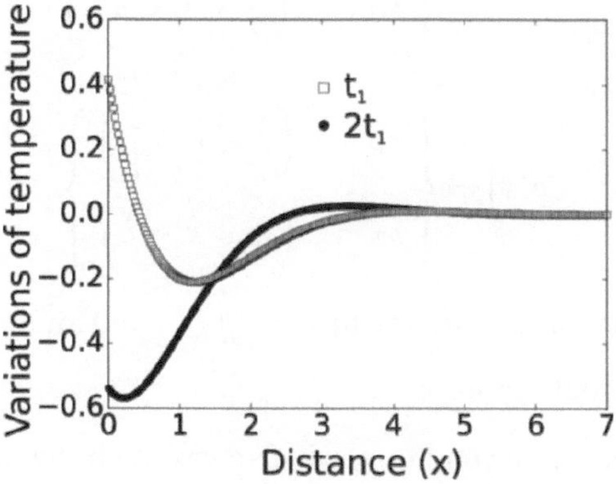

Figure 2.19

2.15 Case Study: Vibrations of Two United Bars

Consider a non-homogeneous bar ($0 < x < L$) with a uniform cross section (S). This bar is created by uniting two homogeneous bars of length $L/2$ with densities ρ_1 and ρ_2 and Young's moduli E_1 and E_2, but with the same speed of sound in both materials: $E_1/\rho_1 = E_2/\rho_2$. The end at $x = 0$ can't move and a constant force F is applied to the end at $x = L$ up until $t = 0$. At that instant, the force is no longer applied. Find the eigenfrequencies and the profile of the principal tone of the vibrations.

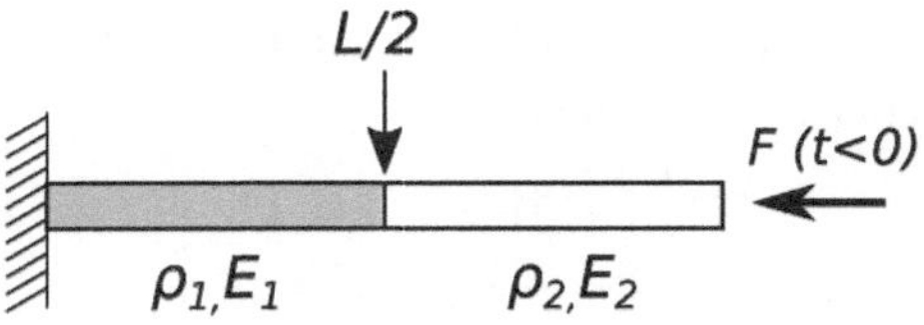

Figure 2.20

Mathematical formulation

$$\rho(x)\frac{\partial^2 u}{\partial t^2} - \frac{\partial}{\partial x}\left[E(x)\frac{\partial u}{\partial x}\right] = 0 \qquad (t > 0) \tag{2.167}$$

$$E(x) = \begin{Bmatrix} E_1 & \left(0 < x < \frac{L}{2}\right) \\ E_2 & \left(\frac{L}{2} < x < L\right) \end{Bmatrix} \tag{2.168}$$

$$\rho(x) = \begin{Bmatrix} \rho_1 & \left(0 < x < \frac{L}{2}\right) \\ \rho_2 & \left(\frac{L}{2} < x < L\right) \end{Bmatrix} \tag{2.169}$$

The boundary conditions are: $u(0) = \left.\frac{\partial u}{\partial x}\right|_{x=L} = 0$ for $t > 0$.

One of the initial conditions is: $\left.\frac{\partial u}{\partial t}\right|_{t=0} = 0$

For the other initial condition, which gives the initial displacement $u_{1,2}(x, 0)$ in one of two parts of the bar, we start from the definition of tension in the bar:

$$T_1 = \frac{F}{S} = E_1 \frac{\partial u_1(x)}{\partial x} \rightarrow u_1(x, 0) = \frac{F}{SE_1} x \tag{2.170}$$

In the same manner it is obtained:

$$u_2(x, 0) = \frac{F}{SE_2} x + \text{Const} \tag{2.171}$$

From the continuity equation we have:

$$u_1 \left(\frac{L}{2}, 0 \right) = u_2 \left(\frac{L}{2}, 0 \right) \tag{2.172}$$

with this, we can isolate the unknown constant:

$$Const = \frac{FL}{2S} \left(\frac{1}{E_1} - \frac{1}{E_2} \right) \tag{2.173}$$

Then:

$$u(x, t = 0) = \begin{Bmatrix} \frac{F}{SE_1} x & \left(0 < x < \frac{L}{2} \right) \\ \frac{F}{SE_2} x + \frac{FL}{2S} \left(\frac{1}{E_1} - \frac{1}{E_2} \right) & \left(\frac{L}{2} < x < L \right) \end{Bmatrix} \tag{2.174}$$

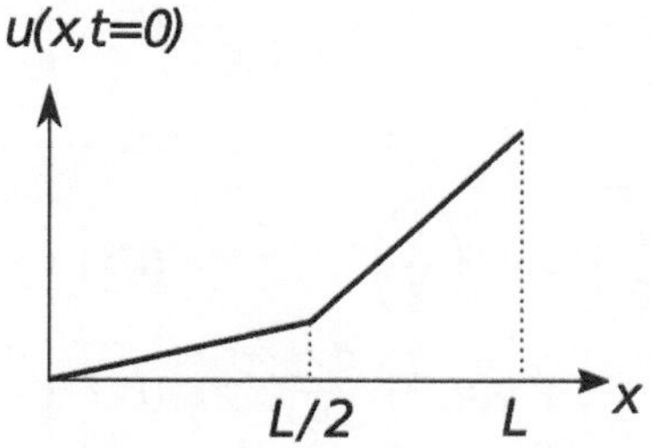

Figure 2.21

Sturm–Liouville problem To seek the solution of the problem we will use the method of separation of variables in the with $u = T(t)v(x)$:

$$\frac{1}{T} \frac{d^2 T}{dt^2} = \frac{1}{\rho(x) v(x)} \frac{d[E(x) \frac{dv}{dx}]}{dx} = -\lambda = -\omega^2 \tag{2.175}$$

We arrive at the following Sturm–Liouville problem for $v(x)$:

$$\frac{d[E(x)\frac{dv}{dx}]}{dx} + \lambda\rho(x)v(x) = 0 \tag{2.176}$$

With the boundary conditions: $v(0) = 0$ and $\frac{dv}{dx}\Big|_{x=L} = 0$.

Note: the orthogonality condition is:

$$(\lambda_n - \lambda_m)\int_0^L \rho(x)v_n(x)v_m(x)dx = 0 \tag{2.177}$$

General solution Separating equation (2.176) and its solutions in two parts:

$$\frac{d^2v_1}{dx^2} + \frac{\omega^2}{a_1^2}v_1 = 0 \qquad \left(x < \frac{L}{2}\right) \tag{2.178}$$

$$\frac{d^2v_2}{dx^2} + \frac{\omega^2}{a_2^2}v_2 = 0 \qquad \left(x > \frac{L}{2}\right) \tag{2.179}$$

with $a_{1,2}^2 = \frac{E_{1,2}}{\rho_{1,2}}$

We find the boundary conditions and the union of $v_{1,2}$ and its derivatives:

$$\left\{\begin{array}{ll} v_1(0) = 0 & (i) \\ \left.\frac{dv_2}{dx}\right|_{x=L} = 0 & (ii) \\ v_1\left(\frac{L}{2}\right) = v_2\left(\frac{L}{2}\right) & (iii) \\ E_1\left.\frac{dv_1}{dx}\right|_{x=\frac{L}{2}} = E_2\left.\frac{dv_2}{dx}\right|_{x=\frac{L}{2}} & (iv) \end{array}\right\} \tag{2.180}$$

Final solution The last condition is found integrating the wave equation near $L/2$.

$$\int_{\frac{L}{2}-\epsilon}^{\frac{L}{2}+\epsilon} \frac{d}{dx}\left[E\frac{dv}{dx}\right]dx + \lambda\int_{\frac{L}{2}-\epsilon}^{\frac{L}{2}+\epsilon} \rho(x)v(x)dx = 0 \tag{2.181}$$

$$\text{From condition (i)} \rightarrow v_1(x) = A\sin\left(\frac{\omega}{a_1}x\right) \tag{2.182}$$

$$\text{From condition (ii)} \rightarrow v_2(x) = B\cos\left(\frac{\omega}{a_2}(x-L)\right) \tag{2.183}$$

$$\text{From condition (iii)} \rightarrow A\sin\left(\frac{\omega}{a_1}\frac{L}{2}\right) = B\cos\left(\frac{\omega}{a_2}\frac{L}{2}\right) \tag{2.184}$$

$$\text{From condition (iv)} \rightarrow AE_1\frac{\omega}{a_1}\cos\left(\frac{\omega}{a_1}\frac{L}{2}\right) = -BE_2\frac{\omega}{a_2}\sin\left(-\frac{\omega}{a_2}\frac{L}{2}\right) \tag{2.185}$$

Using conditions (iii) and (iv) we have a system of two equations with two unknown, A and B. Expressing this system in matrix form we can obtain the values of A and B by equating the determinant of the coefficients to 0:

$$-\frac{E_2}{a_2}\sin\left(\frac{\omega}{a_1}\frac{L}{2}\right)\sin\left(\frac{\omega}{a_2}\frac{L}{2}\right) + \frac{E_1}{a_1}\cos\left(\frac{\omega}{a_1}\frac{L}{2}\right)\cos\left(\frac{\omega}{a_2}\frac{L}{2}\right) = 0 \tag{2.186}$$

From this equation we find the eigenvalues (related to the resonant frequencies). In the limit of a homogeneous bar ($\rho_1 = \rho_2$; $E_1 = E_2$) we arrive to the condition for eigenvalues corresponding to the bar with the right end free and the left one immobile.

$$\sin^2\left(\frac{\omega}{a}\frac{L}{2}\right) - \cos^2\left(\frac{\omega}{a}\frac{L}{2}\right) = -\cos\left(\frac{\omega L}{a}\right) = 0 \tag{2.187}$$

Finally, the frequencies ω_n are obtained in the case of a non-homogeneous bar, but with materials of the same speed of sound:

$$E_1/\rho_1 = E_2/\rho_2 \tag{2.188}$$

$$-E_2\sin\left(\frac{\omega}{a}\frac{L}{2}\right) \times \sin\left(\frac{\omega}{a}\frac{L}{2}\right) + E_1\cos\left(\frac{\omega}{a}\frac{L}{2}\right) \times \cos\left(\frac{\omega}{a}\frac{L}{2}\right) = 0 \tag{2.189}$$

Then:

$$\tan^2\left(\frac{\omega_n}{a}\frac{L}{2}\right) = \frac{E_1}{E_2} \tag{2.190}$$

The relation of amplitudes B_n/A_n will be, when the speeds of sounds are equal:

$$\text{From condition (iii)} \rightarrow B_n/A_n = \frac{\sin\left(\frac{\omega_n}{a_1}\frac{L}{2}\right)}{\cos\left(\frac{\omega_n}{a_2}\frac{L}{2}\right)} = \tan\left(\frac{\omega_n}{a}\frac{L}{2}\right) = \sqrt{\frac{E_1}{E_2}} \tag{2.191}$$

Then with $A_n = 1$ (we are only interested in the shape of the mode, this value simplifies the result), the profile of the first harmonic is:

$$v(x) = \begin{cases} v_1(x) = \sin\left(\frac{\omega_n}{a}x\right) & \left(0 < x < \frac{L}{2}\right) \\ v_2(x) = \sqrt{\frac{E_1}{E_2}}\cos\left[\frac{\omega_n}{a}(x - L)\right] & \left(\frac{L}{2} < x < L\right) \end{cases} \tag{2.192}$$

To find the coefficient $T_1(t)$ of the solution we start from the general solution $u(x, t) = \sum T_n(t)v_n(x)$. Using the initial conditions for an initial displacement $u(x, 0)$, using the properties of orthogonality and integrating, we arrive at:

$$T_1(0) = \frac{\int_0^L u(x, 0)\rho(x)v_{(1)}dx}{\int_0^L \rho(x)v_{(1)}^2 dx} \tag{2.193}$$

Note. $v_{(1)}$ is the profile of the mode corresponding to the main tone $(n = 1)$.

Finally the mode of the lowest frequency is found:

$$u_1(x, t) = T_1(0)v_{(1)}\cos(\omega_1 t) \tag{2.194}$$

(using the condition that the initial velocity is zero).

2.16 Distribution of Temperature in a Non-Homogeneous Bar

A unidimensional bar of length L is made by joining two homogeneous bars $(L = L_1 + L_2)$ with different thermal conductivity coefficients $(k_{1,2})$ and thermal diffusion coefficients $a_{1,2}^2 = k_{1,2}/\rho_{1,2}C_{1,2}$, where $\rho_{1,2}$ are the densities of each material and $C_{1,2}$ are the heat capacities. The lateral surface and the right end of the bar are insulated (see figure), while the left end is in contact with a reservoir at zero temperature. In the initial moment $(t = 0)$ the temperature at every point of the bar is equal to $T = 0$. Find the temperature of the bar in the neighborhood of the thermally insulated end as a function of time.

Note: Only write the integrals for the temporal coefficients, there is no need to solve them. See also problem (4.10) from [1].

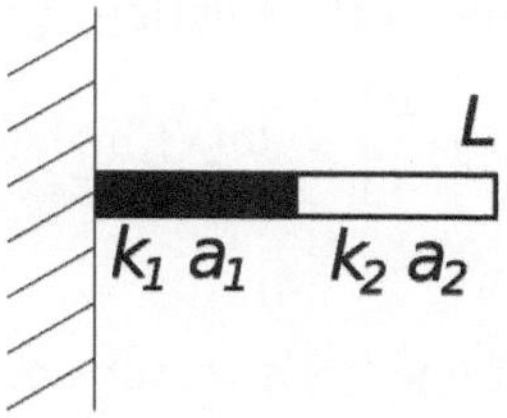

Figure 2.22

Mathematical formulation:

It's important to write from the beginning a single heat equation to describe the whole system (it's incorrect to have two equations, one for each material).

$$\rho(x)C(x)\frac{\partial u}{\partial t} - \frac{\partial}{\partial x}\left[k(x)\frac{\partial u}{\partial x}\right] = 0 \tag{2.195}$$

$$k(x) = \begin{Bmatrix} k_1 \ (0 < x < L_1) \\ k_2 \ (L_1 < x < L) \end{Bmatrix} \tag{2.196}$$

$$\rho(x) = \begin{Bmatrix} \rho_1 \ (0 < x < L_1) \\ \rho_2 \ (L_1 < x < L) \end{Bmatrix} \tag{2.197}$$

$$C(x) = \begin{Bmatrix} C_1 \ (0 < x < L_1) \\ C_2 \ (L_1 < x < L) \end{Bmatrix} \tag{2.198}$$

Boundary conditions: $u(0) = \frac{\partial u}{\partial x}\Big|_{x=L} = 0$

Initial conditions: $u(x, 0) = T_0$

Solution We separate variables to find the solution:

$$u = T(t) \cdot v(x) \tag{2.199}$$

$$\rho(x)C(x)\frac{\partial u}{\partial t} - \frac{\partial}{\partial x}\left[k(x)\frac{\partial u}{\partial x}\right] = 0 \tag{2.200}$$

$$\frac{1}{T}\frac{dT}{dt} = \frac{\frac{d}{dx}\left[k(x)\frac{dv}{dx}\right]}{\rho(x)C(x)v(x)} = -\lambda \tag{2.201}$$

Sturm–Liouville problem for the solution $v(x)$:

$$\frac{\partial}{\partial x}\left[k(x)\frac{\partial v}{\partial x}\right] + \lambda\rho(x)C(x)v(x) = 0 \tag{2.202}$$

Boundary conditions: $v(0) = \left.\frac{\partial v}{\partial x}\right|_{x=L} = 0$

Note: as already mentioned, it's important to formulate a single Sturm–Liouville problem from the beginning, since the orthogonality will be applicable to the corresponding function $v(x)$ and not to each of the parts in which this function will be later divided. The orthogonality condition (see section 4.1.3 from [1]) is:

$$(\lambda_n - \lambda_m)\int_0^L \rho(x)C(x)v_n v_m dx = 0 \tag{2.203}$$

Separating the equation (2.202) in two parts:

$$\frac{d^2v_1}{dx^2} + \lambda\left(\frac{\rho_1 C_1}{k_1}\right)v_1 = 0 \tag{2.204}$$

$$\frac{d^2v_2}{dx^2} + \lambda\left(\frac{\rho_2 C_2}{k_2}\right)v_2 = 0 \tag{2.205}$$

or in another manner:

$$\frac{d^2v_1}{dx^2} + \lambda\left(\frac{1}{a_1^2}\right)v_1 = 0 \tag{2.206}$$

$$\frac{d^2v_2}{dx^2} + \lambda\left(\frac{1}{a_2^2}\right)v_2 = 0 \tag{2.207}$$

With $a_{1,2}^2 = \frac{k_{1,2}}{\rho_{1,2}C_{1,2}}$

We find the boundary conditions and continuity conditions of the functions $v_{1,2}$:

(1) $v_1(0) = 0 \rightarrow v_1(x) = A\sin\left(\frac{\sqrt{\lambda}}{a_1}x\right)$

(2) $\left.\frac{dv_2}{dx}\right|_{x=L} = 0 \rightarrow v_2(x) = B\cos\left(\frac{\sqrt{\lambda}}{a_2}(x-L)\right)$

(3) $v_1(L_1) = v_2(L_1)$

(4) $k_1\left.\frac{dv_1}{dx}\right|_{x=L_1} = k_2\left.\frac{dv_2}{dx}\right|_{x=L_1}$

The last condition has been found by integrating the heat equation near L_1:

$$\int_{L_1-\epsilon}^{L_1+\epsilon} \frac{d}{dx}\left[k\frac{dv}{dx}\right]dx + \lambda \int_{L_1-\epsilon}^{L_1+\epsilon} \rho(x)C(x)v(x)dx = 0 \qquad (2.208)$$

Using the eigenfunctions and conditions (3, 4) we need to solve a system with two equations and two unknowns $(A,\ B)$. From the condition that the determinant of this equation in matrix form must be zero, we find the eigenvalues λ_i.

From (3) we have: $A\sin\left(\frac{\sqrt{\lambda}}{a_1}L_1\right) - B\cos\left[\frac{\sqrt{\lambda}}{a_2}(L_1 - L)\right] = 0$

From (4) we have:

$$Ak_1\frac{\sqrt{\lambda}}{a_1}\cos\left(\frac{\sqrt{\lambda}}{a_1}L_1\right) + Bk_2\frac{\sqrt{\lambda}}{a_2}\sin\left[\frac{\sqrt{\lambda}}{a_2}(L_1 - L)\right] = 0$$

Equating the determinant of the coefficient matrix (which represents equations 3 and 4) to zero, we get the equation to obtain the eigenvalues λ_n:

$$k_1\frac{\sqrt{\lambda}}{a_1}\cos\left(\frac{\sqrt{\lambda}}{a_1}L_1\right)\cos\left(\frac{\sqrt{\lambda}}{a_2}(L_1 - L)\right) + \sin\left(\frac{\sqrt{\lambda}}{a_1}L_1\right)k_2\frac{\sqrt{\lambda}}{a_2}\sin\left(\frac{\sqrt{\lambda}}{a_2}(L_1 - L)\right) = 0 \qquad (2.209)$$

Which is simplified to:

$$\tan\left(\frac{\sqrt{\lambda_n}}{a_2}(L_1 - L)\right)\tan\left(\frac{\sqrt{\lambda_n}}{a_1}L_1\right) = -\frac{k_1a_2}{k_2a_1} \qquad (2.210)$$

We find the ratio between the amplitudes B_n/A_n:

Since $B = \frac{A\sin\left(\frac{\sqrt{\lambda}}{a_1}L_1\right)}{\cos\left(\frac{\sqrt{\lambda}}{a_2}(L_1-L)\right)}$

Then the form of the solutions $v_{1,2}$ (for $A = 1$) could be:

$$v_1(x) = \sin\left(\frac{\sqrt{\lambda_n}}{a_1}x\right) \qquad (0 < x < L_1) \qquad (2.211)$$

$$v_2(x) = \frac{\sin(\frac{\sqrt{\lambda_n}}{a_1}L_1)}{\cos\left(\frac{\sqrt{\lambda_n}}{a_2}(L_1 - L)\right)}\cos\left(\frac{\sqrt{\lambda_n}}{a_2}(x - L)\right) \qquad (L_1 < x < L) \qquad (2.212)$$

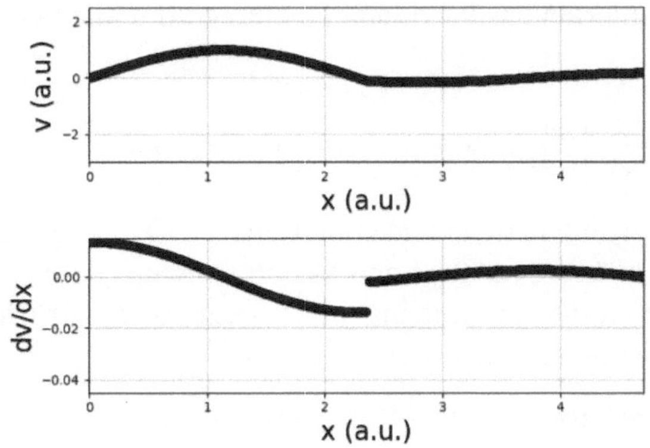

Figure 2.23 Example of eigenfunction for the spatial problem.

Final solution We calculate now the coefficients $T_n(t)$ of the solution. Solution for $x = L$:

$$u(x = L) = \sum_n T_n(t)v_n(L) \tag{2.213}$$

Using the initial conditions and the orthogonality:

$$T_n(0) = \frac{T_0 \int_0^L \rho(x)C(x)v_n(x)dx}{\int_0^L \rho(x)C(x)v_n^2(x)dx} \tag{2.214}$$

Solution of the equation for the temporal coefficients:

$$\frac{dT_n}{dt} + \lambda_n T_n = 0 \tag{2.215}$$

$$T_n(t) = T_n(0)e^{-\lambda_n t} \tag{2.216}$$

Finally:

$$u(x = L) = \sum_n T_n(0)e^{-\lambda_n t} \frac{\sin\left(\frac{\sqrt{\lambda}}{a_1}L_1\right)}{\cos\left(\frac{\sqrt{\lambda}}{a_2}(L_1 - L)\right)} \tag{2.217}$$

2.17 Case Study: Variation in the Ion Concentration in a Rod with Flux across Its Ends

A one-dimensional rod of length L has an initial distribution of ion concentrations at $t = 0$ which can be described by the function $f(x)$: Starting at $t = 0$ ion fluxes J_1 and J_2 cross across the ends of the rod $x = 0$ and $x = L$. Along its length, the rod exchanges ions with the outer medium (which has zero ion concentration), proportionally to the concentration on the surface $q(x) = -hu$ (h is a positive constant). Find the distribution of the concentrations of ions as a function of the position x and time t if the diffusion coefficient is D (another positive constant).

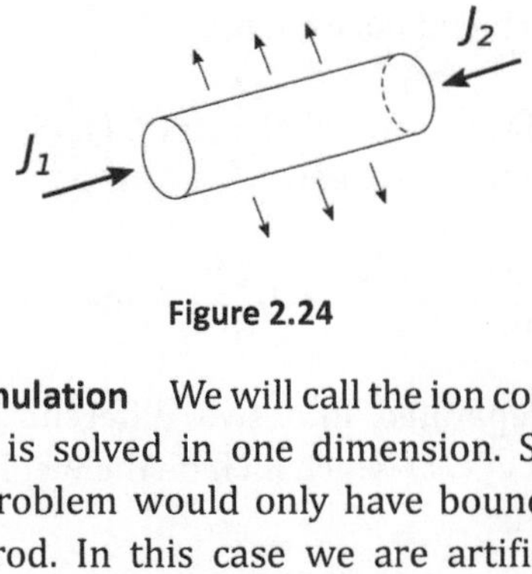

Figure 2.24

Mathematical formulation We will call the ion concentration $u(x, t)$ and the problem is solved in one dimension. Strictly speaking a unidimensional problem would only have boundary conditions at the ends of the rod. In this case we are artificially treating the length of the rod as a boundary, since it exchanges heat with its medium. But instead of including this information as a boundary (to do that we would need to solve the problem in 3D with cylindrical coordinates and apply a boundary condition at the cylinder radius), we will include it as a non-homogeneous term in the diffusion equation:

$$\left\{\begin{array}{l} \dfrac{\partial u}{\partial t} = D\dfrac{\partial^2 u}{\partial x^2} - hu \\ D\left.\dfrac{\partial u}{\partial x}\right|_{x=0} = J_1 \\ D\left.\dfrac{\partial u}{\partial x}\right|_{x=L} = J_2 \\ u(x, t=0) = f(x) \end{array}\right\} \tag{2.218}$$

The sign of the term linear in temperature of the differential equation is determined by the condition that the system loses energy to its surroundings when an excess of temperature is present.

As this is a non-homogeneous equation with non-homogeneous boundary conditions, we will search for the solution as the sum of two functions $u(x, t) = w(x) + v(x, t)$.

(i) The function $w(x)$ describes the stationary part of the solution $w(x)$ (that is, it's independent of time and it would be the solution in the limit $t \to \infty$). This solution will take into account the non-homogeneous boundaries, which persist at all times.
(ii) The other part of the solution would describe the transient part $v(x, t)$ in the presence of homogeneous boundaries (the non-homogeneous contribution of these is already included in w, it must not be included once again).

The expression $u(x, t) = w(x) + v(x, t)$ is substituted in the differential equation, and we get:

$$\frac{\partial v}{\partial t} = D\frac{\partial^2 w}{\partial x^2} + D\frac{\partial^2 v}{\partial x^2} - hw - hv \tag{2.219}$$

The problem is separated into two different problems for the functions $w(x)$ and $v(x, t)$ which include the whole of the initial and boundary conditions:

$$\left\{\begin{array}{l} \left.\dfrac{\partial u}{\partial x}\right|_{x=0} = Q_1 \\ \left.\dfrac{\partial u}{\partial x}\right|_{x=L} = Q_2 \\ u(x, 0) = f(x) \end{array}\right\} \tag{2.220}$$

With $Q_1 = J_1/D$ and $Q_2 = J_2/D$. The non-homogeneous boundary conditions for $w(x)$ are:

$$\left\{\begin{array}{l} \left.\dfrac{\partial w}{\partial x}\right|_{x=0} = Q_1 \\ \left.\dfrac{\partial w}{\partial x}\right|_{x=L} = Q_2 \end{array}\right\} \tag{2.221}$$

Then, to keep the global conditions for the solution $u = v + w$, the conditions for $v(x, t)$ need to change to:

$$\left\{ \begin{array}{l} \left.\dfrac{\partial v}{\partial x}\right|_{x=0} = 0 \\ \left.\dfrac{\partial v}{\partial x}\right|_{x=L} = 0 \\ v(x, 0) = f(x) - w(x) \end{array} \right\} \tag{2.222}$$

The complete formulation for both partial problems is then:

Problem (1) for the transient part

$$\left\{ \begin{array}{l} \dfrac{\partial v}{\partial t} = a^2 \dfrac{\partial^2 v}{\partial x^2} - hv \\ \left.\dfrac{\partial v}{\partial x}\right|_{x=0} = 0 \\ \left.\dfrac{\partial v}{\partial x}\right|_{x=L} = 0 \\ v(x, 0) = f(x) - w(x) \end{array} \right\} \tag{2.223}$$

Problem (2) for the stationary part

$$\left\{ \begin{array}{l} D\dfrac{\partial^2 w}{\partial x^2} - hw = 0 \\ \left.\dfrac{\partial w}{\partial x}\right|_{x=0} = Q_1 \\ \left.\dfrac{\partial w}{\partial x}\right|_{x=L} = Q_2 \end{array} \right\} \tag{2.224}$$

The general solution for problem 2 is:

$$w(x) = A \sinh\left(\sqrt{\frac{h}{D}}x\right) + B \cosh\left(\sqrt{\frac{h}{D}}x\right)$$

Applying the first boundary condition $\left.\frac{\partial w}{\partial x}\right|_{x=0} = Q_1$ we get:

$$A\sqrt{\frac{h}{D}} \cosh\left(\sqrt{\frac{h}{D}} \cdot 0\right) = Q_1 \rightarrow A = Q_1\sqrt{\frac{D}{h}} \tag{2.225}$$

Applying the second boundary condition $\left.\frac{\partial w}{\partial x}\right|_{x=L} = Q_2$ we arrive at the expression:

$$Q_1\sqrt{\frac{D}{h}}\sqrt{\frac{h}{D}} \cosh\left(\sqrt{\frac{h}{D}}L\right) + B\sqrt{\frac{h}{D}} \sinh\left(\sqrt{\frac{h}{D}}L\right) = Q_2 \tag{2.226}$$

Then:

$$B = \frac{Q_2 - Q_1 \cosh\left(\sqrt{\frac{h}{D}}L\right)}{\sqrt{\frac{h}{D}} \sinh\left(\sqrt{\frac{h}{D}}L\right)} \tag{2.227}$$

The solution of problem 2 is:

$$w(x) = Q_1\sqrt{\frac{D}{h}} \sinh\left(\sqrt{\frac{h}{D}}x\right) + \frac{Q_2 - Q_1 \cosh\left(\sqrt{\frac{h}{D}}L\right)}{\sqrt{\frac{h}{D}} \sinh\left(\sqrt{\frac{h}{D}}L\right)} \cosh\left(\sqrt{\frac{h}{D}}x\right) \tag{2.228}$$

Problem (1) for the transient part

To solve the transient part we use the method of separation of variables:

$$v(x, t) = T(t)X(x) \tag{2.229}$$

Separating variables we get:

$$\frac{h + \frac{1}{T}\frac{dT}{dt}}{D} = \frac{1}{X}\frac{dX}{dx} = -\lambda \tag{2.230}$$

We take $\lambda > 0$ to arrive to a Sturm–Liouville with eigenfunctions:

Eigenfunctions: $X_n(x) = \cos\left(\frac{\pi n}{L}x\right)$

Eigenvalues: $\lambda_n = \left(\frac{\pi n}{L}\right)^2$

The differential equation for $T(t)$ is:

$$\frac{dT}{dt} = -\left[D\left(\frac{\pi n}{L}\right)^2 + h\right] T \tag{2.231}$$

The amplitudes of the temporal coefficients are:

$$T_n(t) = A_n e^{-\left(D\left(\frac{\pi n}{L}\right)^2 + h\right)t} \tag{2.232}$$

Then, the general solution is:

$$v(x, t) = \sum_{n=0}^{\infty} A_n e^{-\left(D\left(\frac{\pi n}{L}\right)^2 + h\right)t} \cos\left(\frac{\pi n}{L}x\right) \tag{2.233}$$

Applying the initial conditions:

$$v(x, 0) = f(x) - w(x) = \sum_{n=0}^{\infty} A_n \cos\left(\frac{\pi n}{L}x\right) \tag{2.234}$$

Finally, using the orthogonality of the eigenfunctions $\cos\left(\frac{\pi m}{L}x\right)$ the coefficients of the expansion of the function $v(x, t)$ are found:

$$A_n = \frac{2}{L}\int_0^L (f(x) - w(x)) \cos\left(\frac{\pi n}{L}x\right) dx \tag{2.235}$$

2.18 Oscillations of a Non-Homogeneous String

A non-homogeneous string of length L and fixed at its ends is made of two strings of the same length and different densities ρ_1 and $2\rho_1$ (see figure) and is initially at rest. At $t = 0$ it is hit at its central point. The hit transfers a momentum equal to I. Find the displacement as a function of time.

See also problem (4.6) in [1] and section 4.1.3.

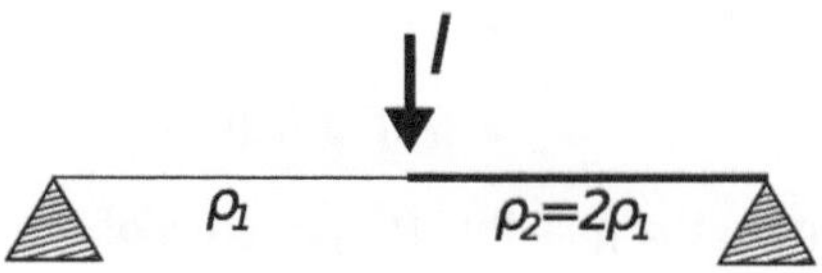

Figure 2.25

Mathematical formulation

$$\rho(x)\frac{\partial^2 u}{\partial t^2} - T\frac{\partial^2 u}{\partial x^2} = 0 \tag{2.236}$$

$$\rho(x) = \left\{\begin{array}{ll} \rho_1 & \left(0 < x < \dfrac{L}{2}\right) \\ 2\rho_1 & \left(\dfrac{L}{2} < x < L\right) \end{array}\right\} \tag{2.237}$$

T is the tension, and it is assumed to be constant along the whole string.

Boundary conditions: $u(0) = u(L) = 0$

Initial condition 1: $u(x, t = 0) = 0$

Initial condition 2, considering that the momentum transferred by the hit is received by a fragment of string of length 2ε, centered at $x = L/2$:

$$\left.\frac{\partial u}{\partial t}\right|_{t=0} = \left\{\begin{array}{ll} 0 & \left(x < \dfrac{L}{2} - \varepsilon\right) \\ \dfrac{I}{\varepsilon(\rho_1 + 2\rho_1)} & \left(\dfrac{L}{2} - \varepsilon < x < \dfrac{L}{2} + \varepsilon\right) \\ 0 & \left(x > \dfrac{L}{2} + \varepsilon\right) \end{array}\right\} \tag{2.238}$$

Sturm–Liouville problem Considering the initial conditions (zero initial displacement, non-zero initial velocity), the shape of the solution as a sum of eigenfunctions will be:

$$u(x,t) = \sum_n Q_n(t)X_n(x) = \sum_n A_n \sin(\omega_n t)X_n(x) \qquad (2.239)$$

Replacing that in the wave equation $u = Q(t) \cdot X(x)$, we separate variables:

$$\frac{1}{QT}\frac{d^2Q}{dt^2} = \frac{1}{\rho(x)X}\frac{d^2X}{dx^2} = -\lambda_n \qquad (2.240)$$

The temporal part is:

$$\frac{d^2Q}{dt^2} + \lambda_n T\, Q = 0 \qquad (2.241)$$

From here we obtain the possible frequencies of the oscillations:

$$\omega_n^2 = \lambda_n T \qquad (2.242)$$

The equation for the spatial form of each mode of oscillation is:

$$\frac{d^2X}{dx^2} + \lambda_n \rho(x)X = 0 \qquad (2.243)$$

Taking into account the boundary conditions, the solution of this equation is presented as a function in two parts $X_{1,2}$ with boundary conditions $X_1(0) = X_2(L) = 0$

$$X_n(x) = \left\{ \begin{array}{ll} X_1 = B_n \sin(\sqrt{\rho_1\lambda_n}x) & \left(0 < x < \frac{L}{2}\right) \\ X_2 = C_n \sin(\sqrt{2\rho_1\lambda_n}(L-x)) & \left(\frac{L}{2} < x < L\right) \end{array} \right\} \qquad (2.244)$$

The relation between B_n and C_n is obtained from the continuity equation:

$$B_n \sin\left(\sqrt{\rho_1\lambda_n}\frac{L}{2}\right) = C_n \sin\left(\sqrt{2\rho_1\lambda_n}\frac{L}{2}\right) \qquad (2.245)$$

Then the solution can be expressed in the form:

$$X_n(x) = \left\{ \begin{array}{ll} X_{1n} = B_n \sin\left(\sqrt{\rho_1\lambda_n}x\right) & \left(0 < x < \frac{L}{2}\right) \\ X_{2n} = B_n \dfrac{\sin\left(\sqrt{\rho_1\lambda_n}\frac{L}{2}\right)}{\sin\left(\sqrt{2\rho_1\lambda_n}\frac{L}{2}\right)} \sin\left(\sqrt{2\rho_1\lambda_n}(L-x)\right) & \left(\frac{L}{2} < x < L\right) \end{array} \right\} \qquad (2.246)$$

Integrating equation (2.243) for every mode in an interval (ε) around the central point:

$$\int_{\frac{L}{2}-\epsilon}^{\frac{L}{2}+\epsilon} \frac{d^2X_n}{dx^2}dx + \int_{\frac{L}{2}-\epsilon}^{\frac{L}{2}+\epsilon} \lambda_n\rho(x)X_n dx = \left.\frac{dX_{n2}}{dx}\right|_{\frac{L}{2}+\epsilon} - \left.\frac{dX_{n1}}{dx}\right|_{\frac{L}{2}-\epsilon}$$
$$+ \int_{\frac{L}{2}-\epsilon}^{\frac{L}{2}} \lambda_n\rho_1 X_{n1}dx + \int_{\frac{L}{2}}^{\frac{L}{2}+\epsilon} \lambda_n\rho_2 X_{n2}dx = 0 \quad (2.247)$$

The first term is the integral of a second derivative, which directly yields a first derivative. The second term is the integral of an integrable function. When we make $\epsilon \rightarrow 0$ these integrals become zero.

Then, letting $\epsilon \rightarrow 0$,

$$\left.\frac{dX_{n1}}{dx}\right|_{x=L/2} = \left.\frac{dX_{n2}}{dx}\right|_{x=L/2} \quad (2.248)$$

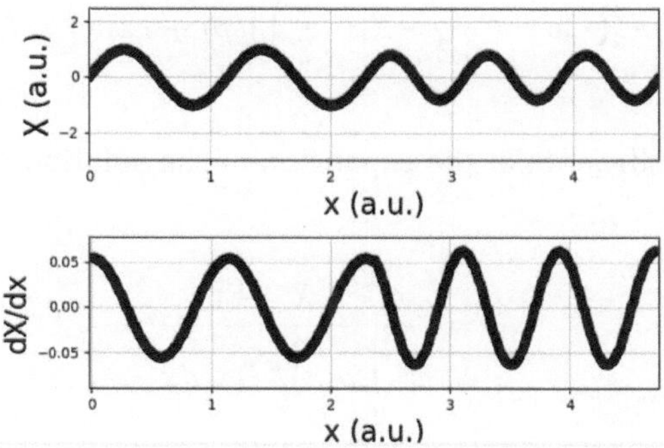

Figure 2.26 Example of eigenfunctions and their derivatives

That gives a transcendental equation to obtain the eigenvalues λ_n:

$$\sqrt{\rho_1\lambda_n}\cos\left[\sqrt{\rho_1\lambda_n}\left(\frac{L}{2}\right)\right]$$
$$= \sqrt{2\rho_1\lambda_n}\frac{\sin(\sqrt{\rho_1\lambda_n}\frac{L}{2})}{\sin(\sqrt{2\rho_1\lambda_n}\frac{L}{2}]}\cos\left[\sqrt{2\rho_1\lambda_n}\left(\frac{L}{2}\right)\right] \quad (2.249)$$

Note: the orthogonality condition (see 4.30 from [1]) is:

$$(\lambda_n - \lambda_m) \int_0^L \rho(x) X_n(x) X_m(x) dx = 0 \tag{2.250}$$

Final solution We will use now the second initial condition to find the coefficients A_n

$$\left.\frac{\partial u}{\partial t}\right|_{t=0} = \sum_n A_n \omega_n X_n(x) = \begin{Bmatrix} 0 & \left(x < \frac{L}{2} - \varepsilon\right) \\ \frac{I}{\varepsilon[\rho_1 + 2\rho_1]} & \left(\frac{L}{2} - \varepsilon < x < \frac{L}{2} + \varepsilon\right) \\ 0 & \left(x > \frac{L}{2} + \varepsilon\right) \end{Bmatrix} \tag{2.251}$$

Both parts are multiplied by $\rho(x) \times X_m(x)$ and are integrated from 0 to L.

$$A_n \omega_n \int_0^L \rho(x) X_n^2(x) dx = \frac{I}{\varepsilon 3\rho_1} X_n\left(\frac{L}{2}\right)[\varepsilon\rho_1 + \varepsilon 2\rho_1] = I \times X_n\left(\frac{L}{2}\right) \tag{2.252}$$

Then the coefficients of the expansion of the solution as a Fourier series are:

$$A_n = \frac{I \times X_n\left(\frac{L}{2}\right)}{\omega_n \int_0^L \rho(x) X_n^2(x) dx} \tag{2.253}$$

2.19 Forced Oscillations of a String

Find the oscillations of a string with length L, fixed at its ends, which at $t > 0$ is subject to a force with density: $f(x, t) = Ae^{-t} \sin(\pi x/L)$. The ration between the tension of the string T and its linear density ρ is $a^2 = T/\rho$. Consider that the string wasn't oscillating at $t = 0$ and that its linear density is $\rho = 1$.

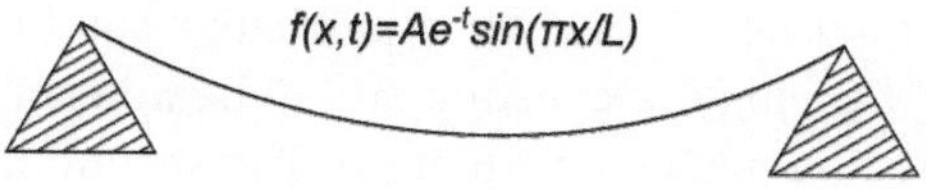

Figure 2.27

Mathematical formulation

$$\left\{\begin{array}{l}\frac{\partial^2 u}{\partial t^2} - a^2\frac{\partial^2 u}{\partial x^2} = Ae^{-t}\sin\left(\frac{\pi x}{L}\right)\ (t>0)\\ u(0,t)=0\\ u(L,t)=0\\ u(x,0)=0\\ \left.\frac{\partial u}{\partial t}\right|_{t=0}=0\end{array}\right\} \tag{2.254}$$

Solution We separate variables to expand the solution of the non-homogeneous equation in eigenfunctions of the homogeneous problem with boundary conditions:

$$u(x,t)=\sum_{n=1}^{\infty} T_n(t)X_n(x) \tag{2.255}$$

$$X_n=\sin\left(\frac{\pi n}{L}x\right);\quad \lambda_n=\left(\frac{\pi n}{L}\right)^2 \tag{2.256}$$

$$\sum_{n=1}^{\infty}\left[\frac{d^2T_n}{dt^2}+a^2\lambda_n T_n\right]X_n = Ae^{-t}\sin\left(\frac{\pi x}{L}\right) \tag{2.257}$$

Taking advantage of the orthogonality of the eigenfunctions we arrive at the equation:

$$\frac{d^2T_n}{dt^2}+a^2\lambda_n T_n = f_n(t)=\frac{2Ae^{-t}}{L}\int_0^L \sin\left(\frac{\pi x}{L}\right)\sin\left(\frac{\pi n}{L}x\right)dx \tag{2.258}$$

$$f_n(t)=\left\{A\,e^{-t}\delta_{1,n}\right\}=\left\{\begin{array}{ll}Ae^{-t} & (n=1)\\ 0 & (n\neq 1)\end{array}\right\} \tag{2.259}$$

We obtain the equation of a forced oscillator (with null initial conditions), corresponding to the excitation of a single eigenmode (main mode). The solution of the temporal part can be found by two alternative methods.

The first could be to consider a particular solution of the type $T_{1,p}(t) = C\,e^{-t}$ (being C a constant still to be determined) plus the solution of the homogeneous equation. The second method would consist in finding the particular solution from Green's equation for a forced oscillator.

$$T_{1p}(t) = \frac{AL}{\pi a}\int_0^t e^{-t'} \sin\left[\frac{\pi}{L}(t-t')\right] dt' \qquad (2.260)$$

The solution is:

$$\begin{aligned} u_p(x,t) &= T_{1p}(t)X_1(x) \\ &= \frac{AL}{\pi a}\sin\left(\frac{\pi}{L}x\right)\int_0^t e^{-x}\sin\left[\frac{\pi}{L}(t-x)\right]dx \\ &= \frac{A}{\left[1+(\frac{\pi a}{L})^2\right]}\left\{e^{-t} - \cos\left(\frac{\pi a}{L}t\right) + \frac{L}{\pi a}\sin\left(\frac{\pi a}{L}t\right)\right\}\sin\left(\frac{\pi}{L}x\right) \end{aligned} \qquad (2.261)$$

It can be shown that from the second way of finding the particular solution, it fulfills the null initial conditions and then is also the final solution $u(x,t) = u_p(x,t)$.

2.20 Case Study: Oscillations of a String Subject to an External Force

A string of length L, with both ends fixed ant initially at rest, is acted upon by an external force of density $f(x,t) = \sinh(x)$ starting at $t = 0$. Find the oscillations of the string at times later than $t = 0$. Consider that the speed of sound in the string (c) and the density (ρ) are equal to one.

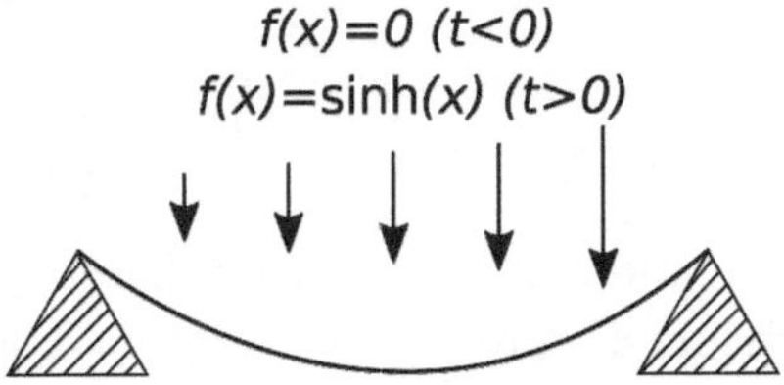

Figure 2.28

Mathematical formulation

$$\frac{\partial^2 u}{\partial t^2} - c^2 \frac{\partial^2 u}{\partial x^2} = \frac{1}{\rho} \sinh(x) \rightarrow \frac{\partial^2 u}{\partial t^2} - \frac{\partial^2 u}{\partial x^2} = \sinh(x) \qquad (t > 0) \qquad (2.262)$$

Boundary conditions: $u(x = 0) = u(x = L) = 0$

Initial conditions: $u(x, 0) = \left.\frac{\partial u}{\partial t}\right|_{t=0} = 0$

This problem can be solved by three methods. One is more "physical" and the other is more formal. The third one uses Green's functions.

Method 1

The string will change its shape as a result of the application of the external force. In the limit of long times the small oscillations will occur on a string of the new equilibrium shape. At infinite times the oscillations will have disappeared due to friction (that is, $\frac{\partial^2 u}{\partial x^2} = -\sinh(x)$). The static shape of the string is only a function of the spatial variable. Then, the solution can be sought as the sum of two functions:

$$u(x, t) = v(x, t) + w(x) \qquad (2.263)$$

The problem is separated into two, so that the summing the corresponding solutions the initial equation and conditions are respected.

$$\frac{\partial^2 v}{\partial t^2} - \frac{\partial^2 v}{\partial x^2} - \frac{\partial^2 w}{\partial x^2} = \sinh(x) \qquad (2.264)$$

From the first boundary condition:

$$v(0) + w(0) = 0 \rightarrow \text{we will assume: } v(0) = w(0) = 0 \qquad (2.265)$$

From the second boundary condition:

$$v(L) + w(L) = 0 \rightarrow \text{we will assume: } v(L) = w(L) = 0 \qquad (2.266)$$

From the first initial condition:

$$v(x, t = 0) + w(x) = 0 \rightarrow \text{we have: } v(x, 0) = -w(x) \qquad (2.267)$$

From the second initial condition:

$$\left.\frac{\partial v}{\partial t}\right|_{t=0} + 0 = 0 \rightarrow \text{we have: } \left.\frac{\partial v}{\partial t}\right|_{t=0} = 0 \qquad (2.268)$$

The problem is separated into two (a,b):

To formulate the problem for the function $w(x)$ we will use the result of replacing $u(x, t) = v(x, t) + w(x)$ into the wave equation:

$$\frac{\partial^2 v}{\partial t^2} - \frac{\partial^2 v}{\partial x^2} - \frac{\partial^2 w}{\partial x^2} = \sinh(x) \tag{2.269}$$

a) The equation to be solved for the function describing the static shape of the string under the external force is:

$$\frac{d^2 w}{dx^2} = -\sinh(x) \tag{2.270}$$

The boundary conditions for $w(x)$ are:

$$w(0) = w(L) = 0 \tag{2.271}$$

The solution of the problem for $w(x)$ can be found by integrating twice the equation:

$$w = -\sinh(x) + Cx + D \tag{2.272}$$

Alternatively, the same general form can be obtained by presenting the solution as the sum of a particular solution: $w_p = -\sinh(x)$ and a general solution for the homogeneous problem $\frac{d^2 w}{dx^2} = 0$.

Applying the conditions $w(0) = 0$ and $w(L) = 0$ we get respectively $D = 0$ and $C = \frac{\sinh(L)}{L}$. Then the equilibrium position of the string under the action of the external force is:

$$w(x) = -\sinh(x) + \frac{\sinh(L)}{L}x \tag{2.273}$$

b) Now we will get the solution of the time dependent problem.

$$\frac{\partial^2 v}{\partial t^2} - \frac{\partial^2 v}{\partial x^2} = 0 \tag{2.274}$$

With boundary conditions: $v(0) = v(L) = 0$.

The initial conditions for $v(x)$ are:

Condition 1: $v(x, 0) = -w(x) = \sinh(x) - \frac{\sinh(L)}{L}x$

Condition 2: $\frac{\partial v}{\partial t}\Big|_{t=0} = 0$

General solution The general solution is known for v:

$$v(x, t) = \sum A_n \cos\left(\frac{\pi n}{L}t\right) \sin\left(\frac{\pi n}{L}x\right) \tag{2.275}$$

Final solution Applying the initial conditions and using the orthogonality properties we get the coefficients:

$$v(x,0)=\sum A_n \sin\left(\frac{\pi n}{L}x\right)=\sinh(x)-\frac{\sinh(L)}{L}x \tag{2.276}$$

$$A_n=\frac{2}{L}\int_0^L\left[\sinh(x)-\frac{\sinh(L)}{L}x\right]\sin\left(\frac{\pi n}{L}x\right)dx \tag{2.277}$$

The integrals can be solved, for example, with Wolfram-Alpha:

$$\int \sinh(x)\sin\left(\frac{\pi n}{L}x\right)dx = \frac{L[L\cosh(x)\sin\left(\frac{\pi n}{L}x\right)-\pi n\sinh(x)\cos(\frac{\pi n}{L}x)]}{L^2+(\pi n)^2}+\text{Const} \tag{2.278}$$

$$\int x\sin\left(\frac{\pi n}{L}x\right)dx=\frac{L[L\sin\left(\frac{\pi n}{L}x\right)-\pi n x\cos\left(\frac{\pi n}{L}x\right)]}{(\pi n)^2}+\text{Const} \tag{2.279}$$

$$\begin{aligned}A_n&=\frac{2}{L}\left[\frac{\sinh(L)}{L(\frac{\pi n}{L})^2}[\pi n\cos(\pi n)-\sin(\pi n)]\right.\\&\quad\left.+\frac{\cosh(L)\sin(\pi n)-\frac{\pi n}{L}\sinh(L)\cos(\pi n)}{(\frac{\pi n}{L})^2+1}\right]\\&=\frac{2\cos(\pi n)\sinh(L)\pi n}{L^2}\left[\frac{1}{(\frac{\pi n}{L})^2}-\frac{1}{(\frac{\pi n}{L})^2+1}\right]\\&=\frac{2\cos(\pi n)\sinh(L)\pi n}{L^2}\left[\frac{1}{[(\frac{\pi n}{L})^2+1](\frac{\pi n}{L})^2}\right]\\&=\frac{2\cos(\pi n)\sinh(L)}{\pi n}\left[\frac{L^2}{[(\pi n)^2+L^2]}\right]=\frac{2L^2}{\pi n}\frac{(-1)^n\sinh(L)}{[(\pi n)^2+L^2]}\end{aligned} \tag{2.280}$$

Method 2

While the Method 1 could be described as a "physical approach" as it introduces the new equilibrium position respect which free oscillations are found, we consider below also more formal approach to solve the same problem.

Sturm–Liouville problem

We separate variables: $u(x, t) = A(x)B(t)$. The spatial part is expanded into the orthogonal eigenfunctions of the Sturm–Liouville problem:

$$\frac{d^2A(x)}{dx^2} + \lambda A(x) = 0 \tag{2.281}$$

$$A(0) = 0; \; A(L) = 0 \tag{2.282}$$

The solution is then:

$$u(x, t) = \sum_n B_n(t) \sin\left(\frac{\pi n}{L}x\right) \tag{2.283}$$

We can replace this solution in: $\frac{\partial^2 u}{\partial t^2} - \frac{\partial^2 u}{\partial x^2} = \sinh(x)$ Using the orthogonality of the eigenfunctions $\sin\left(\frac{\pi n}{L}x\right)$ to arrive at the non-homogeneous equation for the coefficients $B_n(t)$.

$$\frac{d^2B_n(t)}{dt^2} + \left(\frac{\pi n}{L}\right)^2 B_n(t) = -2\pi n\frac{(-1)^n \sinh(L)}{L^2 + (\pi n)^2} \tag{2.284}$$

Initial conditions: $B_n(0) = 0$; $\left.\frac{dB_n}{dt}\right|_{t=0} = 0$

The general solution is the sum of the solution of the homogeneous equation and the particular solution.

$$B_n(t)_{part} = -\frac{2L^2}{\pi n}\frac{(-1)^n \sinh(L)}{L^2 + (\pi n)^2} \tag{2.285}$$

$$B_n(t) = C\sin\left(\frac{\pi n}{L}t\right) + D\cos\left(\frac{\pi n}{L}t\right) - \frac{2L^2}{\pi n}\frac{(-1)^n \sinh(L)}{L^2 + (\pi n)^2} \tag{2.286}$$

First initial condition:

$$B_n(0) = 0 \to D = \frac{2L^2}{\pi n}\frac{(-1)^n \sinh(L)}{L^2 + (\pi n)^2} \tag{2.287}$$

Second initial condition:

$$\left.\frac{dB_n}{dt}\right|_{t=0} = 0 \to C\frac{\pi n}{L} = 0 \to C = 0 \tag{2.288}$$

General solution

$$B_n(t) = \frac{2L^2}{\pi n}\frac{(-1)^n \sinh(L)}{L^2 + (\pi n)^2}\cos\left(\frac{\pi n}{L}t\right) - \frac{2L^2}{\pi n}\frac{(-1)^n \sinh(L)}{L^2 + (\pi n)^2} \tag{2.289}$$

and the solution is:

$$u(x, t) = \sum_{n=1}^{\infty} \frac{2L^2}{\pi n} \frac{(-1)^n \sinh(L)}{L^2 + (\pi n)^2} \left(\cos\left(\frac{\pi n}{L} t\right) - 1 \right) \sin\left(\frac{\pi n}{L} x\right) \tag{2.290}$$

Method 3 (using Green's function of an oscillator)

The problem of the coefficients $B_n(t)$ can be solved using Green's function that was previously used to solve the equation of a forced oscillator:

$$\frac{d^2 B_n(t)}{dt^2} + \left(\frac{\pi n}{L}\right)^2 B_n(t) = -2\pi n \frac{(-1)^n \sinh(L)}{L^2 + (\pi n)^2} = f_n \tag{2.291}$$

$$B_n(t) = C \sin\left(\frac{\pi n}{L} t\right) + D \cos\left(\frac{\pi n}{L} t\right) + \frac{L}{\pi n} \int_0^t f_n \sin\left[\frac{\pi n}{L}(t - \tau)\right] d\tau \tag{2.292}$$

Since both initial conditions are null the solution of the homogeneous equation is trivial ($C = D = 0$). Then we will get:

$$\begin{aligned} B_n(t) &= -2L \frac{(-1)^n \sinh(L)}{L^2 + (\pi n)^2} \int_0^t \sin\left[\frac{\pi n}{L}(t - \tau)\right] d\tau \\ &= \frac{2L^2}{\pi n} \frac{(-1)^n \sinh(L)}{L^2 + (\pi n)^2} \left(\cos\left(\frac{\pi n}{L} t\right) - 1 \right) \end{aligned} \tag{2.293}$$

Note that the form of the solutions obtained with methods 2 and 3 (expanding by orthogonal functions) are equal, but they are different to the solution obtained by the method 1, in which part of the solution has analytic form.

2.21 Case Study: Oscillations of the Gas in a Semi-Open Tube

A tube which is open on one end moves along its axial direction with a constant velocity V. At $t = 0$ the tube stops suddenly. Determine the longitudinal gas vibrations inside the tube as a function of the distance to the closed end (x) and time (t). Note: see also problem 5.6 b) from [1].

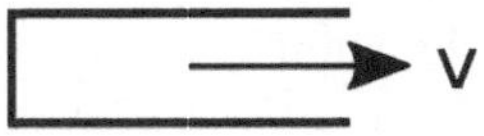

Figure 2.29

Mathematical formulation The problem is formulated in terms of "condensation" $u = \rho - \rho_0$, being ρ_0 the density of gas in equilibrium.

$$\frac{\partial^2 u}{\partial t^2} - a^2 \frac{\partial^2 u}{\partial x^2} = 0 \qquad (0 < x < L;\ t > 0) \tag{2.294}$$

Boundary condition 1: $\left.\frac{\partial u}{\partial x}\right|_{x=0} = 0$

Boundary condition 2: $u(L, t) = 0$

Initial condition 1: $u(x, 0) = 0$

To find a second initial condition for the time derivative of the gas density at $t = 0$ we will use the simplified continuity equation (chapter 5 of [1]):

$$\left.\frac{\partial \rho}{\partial t}\right|_{t=0} = -\rho_0 \frac{dv}{dx} \tag{2.295}$$

where $v(x)$ is the velocity of the molecules, which at $t = 0$ is constant for $x > 0$ and zero for $x < 0$ (that is, is the Heaviside function). Considering that $v(x)$ has the form of the Heaviside function multiplied by V, where V is the velocity of the molecules, and using the properties of the derivative of this function (Appendix 1 of [1]), we can write:

$$\frac{d\rho}{dt}(x, t = 0) = -\rho_0 V \delta(x) \tag{2.296}$$

Then the second boundary condition is:

$$\left.\frac{\partial u}{\partial t}\right|_{t=0} = -\rho_0 V \delta(x) \tag{2.297}$$

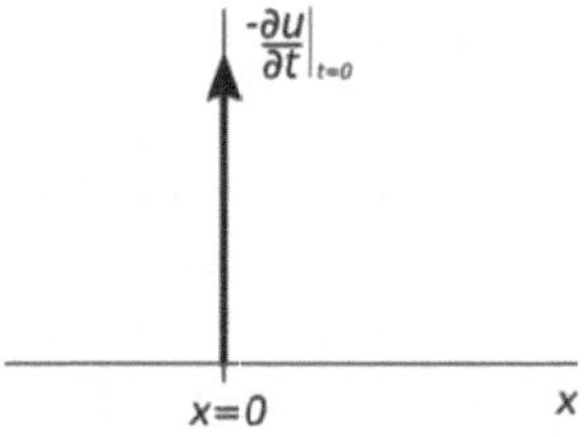

Figure 2.30

Solution The general solution is analogous to the oscillations of a string with the left end free, and the right one fixed.

$$u(x,t)=\sum_{n} T_n(t)\cdot X_n(x)=\sum_{n=0}^{\infty} C_n \sin(\omega_n t)\cdot\cos\left(\frac{\pi[2n+1]}{2L}x\right)$$
$$=\sum_{n=0}^{\infty} C_n \sin(\omega_n t)\cdot\cos\left(\frac{\pi[2n+1]}{2L}x\right) \quad (2.298)$$

This result is obtained by expanding the solution in orthogonal functions which are solutions of the Sturm–Liouville problem:

$$\left\{\begin{array}{l}\dfrac{d^2X}{dx^2}+\lambda X=0\\ \left.\dfrac{dX}{dx}\right|_{x=0}=X(L)=0\\ X_n(x)=\cos\left(\dfrac{\pi(2n+1)}{2L}x\right)\\ \lambda_n=\left[\dfrac{\pi(2n+1)}{2L}\right]^2\end{array}\right\} \quad (2.299)$$

We have also used already the first initial condition $[u(x,0)=0]$ to find the form of the temporal solutions. Finally we apply the second initial condition to find the coefficients C_n:

$$\left.\frac{\partial u}{\partial t}\right|_{t=0}=-\rho_0 V\delta(x)=\sum_{n=0}^{\infty} C_n\omega_n\cdot\cos\left(\frac{\pi[2n+1]}{2L}x\right) \quad (2.300)$$

We use the orthogonality of the functions $X_n(x)$:

$$C_n\omega_n\times\frac{L}{2}=-\rho_0 V \quad (2.301)$$

$$C_n=-\frac{2\rho_0 V}{\pi(2n+1)} \quad (2.302)$$

The final solution will be:

$$u(x,t)=-\frac{2\rho_0 V}{\pi}\sum_{n=0}^{\infty}\frac{1}{(2n+1)}\sin(\omega_n t)\cos\left(\frac{\pi[2n+1]}{2L}x\right) \quad (2.303)$$

2.22 Variation of the Temperature in a Thin Rod Exchanging Heat through Its Surface

The initial temperature of a rod of length L and neglectable cross section is described by a function $f(x)$. From $t = 0$ both ends $x = 0, L$ are connected to a thermal reservoir at $T = 0$. The lateral surface of the rod exchanges heat with an outer medium at a temperature u_0 according to Newton's law (this is, a heat exchange of the form $h(u - u_0)$ where h is a constant). Find the distribution of temperature as a function of time.

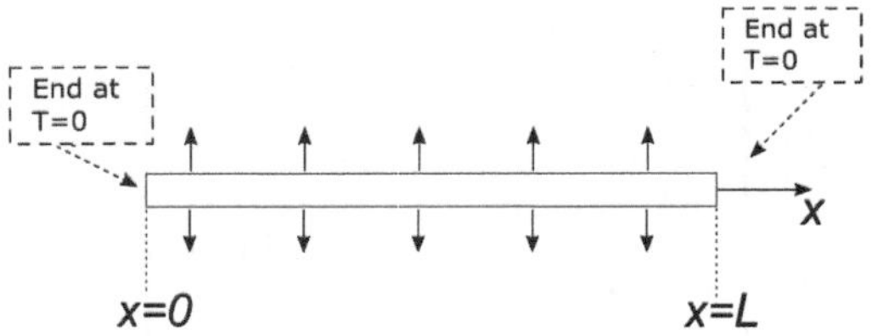

Figure 2.31

Mathematical formulation

$$\begin{cases} \dfrac{\partial u}{\partial t} = a^2 \dfrac{\partial^2 u}{\partial x^2} - h(u - u_0) \\ u(0, t) = u(L, t) = 0 \\ u(x, t = 0) = f(x) \end{cases} \tag{2.304}$$

We will find the solution as the sum of two functions: one corresponds to a stationary solution $u_0 + w(x)$, independent of time, which is the solution at $t \to \infty$), and another transient solution $v(x, t)$.
Replacing in the wave equation:

$$u(x, t) = u_0 + w(x) + v(x, t) \tag{2.305}$$

$$\frac{\partial v}{\partial t} = a^2 \frac{d^2 w}{dx^2} + a^2 \frac{\partial^2 v}{\partial x^2} - hw(x) - hv(x, t) \tag{2.306}$$

We separate the problem in two equations for the $w(x)$ and $v(x, t)$ functions.

Equation for the temporal part:

$$\frac{\partial v}{\partial t} = a^2 \frac{\partial^2 v}{\partial x^2} - hv(x, t) \tag{2.307}$$

With the boundary and initial conditions:

$$\left\{ \begin{array}{l} u(0, t) = u(L, t) = 0 \\ u(x, 0) = f(x) \end{array} \right\} \rightarrow \left\{ \begin{array}{l} v(0, t) = v(L, t) = 0 \\ v(x, 0) = f(x) - u_0 - w(x) \end{array} \right\} \tag{2.308}$$

Equation for the stationary part:

$$a^2 \frac{d^2 w}{dx^2} - hw(x) = 0 \tag{2.309}$$

Boundary conditions for $w(x)$:

$$u(0, t) = u(L, t) = v(0, t) = v(L, t) = 0 \tag{2.310}$$

Then $u_0 + w(0) = u_0 + w(L) = 0$
and we will have $w(0) = w(L) = -u_0$

General solution for equation (2.309)

$$w(x) = A \sinh\left(\frac{\sqrt{h}}{a} x\right) + B \sinh\left(\frac{\sqrt{h}}{a}(L - x)\right) \tag{2.311}$$

Boundary condition 1: $w(0) = -u_0 \rightarrow B = -\frac{u_0}{\sinh[\frac{\sqrt{h}}{a} L]}$

Boundary condition 2: $w(L) = -u_0 \rightarrow A = -\frac{u_0}{\sinh[\frac{\sqrt{h}}{a} L]}$

Solution:

$$w(x) = -\frac{u_0}{\sinh\left(\frac{\sqrt{h}}{a} L\right)} \sinh\left(\frac{\sqrt{h}}{a} x\right) - \frac{u_0}{\sinh\left(\frac{\sqrt{h}}{a} L\right)} \sinh\left(\frac{\sqrt{h}}{a}(L - x)\right) \tag{2.312}$$

On the other hand, to solve equation (2.307) we separate variables:

$$v(x, t) = T(t) X(x) \tag{2.313}$$

Replacing in (2.307):

$$\frac{h + \frac{1}{T}\frac{dT}{dt}}{a^2} = \frac{\frac{d^2 X}{dx^2}}{X} = -\lambda \tag{2.314}$$

We operate in this manner to get to a Sturm–Liouville problem for the spatial part and choose $\lambda > 0$ to achieve oscillating eigenfunctions.

Eigenfunctions and eigenvalues:

$$X(x) = \sin\left(\frac{\pi n}{L}x\right) \tag{2.315}$$

$$\lambda = \left[\frac{\pi n}{L}\right]^2 \tag{2.316}$$

Differential equation for $T(t)$:

$$\frac{dT}{dt} = -\left(\left[\frac{\pi n}{L}\right]^2 + h\right)T \tag{2.317}$$

The time dependent coefficients of the solution are:

$$T_n(t) = A_n e^{-\left(\left[\frac{\pi n}{L}\right]^2 + h\right)t} \tag{2.318}$$

General solution The general solution is:

$$v(x,t) = \sum A_n e^{-\left(\left[\frac{\pi n}{L}\right]^2 + h\right)t} \sin\left(\frac{\pi n}{L}x\right) \tag{2.319}$$

Final solution Applying the initial conditions:

$$v(x,0) = f(x) - u_0 - w(x) = \sum A_n \sin\left(\frac{\pi n}{L}x\right) \tag{2.320}$$

and using the orthogonality of the $\sin(\frac{\pi n}{L}x)$ eigenfunctions we get the coefficients of the Fourier expansion:

$$A_n = \frac{2}{L}\int_0^L [f(x) - u_0 - w(x)] \sin\left(\frac{\pi n}{L}x\right) dx \tag{2.321}$$

2.23 Distribution of Temperature in a Thin Wire with Losses on Its Surface

Find the variations in temperature as a function of time in a wire (length L, thermal diffusion coefficient k, linear density of mass ρ and heat capacity $C = 1$) if it is thin (that is, treat it as if it exchanges heat along its whole length, unlike what would really happen if it was a purely 1D problem, where heat exchange can only occur at $x = 0$ and $x = L$. The wire losses heat through its surface by unit length and by unit time with a constant h, according to Newton's law.

The right end is thermally insulated and the left one is in contact with a thermic reservoir at temperature $T = 0$. The temperature of the outer medium is also $T = 0$. At the initial moment ($t = 0$) the temperature of the wire is a linear function of the distance to the left end: $T(x, 0) = Ax$.

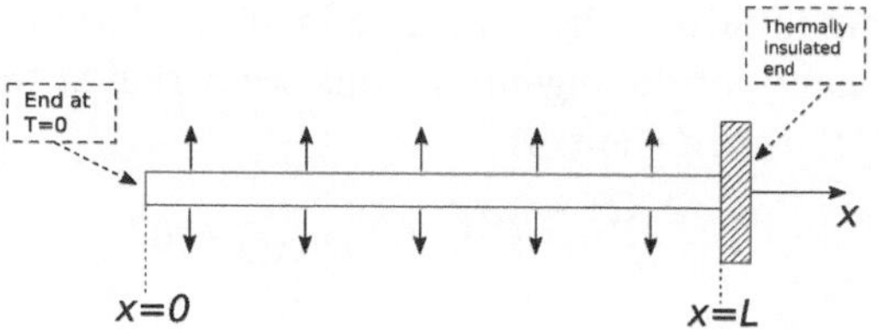

Figure 2.32

Mathematical formulation The density of heat sources acting on the wire can be described by:

$$f(x,t) = \frac{(dq/dx)}{\delta t} = -h[u(x,t) - u_0] \qquad (2.322)$$

where dq is the heat dissipated by the interval dx, while u_0 is the temperature of the outer medium, towards which the heat is dissipated; $\delta t = 1$ second. The problem to be solved consists of:

$$\begin{cases} \dfrac{\partial u(x,t)}{\partial t} = k\dfrac{\partial^2 u(x,t)}{\partial x^2} - h \cdot u(x,t) \\ \text{Initial condition:} \quad u(x, t=0) = Ax \\ \text{Boundary conditions:} \quad u(x=0,t) = \left.\dfrac{\partial u(x,t)}{\partial x}\right|_{x=L} = 0 \end{cases} \qquad (2.323)$$

Sturm–Liouville problem We seek the solution by expanding in eigenfunctions of the homogeneous problem, of known value for the spatial part:

$$v_n(x) = \sin\left(\frac{\pi(2n+1)}{2L}x\right) \qquad (2.324)$$

The eigenvalues are: $\lambda_n = \left(\frac{\pi(2n+1)}{2L}\right)^2$

The general solution is replaced in the heat equation, and the eigenvalues are used as the result of applying the differential operator $\frac{\partial^2}{\partial x^2}$:

$$u(x,t) = \sum_n A_n(t) v_n(x) \qquad (2.325)$$

$$\sum_n \frac{\partial A_n}{\partial t} v_n(x) = -k\sum_n A_n(t)\lambda_n v_n(x) - h\sum_n A_n(t) v_n(x) \qquad (2.326)$$

Both sides are multiplied by $v_m(x)$ and integrated from 0 to L. Using the orthogonality of the eigenfunctions we arrive at the following equation for the coefficients A_n:

$$\frac{\partial A_n(t)}{\partial t} + (k\lambda_n + h)A_n(t) = 0 \tag{2.327}$$

And:

$$A_n(t) = C_n e^{-(k\lambda_n + h)t} \tag{2.328}$$

General solution The general solution, separating the temporal and spatial variables, is:

$$\begin{aligned} u(x,t) &= \sum_n C_n e^{-(k\lambda_n+h)t} \sin\left(\frac{\pi(2n+1)}{2L}x\right) \\ &= e^{-ht}\sum_n C_n e^{-k\lambda_n t} \sin\left(\frac{\pi(2n+1)}{2L}x\right) \end{aligned} \tag{2.329}$$

Final solution Applying the initial conditions:

$$u(x,0) = \sum_n C_n \sin\left(\frac{\pi(2n+1)}{2L}x\right) = Ax \tag{2.330}$$

We have the identities:

$$\begin{aligned} A\int_0^L x \sin\left(\frac{\pi(2n+1)}{2L}x\right) dx &= A\frac{4L^2}{[\pi(2n+1)]^2}\cos(\pi n) \\ &= A\frac{4L^2}{[\pi(2n+1)]^2}(-1)^n \end{aligned} \tag{2.331}$$

$$\int_0^L \left[\sin\left(\frac{\pi(2n+1)}{2L}x\right)\right]^2 dx = \frac{\pi L(2n+1)}{[2\pi(2n+1)]} = \frac{L}{2} \tag{2.332}$$

Finally we arrive at the coefficients of the expansion used as solution:

$$C_n = \frac{8AL}{[\pi(2n+1)]^2}(-1)^n \tag{2.333}$$

Note: only due to the specific form of the non-homogeneous term (proportional to the solution) in this case it would be possible to solve the problem also using the separation of variables method, but applied to the whole PDE, and not only to the homogeneous part.

2.24 Case Study: Oscillations of a Finite String with Friction

A string of length L with both ends fixed is placed in a viscous medium with friction proportional to the transversal component of the velocity (with proportionality coefficient k). Consider that initial displacement of the string has the form of the function $f(x)$ and that the initial transversal velocity is $\psi(x)$. The sound speed is equal to c. Find the transversal oscillations from $t = 0$ if the linear density of mass is equal to 1.

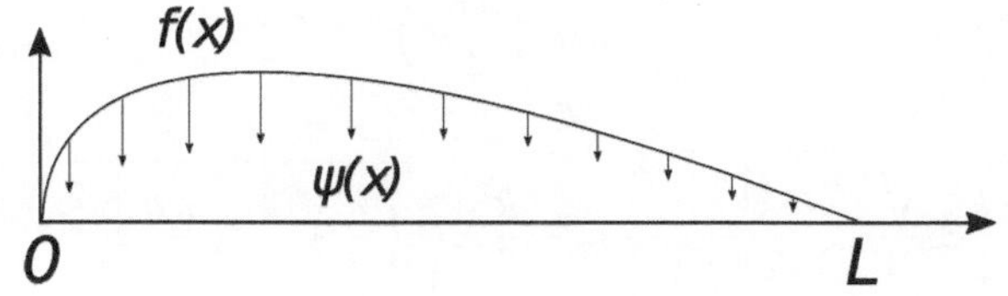

Figure 2.33

Mathematical formulation

$$\left\{\begin{array}{ll} \dfrac{\partial^2 u}{\partial t^2} - c^2\dfrac{\partial^2 u}{\partial x^2} = -k\dfrac{\partial u}{\partial t} & c, k > 0 \\ u(0, t) = u(L, t) = 0 & \\ u(x, t = 0) = f(x); \quad \left.\dfrac{\partial u}{\partial t}\right|_{t=0} = \psi(x) & \end{array}\right\} \tag{2.334}$$

Sturm–Liouville problem

The method of separation of variables is used: $u(x, t) = \sum T_n(t)X_n(x)$.

$X_n(x)$ are orthogonal eigenfunctions, which are solutions of the Sturm–Liouville problem with boundary conditions:

$$\left\{\begin{array}{l} \dfrac{d^2X}{dx^2} + \lambda_n X_n = 0 \\ X(0) = X(L) = 0 \end{array}\right\} \tag{2.335}$$

The eigenfunctions are known:

$$X_n = \sin\left(\frac{\pi n}{L}x\right); \qquad \lambda_n = \left(\frac{\pi n}{L}\right)^2 \quad (n = 1, 2, 3 \ldots) \tag{2.336}$$

General solution Replacing $X(x)$ in the wave equation, we obtain the equation for the temporal part.

$$\frac{d^2T_n}{dt^2} + k\frac{dT_n}{dt} + \left(\frac{\pi n}{L}c\right)^2 T_n(t) = 0 \tag{2.337}$$

We seek a solution of the form $T_n(t) = e^{\alpha t}$, which replaced in the previous equation gives:

$$\alpha^2 + k\alpha + \left(\frac{\pi n}{L}c\right)^2 = 0 \tag{2.338}$$

$$\alpha_{1,2} = -\frac{k}{2} \pm \frac{\sqrt{k^2 - 4(\frac{\pi n}{L}c)^2}}{2} = -\frac{k}{2} \pm \sqrt{\left(\frac{k}{2}\right)^2 - \left(\frac{\pi n}{L}c\right)^2} = -\frac{k}{2} \pm \beta_n \tag{2.339}$$

The values of $\alpha_{1,2}$ will be real when:

$$k^2 - 4\left(\frac{\pi n}{L}c\right)^2 > 0; \quad n \leq \frac{kL}{2\pi c} \tag{2.340}$$

There will be two independent solutions:

$$\begin{Bmatrix} T_{1,n}(t) = e^{\alpha_1 t} \\ T_{2,n}(t) = e^{\alpha_2 t} \end{Bmatrix} \tag{2.341}$$

The solution can be simplified by creating two independent combinations:

$$f_1 = \frac{T_{1,n}(t) + T_{2,n}(t)}{2} = e^{-\frac{kt}{2}} \frac{e^{\beta t} + e^{-\beta t}}{2} = e^{-\frac{k}{2}t} \cosh(\beta_n t) \tag{2.342}$$

$$f_2 = \frac{T_{1,n}(t) - T_{2,n}(t)}{2} = e^{-\frac{kt}{2}} \frac{e^{\beta t} - e^{-\beta t}}{2} = e^{-\frac{k}{2}t} \sinh(\beta_n t) \tag{2.343}$$

The solution for the range of indices $n \leq \frac{kL}{2\pi c}$ (n is compared to the maximum integer value of $\frac{kL}{2\pi c}$ for which the solutions of $\alpha_{1,2}$ are still real) must be presented in the form of a combination of two independent solutions in general, with different weights:

$$T_n(t) = e^{-\frac{k}{2}t}[A_n \cosh(\beta_n t) + B_n \sinh(\beta_n t)] \tag{2.344}$$

Finally the indices $n > \frac{kL}{2\pi c} \rightarrow \alpha_{1,2}$ will have complex values (we try to compare n to the minimum integer of $\frac{kL}{2\pi c}$ for which the solutions of $\alpha_{1,2}$ start to be imaginary):

$$\alpha_{1,2} = -\frac{k}{2} \pm \sqrt{\left(\frac{k}{2}\right)^2 - \left(\frac{\pi n}{L}c\right)^2} = -\frac{k}{2} \pm i\beta_n \tag{2.345}$$

Analogously we have:

$$T_n(t) = e^{-\frac{k}{2}t}[A_n \cos(\beta_n t) + B_n \sin(\beta_n t)] \tag{2.346}$$

The general solution for the problem will be:

$$\begin{aligned} u(x, t) = e^{-\frac{kt}{2}} \sum_{n \leq \frac{kL}{2\pi c}} [A_n \cosh(\beta_n t) + B_n \sinh(\beta_n t)] \sin\left(\frac{\pi n}{L}x\right) \\ +e^{-\frac{kt}{2}} \sum_{n > \frac{kL}{2\pi c}} [A_n \cos(\beta_n t) + B_n \sin(\beta_n t)] \sin\left(\frac{\pi n}{L}x\right) \end{aligned} \tag{2.347}$$

Final solution The coefficients A_n and B_n are found by applying the initial conditions:

Note: we can use the same names of the indices A_n and B_n of the sum, since they are separated in two sums with two different ranges of n that don't overlap.
From the first initial condition:

$$u(x, 0) = f(x) = \sum_{1 \leq n < \infty} A_n \sin\left(\frac{\pi n}{L}x\right) \tag{2.348}$$

Using the orthogonality of the eigenfunctions $X_n = \sin\left(\frac{\pi n}{L}x\right)$

$$A_n = \frac{2}{L} \int_0^L f(x) \sin\left(\frac{\pi n}{L}x\right) dx \tag{2.349}$$

Analogously, applying the second initial condition:

$$\left.\frac{\partial u}{\partial t}\right|_{t=0} = \psi(x) = \sum_{1 \leq n < \infty} \left(-\frac{k}{2}\right) A_n \sin\left(\frac{\pi n}{L}x\right) + \sum_{1 \leq n < \infty} \beta B_n \sin\left(\frac{\pi n}{L}x\right) \tag{2.350}$$

$$\sum_{1 \leq n < \infty} \left(-\frac{k}{2}A_n + \beta_n B_n\right) \times \sin\left(\frac{\pi n}{L}x\right) = \psi(x) \tag{2.351}$$

We arrive at:

$$\left(-\frac{k}{2}A_n + \beta_n B_n\right) = \frac{2}{L}\int_0^L \psi(x)\sin\left(\frac{\pi n}{L}x\right)dx$$

$$B_n = \frac{1}{\beta_n}\left[\frac{k}{2}A_n + \frac{2}{L}\int_0^L \psi(x)\sin\left(\frac{\pi n}{L}x\right)dx\right]$$
$$= \frac{1}{\beta_n}\left[\frac{k}{L}\int_0^L f(x)\sin\left(\frac{\pi n}{L}x\right)dx + \frac{2}{L}\int_0^L \psi(x)\sin\left(\frac{\pi n}{L}x\right)dx\right] \tag{2.352}$$

2.25 Propagation of a Thermal Pulse in a Thin Bar with Insulated Ends

At $t = 0$ the temperature distribution in a very thin rod whose surface and its ends stay thermally insulated is $T(x) = A\delta(x - x_0)$.

Find the distribution of temperature along the rod as a function of time.

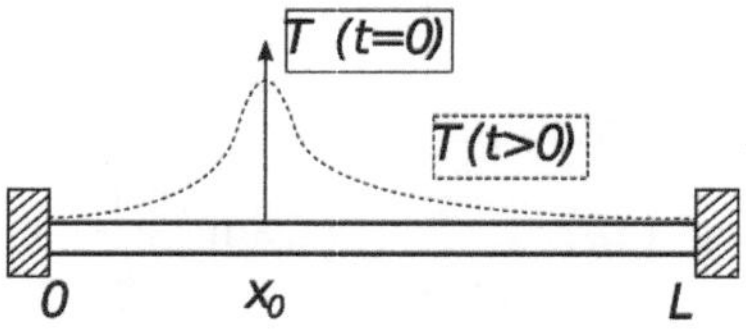

Figure 2.34

Mathematical formulation

$$\begin{Bmatrix} \dfrac{\partial u}{\partial t} = c^2\dfrac{\partial^2 u}{\partial x^2} \\ \left.\dfrac{\partial u}{\partial x}\right|_{x=0} = \left.\dfrac{\partial u}{\partial x}\right|_{x=L} = 0 \\ u(x, t = 0) = A\delta(x - x_0) \end{Bmatrix} \tag{2.353}$$

Sturm–Liouville problem Using the method of separation of variables:

$$u(x, t) = T(t)X(x) \tag{2.354}$$

To find the eigenfunctions of the spatial part $X_n(x)$ we need to solve a problem similar to that of a string with both ends free:

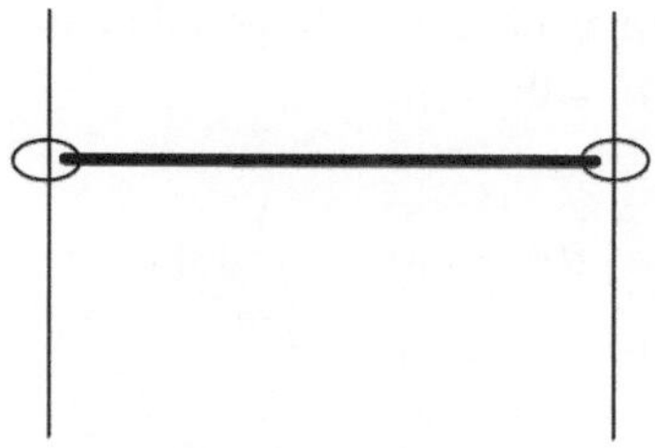

Figure 2.35

The Sturm–Liouville problem for $X(x)$ is:

$$\left\{ \begin{array}{c} \dfrac{d^2X}{dx^2} + \lambda X = 0 \\ \left.\dfrac{dX}{dx}\right|_{x=0} = \left.\dfrac{dX}{dx}\right|_{x=L} = 0 \end{array} \right\} \tag{2.355}$$

The eigenfunctions and eigenvalues for this problem are:

$$X_n(x) = \cos(\sqrt{\lambda_n} x); \tag{2.356}$$

$$\lambda_n = \left(\frac{\pi n}{L}\right)^2 \quad (n = 0, 1, 2, \ldots) \tag{2.357}$$

Including this value for the $X_n(x)$ in the solution $u(x, t) = \sum T_n(t)X_n(x)$ and replacing it in the heat equation, and taking advantage of the orthogonality of the eigenfunctions, we arrive at an equation for the coefficients $T_n(t)$. They are determined by using the initial conditions:

$$T_n(t) = B_n e^{-c^2\left(\frac{\pi n}{L}\right)^2 t} \tag{2.358}$$

General solution The solution is:

$$u(x, t) = B_0 + \sum_{n=1}^{\infty} B_n e^{-c^2\left(\frac{\pi n}{L}\right)^2 t} \cos\left(\frac{\pi n}{L} x\right) \tag{2.359}$$

Final solution Using the initial conditions:

$$u(x,0) = B_0 + \sum_{n=1}^{\infty} B_n \cos\left(\frac{\pi n}{L}x\right) = A\delta(x - x_0) \qquad (2.360)$$

And taking advantage of the orthogonality of the eigenfunctions $X_n(x)$, we have for $n = 0$:

$$B_0 \int_0^L \cos\left(\frac{\pi 0}{L}x\right) dx = B_0 L = \int_0^L A\delta(x - x_0)\cos\left(\frac{\pi 0}{L}x\right) dx$$
$$= \int_0^L A\delta(x - x_0) dx = A \qquad (2.361)$$

$$B_0 = \frac{A}{L} \qquad (2.362)$$

For n$\neq$ 0 we have:

$$B_n \int_0^L \cos^2\left(\frac{\pi n}{L}x\right) dx = B_n \frac{L}{2} = \int_0^L A\delta(x - x_0)\cos\left(\frac{\pi n}{L}x\right) dx$$
$$= A\cos\left(\frac{\pi n}{L}x_0\right) \qquad (2.363)$$

Then the coefficients of the series are:

$$B_n = \int_0^L A\delta\,(x - x_0)\cos\left(\frac{\pi n}{L}x\right) dx = \frac{2A}{L}\cos\left(\frac{\pi n}{L}x_0\right) \qquad (2.364)$$

2.26 Forced Oscillations of a Hanging String in a Gravitational Field

Consider a heavy string of length L and linear mass density $\rho = 1$ (in the presence of gravity), hanging from a point which oscillates (transversally to the string) like this: $u_0 \sin(\omega t)$.

Determine the displacement of the string in the limit of small oscillations (see also problem 4.9 and section 4.1.4 from [1]).

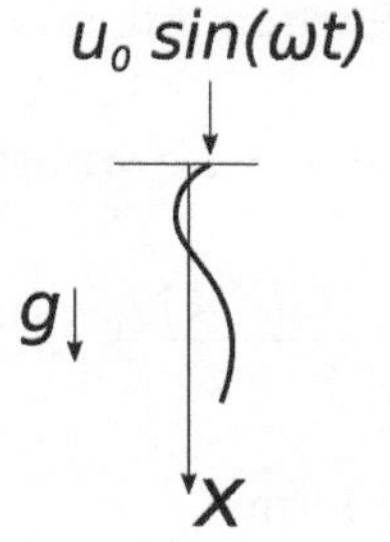

Figure 2.36

Mathematical formulation

$$\rho\frac{\partial^2 u(x,t)}{\partial t^2}-\frac{\partial}{\partial x}\left[T(x)\frac{\partial u(x,t)}{\partial x}\right]=0 \tag{2.365}$$

The oscillating end of the string will be described with the boundary condition $u(0,t) = u_0\sin(\omega t)$. The equation will simplify by considering that the tension is due to gravity.

$$T(x)=\rho g x \tag{2.366}$$

Then:

$$\frac{\partial^2 u(x,t)}{\partial t^2}-g\frac{\partial}{\partial x}\left[x\frac{\partial u(x,t)}{\partial x}\right]=0 \tag{2.367}$$

Sturm–Liouville problem and general solution We will seek the solution by separating space and time variables as:

$$u(x)=C\,X(x)\sin(\omega t) \tag{2.368}$$

Replacing that into the wave equation:

$$-\omega^2 X(x)\,\sin(\omega t)-g\frac{d}{dx}\left[x\frac{dX(x)}{dx}\right]\sin(\omega t)=0 \tag{2.369}$$

$$-\frac{\omega^2}{g}-\frac{1}{X(x)}\frac{d}{dx}\left[x\frac{dX(x)}{dx}\right]=0 \tag{2.370}$$

We seek the solution of the equation:

$$\frac{1}{X(x)}\frac{d}{dx}\left[x\frac{dX(x)}{dx}\right]=-\lambda\ (\lambda>0) \tag{2.371}$$

The solution of the equation:

$$\frac{d}{dx}\left[x\frac{dX(x)}{dx}\right]+\lambda X(x)=0 \tag{2.372}$$

is the zero order Bessel function if the function has no discontinuities.
With the change of variable: $\xi = \sqrt{x}$ we arrive at the equation:

$$\frac{d^2X(\xi)}{d\xi^2} + \frac{1}{\xi}\frac{dX(\xi)}{d\xi} + 4\lambda X(\xi) = 0 \tag{2.373}$$

which has a solution of the form:

$$X(x) = J_0(2\sqrt{\lambda x}) \tag{2.374}$$

Since $J_0(0) = 1 \to C = u_0$

Final solution The non-trivial solutions correspond to the values:

$$\lambda = \frac{\omega^2}{g} \tag{2.375}$$

Then the solution will be:

$$u(x, t) = u_0 J_0(2\sqrt{\lambda x}) \sin(\omega t) \tag{2.376}$$

Note: the shape of the string (the number of nodes) will change with the applied frequency.

2.27 Case Study: Temperature Equilibrium in a Bar with Heat Sources

Consider a bar of length L with heat capacity C, density ρ and heat conduction coefficient k. The left end of the bar ($x = 0$) is insulated, while the right end ($x = L$) is in contact with the outer medium, which is at T_0 and exchanges heat according to Newton's law (with negative constant H). At the initial moment the temperature of the bar is equal to T_0. Find the variation of temperature as a function of time if starting at $t = 0$ in the central part of the bar ($L/4 < x < 3L/4$) are heat sources with a constant density F.

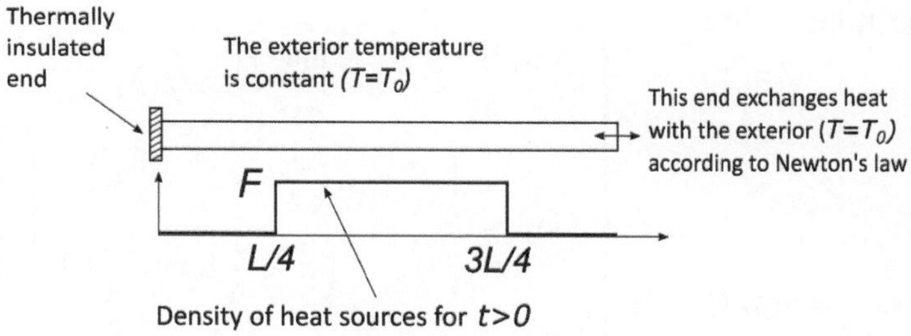

Figure 2.37

Mathematical formulation

$$\left\{\begin{array}{l} C\rho\dfrac{\partial u(x,t)}{\partial t} - k\dfrac{\partial^2 u(x,t)}{\partial x^2} = f(x) \\ f(x) = \left\{\begin{array}{ll} F & \left(\dfrac{1}{4}L < x < \dfrac{3}{4}L\right) \\ 0 & \left(x < \dfrac{1}{4}L; x > \dfrac{3}{4}L\right) \end{array}\right\} \\ u(x,0) = T_0 \\ \left.\dfrac{du}{dx}\right|_{x=0} = 0 \\ -k\left.\dfrac{du}{dx}\right|_{x=L} = H(u(L,t) - T_0) \end{array}\right\} \quad (2.377)$$

We have chosen the sign for the heat exchange in the right end so that the negative gradient of temperature describes the heat flux in the positive direction.

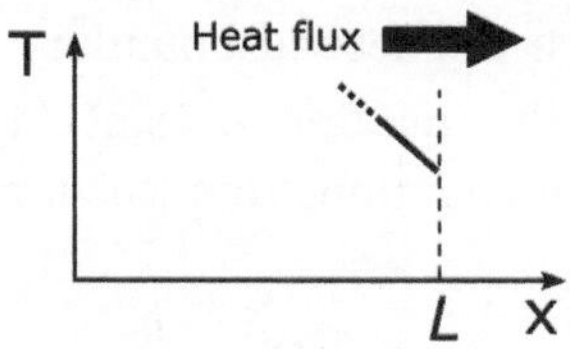

Figure 2.38 Heat flux direction at $x = L$

Main method of resolution We seek the solution subtracting first the thermal background T_0 to have homogeneous boundary conditions.

$$v(x,t)=u(x,t)-T_0\left\{\begin{array}{l} C\rho\dfrac{\partial v(x,t)}{\partial t}-k\dfrac{\partial^2 v(x,t)}{\partial x^2}=f(x) \\ f(x)=\left\{\begin{array}{ll} F & \left(\dfrac{1}{4}L<x<\dfrac{3}{4}L\right) \\ 0 & \left(x<\dfrac{1}{4}L; x>\dfrac{3}{4}L\right)\end{array}\right\} \\ v(x,t=0)=0 \\ \left.\dfrac{dv}{dx}\right|_{x=0}=0 \\ \left.\dfrac{dv}{dx}\right|_{x=L}=-hv(L,t);\quad h=\dfrac{H}{k}\end{array}\right\} \tag{2.378}$$

Sturm–Liouville problem We seek the solution as a summation:

$$v(x,t)=\sum_n A_n(t)X_n(x) \tag{2.379}$$

The spatial part is expanded in orthogonal eigenfunctions which solve the Sturm–Liouville problem:

$$\left\{\begin{array}{l} \dfrac{d^2X_n}{dx^2}+\lambda X_n=0 \\ \left.\dfrac{dX_n}{dx}\right|_{x=0}=0 \\ \left.\dfrac{dX_n}{dx}\right|_{x=L}=-hX_n(L)\end{array}\right\} \tag{2.380}$$

$$X_n(x)=A\sin(\sqrt{\lambda_n}x)+B\cos(\sqrt{\lambda_n}x) \tag{2.381}$$

From the first boundary condition: $\frac{dX_n}{dx}\big|_{x=0}=0=A\sqrt{\lambda_n}\cos(\sqrt{\lambda_n}0)$ we have $A=0$. From the second boundary condition: $\frac{dX}{dx}\big|_{x=L}+hX(L)=0$ we deduce the equation: $\sqrt{\lambda}\sin(\sqrt{\lambda}L)=h\cos(\sqrt{\lambda}L)$. In a more compact form this equation, from which we get the eigenvalues λ_n is:

$$\tan(\sqrt{\lambda}L)=\frac{h}{\sqrt{\lambda}} \tag{2.382}$$

Then we get the eigenfunctions:

$$X_n(x)=B_n\cos[\sqrt{\lambda_n}x] \tag{2.383}$$

Note: here $n=1,2,3\ldots$ indicates the numeration of eigenvalues and the corresponding eigenfunctions.

General solution To analyze the temporal part, we replace the expansion in eigenfunctions into equation (2.378):

$$C\rho\frac{\partial}{\partial t}\left[\sum_n A_n(t)X_n(x)\right]-k\frac{d^2}{dx^2}\left[\sum_n A_n(t)X_n(x)\right]=f(x) \quad (2.384)$$

$$C\rho\sum_n\frac{dA_n(t)}{dt}X_n(x)-k\sum_n A_n(t)\frac{d^2X_n(x)}{dx^2}=f(x) \quad (2.385)$$

$$\sum_n\left[\frac{dA_n(t)}{dt}+\chi\lambda_nA_n(t)\right]X_n(x)=\frac{f(x)}{C\rho} \quad (2.386)$$

(where $\chi=\frac{k}{C\rho}$). Both parts of the previous expression are multiplied by $X_m(x)$ and integrated between 0 and L. Using the orthogonality of the eigenfunctions $X_n(x)$ we arrive at the equation:

$$\left[\frac{dA_n(t)}{dt}+\chi\lambda_nA_n(t)\right]\int_0^L[X_n(x)]^2dx=\frac{1}{C\rho}\int_0^L f(x)X_n(x)dx$$

$$=\frac{F}{C\rho}\int_{\frac{L}{4}}^{\frac{3L}{4}}\cos[\sqrt{\lambda_n}x]dx \quad (2.387)$$

$$\frac{dA_n(t)}{dt}+\chi\lambda_nA_n(t)=\frac{F}{C\rho\int_0^L[\cos\sqrt{\lambda_n}x]^2dx}\int_{\frac{L}{4}}^{\frac{3L}{4}}\cos[\sqrt{\lambda_n}x]dx \quad (2.388)$$

$$\frac{dA_n(t)}{dt}+\chi\lambda_nA_n(t)=\frac{F[\sin\sqrt{\lambda_n}\frac{3L}{4}-\sin\sqrt{\lambda_n}\frac{L}{4}]}{\sqrt{\lambda_n}C\rho\int_0^L[\cos\sqrt{\lambda_n}x]^2dx}=f_n \quad (2.389)$$

Note about the norm:

$$\int_0^L[\cos(ax)]^2dx=\frac{2aL+\sin(2aL)}{4a}\neq\frac{L}{2}\quad\text{(in general)} \quad (2.390)$$

The general solution of the obtained equation:

$$\frac{dA_n(t)}{dt}+\chi\lambda_nA_n(t)=f_n \quad (2.391)$$

is the sum of the solutions of the homogeneous equation:

$$A_{n,hom}(t) = A_n(0)e^{-\chi\lambda_n t} \tag{2.392}$$

and the particular solution of the non-homogeneous equation:

$$A_{n,part} = \frac{f_n}{\chi\lambda_n} \tag{2.393}$$

Final solution Finally the solution is:

$$A_n(t) = A_n(0)e^{-\chi\lambda_n t} + \frac{f_n}{\chi\lambda_n} \tag{2.394}$$

All is left to do is using the initial condition to find the coefficients $A_n(0)$:

$$v(x, 0) = \sum_n A_n(0)X_n(x) = \sum_n \left[A_n(0)e^0 + \frac{f_n}{\chi\lambda_n}\right] X_n(x) = 0 \tag{2.395}$$

From here we get the values $A_n(0) = -\frac{f_n}{\chi\lambda_n}$

Finally the solution is:

$$u(x, t) = T_0 + \sum_n \frac{f_n}{\chi\lambda_n}\left[1 - e^{-\chi\lambda_n t}\right]\cos\left(\sqrt{\lambda_n}x\right) \tag{2.396}$$

Alternative method In this alternative method we subtract a thermal background T_0 (this method is more "physical"). The solution can be separated into two, one corresponding to the profile of temperature $w(x)$ reached in the limit $t = \infty$ and another transitory one $v(x, t)$

$$u(x, t) = v(x, t) + w(x) \tag{2.397}$$

Complete problem:

$$\left\{\begin{array}{l} C\rho\dfrac{\partial u(x, t)}{\partial t} - k\dfrac{d^2u(x, t)}{dx^2} = f(x) \\ f(x) = \left\{\begin{array}{ll} F & \left(\dfrac{1}{4}L < x < \dfrac{3}{4}L\right) \\ 0 & \left(x < \dfrac{1}{4}L; x > \dfrac{3}{4}L\right) \end{array}\right\} \\ u(x, 0) = 0 \\ \left.\dfrac{du}{dx}\right|_{x=0} = 0 \\ \left.\dfrac{du}{dx}\right|_{x=L} = -hu(L, t) \end{array}\right\} \tag{2.398}$$

Problem 1 for $w(x)$:

$$\left\{\begin{array}{l} -k\dfrac{d^2w(x)}{dx^2} = f(x) \\ f(x) = \left\{\begin{array}{ll} F & \left(\dfrac{1}{4}L < x < \dfrac{3}{4}L\right) \\ 0 & \left(x < \dfrac{1}{4}L; x > \dfrac{3}{4}L\right) \end{array}\right\} \\ \left.\dfrac{dw}{dx}\right|_{x=0} = 0 \\ \left.\dfrac{dw}{dx}\right|_{x=L} = -hw(L) \end{array}\right\} \tag{2.399}$$

Problem 2 for $v(x, t)$

$$\left\{\begin{array}{l} C\rho\dfrac{\partial v(x,t)}{\partial t} - k\dfrac{d^2v(x,t)}{dx^2} = 0 \\ u(x, t = 0) = 0 \rightarrow v(x, 0) = -w(x) \\ \left.\dfrac{dv}{dx}\right|_{x=0} = 0 \\ \left.\dfrac{dv}{dx}\right|_{x=L} = -hv(L, t) \end{array}\right\} \tag{2.400}$$

With this separation we reach the solution by solving two simpler problems. To find $w(x)$ we can find the solution in three intervals of x:

$$w(x) = \left\{\begin{array}{ll} w_1(x) & \left(0 < x < \dfrac{1}{4}L\right) \\ w_2(x) & \left(\dfrac{1}{4}L < x < \frac{3}{4}L\right) \\ w_3(x) & \left(\dfrac{3}{4}L < x < L\right) \end{array}\right\} \tag{2.401}$$

Problem for $w_1(x)$

$$\left\{\begin{array}{ll} -k\dfrac{d^2w_1(x)}{dx^2} = 0 & (0 < x < \dfrac{1}{4}L) \\ \left.\dfrac{dw_1}{dx}\right|_{x=0} = 0 & \end{array}\right\} \tag{2.402}$$

Since $w_1(x) = A_1x + B_1$

$$\left.\frac{dw}{dx}\right|_{x=0} = 0 \rightarrow A_1 = 0 \rightarrow w_1(x) = B_1 \tag{2.403}$$

Problem for $w_3(x)$

$$\begin{Bmatrix} -k\dfrac{d^2w_3(x)}{dx^2} = 0 & \left(\dfrac{3}{4}L < x < L\right) \\ \left.\dfrac{dw_3}{dx}\right|_{x=L} = -hw_3(L) & \end{Bmatrix} \tag{2.404}$$

Since $w_3(x) = A_3x + B_3$

$$\left.\frac{dw_3}{dx}\right|_{x=L} + hw_3(L) = 0 \to A_3 + h(A_3L + B_3)$$
$$= 0 \to A_3(1 + hL) + B_3h = 0 \tag{2.405}$$

In this way we can relate the constants A_3 and B_3.

Finally we formulate the problem for $w_2(x)$.

$$\frac{d^2w_2(x)}{dx^2} = -\frac{F}{k} \quad \left(\frac{1}{4}L < x < \frac{3}{4}L\right) \tag{2.406}$$

$$w_2(x) = -\frac{F}{2k}x^2 + A_2x + B_2 \tag{2.407}$$

From the continuity conditions in the boundary, and their derivatives, between $w_1(x)$, $w_2(x)$ and $w_3(x)$ we find the values of the coefficients. Un this way we will have five conditions for the five unknown constants.

$$\begin{Bmatrix} (1)\ A_3(1 + hL) + B_3h = 0 \\ (2)\ w_2\left(\dfrac{L}{4}\right) = -\dfrac{F}{2k}\left(\dfrac{L}{4}\right)^2 + A_2\left(\dfrac{L}{4}\right) + B_2 = w_1\left(\dfrac{L}{4}\right) = B_1 \\ (3)\ \left.\dfrac{dw_2}{dx}\right|_{x=\frac{L}{4}} = -\dfrac{F}{k}\left(\dfrac{L}{4}\right) + A_2 = \left.\dfrac{dw_1}{dx}\right|_{x=\frac{L}{4}} = 0 \\ (4)\ w_2\left(\dfrac{3L}{4}\right) = -\dfrac{F}{2k}\left(\dfrac{3L}{4}\right)^2 + A_2\left(\dfrac{3L}{4}\right) + B_2 = w_3\left(\dfrac{3L}{4}\right) = A_3\left(\dfrac{3L}{4}\right) + B_3 \\ (5)\ \left.\dfrac{dw_2}{dx}\right|_{x=\frac{3L}{4}} = -\dfrac{F}{k}\left(\dfrac{3L}{4}\right) + A_2 = \left.\dfrac{dw_3}{dx}\right|_{x=\frac{3L}{4}} = A_3 \end{Bmatrix} \tag{2.408}$$

$$\left\{\begin{array}{l} A_2 = \dfrac{FL}{4k} \\ A_3 = -\dfrac{FL}{2k} \\ B_1 = \dfrac{F}{2k}\left(\dfrac{L^2}{h}\left(h+\dfrac{1}{L}\right) - 2\left(\dfrac{L}{2}\right)^2\right) \\ B_2 = \dfrac{F}{2k}\left(\dfrac{L^2}{h}\left(h+\dfrac{1}{L}\right) - \left(\dfrac{3L}{4}\right)^2\right) \\ B_3 = \dfrac{FL}{2k}\dfrac{1+hL}{h} \end{array}\right\} \tag{2.409}$$

Equations (1-5) allow us to find the form of $w(x)$. The problem 2 for $v(x, t)$ is homogeneous with homogeneous boundary conditions of the second and third type and with known initial conditions. We solve it by expanding into orthogonal functions, which we have found in the first method of solving the problem.

$$v(x, t) = \sum_n A_n e^{-\chi \lambda_n t} \cos(\sqrt{\lambda_n} x) \tag{2.410}$$

Applying the initial conditions $u(x, 0) = 0 \to v(x, 0) = -w(x)$

$$-w(x) = \sum_n A_n \cos[\sqrt{\lambda_n} x] \tag{2.411}$$

and using the orthogonality of the eigenfunctions $\cos(\sqrt{\lambda_n} x)$ between 0 and L we solve the A_n coefficients.

$$A_n = \frac{-\int_0^L w(x) \cos[\sqrt{\lambda_n} x] dx}{\int_0^L [\cos \sqrt{\lambda_n} x]^2 dx} \tag{2.412}$$

While the first way of solving the problem expands the function $w(x)$ of the second part in a summation of orthogonal functions, the second method obtains explicitly their form as a function in parts (see figure).

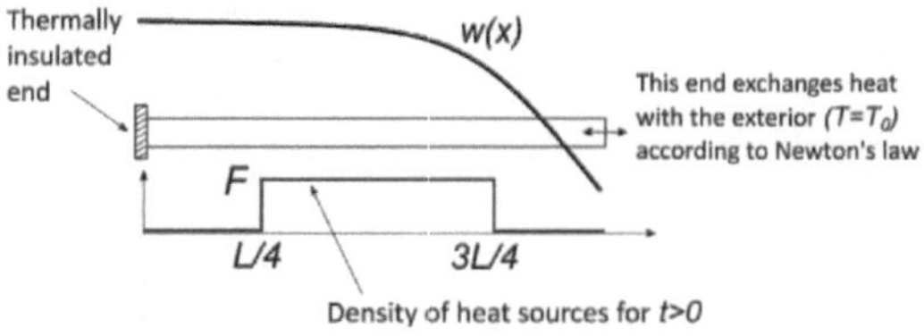

Figure 2.39

2.28 Case Study: String under a Gravitational Field

Consider a string of length L, tension T and linear density of mass ρ. The string has its left end connected to a spring (with constant β) and the right end can move freely in the transversal direction. At the instant $t = 0$ the string is at rest in the horizontal position, since there is no gravitational field. From $t = 0$ onwards the string becomes subject to the Earth's gravitational field (g).

Determine the form of the string as a function of time for $t > 0$.

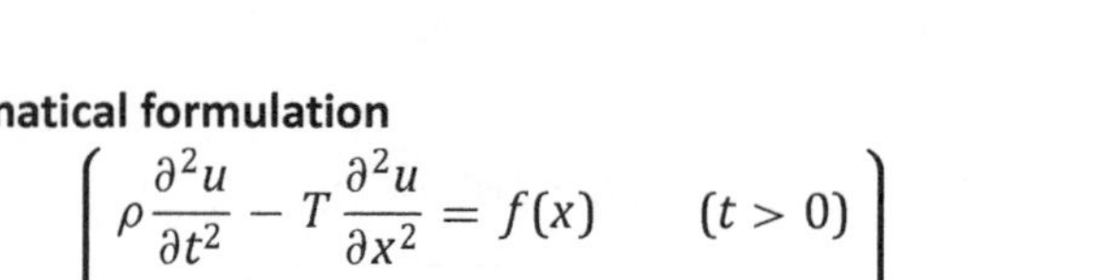

Figure 2.40

Mathematical formulation

$$\left\{\begin{array}{l} \rho\dfrac{\partial^2 u}{\partial t^2} - T\dfrac{\partial^2 u}{\partial x^2} = f(x) \qquad (t > 0) \\ T\left.\dfrac{\partial u}{\partial x}\right|_{x=0} - \beta u(0,t) = 0 \\ \left.\dfrac{\partial u}{\partial x}\right|_{x=L} = 0 \\ u(x, t=0) = 0 \\ \left.\dfrac{\partial u}{\partial t}\right|_{t=0} = 0 \end{array}\right\} \tag{2.413}$$

We will discuss first how to find the density of force. The force applied to an element of length l in the string, which is directed in the negative direction, is:

$$F = -\rho l g; \quad f(x) = \frac{F}{l} = -\rho g \tag{2.414}$$

Then, the problem to be solved is:

$$\left\{ \begin{array}{l} \dfrac{\partial^2 u}{\partial t^2} - c^2 \dfrac{\partial^2 u}{\partial x^2} = -g \qquad (t > 0) \\ \left.\dfrac{\partial u}{\partial x}\right|_{x=0} - hu(0, t) = 0 \\ \left.\dfrac{\partial u}{\partial x}\right|_{x=L} = 0 \\ u(x, t = 0) = 0 \\ \left.\dfrac{\partial u}{\partial t}\right|_{t=0} = 0 \end{array} \right\} \tag{2.415}$$

Being $c^2 = \frac{T}{\rho}$ and $h = \frac{\beta}{T}$

The solution of problem (2.415) can be decomposed into two functions: one which corresponds to the new equilibrium of the string (at rest and without oscillations) and a transient part towards that new equilibrium.

$$u(x, t) = v(x, t) + w(x) \tag{2.416}$$

The function $w(x)$ is the solution of the equation that includes the gravitational field but not the temporal evolution:

$$\left\{ \begin{array}{l} c^2 \dfrac{d^2 w}{dx^2} = g \qquad (t > 0) \\ \left.\dfrac{dw}{dx}\right|_{x=0} - hw(0) = 0 \\ \left.\dfrac{dw}{dx}\right|_{x=L} = 0 \end{array} \right\} \tag{2.417}$$

On the other hand, the function $v(x, t)$ is the solution of the transient homogeneous equation:

$$\left\{\begin{array}{c} \dfrac{d^2 v}{dt^2} - c^2 \dfrac{d^2 v}{dx^2} = 0 \ (t > 0) \\ \left.\dfrac{\partial v}{\partial x}\right|_{x=0} - hv(0, t) = 0 \\ \left.\dfrac{\partial v}{\partial x}\right|_{x=L} = 0 \\ v(x, t = 0) = -w(x) \\ \left.\dfrac{\partial v}{\partial t}\right|_{t=0} = 0 \end{array}\right\} \quad (2.418)$$

We look for the solution for $w(x)$:

$$\frac{d^2 w}{dx^2} = \frac{g}{c^2} \to w(x) = \frac{g}{2c^2} x^2 + Ax + B \quad (2.419)$$

$$\left.\frac{dw}{dx}\right|_{x=L} = 0 \to \frac{g}{c^2} L + A = 0 \quad (2.420)$$

$$\left.\frac{dw}{dx}\right|_{x=0} - hw(0) = 0 \to A - hB = 0 \quad (2.421)$$

The coefficients are found like this:

$$A = -\frac{g}{c^2} L; \qquad B = -\frac{g}{hc^2} L \quad (2.422)$$

The stationary solution will be:

$$w(x) = \frac{g}{2c^2} x^2 - \frac{g}{c^2} Lx - \frac{g}{hc^2} L = \frac{g}{2c^2} \left(x^2 - 2Lx - \frac{2L}{h} \right) \quad (2.423)$$

Sturm–Liouville problem To solve the transient part, we will seek a solution by expanding in eigenfunctions of the spatial Sturm–Liouville problem with second and third type boundaries.

$$v(x, t) = \sum_n Q_n(t)\, X_n(x) \quad (2.424)$$

The Sturm–Liouville problem, adapting the boundary conditions of $u(x, t)$ to the spatial part $X(x)$, is:

$$\left\{\begin{array}{l} \dfrac{d^2 X}{dx^2} + \lambda X = 0 \\ \left.\dfrac{dX}{dx}\right|_{x=0} - hX(0) = 0 \\ \left.\dfrac{dX}{dx}\right|_{x=L} = 0 \end{array}\right\} \quad (2.425)$$

Applying the second boundary condition:

$$X_n(x) = \cos[\sqrt{\lambda}(x - L)] \tag{2.426}$$

And applying the first:

$$-\sqrt{\lambda}\sin(-\sqrt{\lambda}L) - h\cos(\sqrt{\lambda}L) = 0 \tag{2.427}$$

The eigenvalues λ_n satisfy then:

$$\tan(\sqrt{\lambda_n}L) = \frac{h}{\sqrt{\lambda_n}} \tag{2.428}$$

Being $(n = 1, 2, 3\ldots)$. These eigenvalues are marked with the subindex n, but they are unrelated to n in such a simple fashion as in the case of a string with fixed ends. Their values are not as "predictable" and they must be determined numerically or graphically from the transcendental equation:

General solution General solution and initial conditions:

$$v(x, t) = \sum_n [A_n \cos(\omega_n t) + B_n \sin(\omega_n t)] \cos\left[\sqrt{\lambda_n}(x - L)\right] \tag{2.429}$$

Being $\omega_n = c\sqrt{\lambda_n}$

Final solution From the second initial condition $\frac{\partial v}{\partial t}\big|_{t=0} = 0$ we have $B_n = 0$.

From the first initial condition:

$$\begin{aligned} v(x, t = 0) = -w(x) &= -\frac{g}{2c^2}\left(x^2 - 2Lx - \frac{2L}{h}\right) \\ &= \sum_n A_n \cos[\sqrt{\lambda_n}(x - L)] \end{aligned} \tag{2.430}$$

Using the orthogonality of the eigenfunctions:

$$\begin{aligned} &-\frac{g}{2c^2}\int_0^L \left(x^2 - 2Lx - \frac{2L}{h}\right)\cos[\sqrt{\lambda_n}(x - L)]dx \\ &= A_n \int_0^L \left(\cos[\sqrt{\lambda_n}(x - L)]\right)^2 dx \end{aligned} \tag{2.431}$$

The coefficients of the expansion are then:

$$A_n = -\frac{g}{2c^2} \frac{\int\limits_0^L \left(x^2 - 2Lx - \frac{2L}{h}\right) \cos[\sqrt{\lambda_n}(x - L)] dx}{\int\limits_0^L \left(\cos[\sqrt{\lambda_n}(x - L)]\right)^2 dx} \tag{2.432}$$

The final solution is:

$$u(x, t) = \frac{g}{2c^2} \left(x^2 - 2Lx - \frac{2L}{h}\right) + \sum_n A_n \cos(\omega_n t) \cos[\sqrt{\lambda_n}(x - L)] \tag{2.433}$$

Alternative solution

Sturm–Liouville problem We seek the solution of the problem (2.415) as a sum:

$$u(x, t) = \sum_n Q_n(t)\, X_n(x) \tag{2.434}$$

where $X_n(x)$ is a set of orthogonal eigenfunctions of the previous Sturm–Liouville problem.

$$\left\{ \begin{array}{l} \dfrac{d^2X}{dx^2} + \lambda X = 0 \\ \left.\dfrac{dX}{dx}\right|_{x=0} - hX(0) = 0 \\ \left.\dfrac{dX}{dx}\right|_{x=L} = 0 \end{array} \right\} \tag{2.435}$$

General solution Substituting the general solution (2.434) in the wave equation gives:

$$\left\{ \begin{array}{l} \sum\limits_n \dfrac{d^2Q_n}{dt^2} X_n(x) - c^2 \sum\limits_n Q_n(t) \dfrac{d^2X_n}{dx^2} = -g \rightarrow \\ \rightarrow \sum\limits_n \left[\dfrac{d^2Q_n}{dt^2} + c^2\lambda_n Q_n(t)\right] X_n(x) = -g \\ u(x, 0) = 0 \\ \left.\dfrac{\partial u}{\partial t}\right|_{t=0} = 0 \end{array} \right\} \tag{2.436}$$

Final solution Using the orthogonality of the eigenfunctions $X_n(x)$ we multiply both sides of:

$$\sum_n \left[\frac{d^2 Q_n}{dt^2} + c^2 \lambda_n Q_n\right] X_n(x) = -g \tag{2.437}$$

$$\left[\frac{d^2 Q_n}{dt^2} + c^2 \lambda_n Q_n\right] \int_0^L \left(\cos[\sqrt{\lambda_n}(x - L)]\right)^2 dx$$

$$= -g \int_0^L \cos[\sqrt{\lambda_n}(x - L)]dx \tag{2.438}$$

We are left with the equation:

$$\frac{d^2 Q_n}{dt^2} + c^2 \lambda_n Q_n = E_n \tag{2.439}$$

$$E_n = -g \frac{\int_0^L \cos[\sqrt{\lambda_n}(x - L)]dx}{\int_0^L \left(\cos[\sqrt{\lambda_n}(x - L)]\right)^2 dx}$$

$$= \frac{g}{\sqrt{\lambda_n}} \frac{\sin[\sqrt{\lambda_n} L]}{\left[\frac{L}{2} + \frac{1}{\sqrt{\lambda_n}} \sin(\sqrt{\lambda_n} L) \cos(\sqrt{\lambda_n} L)\right]} \tag{2.440}$$

$$\int_0^L \cos\left[\sqrt{\lambda_n}(x - L)\right] dx = -\frac{1}{\sqrt{\lambda_n}} \sin\left[\sqrt{\lambda_n} L\right] \tag{2.441}$$

$$\int_0^L \left(\cos\left[\sqrt{\lambda_n}(x - L)\right]\right)^2 dx = \left[\frac{L}{2} + \frac{1}{\sqrt{\lambda_n}} \sin(\sqrt{\lambda_n} L) \cos(\sqrt{\lambda_n} L)\right] \tag{2.442}$$

This is the equation of an oscillator at rest and without gravity, to which a constant gravitational force is applied at $t = 0$. We will seek the solution as the sum of the homogeneous and particular solution:

$$Q_{n,hom}(t) = A_n \cos(\omega_n t) + B_n \sin(\omega_n t) \tag{2.443}$$

$$Q_{n,part}(t) = \frac{E_n}{c^2 \lambda_n} \tag{2.444}$$

With the general solution, we can apply the initial conditions to find the coefficients:

$$u(x,t) = \sum_n \left[A_n \cos(\omega_n t) + B_n \sin(\omega_n t) + \frac{E_n}{c^2\lambda_n}\right] X_n(x) \quad (2.445)$$

From the first initial condition:

$$u(x,0) = \sum_n \left[A_n + \frac{E_n}{c^2\lambda_n}\right] X_n(x) = 0 \quad (2.446)$$

we get:

$$A_n = -\frac{E_n}{c^2\lambda_n} \quad (2.447)$$

From the second initial condition:

$$\left.\frac{\partial u}{\partial t}\right|_{t=0} = \sum_n [\omega_n B_n]\, X_n(x) = 0 \quad (2.448)$$

We get:

$$B_n = 0 \quad (2.449)$$

Finally:

$$u(x,t) = \sum_n \frac{E_n}{c^2\lambda_n} [1 - \cos(\omega_n t)] \cos\left[\sqrt{\lambda_n}(x-L)\right] \quad (2.450)$$

2.29 String with Oscillations Forced in One of Its Ends

Find the forced oscillations in a mass of length L and speed of sound c if it's fixed in an end ($x = L$) and the other one moves periodically according to $u(0,t) = \sin(\omega t)$. Consider the case of a string at rest up until $t = 0$.

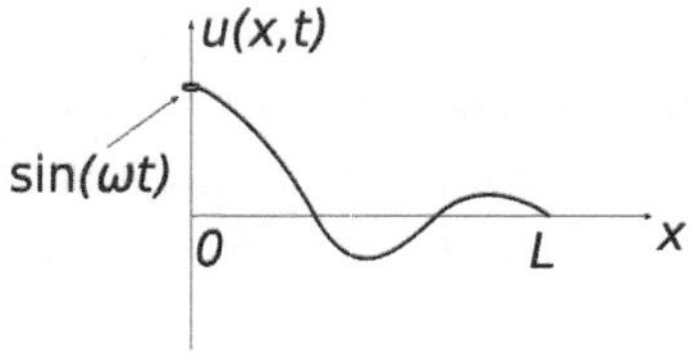

Figure 2.41

Mathematical formulation

$$\left\{\begin{array}{ll} \dfrac{d^2u}{dt^2} - c^2\dfrac{d^2u}{dx^2} = 0 & \\ u(0,t) = \sin(\omega t) & (t > 0) \\ u(L,t) = 0 & (t > 0) \\ u(x,0) = \left.\dfrac{\partial u}{\partial t}\right|_{t=0} = 0 & \end{array}\right\} \quad (2.451)$$

We look for the solution as the sum of a stationary and a transient one.

$$u(x,t) = w(x,t) + v(x,t) \quad (2.452)$$

In this case, the stationary part is not independent of time, but it corresponds to a regime of periodic oscillations which follow the external excitation with the same frequency. The transient part describes how the string goes from the initial to the stationary state at long times.

Formulation of the stationary problem:

$$\left\{\begin{array}{ll} \dfrac{d^2w}{dt^2} - c^2\dfrac{d^2w}{dx^2} = 0\ (*) & \\ u(0,t) = \sin(\omega t) & (-\infty < t < +\infty) \\ u(L,t) = 0 & (-\infty < t < +\infty) \end{array}\right\} \quad (2.453)$$

The stationary solution must be proportional to $\sin(\omega t)$, as previously indicated, since the string at long times will only oscillate with the external frequency. Separating variables:

$$w(x,t) = A\sin(\omega t)X(x) \quad (2.454)$$

Replacing this form of the solution into equation (*) we have:

$$-\omega^2 X - c^2\frac{d^2X}{dx^2} = 0 \quad (2.455)$$

The general solution for $X(x)$, which already assumes the boundary condition for $x = L$, is:

$$X(x) = B\sin\left[\frac{c}{\omega}(L-x)\right] \quad (2.456)$$

In problems of this kind, in which only the end which is <u>not</u> at $x = 0$ is fixed, it is useful to change the origin of coordinates of the argument of the sine function so that it is zero at $x = L$. What is left is to apply the boundary condition at $x = 0$:

$$\sin(\omega t)X(0) = B\sin(\omega t)\sin\left[\frac{c}{\omega}(L)\right] \tag{2.457}$$

from where we have

$$B = \frac{1}{\sin\left[\frac{cL}{\omega}\right]} \tag{2.458}$$

$$w(x,t) = \frac{\sin(\omega t)\sin\left[\frac{c}{\omega}(L-x)\right]}{\sin\left[\frac{cL}{\omega}\right]} \tag{2.459}$$

Mathematical formulation of the transient problem, taking into account, because of the time dependence of the stationary part, that at the initial instant we have $w(x,0)$ in all the string

$$\left\{\begin{array}{ll} \dfrac{d^2v}{dt^2} - c^2\dfrac{d^2v}{dx^2} = 0 & \\ v(0,t) = 0 & (t > 0) \\ u(L,t) = 0 & (t > 0) \\ v(x,t=0) = 0 & \\ \left.\dfrac{\partial v}{\partial t}\right|_{t=0} = -\left.\dfrac{\partial w}{\partial t}\right|_{t=0} & \end{array}\right\} \tag{2.460}$$

From the previous result we have:

$$\left.\frac{\partial w}{\partial t}\right|_{t=0} = \frac{\omega\sin\left[\frac{c}{\omega}(L-x)\right]}{\sin\left[\frac{c}{\omega}L\right]} \tag{2.461}$$

The general solution corresponds to the problem of a string which is free and fixed at the left and right borders respectively.

$$v(x,t) = \sum_{n=1}^{\infty}[C_n\sin(\omega_n t) + D_n\cos(\omega_n t)]\sin\left[\frac{\pi n(x-L)}{L}\right] \tag{2.462}$$

with $\omega_n = c\frac{\pi n}{L}$. It is evident that the first initial condition $v(x,t=0) = 0$ forces us to impose $D_n = 0$. To find the C_n we use the second initial condition:

$$v_t(x) = \omega_n\sum_{n=1}^{\infty}C_n\sin\left[\frac{\pi n(x-L)}{L}\right] = -w_t(x,0) \tag{2.463}$$

Final solution All that is left is using the orthogonality of the eigenfunctions $\sin\left[\frac{\pi n(x-L)}{L}\right]$ to find the coefficients of the expansion of the transient solution.

$$C_n = \frac{-1}{\omega_n}\frac{\int_0^L \frac{\partial w}{\partial t}\Big|_{t=0}\sin[\frac{\pi n(x-L)}{L}]dx}{\int_0^L \left[\sin\left[\frac{\pi n(x-L)}{L}\right]\right]^2 dx} \tag{2.464}$$

2.30 Oscillations of a String with a Force That Increases Linearly in Time

Find the forced oscillations of a string of length π that is initially at rest if, from $t = 0$ a force starts acting on it, with a distributed density of force $t \times \sin(x)$. The speed of sound is a. Suppose that the linear density of mass of the string is $\rho = 1$.

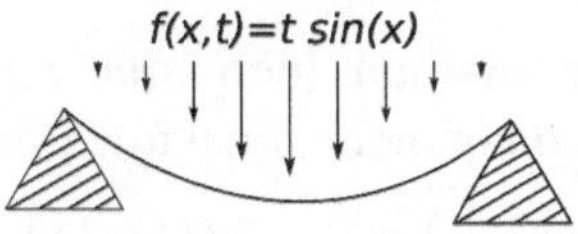

Figure 2.42

Mathematical formulation

$$\begin{Bmatrix} \dfrac{\partial^2 u}{\partial t^2} - a^2\dfrac{\partial^2 u}{\partial x^2} = t\sin(x) \\ u(x, 0) = u_t(x, 0) = 0 \\ u(0, t) = u(\pi, t) = 0 \end{Bmatrix} \tag{2.465}$$

Since the profile of the force corresponds to one of the eigenmodes of the string, we will look for the solution of this non-homogeneous equation as the product of the sought-after amplitude and the profile of this only mode excited by the forced vibrations.

$$u(x, t) = v(t)\sin(x) \tag{2.466}$$

Replacing this solution into the wave equation:

$$\frac{d^2 v}{dt^2}\sin(x) + a^2 v(t)\sin(x) = t\sin(x) \rightarrow \tag{2.467}$$

$$\rightarrow \frac{d^2v}{dt^2} + a^2v(t) = t \tag{2.468}$$

with initial conditions: $v(0) = \frac{dv}{dt}\Big|_{t=0} = 0$

Seeking the solution as the sum of solutions of the homogeneous equation and a particular solution, we get to the general solution:

$$v(t) = C_1 \cos(at) + C_2 \sin(at) + \frac{t}{a^2} \tag{2.469}$$

From the initial conditions we have:

$$C_1 = 0, C_2 = -\frac{1}{a^3} \tag{2.470}$$

$$v(t) = \frac{1}{a^2}\left[t - \frac{1}{a}\sin(at)\right] \tag{2.471}$$

The final solution is:

$$u(x, t)) = \frac{\sin x}{a^2}\left[t - \frac{1}{a}\sin(at)\right] \tag{2.472}$$

Alternative resolution method (non-intuitive) We can seek the solution as an expansion of orthogonal functions:

$$u(x, t) = \sum_n v_n(t)X_n(x) \tag{2.473}$$

being $X_n(x)$ the solutions of the Sturm–Liouville problem:

$$\left\{\begin{array}{l} \dfrac{d^2X}{dx^2} + \lambda X = 0 \\ X(0) = X(\pi) = 0 \end{array}\right\} \tag{2.474}$$

$$X_n(x) = \sin(nx); \qquad \lambda_n = (n)^2 \tag{2.475}$$

Replacing the summation in the wave equation:

$$\sum_n \left[\frac{d^2v_n}{dt^2} + a^2n^2v_n\right] X_n(x) = t\sin(x) \tag{2.476}$$

We use the orthogonality properties of the eigenfunctions, multiplying by $X_m(x)$ and integrating in the range $0 \leq x \leq \pi$. In the summation on the left side only the term with $n = m$ is different from zero:

$$\frac{d^2v_n}{dt^2} + a^2n^2v_n = \frac{2t}{\pi}\int_0^{\pi} \sin(x)\sin(nx)dx \tag{2.477}$$

$$\int_0^{\pi} \sin(x)\sin(nx)dx = \begin{Bmatrix} \frac{\pi}{2} & (n = 1) \\ 0 & (n \neq 1) \end{Bmatrix} \tag{2.478}$$

In this way we get to the same non-homogeneous equation that we obtained with the intuitive method:

$$\frac{\partial^2 v_1(t)}{\partial t^2} + a^2 v_1(t) = t \tag{2.479}$$

2.31 Case Study: Lateral Photoeffect

Consider a thin rod (length L). Around x_0, where a laser light is directed, mobile charge carriers are generated, and they diffuse along the rod with a diffusion coefficient D.

Find the variation of the concentration of carriers as a function of time and position if, starting at $t = 0$ the laser starts generating P particles by unit time and length. At the initial moment the carrier concentration generated by the laser was equal to zero.

Consider that the particles generated by the laser, when they diffuse along the rod (consider only the coordinate x) are annihilated by their corresponding antiparticles (electrons "annihilate" with holes) at a rate proportional to the local concentration (that is, the annihilation is proportional to a constant H). Furthermore, consider that there is a leak of particles at the ends of the bar, with fluxes proportional to the concentration (with constant h), due to the rod ends being connected through a resistor.

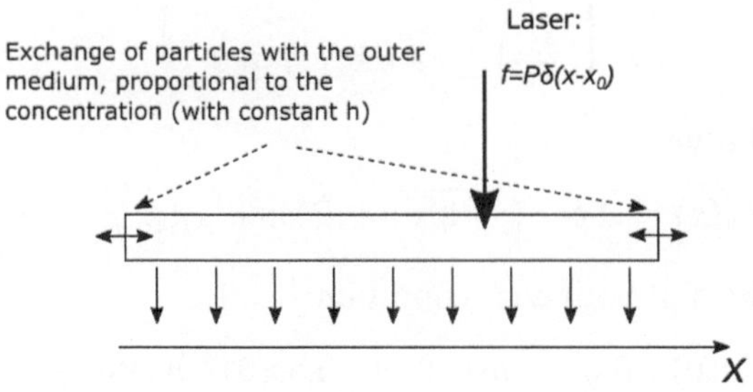

Figure 2.43

Mathematical formulation

$$\left\{\begin{array}{l}\dfrac{\partial u(x,t)}{\partial t}-D\dfrac{\partial^2 u(x,t)}{\partial x^2}=P\delta(x-x_0)-Hu\\ u(x,0)=0\\ \left.\dfrac{du}{dx}\right|_{x=0}-hu(0,t)=0\\ \left.\dfrac{du}{dx}\right|_{x=L}+hu(L,t)=0\end{array}\right\} \quad (2.480)$$

The function $u(x, t)$ represents the particle concentration.

D is the diffusion coefficient.

P is a coefficient proportional to the laser power by unit length.

H is a coefficient inversely proportional to the recombination time.

h is a coefficient proportional to the leak of particles through the rod ends.

We seek the solution as a summation of eigenfunctions of the Sturm–Liouville problem for homogeneous boundaries:

$$u(x,t)=\sum_n Q_n(t)X_n(x) \quad (2.481)$$

Sturm–Liouville problem The Sturm–Liouville problem is:

$$\left\{\begin{array}{l}\dfrac{d^2Xn}{dx^2}+\lambda_n X=0\\ \left.\dfrac{dX_n}{dx}\right|_{x=0}-hX_n(0)=0\\ \left.\frac{dX_n}{dx}\right|_{x=L}+hX_n(L)=0\end{array}\right\} \quad (2.482)$$

General solution

$$X_n(x)=A\cos\left(\sqrt{\lambda_n}x\right)+B\sin\left(\sqrt{\lambda_n}x\right) \quad (2.483)$$

Applying the first boundary condition:

$$-A\sqrt{\lambda_n}\sin(0)+B\sqrt{\lambda_n}\cos(0)-h[A\cos(0)+B\sin(0)]=0 \quad (2.484)$$

$$B\sqrt{\lambda_n}-hA=0\rightarrow A=B\frac{\sqrt{\lambda_n}}{h} \quad (2.485)$$

Then:

$$X_n(x) = \frac{\sqrt{\lambda_n}}{h}\cos(\sqrt{\lambda_n}x) + \sin(\sqrt{\lambda_n}x) \tag{2.486}$$

We can see that if the second boundary condition is applied, we won't be able to determine the value of B, and A and B will always depend on each other and will never be fully determined. In principle some other condition would be necessary, but we can set $B = 1$ to simplify matters. What is really important is the form of the eigenfunctions, not so much their absolute value, since we still need to apply the initial condition, where another coefficient will solve the uncertainty that we now face. This is a consequence of the boundary conditions of this problem. Applying the second boundary condition (and considering the ratio between the A and B coefficients):

$$-\frac{\sqrt{\lambda_n}}{h}\sqrt{\lambda_n}\sin\left(\sqrt{\lambda_n}L\right) + \sqrt{\lambda_n}\cos\left(\sqrt{\lambda_n}L\right)$$
$$+h\left[\frac{\sqrt{\lambda_n}}{h}\cos\left(\sqrt{\lambda_n}L\right) + \sin\left(\sqrt{\lambda_n}L\right)\right] = 0 \tag{2.487}$$

Operating, we arrive at the equation that determines the λ_n eigenvalues:

$$\tan\left(\sqrt{\lambda_n}L\right) = \frac{2h\sqrt{\lambda_n}}{\lambda_n - h^2} \tag{2.488}$$

Final solution To analyze the temporal part, we replace the expansion of the solution in orthogonal eigenfunctions into equation (2.480):

$$\sum_n \frac{dQ_n(t)}{dt}X_n(x) - D\sum_n Q_n(t)\frac{d^2X_n(x)}{dx^2} = P\delta(x - x_0) - Hu(x,t) \tag{2.489}$$

$$\sum_n \left[\frac{dQ_n(t)}{dt} + [D\lambda_n + H]Q_n(t)\right] X_n(x) = P\delta(x - x_0) \tag{2.490}$$

Multiplying both sides by $X_n(x)$ and integrating between 0 and L, using the orthogonality of the $X_n(x)$ eigenfunctions, we arrive at an equation for the amplitudes of the $Q_n(t)$ modes:

$$\left[\frac{dQ_n(t)}{dt} + [D\lambda_n + H]Q_n(t)\right]\int_0^L [X_n(x)]^2 dx = P\int_0^L \delta(x-x_0)X_n(x)dx \tag{2.491}$$

$$\frac{dQ_n(t)}{dt}+[D\lambda_n+H]Q_n(t)=\frac{P\left[\frac{\sqrt{\lambda_n}}{h}\cos(\sqrt{\lambda_n}x_0)+\sin(\sqrt{\lambda_n}x_0)\right]}{\int\limits_0^L\left[\frac{\sqrt{\lambda_n}}{h}\cos(\sqrt{\lambda_n}x)+\sin\left(\sqrt{\lambda_n}x\right)\right]^2dx}=f_n \tag{2.492}$$

Note: In the case that the laser had a more realistic cross section, for instance gaussian, $P\delta(x-x_0)$ would need to be changed by

$$Pe^{-(x-x_0)^2/c^2}$$

where c will be related to the width of the laser beam.

In the case of a gaussian cross section:

$$f_n=\frac{P\int\limits_0^L e^{-\frac{(x-x_0)^2}{c^2}}\left[\frac{\sqrt{\lambda_n}}{h}\cos(\sqrt{\lambda_n}x)+\sin(\sqrt{\lambda_n}x)\right]dx}{\int\limits_0^L\left[\frac{\sqrt{\lambda_n}}{h}\cos(\sqrt{\lambda_n}x)+\sin(\sqrt{\lambda_n}x)\right]^2dx} \tag{2.493}$$

General solution The general solution of the equation is:

$$\frac{dQ_n(t)}{dt}+[D\lambda_n+H]Q_n(t)=f_n \tag{2.494}$$

Final solution The amplitudes $Q_n(t)$ are the solution of the homogeneous equation:

$$Q_{n,hom}(t)=Q_n(0)e^{-(D\lambda_n+H)t} \tag{2.495}$$

plus the particular solution of the non-homogeneous equation:

$$Q_{n,\,part}=\frac{f_n}{[D\lambda_n+H]} \tag{2.496}$$

Then the solution for $Q_n(t)$ is:

$$Q_n(t)=Q_n(0)e^{-(D\lambda_n+H)t}+\frac{f_n}{D\lambda_n+H} \tag{2.497}$$

We just need to impose the initial condition to find the $Q_n(0)$ coefficients.

$$u(x,0)=\sum_n Q_n(0)X_n(x)=\sum_n\left[Q_n(0)+\frac{f_n}{D\lambda_n+H}\right]X_n(x)=0 \tag{2.498}$$

From here we get the values $Q_n(0)=-\frac{f_n}{D\lambda_n+H}$

Finally the solution of the problem of the variation of the lateral photoeffect as a function of time is:

$$u(x,t) = \sum_n \frac{f_n}{D\lambda_n + H}\left[1 - e^{-(D\lambda_n+H)t}\right] \times \left[\frac{\sqrt{\lambda_n}}{h}\cos\left(\sqrt{\lambda_n}x\right) + \sin\left(\sqrt{\lambda_n}x\right)\right] \quad (2.499)$$

The distribution of the concentration of carriers at t_0, after applying the laser pulse is:

$$u(x,t_0) = \sum_n \frac{f_n}{D\lambda_n + H}\left[1 - e^{-(D\lambda_n+H)t_0}\right] \times \left[\frac{\sqrt{\lambda_n}}{h}\cos\left(\sqrt{\lambda_n}x\right) + \sin\left(\sqrt{\lambda_n}x\right)\right] \quad (2.500)$$

The distribution $u(x)$ of the carrier concentration in the limit $t = \infty$ is:

$$u(x,\infty) = \sum_n \frac{f_n}{D\lambda_n + H}\left[\frac{\sqrt{\lambda_n}}{h}\cos(\sqrt{\lambda_n}x) + \sin(\sqrt{\lambda_n}x)\right] \quad (2.501)$$

Note: there is an alternative method to find the stationary distribution $u(x,\infty)$ (that is, when $\frac{\partial u(x,t)}{\partial t} = 0$) by solving the problem analytically:

$$\left\{\begin{array}{l} -D\dfrac{d^2u(x,t)}{dx^2} = A\delta(x-x_0) - Hu \\ \left.\dfrac{du}{dx}\right|_{x=0} - hu(L,t) = 0 \\ \left.\dfrac{du}{dx}\right|_{x=L} + hu(L,t) = 0 \end{array}\right\} \quad (2.502)$$

The solution is separated in two intervals $(x < x_0)$ and $(x > x_0)$, where the equation becomes homogeneous and, in this way, we get a solution by parts (two exponential functions) that must be continuous at $(x = x_0)$, as well as using the boundary conditions at the ends to find the coefficients of the linearly independent solutions.

Now we will see how to find the temporal variation of the distribution of carriers after turning off the illumination.

Mathematical formulation

$$\left\{\begin{array}{l} \dfrac{\partial w(x,t)}{\partial t} - D\dfrac{d^2w(x,t)}{dx^2} = -Hw \\ (t = t' - t_0),\ t \text{ is now the time after turning off the laser.} \\ w(x,0) = u(x,t_0) \\ \frac{dw}{dx}\Big|_{x=0} - hw(0,t) = 0 \\ \frac{dw}{dx}\Big|_{x=L} + hw(L,t) = 0 \end{array}\right\} \tag{2.503}$$

General solution Once again we seek the solution as an expansion in orthogonal eigenfunctions, solutions of the Sturm–Liouville that correspond to the homogeneous boundaries.

$$w(x,t) = \sum_n Q_n(t)X_n(x) \tag{2.504}$$

obviously the $X_n(x)$; λ_n are not the same as before.

Replacing $w(x,t)$ into (2.503)

$$\sum_n \left[\frac{dQ_n(t)}{dt} + [D\lambda_n + H]Q_n(t)\right] X_n(x) = 0 \tag{2.505}$$

$$\frac{dQ_n(t)}{dt} + [D\lambda_n + H]Q_n(t) = 0 \tag{2.506}$$

This equation has a solution:

$$Q_n(t) = A_n e^{-(D\lambda_n+H)t} \tag{2.507}$$

Final solution Then the solution is:

$$w(x,t) = \sum_n A_n e^{-(D\lambda_n+H)t}\left[\frac{\sqrt{\lambda_n}}{h}\cos\left(\sqrt{\lambda_n}x\right) + \sin\left(\sqrt{\lambda_n}x\right)\right] \tag{2.508}$$

Applying the initial conditions:

$$u(x,t_0) = \sum_n A_n\left[\frac{\sqrt{\lambda_n}}{h}\cos\left(\sqrt{\lambda_n}x\right) + \sin\left(\sqrt{\lambda_n}x\right)\right] \tag{2.509}$$

and using the orthogonality of the $X_n(x)$ we arrive at:

$$A_n = \frac{\int_0^L u(x,t_0)\left[\frac{\sqrt{\lambda_n}}{h}\cos(\sqrt{\lambda_n}x) + \sin(\sqrt{\lambda_n}x)\right]dx}{\int_0^L \left[\frac{\sqrt{\lambda_n}}{h}\cos(\sqrt{\lambda_n}x) + \sin(\sqrt{\lambda_n}x)\right]^2 dx} \tag{2.510}$$

which solves the problem.

2.32 Oscillations of a String under the Influence of a Gravitational Field

A string of length L with both ends fixed and initially at rest, from $t = 0$ onwards is subject to the action of the gravitational field. Find the oscillations of the string starting at $t = 0$. Consider that the speed of sound of the string is (c) and the density (ρ) are equal to one.

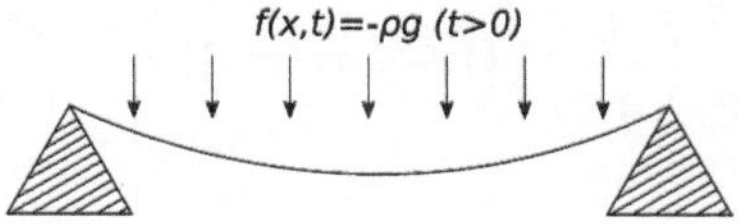

Figure 2.44

Mathematical formulation

$$\left\{\begin{array}{l} \dfrac{\partial^2 u}{\partial t^2} - c^2 \dfrac{\partial^2 u}{\partial x^2} = \dfrac{-\rho g}{\rho} = -g \\ u(0) = 0; \quad u(L) = 0 \\ u(x,0) = 0; \quad \left.\dfrac{\partial u}{\partial t}\right|_{t=0} = 0 \end{array}\right\} \tag{2.511}$$

Sturm–Liouville problem We seek the solution by expanding it in orthogonal eigenfunctions that are the solutions of the homogeneous Sturm–Liouville problem:

$$\left\{\begin{array}{l} \dfrac{\partial^2 X}{\partial x^2} + \lambda X = 0 \\ u(0) = 0 \\ u(L) = 0 \end{array}\right\} \tag{2.512}$$

With known eigenfunctions, that yield the solution:

$$u(x,t) = \sum B_n(t) \sin\left(\frac{\pi n}{L} x\right) \tag{2.513}$$

General solution We can replace this solution into the waves equation. We use the orthogonality of the $\sin(\frac{\pi n}{L} x)$ eigenfunctions to arrive at the non-homogeneous equation for the $B_n(t)$ coefficients.

$$\frac{d^2 B_n}{dt^2} + \left(\frac{\pi n}{L}\right)^2 B_n(t) = -g \frac{2}{n\pi}(1 - \cos \pi n) \tag{2.514}$$

Final solution with the initial conditions:

$$B_n(0) = 0; \; \left.\frac{dB_n}{dt}\right|_{x=0} = 0$$

The general solution is the sum of the solution to the homogeneous equation and the particular solution. The latter can be considered as $B_{n,\, part}$ =Constant:

$$B_{n,\, part}(t) = -\frac{2g}{n\pi}\left(\frac{L}{\pi n}\right)^2 [1 - \cos(\pi n)] = -\frac{2gL^2}{(n\pi)^3}[1 - (-1)^n] \tag{2.515}$$

$$B_n(t) = C \sin\left(\frac{\pi n}{L}t\right) + D \cos\left(\frac{\pi n}{L}t\right) - \frac{2gL^2}{(n\pi)^3}[1 - (-1)^n] \tag{2.516}$$

First initial condition $B_n(0) = 0 \rightarrow$

$$D = \frac{2gL^2}{(n\pi)^3}[1 - (-1)^n] \tag{2.517}$$

Second initial condition: $\left.\frac{dB_n}{dt}\right|_{t=0} = 0 \rightarrow C\frac{\pi n}{L} = 0C = 0$
Then:

$$B_n(t) = \frac{2gL^2}{(n\pi)^3}[1 - (-1)^n]\left[\cos\left(\frac{\pi n}{L}t\right) - 1\right] \tag{2.518}$$

The solution (with $[1 - (-1)^n] = 0$ for even n) is:

$$u(x,\, t) = \sum_{k=0}^{\infty} \frac{4gL^2}{[(2k+1)\pi]^3}\left[\cos\left(\frac{\pi n}{L}t\right) - 1\right] \sin\left(\frac{\pi n}{L}x\right) \tag{2.519}$$

2.33 Dynamic String with Free Ends and a Point Mass at $x = x_0$

A string of length L, tension T and linear density ρ has a point mass m at $x = x_0$. Both ends of the string are free and the string is initially at rest and without any external fields applied. Find the movement of the string from $t = 0$, when it becomes subject to the gravitational field.

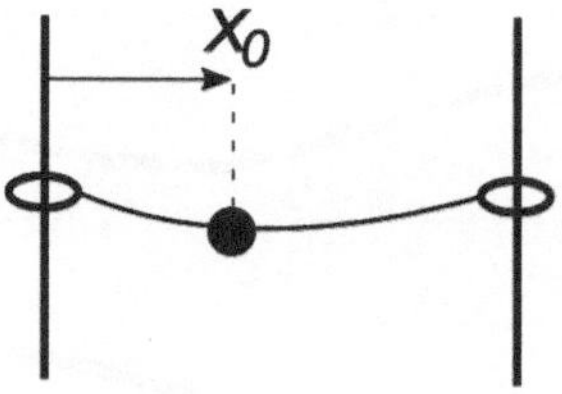

Figure 2.45

Mathematical formulation ($t < 0$)

$$\left\{\begin{array}{l} \rho(x)\dfrac{\partial^2 u}{\partial t^2} - T\dfrac{\partial^2 u}{\partial x^2} = 0 \\ \left.\dfrac{\partial u}{\partial x}\right|_{x=0} = 0 \\ \left.\dfrac{\partial u}{\partial x}\right|_{x=L} = 0 \end{array}\right\} \quad (2.520)$$

with

$$\rho(x) = \rho + m\delta(x - x_0) \rightarrow \int_0^L \rho(x)dx = \rho L + m$$

Sturm–Liouville problem Separating variables we arrive at the Sturm–Liouville problem:

$$\left\{\begin{array}{l} \dfrac{dX^2}{dx^2} + \lambda\rho(x)X = 0 \\ \left.\dfrac{dX}{dx}\right|_{x=0} = 0 \\ \left.\dfrac{dX}{dx}\right|_{x=L} = 0 \end{array}\right\} \quad (2.521)$$

To find the eigenfunctions while avoiding the anomalous point where the linear density of mass diverges, we must separate the eigenfunctions in two parts:

$$X_1(x) = A\cos(\sqrt{\lambda\rho}x) \quad (2.522)$$

(having applied the first boundary condition).

$$X_2(x) = B\cos(\sqrt{\lambda\rho}(x - L)) \quad (2.523)$$

(having applied the second boundary condition).

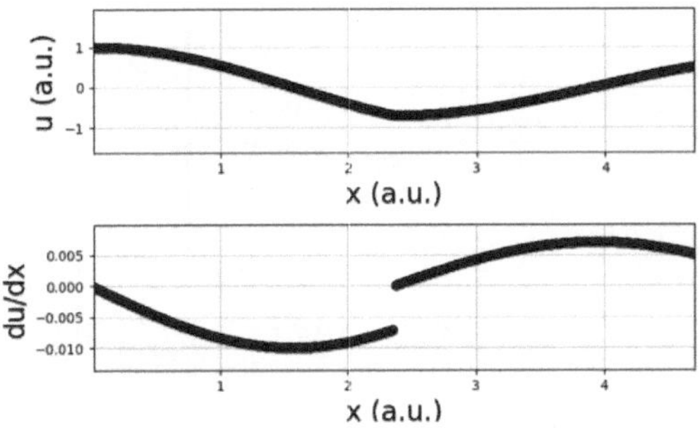

Figure 2.46 Numeric example of eigenfunction and its derivative

Now we apply the condition of continuity of the eigenfunctions at $x = x_0$:

$$A\cos\left(\sqrt{\lambda\rho}x_0\right) - B\cos\left(\sqrt{\lambda\rho}(x_0 - L)\right) = 0 \tag{2.524}$$

Orthogonal eigenfunctions:

$$X_n(x) = \left\{ \begin{array}{l} A\cos(\sqrt{\lambda_n\rho}x) \\ A\dfrac{\cos\left(\sqrt{\lambda_n\rho}x_0\right)}{\cos\left(\sqrt{\lambda_n\rho}(x_0 - L)\right)}\cos\left(\sqrt{\lambda_n\rho}(x - L)\right) \end{array} \right\} \tag{2.525}$$

where λ_n are the eigenfunctions to be determined. Now we apply the condition of discontinuity of the first derivatives of the eigenfunctions, integrating equation (2.521) around $x = x_0$:

$$\left.\frac{dX_1}{dx}\right|_{x=x_0} - \left.\frac{dX_2}{dx}\right|_{x=x_0} = \lambda m X_1(x_0) = A\lambda m\cos\left(\sqrt{\lambda\rho}x_0\right) \tag{2.526}$$

$$\begin{aligned} -A\left[\sqrt{\lambda\rho}\sin(\sqrt{\lambda\rho}x_0) - \lambda m\cos(\sqrt{\lambda\rho}x_0)\right] \\ +B\sqrt{\lambda\rho}\sin\left[\sqrt{\lambda\rho}(x_0 - L)\right] = 0 \end{aligned} \tag{2.527}$$

The eigenvalues λ_n can be obtained from the result of equating to zero the determinant of coefficients that results from expressing equations 2.524 and 2.527 in matrix form. Furthermore, we see that $\lambda_0 = 0$ is also an eigenvalue, since it satisfies both equations 2.524 and 2.527 and describes the $X_0(x) = A$ eigenfunction. We now solve the problem of the oscillations of the string for $t > 0$:

$$\begin{Bmatrix} \dfrac{du^2}{dt^2} - \dfrac{T}{\rho(x)}\dfrac{d^2u}{dx^2} = -g \\ \left.\dfrac{du}{dx}\right|_{x=0} = 0 \\ \left.\dfrac{du}{dx}\right|_{x=L} = 0 \\ u(x, t = 0) = \left.\dfrac{\partial u}{\partial t}\right|_{t=0} = 0 \end{Bmatrix} \tag{2.528}$$

General solution We seek a solution $u(x, t) = \sum_n Q_n(t)X_n(x)$ with $X_n(x)$ being the previously found eigenmodes (the spatial dependency does not change in the whole problem). Replacing into the wave equation we have:

$$\sum_n \left[\frac{d^2Q_n(t)}{dt^2}X_n(x) - Q_n(t)T\frac{1}{\rho(x)}\frac{d^2X_n(x)}{dx^2}\right] = -g \tag{2.529}$$

$$\sum_n \left[\frac{d^2Q_n(t)}{dt^2} + Q_n(t)T\lambda_n\right] X_n(x) = -g \tag{2.530}$$

Applying the orthogonality of the $X_n(x)$ and integrating from 0 to L with a weight $\rho(x)$:

$$\frac{d^2Q_n(t)}{dt^2} + T\lambda_nQ_n(t) = -g\frac{\int_0^L \rho(x)X_n(x)dx}{\int_0^L \rho(x)[X_n(x)]^2dx}$$

$$= -g\frac{\rho\int_0^L X_n(x)dx + mX_n(x_0)}{\int_0^L \rho(x)[X_n(x)]^2dx} = f_n \tag{2.531}$$

The general solution for $Q_n(t)$ is composed of the solution of the homogeneous equation and the particular solution:

$$Q_n(t) = C_n\cos(\omega_nt) + D_n\sin(\omega_nt) + \frac{f_n}{T\lambda_n} \tag{2.532}$$

Final solution Applying both initial conditions:

$$C_n = -\frac{f_n}{T\lambda_n}$$

$$D_n = 0$$

The final solution is:

$$u(x, t) = \sum_{n=1}^{\infty} \frac{f_n}{T\lambda_n}(1 - \cos\omega_n t)X_n(x) \qquad (2.533)$$

with $\omega_n^2 = T\lambda_n$
To find the inhomogeneous part of the solution that corresponds to $\lambda_0 = 0$ (free fall of the string) we need to solve the equation:

$$\frac{d^2 Q_0(t)}{dt^2} X_0(x) = -g \qquad (2.534)$$

which will give us the displacement due to the acceleration of the string with the point mass in the presence of the gravitational field (with $X_0 = 1$): $\frac{-gt^2}{2}$

Alternative method Separating the solution in two: one of them describes the static form of the string in under the gravitational field, $w(x)$. The other is a transient function $v(x.t)$. We replace $u(x.t) = v(x.t) + w(x)$ into the wave equation:

$$\frac{d^2 v}{dt^2} - \frac{T}{\rho(x)}\frac{d^2 v}{dx^2} - \frac{T}{\rho(x)}\frac{d^2 w}{dx^2} = -g \qquad (2.535)$$

Problem for $w(x)$:

$$\begin{pmatrix} \dfrac{d^2 w(x)}{dx^2} = \rho(x)\dfrac{g}{T} \\ \left.\dfrac{dw}{dx}\right|_{x=0} = 0 \\ \left.\dfrac{dw}{dx}\right|_{x=L} = 0 \end{pmatrix} \qquad (2.536)$$

We are left with a defined solution (for which a constant is still to be found), consisting in two inverted parabolic functions, united at $x = x_0$.

The problem for $v(x.t)$ is:

$$\begin{cases} \dfrac{\partial^2 v}{\partial t^2} - \dfrac{T}{\rho(x)} \dfrac{\partial^2 v}{\partial x^2} = 0 \\ \left.\dfrac{dv}{dx}\right|_{x=0} = 0 \\ \left.\dfrac{dv}{dx}\right|_{x=L} = 0 \\ v(x, t = 0) = -w(x) \\ \left.\dfrac{\partial v}{\partial t}\right|_{t=0} = 0 \end{cases} \quad (2.537)$$

We seek the solution by separating by the $X_n(x)$ eigenfunctions obtained earlier at different regions for the transient part:

$$v(x, t) = \sum_n Q_n(t) X_n(x) \quad (2.538)$$

2.34 Oscillations in a String Interrupted by a Spring

A homogeneous string of length L and tension T is connected to a spring with constant β on its mid-point ($L/2$). Its ends can move freely in the direction transversal to the string. From $t = -\infty$ a local, periodic force acts on the string at $x = x_0$. Find the stationary oscillations of the string at the right end ($x = L$).

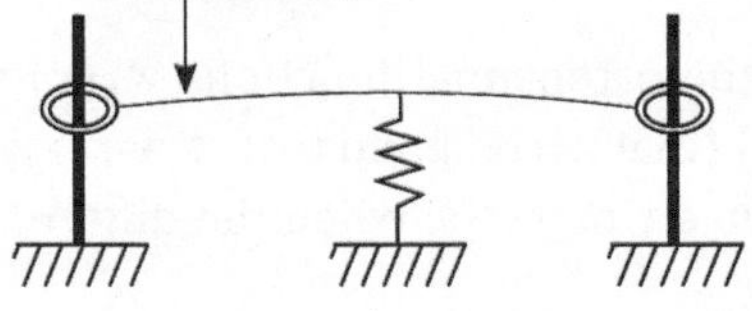

Figure 2.47

Mathematical formulation

$$\rho \frac{\partial^2 u}{\partial t^2} - T \frac{\partial^2 u}{\partial x^2} + \delta(x - L/2)\beta u = \delta(x - x_0)\sin(\omega t) \quad (-\infty < t < +\infty) \quad (2.539)$$

Both terms with Delta functions describe the linear density of local forces.
The boundary conditions are:

$$\left.\frac{\partial u}{\partial x}\right|_{x=0} = \left.\frac{\partial u}{\partial x}\right|_{x=L} = 0$$

Since we are looking for the stationary solution, we do not need any initial condition.

Sturm–Liouville problem We will seek the general solution as a function with a forced temporal variation and eigenfunctions of the system:

$$u(x,t) = \sum_n A_n(t) X_n(x) = \sum_n Q_n sen(\omega t) X_n(x) \qquad (2.540)$$

Replacing this into the wave equation we get:

$$\sum_n Q_n \left[\rho(-\omega^2) X_n(x) - T\frac{d^2 X_n(x)}{dx^2} + \delta(x - L/2)\beta X_n(x)\right] = \delta(x - x_0) \qquad (2.541)$$

To find $X_n(x)$ we first need to formulate the Sturm–Liouville problem separating the variables of the homogeneous equation. By doing this we arrive at an equation with both boundaries homogeneous of the second type. The condition of the discontinuity of the derivative can be deduced by integrating the Sturm–Liouville problem close to the point of junction of both strings.

$$T\frac{d^2 X_n(x)}{dx^2} - \delta(x - L/2)\beta X_n(x) = -\lambda_n X_n \qquad (2.542)$$

The solution for the orthogonal functions $X_n(x)$ will be sought in two parts $X_{n1,2}(x)$ (that correspond to two parts with respect to the point on which the string rests, when the differential equation can be simplified):

$$\left\{\begin{array}{l} T\dfrac{d^2 X_{n1,2}(x)}{dx^2} + \lambda_n X_{n1,2} = 0 \\ \left.\dfrac{dX_{n1}(x)}{dx}\right|_{x=0} = \left.\dfrac{dX_{n2}}{dx}\right|_{x=L} = 0 \\ \text{+ Condition of continuity of } X_{n1,2} \text{ and} \\ \text{discontinuity of its derivatives at } x = L/2 \end{array}\right\} \qquad (2.543)$$

It's more comfortable to seek the solution as a function by parts of the following form, since it applies the first two boundary conditions:

$$X_n(x) = \begin{cases} X_{n1} = A_n \cos(\sqrt{\lambda_n} x) & \left(0 < x < \dfrac{L}{2}\right) \\ X_{n2} = B_n \cos(\sqrt{\lambda_n}(L - x)) & \left(\dfrac{L}{2} < x < L\right) \end{cases} \tag{2.544}$$

We will have to join two parts of the solution by parts imposing the continuity conditions at $x = L/2$ and the last condition will be for the variation of the derivatives at $x = L/2$.
The application of the continuity conditions entails the following conditions at $x = L/2$:

$$\begin{cases} X_{n1}\left(\dfrac{L}{2}\right) = X_{n2}\left(\dfrac{L}{2}\right) \\ T\left.\dfrac{dX_{n1}}{dx}\right|_{x=L/2} - T\left.\dfrac{dX_{n2}}{dx}\right|_{x=L/2} + \beta X_{n1}\left(\dfrac{L}{2}\right) = 0 \end{cases} \tag{2.545}$$

The condition of variation of the derivatives is obtained by integrating equation (2.542) in the proximities of $x = L/2$

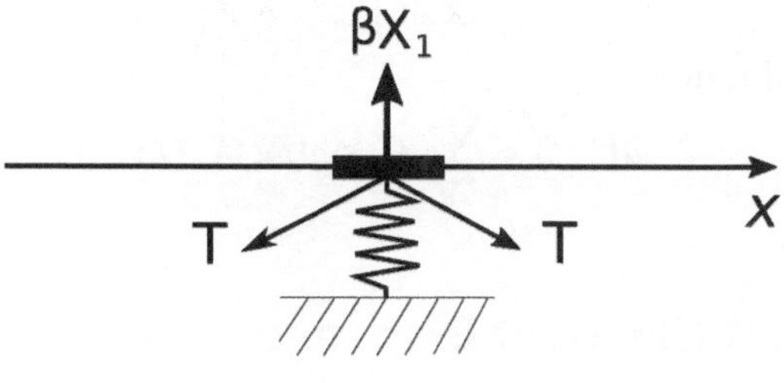

Figure 2.48

To correctly write this boundary condition for the change of the derivatives of each eigenmode we need to add the two projections of the tension on the vertical axis (that act on the spring on each side) and equate them to the tension of the spring due to the vertical displacement (see figure). Applying the form before specifying that the two functions X_1 and X_2 we get relations for the relative amplitude of X_1 and X_2 and the equation to find the eigenvalues of the problem:

$$\left\{\begin{array}{l} A_n - B_n = 0 \\ A_n T\left(-\sqrt{\lambda_n}\sin[(\sqrt{\lambda_n}(L/2)] - B_n T\left(\sqrt{\lambda_n}\sin[(\sqrt{\lambda_n}(L/2)]\right.\right. \\ +\beta\cos[(\sqrt{\lambda_n}(L/2)] = 0 \rightarrow \\ A_n T\left(\sqrt{\lambda_n}\sin[(\sqrt{\lambda_n}(L/2)] + B_n T\left(\sqrt{\lambda_n}\sin[(\sqrt{\lambda_n}(L/2)]\right.\right. \\ = \beta\cos[(\sqrt{\lambda_n}(L/2)] \end{array}\right\} \tag{2.546}$$

The determinant of the first and third equations in matrix form will give us the eigenvalues of the problem. Replacing the general solution into the wave equation (2.539) and eliminating the temporal term we will obtain:

$$\sum_n Q_n[\lambda_n - \rho\omega^2]X_n(x) = \delta(x - x_0) \tag{2.547}$$

Final solution Multiplying both sides by the orthogonal function X_m and integrating between the limits 0 and L we will get the solution for the Q_n coefficients:

$$Q_n = \frac{X_n(x_0)}{|X_n|^2\,[\lambda_n - \rho\omega^2]} \tag{2.548}$$

The final solution is:

$$u(L,.t) = \sum_n Q_n \sin(\omega t)X_n(L) \tag{2.549}$$

2.35 Point Like Heat Exchange

An insulated bar, whose temperature changes as $T_0\cos(\omega t)$ at the point x_0, exchanges heat with an external media at $T = 0$ according to Newton's law at $x = L/2$. Find the stationary distribution of temperature along the bar.

Mathematical formulation

$$C\rho\frac{\partial T(x,t)}{\partial t} - \kappa\frac{\partial^2 T(x,t)}{\partial x^2} + hT(x,t)\delta\left(x - \frac{L}{2}\right) = T_0\cos(\omega t)\delta(x - x_0) \tag{2.550}$$

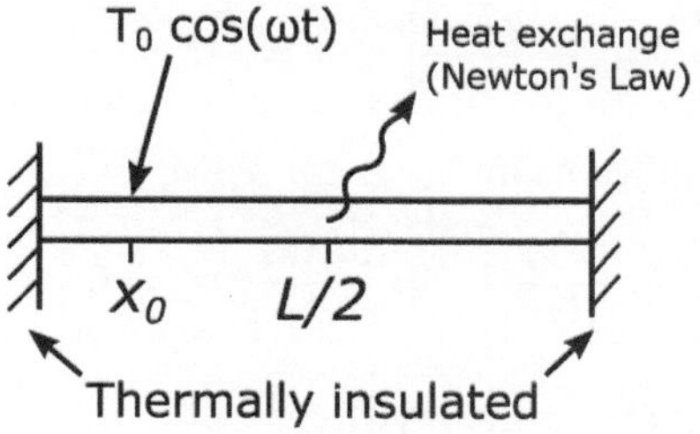

Figure 2.49

We use the method of separation of variables and take only the real part of the solution:

$$T(x,t) = \sum_n Q_n(t) X_n(x) = \sum_n X_n(x) e^{-i\omega t} \tag{2.551}$$

And we have the boundary conditions:

$$\left.\frac{\partial T}{\partial x}\right|_{x=0,\,L} = 0 \tag{2.552}$$

Replacing in the heat equation:

$$-i\omega C\rho X_n(x) - \kappa \frac{\partial^2 X_n(x)}{\partial x^2} + hX_n(x)\delta\left(x - \frac{L}{2}\right) = -T_0\delta(x - x_0) \tag{2.553}$$

$$\kappa \frac{\partial^2 X_n(x)}{\partial x^2} + \left(-h\delta\left(x - \frac{L}{2}\right) - i\omega C\rho\right) X_n(x) = -T_0\delta(x - x_0) \tag{2.554}$$

Sturm–Liouville problem We can formulate the following Sturm–Liouville problem:

$$\frac{d^2 X_n(x)}{dx^2} - h\delta\left(x - \frac{L}{2}\right) + \lambda_n X_n(x) = 0 \tag{2.555}$$

The solution will be defined by parts, to the left and to the right of the singularity at $x = L/2$:

$$X_n(x) = \left\{ \begin{array}{l} X_{n1}(x)\ 0 < x < L/2 \\ X_{n2}(x)\ L/2 < x < L \end{array} \right\} \tag{2.556}$$

The function is continuous at $x = L/2$ but its derivative is not:

$$X_{n1}(L/2) = X_{n2}(L/2) \tag{2.557}$$

$$\left.\frac{dX_{n2}}{dx}\right|_{x=L/2} - \left.\frac{dX_{n1}}{dx}\right|_{x=L/2} = \frac{h}{\kappa}X_n(L/2) \tag{2.558}$$

The boundary conditions are:

$$\left.\frac{dX_{n1}}{dx}\right|_{x=0} = \left.\frac{dX_{n2}}{dx}\right|_{x=L} = 0 \tag{2.559}$$

$$X_n(x) = \left\{ \begin{array}{ll} A_n \cos(\sqrt(\lambda_n)x) & 0 < x < L/2 \\ B_n \cos(\sqrt(\lambda_n)(x-L)) & L/2 < x < L \end{array} \right\} \tag{2.560}$$

$$A_n \cos(\sqrt{\lambda_n}L/2) = B_n \cos(\sqrt{\lambda_n}L/2) \rightarrow A_n = B_n \tag{2.561}$$

$$\sqrt{\lambda_n}B_n \sin(\sqrt{\lambda_n}L/2) + \sqrt{\lambda_n})B_n \sin(\sqrt{\lambda_n}L/2) = \frac{h}{\kappa}B_n \cos(\sqrt{\lambda_n}L/2)$$

$$\rightarrow \tan\left(\frac{\sqrt{\lambda_n}}{2}\right) = \frac{h}{\kappa\sqrt{\lambda_n}} \tag{2.562}$$

This equation gives the eigenvalues λ_n but has no analytical solution, the eigenvalues must be found numerically.
To find the A_n coefficients we replace in the heat equation:

$$\sum[-\kappa\lambda_n X_n(x) + i\omega c\rho X_n(x)] = -T_0\delta(x-x_0) \tag{2.563}$$

$$A_n = \frac{T_0}{\kappa\lambda_n - i\omega C\rho}\frac{2}{L}\cos(\sqrt{\lambda_n}x_0) = \frac{2T_0\cos(\sqrt{\lambda_n}x_0)(\kappa\lambda_n + i\omega C\rho)}{\kappa^2\lambda_n^2 + \omega^2C^2\rho^2} \tag{2.564}$$

The final solution will be (using $e^{-i\omega t} = \cos(\omega t) - i\sin(\omega t)$ and using only the real part of $A_n e^{-i\omega t}$):

$$T(x,t) =$$

$$\left\{ \begin{array}{ll} \sum_n \frac{2T_0\cos(\sqrt{\lambda_n}x_0)}{\kappa^2\lambda_n^2+\omega^2C^2\rho^2}\cos(\sqrt{\lambda_n}x)(\kappa\lambda_n\cos(\omega t) + \omega C\rho\sin(\omega t)) & 0 < x < \frac{L}{2} \\ \sum_n \frac{2T_0\cos(\sqrt{\lambda_n}x_0)}{\kappa^2\lambda_n^2+\omega^2C^2\rho^2}\cos(\sqrt{\lambda_n}(x-L))(\kappa\lambda_n\cos(\omega t) + \omega C\rho\sin(\omega t)) & \frac{L}{2} < x < L \end{array} \right. \tag{2.565}$$

Chapter 3

Bidimensional Problems

This chapter describes in a detailed manner the solutions of different problems in confined systems in two dimensions in Cartesian coordinates. The solutions will be sought by using the method of separation of variables, as functions of two (for example for static Laplace or Poisson problems) or three variables in the case of wave or diffusion problems (when time is considered). The separation of variables will be performed in order to expand the solution in orthogonal functions in one or two dimensions.

In the case of converting the initial problem in several simpler problems we need to take care that the Laplace problem with all boundaries of homogeneous type has trivial solution (zero or a constant). Just like in the previous chapter, when possible, the steps of the solution will be distributed in four stages: (i) General formulation of the problem, including the PDE, the initial conditions (if they exist) and the boundary conditions. (ii) Search for the solution from a partial solution, by solving a Sturm–Liouville problem. (iii) Steps to reach the general solution. (iv) Steps to find the final solution of the problem.

Mathematical Methods for Physics: Problems and Solutions
Farkhad G. Aliev and Antonio Lara

ISBN 978-981-4968-71-3 (Hardcover), 978-1-003-41088-1 (eBook)
www.jennystanford.com

In several cases the intermediate solutions will be analyzed graphically, whereas the final solutions will be presented using the PDE tool from MATLAB.

3.1 Forced Oscillations of a Membrane

A rectangular membrane whose sides have lengths a and b and with two fixed borders the other two (on opposite sides) are free, is subject to a point force, perpendicular to the plane, of value $A\sin(\omega t)$ at the point indicated in the figure ($0 < x < b$ and $0 < y < a$). Find the eigenfunctions of the Sturm–Liouville problem and the frequencies of the vibrations excited by the applied force.

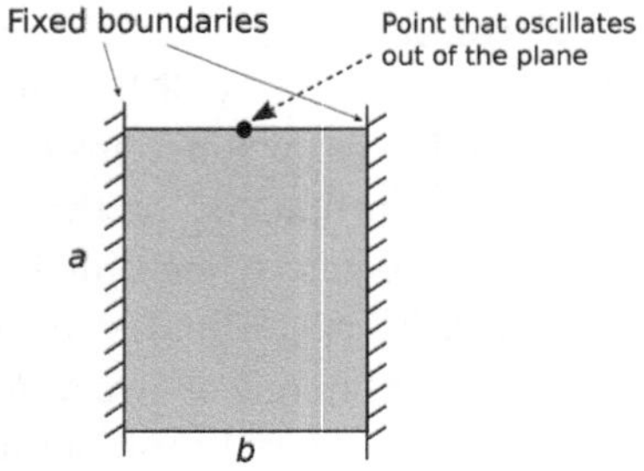

Figure 3.1

Mathematical formulation

Although the point force is applied on the border, it can be represented as applied on a point infinitely close to the border, at a distance ε:

$$f(x, y, t) = A\sin(\omega t)\,\delta\left(x - \frac{b}{2}\right)\delta(y - [a - \varepsilon]) \qquad (3.1)$$

And then consider the solution in the limit $\varepsilon \to 0$. The equation to be solved is:

$$\frac{\partial^2 u}{\partial t^2} - c^2\left(\frac{\partial^2 u}{\partial x^2} + \frac{\partial^2 u}{\partial y^2}\right) = \frac{A}{\rho}\sin(\omega t)\delta\left(x - \frac{b}{2}\right)\delta(y - [a - \varepsilon]) \quad (3.2)$$

being ρ the density of the material of the membrane and c the speed of sound.

Sturm–Liouville problem

We will look for the excited vibrations by expanding the solution:

$$u(x, y, t) = \sum T_{nm}(t)v_{nm}(x, y) \tag{3.3}$$

into orthogonal eigenfunction of the following Sturm–Liouville problem:

$$\left\{\begin{array}{l} \dfrac{\partial^2 v}{\partial x^2} + \dfrac{\partial^2 v}{\partial y^2} + \lambda v = 0 \\ v(0, y) = v(b, y) = 0 \\ \left.\dfrac{\partial v}{\partial y}\right|_{y=0} = \left.\dfrac{\partial v}{\partial y}\right|_{y=a} = 0 \end{array}\right\} \tag{3.4}$$

The eigenfunctions and eigenvalues of this problem are known. The normalized eigenfunctions are:

$$v_{nm} = \frac{2}{\sqrt{ab}} \sin\left(\frac{\pi n}{b}x\right) \cos\left(\frac{\pi m}{a}y\right) \tag{3.5}$$

General solution

Replacing the solution in the non-homogeneous wave equation and using the results of the spatial eigenfunctions we arrive at:

$$\sum_{n,m} \left\{\frac{\partial^2 T_{nm}}{\partial t^2} + c^2\pi^2 \left[\left(\frac{n}{b}\right)^2 + \left(\frac{m}{a}\right)^2\right] T_{nm}\right\} v_{nm}$$
$$= \frac{A}{\rho} \sin(\omega t)\, \delta\left(x - \frac{b}{2}\right) \delta(y - [a - \varepsilon]) \tag{3.6}$$

Final solution

Multiplying the previous relation by $v_{nm}(x, y)$ and integrating between $0 < x < b$ and $0 < y < a$ we have:

$$\frac{\partial^2 T_{nm}}{\partial t^2} + c^2\pi^2 \left[\left(\frac{n}{b}\right)^2 + \left(\frac{m}{a}\right)^2\right] T_{nm}$$
$$= \int_0^a dy \int_0^b dx \frac{2}{\sqrt{ab}} \frac{A}{\rho} \sin(\omega t)\, \delta\left(x - \frac{b}{2}\right) \delta(y - [a - \varepsilon]) \sin\left(\frac{\pi n}{b}x\right) \cos\left(\frac{\pi m}{a}y\right)$$
$$= \frac{2}{\sqrt{ab}} \frac{A}{\rho} \sin(\omega t) \sin\left(\frac{\pi n}{b}\frac{b}{2}\right) \cos\left(\frac{\pi m}{a}(a - \varepsilon)\right)$$
$$\tag{3.7}$$

Note: $\sin\left(\frac{\pi n}{2}\right) = \begin{Bmatrix} 0 & (n = 2k) \\ (-1)^k & (n = 2k+1) \end{Bmatrix}$

Note: $\cos\left(\frac{\pi m}{a}(a-\varepsilon)\right)$ [for $\varepsilon \to 0$] $= \cos(\pi m) = (-1)^m\{m = 0, 1, 2, \ldots\}$

Then the modes excited by the applied force will be:

$$\omega^2_{2k+1,m} = c^2\pi^2\left\{\frac{(2k+1)^2}{b^2} + \frac{m^2}{a^2}\right\} \tag{3.8}$$

Note: modes in the x direction with even value of n that have nodes in the central vertical line of the membrane are not excited, since they are suppressed by the symmetry of the applied force.

3.2 Oscillations of a Membrane Fixed at Two Boundaries

A square membrane, whose sides are of length π have two opposite boundaries free to move (at $y = 0$ and $y = \pi$) and the other two ($x = 0$, $x = \pi$), fixed. Starting at $t = 0$ the membrane is subject to a periodic force with areal density of the form $\sin(t)\sin(x)\cos(y)$. Find the membrane displacement for $t > 0$. Consider that the surface density and the speed of sound in the membrane are equal to 1.

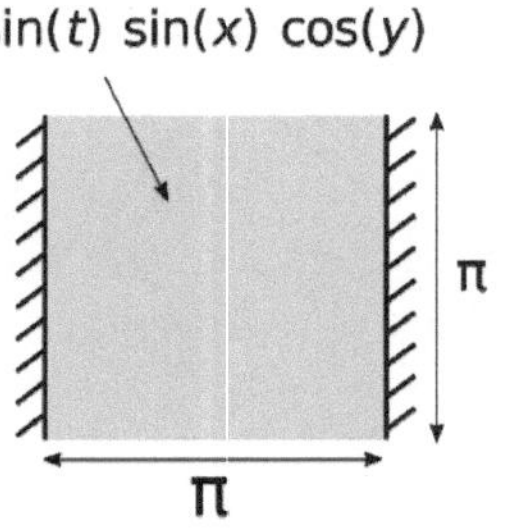

Figure 3.2

Mathematical formulation

$$\left\{\begin{array}{l} \dfrac{\partial^2 u}{\partial t^2} = c^2\left(\dfrac{\partial^2 u}{\partial x^2} + \dfrac{\partial^2 u}{\partial y^2}\right) + \sin(t)\cdot\sin(x)\cdot\cos(y) \\ \text{Boundary conditions: } \left\{\begin{array}{c} \left.\dfrac{\partial u}{\partial y}\right|_{y=0} = 0 \\ \left.\dfrac{\partial u}{\partial y}\right|_{y=\pi} = 0 = 0 \\ u(x=0) = 0 \\ u(x=\pi) = 0 \end{array}\right\} \\ \text{Initial conditions: } \left\{\begin{array}{c} u(x, y, 0) = 0 \\ \left.\dfrac{\partial u}{\partial t}\right|_{t=0} = 0 \end{array}\right\} \end{array}\right\} \tag{3.9}$$

Sturm–Liouville problem

We seek the solution as an expansion of orthogonal functions:

$$u(x, y, t) = \sum_{n,m}^{\infty} A_{nm}(t) v_{nm}(x, y) \tag{3.10}$$

where v_{nm} are the solutions of the Sturm–Liouville problem:

$$\left\{\begin{array}{l} \dfrac{\partial^2 v}{\partial x^2} + \dfrac{\partial^2 v}{\partial y^2} + \lambda v = 0 \\ \text{Boundary conditions: } \left\{\begin{array}{c} \left.\dfrac{\partial v}{\partial y}\right|_{y=0} = 0 \\ \left.\dfrac{\partial v}{\partial y}\right|_{y=\pi} = 0 \\ v(x=0) = 0 \\ v(x=\pi) = 0 \end{array}\right\} \end{array}\right\} \tag{3.11}$$

The eigenfunctions are well known. For the side π of this membrane:

$$v_{nm}(x, y) = \sin(nx)\cos(my) \qquad (n = 1, 2\ldots \quad m = 0, 1, 2\ldots) \tag{3.12}$$

with eigenvalues:

$$\lambda_{nm} = n^2 + m^2 \tag{3.13}$$

General solution Replacing

$$u(x, y, t) = \sum_{n,m}^{\infty} A_{nm}(t) \sin(nx) \cos(my) \tag{3.14}$$

into equation (3.9)

$$\sum_{n,m}^{\infty} \left[\frac{\partial^2 A_{nm}(t)}{\partial t^2} + c^2 \lambda_{nm} A_{nm}(t) \right] \sin(nx) \cos(my) = \sin(t) \cdot \sin(x) \cdot \cos(y) \tag{3.15}$$

Applying the orthogonality of the $v_{nm}(x, y)$ eigenfunctions we arrive at the equation for the amplitudes $A_{nm}(t)$.

Due to the membrane being excited with a force with the spatial profile of a single mode, only this mode gets excited, $A_{11}(t)$. Mathematically it is the consequence that the rest of the eigenfunctions are orthogonal with the function that describes the profile of the applied force. Once the orthogonality conditions have been applied and, integrating both functions in the range $[0, \pi]$ we have:

$$\frac{d^2 A_{11}(t)}{dt^2} + \lambda_{11} A_{11}(t) = \sin(t) \tag{3.16}$$

Since $\lambda_{11} = 1^2 + 1^2 = 2$ we will need to solve the next equation:

$$\left\{ \begin{array}{l} \dfrac{d^2 A_{11}(t)}{dt^2} + 2A_{11}(t) = \sin(t) \\ A_{11}(0) = 0 \\ \left. \dfrac{dA_{11}}{dt} \right|_{t=0} = 0 \end{array} \right\} \tag{3.17}$$

The particular solution is $A_{11,part}(t) = \sin(t)$:

The general solution of the homogeneous equation is:

$$\frac{d^2 A_{11}}{dt^2} + 2A_{11} = 0 \tag{3.18}$$

$$A_{11,hom} = C \sin(\sqrt{2}t) + D \cos(\sqrt{2}t) \tag{3.19}$$

Applying the initial conditions to the solution $A_{11} = A_{11,hom} + A_{11,part}$:

$$\begin{pmatrix} C\sin(\sqrt{2}\,0) + D\cos(\sqrt{2}\,0) + \sin(0) = 0 \\ C\sqrt{2}\cos(\sqrt{2}\,0) - D\sqrt{2}\sin(\sqrt{2}\,0) + \cos(0) = 0 \end{pmatrix} \tag{3.20}$$

We have:

$$\begin{pmatrix} D = 0 \\ C\sqrt{2} + 1 = 0 \end{pmatrix} \tag{3.21}$$

$$A_{11}(t) = -\frac{1}{\sqrt{2}}\sin(\sqrt{2}t) + \sin(t) \tag{3.22}$$

Final solution

$$\begin{aligned} u(x, y, t) &= A_{11}\sin(x)\cos(y) \\ &= \left[\sin(t) - \frac{1}{\sqrt{2}}\sin(\sqrt{2}t)\right]\sin(x)\cos(y) \end{aligned} \tag{3.23}$$

Note: in the case that the membrane oscillates in a viscous medium we could approximate the viscosity by inserting a term proportional to the velocity ($\alpha\frac{du}{dt}$). The solution method is the same with this new term, changing only the form of the equation for $A(t)$, where we will have a term with the first derivative of $A(t)$.

3.3 Electrostatic Field inside a Semi-Infinite Region

Find the electrostatic potential inside a semi-infinite region, limited by conductor plates at ($y = 0, y = b, x = 0$) if the plate at $x = 0$ is connected to a V_0 potential (see figure).

The plates at $y = 0$, $y = b$ are grounded and there are no charges inside the region.

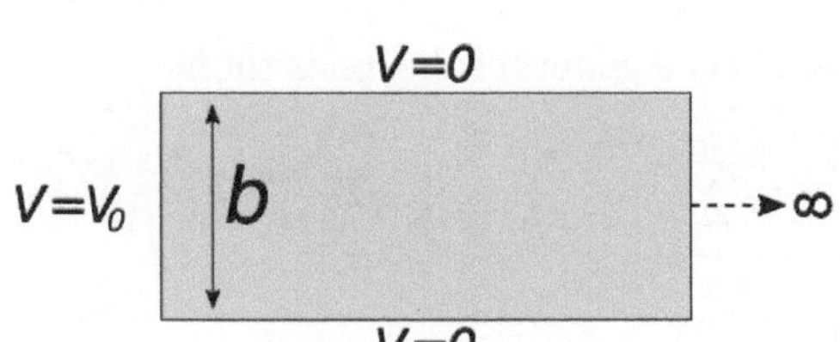

Figure 3.3

Mathematical formulation

$$\left\{\begin{array}{l}\dfrac{\partial^2 u}{\partial x^2}+\dfrac{\partial^2 u}{\partial y^2}=0\\ u(0,y)=V_0\\ u(x,0)=u(x,b)=0\\ u(x\to\infty,y)=0\end{array}\right\} \tag{3.24}$$

Sturm–Liouville problem

Separating variables and taking advantage of the homogeneous boundary conditions in the y direction the solution is expanded in eigenfunctions of the Sturm–Liouville problem. We seek the solution as:

$$u = X(x)Y(y) \tag{3.25}$$

Separating variables we arrive at the Sturm–Liouville problem for $Y(y)$:

$$\left\{\begin{array}{l}\dfrac{d^2 Y}{dy^2}+\lambda Y=0\\ Y(0)=Y(b)=0\end{array}\right\} \tag{3.26}$$

The $Y(y)$ functions, when replaced into the Laplace equation, give rise to the problem for $X(x)$:

$$X'' - \lambda X = 0 \tag{3.27}$$

which has a solution in the form of two exponential solutions:

$$X_n(x) = A_n e^{-\frac{\pi n}{b}x} + B_n e^{+\frac{\pi n}{b}x} \tag{3.28}$$

General solution The general solution is then:

$$u(x,y) = \sum_{n=1}^{\infty} \sin\left(\frac{\pi n}{b}y\right)\left[A_n e^{-\frac{\pi n}{b}x} + B_n e^{\frac{\pi n}{b}x}\right] \tag{3.29}$$

Final solution

From the boundary condition at infinity, we must impose: $u(x \to \infty, y) = 0$ so that the solution does not diverge.

Because of which $B_n = 0$ and the solution simplifies:

$$u(x, y) = \sum_{n=1}^{\infty} A_n \sin\left(\frac{\pi n}{b} y\right) e^{-\frac{\pi n}{b} x} \tag{3.30}$$

Imposing the boundary condition $u(0, y) = V_0$ we have:

$$V_0 = \sum_{n=1}^{\infty} A_n \sin\left(\frac{\pi n}{b} y\right) \tag{3.31}$$

and using the orthogonality of the eigenfunctions (in the y direction):

$$\frac{2V_0}{b} \int_0^b \sin\left(\frac{\pi n}{b} y\right) dy = A_n \tag{3.32}$$

The integral has a value:

$$\int_0^b \sin\left(\frac{\pi n}{b} y\right) dy = \frac{b}{\pi n}[\cos(\pi n) - \cos(0)] = \begin{Bmatrix} -2 & (n = 2k+1) \\ 0 & (n = 2k) \end{Bmatrix} \tag{3.33}$$

where k are integer numbers from $k = 0$.

We finally arrive at the solution in a compact form:

$$u(x, y) = \frac{4V_0}{\pi} \sum_{k=0}^{\infty} e^{-\frac{\pi(2k+1)}{b} x} \frac{\sin\left(\frac{\pi(2k+1)}{b} y\right)}{(2k+1)} \tag{3.34}$$

Graphical representation (using MATLAB's PDE Toolbox with the right boundary at a finite distance and null electric field transversal to the boundary).

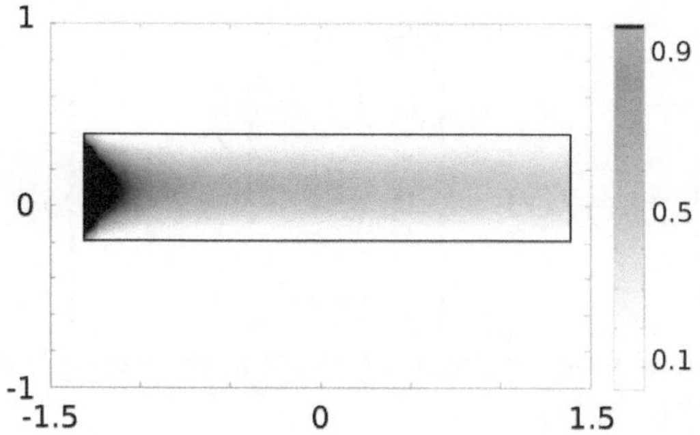

Figure 3.4

3.4 Distribution of Electrostatic Potential in a Rectangle

A metallic prism has a rectangular cross section $L_x \times L_y$ and infinite length (along the z axis). The whole prism is grounded except for the central region of the face at $x = 0$ which, being insulated from the rest, is at electric potential V_0 (the thickness of this region is $L_y/2$. Supposing that $L_x >> L_y$, obtain an approximate expression for the electric potential.

Figure 3.5

Mathematical formulation This problem can be solved with Laplace's equation:

$$\frac{\partial^2 u}{\partial x^2} + \frac{\partial^2 u}{\partial y^2} = 0 \tag{3.35}$$

Boundary conditions 1 and 2:

$$u(x, 0) = 0 \tag{3.36}$$

$$u(x, L_y) = 0 \tag{3.37}$$

Boundary conditions 3 and 4:

$$u(L_x, y) = 0 \tag{3.38}$$

$$u(0, y) = f(y) = \left\{ \begin{array}{l} 0 \ \left(y < \frac{L_y}{4}\right) \\ V_0 \ \left(\frac{L_y}{4} < y < \frac{3L_y}{4}\right) \\ 0 \ \ (y > \frac{3L_y}{4}) \end{array} \right\} \tag{3.39}$$

It can be solved in an analogous fashion to the example described in section 2.2.2 from [1]:

We use separation of spatial variables: $u = X \cdot Y$.

Solution Sturm–Liouville problem in the y direction:

$$\frac{\partial^2 Y}{\partial y^2} + \lambda Y = 0 \tag{3.40}$$

With homogeneous boundary conditions of the first type. The corresponding eigenfunctions and eigenvalues are:

$$v_n = \sin\left(\frac{\pi n}{L_y} y\right) \tag{3.41}$$

$$\lambda_n = \left(\frac{\pi n}{L_y}\right)^2 \tag{3.42}$$

We will look for the solution of the problem as an eigenfunction expansion:

$$u(x, y) = \sum X_n(x) \sin\left(\frac{\pi n}{L_y} y\right) \tag{3.43}$$

Replacing this solution into Laplace's equation:

$$\sum_n \left[\frac{d^2 X_n}{dx^2} - \lambda_n X_n\right] \sin\left(\frac{\pi n}{L_y} y\right) = 0 \tag{3.44}$$

Then the equations to find X_n are:

$$\begin{Bmatrix} \dfrac{d^2 X_n}{dx^2} - \lambda_n X_n = 0 \\ X_n(L_x) = 0 \\ X_n(0) = A_n \end{Bmatrix} \tag{3.45}$$

Final solution The A_n coefficients are given by:

$$u(0, y) = \sum X_n(0) \sin\left(\frac{\pi n}{L_y} y\right) = f(y) = \begin{Bmatrix} 0 & \left(y < \dfrac{Ly}{4}\right) \\ V_0 & \left(\dfrac{Ly}{4} < y < \dfrac{3Ly}{4}\right) \\ 0 & \left(y > \dfrac{3Ly}{4}\right) \end{Bmatrix} \tag{3.46}$$

Using the orthogonality of the Y_n:

$$X_n(0) = \frac{2}{L_y} V_0 \int_{\frac{L_y}{4}}^{\frac{3L_y}{4}} \sin\left(\frac{\pi n}{L_y} y\right) dy = \frac{2}{L_y} V_0 \left[\cos\left(\frac{\pi n}{4}\right) - \cos\left(\frac{3\pi n}{4}\right)\right] \tag{3.47}$$

Solution for the X_n:

$$X_n(x) = C_n e^{-\sqrt{\lambda_n}x} + D_n e^{\sqrt{\lambda_n}x} \tag{3.48}$$

Imposing the boundary conditions:

$$C_n e^{-\sqrt{\lambda_n}L_x} + D_n e^{\sqrt{\lambda_n}L_x} = 0 \tag{3.49}$$

$$C_n + D_n = A_n \tag{3.50}$$

$$D_n = \frac{A_n}{1 - e^{2\sqrt{\lambda_n}L_x}} \tag{3.51}$$

$$C_n = \frac{A_n e^{2\sqrt{\lambda_n}L_x}}{e^{2\sqrt{\lambda_n}L_x} - 1} \tag{3.52}$$

Considering the solution in the limit $\frac{L_x}{L_y} \to \infty$ we have: $D_n \to 0$ and $C_n \to A_n$.

3.5 Distribution of Temperature in a Semi-Insulated and Semi-Infinite Slab

a) Find the stationary distribution of temperature $u(x, y)$ inside a box with respect to the temperature of its base (which is constant but is not defined). The box is infinite in the z direction and semi-infinite for $x > 0$. The faces at $y = 0$ and $y = L$ are thermally insulated. The face $(x = 0,\ 0 < y < L)$ receives heat with a density of flux $F_0 \times y \times (L - y)$. The thermal conductivity of the box is k.

b) Solve the same problem supposing that the lower boundary exchanges heat with the outer medium, which is at a temperature T_0, according to the Newton's law, with constant h.

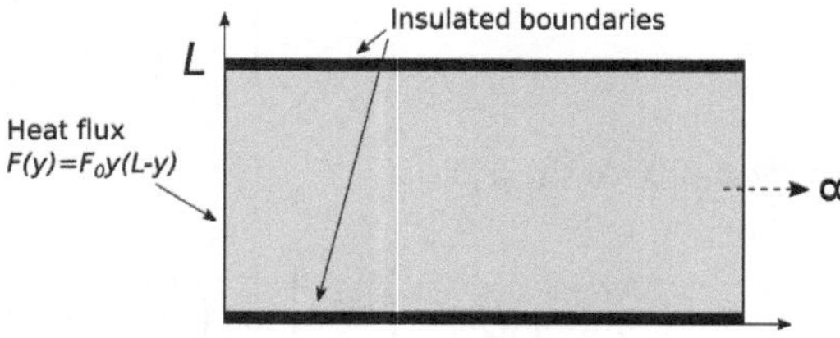

Figure 3.6

a) Mathematical formulation

The variable u represents the temperature with respect to the original temperature of the bar, which is unknown, therefore u represents the variations of temperature. As the bar is infinite and the heat occurs by a source at $x = 0$, the other boundary, at $x \to \infty$ remains at the initial temperature $u = 0$, since the heat would require an infinite time to get there. Since the slab is infinite in the z direction it's enough to solve Laplace's equation in two dimensions (see figure).

$$\left\{\begin{array}{ll}\dfrac{\partial^2 u}{\partial x^2}+\dfrac{\partial^2 u}{\partial y^2}=0 & (0<x<\infty,\ 0<y<L)\\ \left.\dfrac{\partial u}{\partial y}\right|_{y=0}=\left.\dfrac{\partial u}{\partial y}\right|_{y=L}=u(+\infty, y)=0 & \\ \left.\dfrac{\partial u}{\partial x}\right|_{x=0}=-\dfrac{F_0}{k}y(L-y) & \end{array}\right\} \tag{3.53}$$

Sturm–Liouville problem

We seek the general solution by separating variables, with the intention of expanding the solution in a Fourier series with orthogonal functions (in the y direction, which has second type boundary conditions).

$$u(x, y) = X(x)Y(y) \tag{3.54}$$

We arrive at two equations:

Equation for the Sturm–Liouville problem for $Y(y)$:

$$\left\{\begin{array}{l}\dfrac{d^2Y}{dy^2}+\lambda Y=0\\ \left.\dfrac{dY}{dy}\right|_{y=0}=\left.\dfrac{dY}{dy}\right|_{y=L}=0\end{array}\right\} \tag{3.55}$$

Which has eigenfunctions and eigenvalues:

$$Y_n(y) = \cos\left(\frac{n\pi}{L}y\right) \qquad (n = 0, 1, 2, \ldots) \tag{3.56}$$

$$\lambda_n = \left(\frac{n\pi}{L}\right)^2 \tag{3.57}$$

The equation and the corresponding solutions for $X(x)$ are:

$$\frac{d^2X}{dx^2} - \lambda X = 0 \tag{3.58}$$

$$X_n(x) = A_n e^{-\left(\frac{n\pi}{L}x\right)} + B_n e^{\left(\frac{n\pi}{L}x\right)} \qquad (n = 0, 1, 2, \ldots) \tag{3.59}$$

General solution

$$u(x, y) = \sum \left[A_n e^{\left(-\frac{n\pi}{L}x\right)} + B_n e^{\left(\frac{n\pi}{L}x\right)}\right] \cos\left(\frac{n\pi}{L}y\right) \quad (n = 0, 1, 2, \ldots) \tag{3.60}$$

Note: A_0 must no appear in the sum, since the base temperature of the problem is not defined. We are only interested in the variations of temperature due to the heat flux. Applying the boundary condition:

$$u(+\infty, y) = 0 \tag{3.61}$$

We arrive at $B_n = 0\ (n = 1, 2, \ldots)$

Imposing the other boundary condition:

$$\left.\frac{du}{dx}\right|_{x=0} = -\frac{F_0}{k}y(L-y) = \sum A_n \left(-\frac{n\pi}{L}\right) e^{-\left(\frac{n\pi}{L}0\right)} \cos\left(\frac{n\pi}{L}y\right) \tag{3.62}$$

Final solution Using the orthogonality of the eigenfunctions Y_n we get the coefficients:

$$A_n (n \geq 1) \tag{3.63}$$

$$A_n = \frac{2F_0}{k\pi n} \int_0^L y(L-y) \cos\left(\frac{n\pi}{L}y\right) dy = \frac{2F_0 L^3}{\pi^3 n^3 k}[(-1)^n + 1] \tag{3.64}$$

Finally the variation of temperature along the x axis, due to the supplied heat flux is:

$$u(x, y) = \sum_{n=\text{even}>0}^{\infty} \frac{2F_0 L^3}{\pi^3 n^3 k}[(-1)^n + 1] e^{-\left(\frac{n\pi}{L}x\right)} \cos\left(\frac{n\pi}{L}y\right) \tag{3.65}$$

The term with $n = 0$ does not exist in the sum, since the solution is defined with respect to the unknown temperature of the object.

b) Now we solve the problem supposing that the lower border losses heat according to Newton's Law, towards the surroundings, with temperature T_0, with constant h.

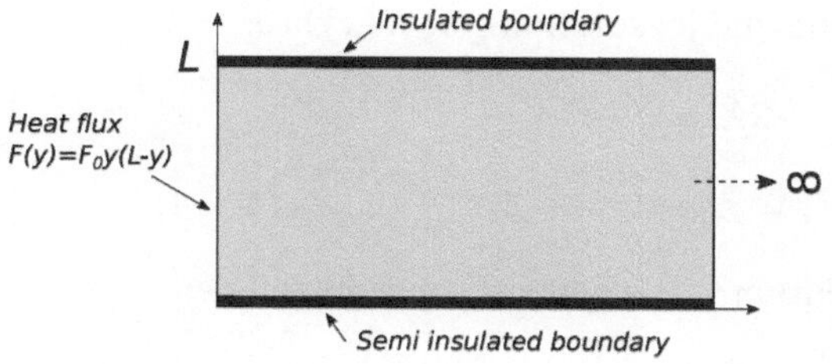

Figure 3.7

In this case, resetting the value of the constant T_0 would not change the boundary conditions (neither left nor upper). The solution with the T_0 background subtracted tends to zero in the limit x equal to infinity.

Sturm–Liouville problem

We formulate the same Laplace's problem, with the same boundary conditions, except for the lower border, which would change from type two to type three. The equation for the Sturm–Liouville problem for $Y(y)$ is, considering the direction of the heat flux:

$$\left\{\begin{array}{l} \dfrac{d^2Y}{dy^2}+\lambda Y=0 \\ -k\left.\dfrac{dY}{dy}\right|_{y=0}+hY(0)=0 \\ \left.\dfrac{dY}{dy}\right|_{y=L}=0 \end{array}\right\} \tag{3.66}$$

Sign of the first boundary condition is chosen to relate correctly the derivative of the solution in vertical coordinate (y) with corresponding direction of the heat flow. A solution which satisfies the second boundary condition is:

$$Y_n(y)=C_n\cos\left(\sqrt{\lambda_n}(y-L)\right) \qquad (n=1,2,\ldots) \tag{3.67}$$

From the first boundary condition:

$$C_n\left[k\sqrt{\lambda_n}\sin\left(\sqrt{\lambda_n}L\right)+h\cos\left(\sqrt{\lambda_n}L\right)\right]=0 \tag{3.68}$$

The solutions of the equation are the eigenvalues:

$$\tan\left(\sqrt{\lambda_n}L\right)=\frac{-h}{k\sqrt{\lambda_n}} \tag{3.69}$$

The equation and its solutions for $X(x)$ are:

$$\left\{\begin{array}{ll}\dfrac{d^2X}{dx^2}-\lambda X=0 & \\ X_n(x)=A_n e^{-\sqrt{\lambda_n}x}+B_n e^{+\sqrt{\lambda_n}x} & (n=1,2,\ldots)\end{array}\right\} \tag{3.70}$$

General solution The general solution will be:

$$u(x,y)=\sum\left[A_n e^{-\sqrt{\lambda_n}x}+B_n e^{+\sqrt{\lambda_n}x}\right]\cos\sqrt{\lambda_n}[y-L] \quad (n=1,2,\ldots) \tag{3.71}$$

Final solution Applying the condition:

$$u(+\infty,y)=0 \tag{3.72}$$

We have $B_n=0$, with $n=1,2,\ldots$

Imposing the condition:

$$\left.\frac{du}{dx}\right|_{x=0}=-\frac{F_0}{k}y(L-y)=\sum[A_n(-\sqrt{\lambda_n})e^0]\cos\left(\sqrt{\lambda_n}(y-L)\right) \tag{3.73}$$

Using the orthogonality of the eigenfunctions Y_n we obtain the coefficients $A_n(n\geq 1)$

$$A_n=\frac{F_0}{k\sqrt{\lambda_n}}\frac{\int_0^L y(L-y)\cos\left(\sqrt{\lambda_n}[y-L]\right)dy}{\int_0^L \cos\left(\sqrt{\lambda_n}[y-L]\right)^2 dy}= \tag{3.74}$$

$$\int_0^L y(L-y)\cos[\sqrt{\lambda_n}(y-L)]dy=\frac{L}{\lambda_n}\left[1-\cos\left(\sqrt{\lambda_n}L\right)\right] -\frac{2\sqrt{\lambda_n}L-\sin(\sqrt{\lambda_n}L)}{(\lambda_n)^{3/2}} \tag{3.75}$$

$$\int_0^L\left[\cos\left(\sqrt{\lambda_n}[y-L]\right)\right]^2 dy=\frac{2\sqrt{\lambda_n}L+\sin(2\sqrt{\lambda_n}L)}{4\sqrt{\lambda_n}} =\frac{L}{2}+\frac{\sin(2\sqrt{\lambda_n}L)}{4\sqrt{\lambda_n}} \tag{3.76}$$

$$A_n=\frac{F_0}{k\sqrt{\lambda_n}}\frac{\frac{L}{\lambda_n}[1-\cos(\sqrt{\lambda_n}L)]-\frac{2\sqrt{\lambda_n}L-\sin(\sqrt{\lambda_n}L)}{(\lambda_n)^{3/2}}}{\frac{L}{2}+\frac{\sin(2\sqrt{\lambda_n}L)}{4\sqrt{\lambda_n}}} \tag{3.77}$$

Finally the variation of the temperature along the x axis due to the supplied heat is:

$$u(x, y) = \sum_{n=1}^{\infty} \left[A_n e^{-\sqrt{\lambda_n} x} \cos\left(\sqrt{\lambda_n} [y - L] \right) + T_0 \right] \tag{3.78}$$

3.6 Oscillations of a Semi-Fixed Membrane

A square membrane of side L has a fixed border, whereas the other three can move transversally. The border opposite to the fixed one is kept at a distance h from its mechanical equilibrium position. At $t = 0$ that border is released. Determine the form of the membrane as a function of time (in the absence of gravity).

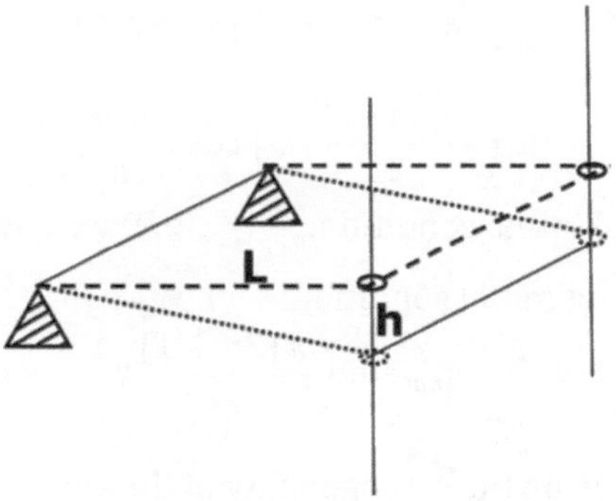

Figure 3.8

Mathematical formulation

$$\left\{ \begin{array}{l} \dfrac{d^2u}{dt^2} - c^2 \Delta u = 0 \\ u(0, y, t) = 0 \\ \left.\dfrac{\partial u}{\partial x}\right|_{x=L} = \left.\dfrac{\partial u}{\partial y}\right|_{y=0} = \left.\dfrac{\partial u}{\partial y}\right|_{y=L} = 0 \\ u(x, y, 0) = \dfrac{h}{L} x \\ \left.\dfrac{\partial u}{\partial t}\right|_{t=0} = 0 \end{array} \right\} \tag{3.79}$$

Sturm–Liouville problem The solution is expanded into eigenfunctions of the Sturm–Liouville problem in 2D:

$$u(x, y, t) = \sum_{nm} T_{nm}(t)X_n(x)Y_n(y) \tag{3.80}$$

Formulating the Sturm–Liouville problem we get the eigenfunctions:

$$X_n(x)Y_n(y) = \sin\left(\frac{\pi(2n+1)}{2L}x\right)\cos\left(\frac{\pi m}{L}y\right) \tag{3.81}$$

where $n = 0, 1, 2 \ldots \infty$ and $m = 0, 1, 2 \ldots \infty$

And the eigenvalues:

$$\lambda_{nm} = \left[\frac{\pi(2n+1)}{2L}\right]^2 + \left[\frac{\pi m}{L}\right]^2 \tag{3.82}$$

General solution The general solution and the initial conditions are:

$$u(x, y, t) = \sum_{nm} [A_{nm}\cos(\omega_{nm}t) + B_{nm}\sin(\omega_{nm}t)]\sin \times \left(\frac{\pi(2n+1)}{2L}x\right)\cos\left(\frac{\pi m}{L}y\right) \tag{3.83}$$

Using the second initial condition: $\frac{\partial u}{\partial t}\Big|_{t=0} = 0$ we have $B_{nm} = 0$

Also, from the first initial condition:

$$u(x, y, 0) = \frac{h}{L}x = \sum_{nm} A_{nm}\sin\left(\frac{\pi(2n+1)}{2L}x\right)\cos\left(\frac{\pi m}{L}y\right) \tag{3.84}$$

Final solution Using the orthogonality of the eigenfunctions:

$$\int_0^L \frac{h}{L}x\sin\left(\frac{\pi(2n+1)}{2L}x\right)dx\int_0^L \cos\left(\frac{\pi m}{L}y\right)dy = A_{nm}\int_0^L \sin^2\left(\frac{\pi(2n+1)}{2L}x\right)dx\int_0^L \cos^2\left(\frac{\pi m}{L}y\right)dy \tag{3.85}$$

Finally we obtain the coefficients A_{nm}:

A sine function is orthogonal to a constant value (which is one of the orthogonal eigenfunctions of the $\cos\left(\frac{\pi 0}{L}y\right)$). The integral:

$$\int_0^L \cos\left(\frac{\pi m}{L}y\right)dy = 0 \tag{3.86}$$

for the indices $m \neq 0$

For $m = 0$ the previous integral equals L. In that case, the A_{nm} coefficients are not zero. We would need these integrals to obtain its value:

$$\int_0^L \cos^2\left(\frac{\pi m}{L}y\right) dy = L \qquad (m = 0) \tag{3.87}$$

$$\int_0^L \sin^2\left(\frac{\pi(2n+1)}{2L}x\right) dx = \frac{L}{2} \tag{3.88}$$

With the two sums only the terms with n indices survive. Finally, the coefficients are:

$$A_{n0} = \frac{2}{L}\frac{h}{L}\int_0^L x \sin\left(\frac{\pi(2n+1)}{2L}x\right) dx = \frac{8h}{\pi^2}\frac{(-1)^n}{(2n+1)^2} \tag{3.89}$$

3.7 Stationary Temperature in a Rectangle with Heat Losses through Its Boundaries

Find the stationary distribution of temperature (T) in a rectangle $(a,\ b)$. Two of its sides ($x = 0, a$) are in contact with a thermal reservoir at $T = 0$. The thermal conductivity is $k = 1$. The boundary at $y = 0$ exchanges a heat flux with the outer medium, which creates a gradient of temperature in the direction perpendicular to the boundary, given by: $\left.\frac{\partial T}{\partial y}\right|_{y=0} = -2T(x, y = 0)$. The opposite border ($y = b$) exchanges a heat flux with a density $-f(x)$ per unit time and unit surface.

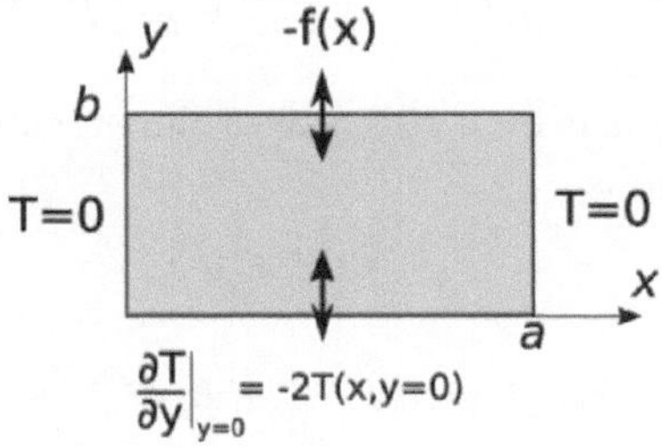

Figure 3.9

Mathematical formulation

$$\left\{\begin{array}{l} \dfrac{\partial^2 u}{\partial x^2}+\dfrac{\partial^2 u}{\partial y^2}=0 \\ u(x=0)=u(x=a)=0 \\ \left.\dfrac{du}{dy}\right|_{y=0}+2u(x,0)=0 \\ \left.\dfrac{du}{dy}\right|_{y=b}=f(x) \end{array}\right\} \tag{3.90}$$

Sturm–Liouville problem

We separate variables:

$$v(x,y)=X(x)\cdot Y(y) \tag{3.91}$$

We arrive at two equations, one for X and another one for Y. The Sturm–Liouville problem for X is:

$$\left\{\begin{array}{ll} \dfrac{d^2X}{dx^2}+\lambda X=0 & \\ X(0)=X(a)=0 & \\ X_n(x)=\sin\left(\frac{n\pi}{a}x\right) & (n=1,2,\ldots) \end{array}\right\} \tag{3.92}$$

The sign of the constant λ is chosen to expand the solution in eigenfunctions, orthogonal in the x direction, where there are homogeneous boundary conditions.

Equation and solution for the y variable:

$$\frac{d^2Y}{dy^2}-\lambda Y=0$$

$$Y_n(y)=A_n\cosh\left(\frac{n\pi}{a}y\right)+B_n\sinh\left(\frac{n\pi}{a}y\right) \quad (n=1,2,\ldots) \tag{3.93}$$

The solution can also be written as a linear combination of two exponential solutions.

$$\frac{d^2Y}{dy^2}-\lambda Y=0$$

$$Y_n(y)=A_n e^{-\frac{n\pi}{a}y}+B_n e^{+\frac{n\pi}{a}y} \tag{3.94}$$

The latter form is the most convenient one when any of the boundaries of the rectangle in the y is not fixed.

General solution

In this case the best form of the general solution is:

$$u(x, y) = \sum_n \left[A_n \cosh\left(\frac{n\pi}{a}y\right) + B_n \sinh\left(\frac{n\pi}{a}y\right)\right] \sin\left(\frac{n\pi}{a}x\right)$$
$$(n = 1, 2, \ldots) \tag{3.95}$$

Applying the boundary condition: $\left.\frac{du}{dy}\right|_{y=0} + 2u = 0 \rightarrow$

$$\sum_n \left[B_n \frac{n\pi}{a} + 2A_n\right] \sin\left(\frac{n\pi}{a}x\right) = 0 \rightarrow B_n \frac{n\pi}{a} + 2A_n$$
$$= 0 \rightarrow B_n = -\frac{2aA_n}{n\pi} \tag{3.96}$$

We arrive at the following general solution:

$$u(x, y) = \sum_n A_n \left[\cosh\left(\frac{n\pi}{a}y\right) - \frac{2a}{n\pi} \sinh\left(\frac{n\pi}{a}y\right)\right] \sin\left(\frac{n\pi}{a}x\right) \tag{3.97}$$

Final solution

Imposing the second boundary condition: $\left.\frac{du}{dy}\right|_{y=b} = f(x)$

$$f(x) = \sum_n A_n \left[\frac{n\pi}{a} \sinh\left(\frac{n\pi}{a}b\right) - 2\cosh\left(\frac{n\pi}{a}b\right)\right] \sin\left(\frac{n\pi}{a}x\right) \tag{3.98}$$

and using the orthogonality of the eigenfunctions, we find the coefficients A_n:

$$A_n = \frac{2}{a\left[\frac{n\pi}{a} \sinh\left(\frac{n\pi}{a}b\right) - 2\cosh\left(\frac{n\pi}{a}b\right)\right]} \int_0^a f(x) \sin\left(\frac{n\pi}{a}x\right) dx \tag{3.99}$$

Next, the graphic solution of the problem for the case $f(x) = 2$ is shown, using the PDE Toolbox module from MATLAB:

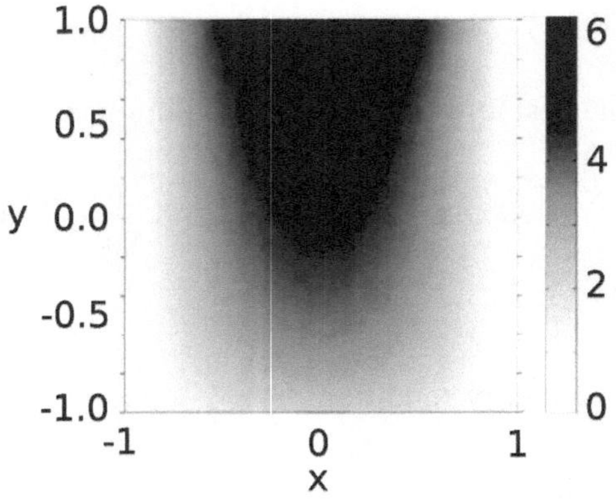

Figure 3.10

3.8 Case Study: Heat Leak from a Rectangle

Find the temporal variations of temperature of a rectangular membrane of sides a and b with thermal conductivity k, heat capacity C and mass density m. Until $t = 0$ two contiguous borders are thermally insulated and the other two are in contact with a thermal bath at T_0, so that the membrane is initially at thermal equilibrium. At $t = 0$ one of the insulated borders changes and starts exchanging heat with the outer medium (which is at $T = 0$), according to Newton's law (with constant h). Initially all the membrane is at temperature $T = T_0$.

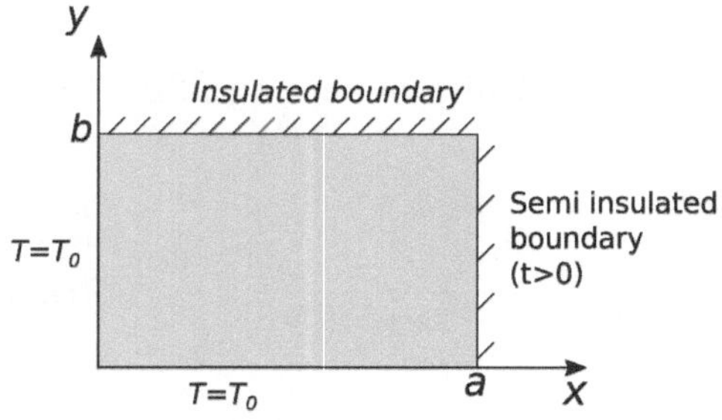

Figure 3.11

Mathematical formulation

$$\begin{Bmatrix} \dfrac{\partial u}{\partial t} - \kappa \Delta u = 0 \\ \text{Boundary condition } (t > 0): \begin{Bmatrix} u(0, y) = T_0 \\ u(x, 0) = T_0 \\ \left.\dfrac{\partial u}{\partial y}\right|_{y=b} = 0 \\ -k \left.\dfrac{\partial u}{\partial x}\right|_{x=a} - hu(x = a) = 0 \end{Bmatrix} \\ \text{Initial condition: } u(x, y, 0) = T_0 \end{Bmatrix} \tag{3.100}$$

Where $\kappa = \frac{k}{C\rho}$. We seek the solution by shifting the origin of temperatures: $v(x, y, t) = u(x, y, 0) - T_0$, to be able to expand the solution in orthogonal functions in the y direction.

$$\begin{Bmatrix} \dfrac{\partial v}{\partial t} - \kappa \Delta v = 0 \\ \text{Boundary condition: } \begin{Bmatrix} v(0, y) = 0 \\ v(x, 0) = 0 \\ \left.\dfrac{\partial v}{\partial y}\right|_{y=b} = 0 \\ k \left.\dfrac{\partial v}{\partial x}\right|_{x=a} + hv(x = a) = -hT_0 \end{Bmatrix} \\ \text{Initial condition: } v(x, y, 0) = 0 \end{Bmatrix} \tag{3.101}$$

The problem is split in two parts: one with a stationary solution $w(x, y)$ and the other with the transient solution $S(x, y, t)$, which will be added.

$$v(x, y, t) = w(x, y) + S(x, y, t) \tag{3.102}$$

The stationary problem (to arrive at a non-trivial solution) must be:

$$\left\{\begin{array}{l}\Delta w = 0 \\ w(0, y) = 0 \\ w(x, 0) = 0 \\ \left.\dfrac{\partial w}{\partial y}\right|_{y=b} = 0 \\ k\left.\dfrac{\partial w}{\partial x}\right|_{x=a} + hw(a, y) = -hT_0\end{array}\right\} \tag{3.103}$$

On the other hand, the transient problem is:

$$\left\{\begin{array}{l}\dfrac{\partial S}{\partial t} - \kappa\Delta S = 0 \\ \left\{\begin{array}{l}S(0, y) = 0 \\ S(x, 0) = 0 \\ \left.\dfrac{\partial S}{\partial y}\right|_{y=b} = 0 \\ k\left.\dfrac{\partial S}{\partial x}\right|_{x=a} + hS(a, y) = 0\end{array}\right\} \\ \text{Initial condition: } S(x, y, 0) = -w(x, y)\end{array}\right\} \tag{3.104}$$

Sturm–Liouville problem

We first seek the solution of the first problem in the form:

$$w(x, y) = \sum_n X_n(x)\, Y_n(y) \tag{3.105}$$

where $Y_n(y)$ are orthogonal eigenfunctions, since both boundaries in the y direction are homogeneous. Solving the corresponding Sturm–Liouville problem we have:

$$Y_n(y) = \sin(\sqrt{\lambda_n}y) \tag{3.106}$$

$$\lambda_n = \left[\frac{\pi(2n+1)}{2b}\right]^2 \tag{3.107}$$

We arrive at the equation for the $X(x)$ function, replacing $w(x, y)$ in the Poisson's equation:

$$\frac{d^2X_n}{dx^2} - \lambda_n X_n = 0 \qquad (3.108)$$

Considering the boundary conditions, it is convenient to write the solution in the form:

$$X_n = A_n \sinh\left(\sqrt{\lambda_n}x\right) + B_n \cosh\left(\sqrt{\lambda_n}x\right) \qquad (3.109)$$

General solution

$$w(x, y) = \sum_n \left[A_n \sinh\left(\sqrt{\lambda_n}x\right) + B_n \cosh\left(\sqrt{\lambda_n}x\right)\right] Y_n(y) \qquad (3.110)$$

Applying the boundary condition $w(0, y) = 0$ we have $B_n = 0$.

Applying the condition $k\left.\frac{\partial w}{\partial x}\right|_{x=a} + hw(x = a) = -hT_0$

$$\sum_n A_n \left[k\sqrt{\lambda_n} \cosh\left(\sqrt{\lambda_n}a\right) + h \sinh\left(\sqrt{\lambda_n}a\right)\right] Y_n(y) = -hT_0 \qquad (3.111)$$

And using the orthogonality of the eigenfunctions:

$$A_n \left[k\sqrt{\lambda_n} \cosh\left(\sqrt{\lambda_n}a\right) + h \sinh\left(\sqrt{\lambda_n}a\right)\right] \int_0^b |Y_n(y)|^2\, dy$$

$$= -hT_0 \int_0^b Y_n(y)dy \qquad (3.112)$$

$$\frac{\int_0^b \sin\left(\sqrt{\lambda_n}y\right) dy}{\int_0^b \left|\sin\left(\sqrt{\lambda_n}y\right)\right|^2 dy} = -\frac{2}{b\sqrt{\lambda_n}}\left[\cos\left(\frac{\pi(2n+1)}{2}\right) - 1\right] = \frac{2}{b\sqrt{\lambda_n}} \qquad (3.113)$$

$$A_n = -\frac{2T_0 h}{b\sqrt{\lambda_n}\left[k\sqrt{\lambda_n}\cosh\left(\sqrt{\lambda_n}a\right) + h\sinh\left(\sqrt{\lambda_n}a\right)\right]} \qquad (3.114)$$

To solve the second homogeneous problem the general solution is quite simple, since it has the form of a summation of the product of both eigenfunctions in two dimensions $X_n(x) \cdot Y_n(y)$ and the temporal part. Contrary to the previous case, here, due to the equation having a temporal part, when doing the separation of variables we get two sets of different eigenvalues, one for each direction:

$$S(x, y, t) = \sum_{mn} T_{mn}(t) X_m(x) Y_n(y) \tag{3.115}$$

$$Y_n(y) = \sin\left(\sqrt{\mu_n} y\right) \tag{3.116}$$

with $\mu_n = \left[\frac{\pi(2n+1)}{2b}\right]^2$

$$X_m(x) = \sin(\sqrt{\nu_m} x) \tag{3.117}$$

with ν_m defined according to the boundary condition:

$$k \left.\frac{dX_m}{dx}\right|_{x=a} + hX_m(a) = 0 \tag{3.118}$$

Which produces the transcendental equation:

$$\tan\left(\sqrt{\nu_m} a\right) = -\frac{k}{h}\sqrt{\nu_m} \tag{3.119}$$

The temporal solutions can be obtained replacing $\sum_{mn} T_{mn}(t) X_m(x)$ $Y_n(y)$ in the homogeneous equation, and we get the problem:

$$\frac{dT_{mn}}{dt} + \kappa(\mu_n + \nu_m) T_{mn} = 0 \tag{3.120}$$

Which has a solution:

$$T_{mn}(t) = T_{mn}(0) e^{-\kappa(\mu_n + \nu_m)t} \tag{3.121}$$

The general solution is:

$$S(x, y, t) = \sum_{mn} T_n(0) e^{-\kappa(\mu_n + \nu_m)t} \sin\left(\sqrt{\nu_m} x\right) \sin\left(\sqrt{\mu_n} y\right) \tag{3.122}$$

Final solution

We solve the problem by including the initial condition:

$$-w(x, y) = \sum_{mn} T_{mn}(0) \sin\left(\sqrt{\nu_m} x\right) \sin\left(\sqrt{\mu_n} y\right) \tag{3.123}$$

And using the orthogonality of $X_m(x)$ and $Y_n(y)$ we get:

$$T_{mn}(0) = -\frac{\int_0^a \int_0^b w(x, y) \sin\left(\sqrt{\nu_m} x\right) \sin\left(\sqrt{\mu_n} y\right) \, dxdy}{\int_0^a \int_0^b \sin^2\left(\sqrt{\nu_m} x\right) \sin^2\left(\sqrt{\mu_n} y\right) \, dxdy} \tag{3.124}$$

3.9 Rectangular Hit on a Square Membrane

Find the movement of a square membrane of side b which is which has two fixed opposite borders, whereas the other two are free. At $t = 0$ a vertical hit transfers a total impulse I to the surface. The hit is uniformly distributed over a rectangle centered in the origin $(b/2 \times b/4)$. The tension is T, the density of the membrane by unit surface is ρ. Consider that the membrane does not vibrate until the initial instant $(t = 0)$.

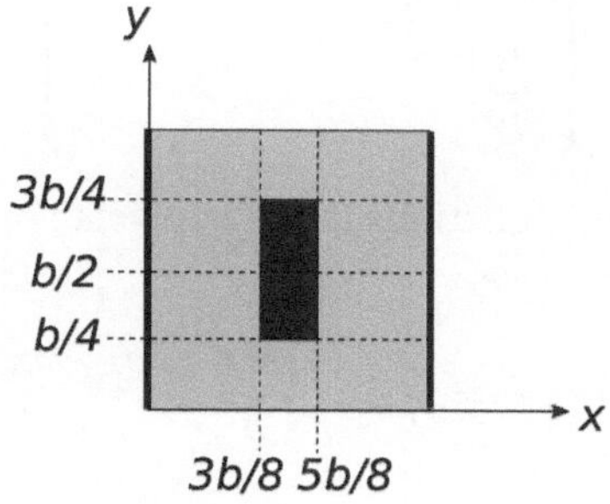

Figure 3.12

Mathematical formulation

$$\left\{\begin{array}{l} \dfrac{\partial^2 u}{\partial t^2} - a^2 \Delta u = 0 \\ \text{Boundary condition: } \left\{\begin{array}{l} u(0, y, t) = 0 \\ u(b, y, t) = 0 \\ \left.\dfrac{\partial u}{\partial y}\right|_{y=0} = 0 \\ \left.\dfrac{\partial u}{\partial y}\right|_{y=b} = 0 \end{array}\right\} \\ \text{Initial conditions: } \left\{\begin{array}{l} u(x, y, 0) = 0 \\ \left.\dfrac{\partial u}{\partial t}\right|_{t=0} = \left\{\begin{array}{ll} \dfrac{8I}{\rho b^2} & \text{(inside the rectangle)} \\ 0 & \text{(outside the rectangle)} \end{array}\right\} \end{array}\right\} \end{array}\right\} \quad (3.125)$$

Being: $a^2 = \frac{T}{\rho}$

We separate variables $u(x, y, t) = T(t)X(x)Y(y) = T(t)v(x, y)$:

Sturm–Liouville problem

$$\begin{Bmatrix} \Delta v(x, y) + \lambda v(x, y) = 0 \\ v(0, y) = 0 \\ v(b, y) = 0 \\ \left.\dfrac{\partial v}{\partial y}\right|_{y=0} = 0 \\ \left.\dfrac{\partial v}{\partial y}\right|_{y=b} = 0 \end{Bmatrix} \tag{3.126}$$

The normalized eigenfunctions of the Laplacian operator, with these boundary conditions, are:

$$v_{nm}(x, y) = \frac{b}{2} \sin\left(\frac{n\pi x}{b}\right) \cos\left(\frac{m\pi y}{b}\right) \tag{3.127}$$

With eigenvalues:

$$\lambda_{nm} = \left(\pi \frac{n}{b}\right)^2 + \left(\pi \frac{m}{b}\right)^2 \qquad (n = 1, 2, \ldots; \; m = 0, 1, 2, \ldots) \tag{3.128}$$

General solution

The equation for the temporal part is:

$$\frac{d^2 T_{nm}}{dt^2} + a^2 \lambda_{nm} T_{nm}(t) = 0 \tag{3.129}$$

with solutions:

$$T_{nm}(t) = A_{nm} \sin\left(a\sqrt{\lambda_{nm}} t\right) + B_{nm} \cos\left(a\sqrt{\lambda_{nm}} t\right) \tag{3.130}$$

Final solution

From the first initial condition we have $B_{nm} = 0$.

Then the general solution can be written (with normalized eigenfunctions) as:

$$u(x, y, t) = \sum_{nm} \left[A_{nm} \frac{b}{2} \sin(a\sqrt{\lambda_{nm}} t) \right] \sin\left(\frac{n\pi x}{b}\right) \cos\left(\frac{m\pi y}{b}\right) \tag{3.131}$$

Applying the second initial condition:

$$u_t(x, y, 0) = \sum_{nm} \left[A_{nm} \frac{b}{2} a\sqrt{\lambda_{nm}} \cos\left(a\sqrt{\lambda_{nm}} 0\right) \right]$$

$$\times \sin\left(\frac{n\pi x}{b}\right) \cos\left(\frac{m\pi y}{b}\right) \tag{3.132}$$

$$= \left\{ \begin{array}{ll} \dfrac{8I}{\rho b^2} & \left(\dfrac{3b}{8} < x < \dfrac{5b}{8}\right), \left(\dfrac{b}{4} < y < \dfrac{3b}{4}\right) \\ 0 & \text{(rest of values of } x, y) \end{array} \right\} \tag{3.133}$$

Using the properties of orthogonality, we multiply by $\frac{2}{b} \sin\left(\frac{n\pi x}{b}\right) \cos\left(\frac{m\pi y}{b}\right)$ both sides and integrate between 0 and b for x and y, to obtain the coefficients of the expansion, which are:

$$A_{nm} a\sqrt{\lambda_{nm}} = \int_{\frac{3b}{8}}^{\frac{5b}{8}} dx \int_{\frac{b}{4}}^{\frac{3b}{4}} dy \, \frac{8I}{\rho b^2} \sin\left(\frac{n\pi x}{b}\right) \cos\left(\frac{m\pi y}{b}\right) \tag{3.134}$$

$$A_{nm} = \frac{8I}{\rho b^2 a\sqrt{\lambda_{nm}}} \int_{\frac{3b}{8}}^{\frac{5b}{8}} dx \int_{\frac{b}{4}}^{\frac{3b}{4}} dy \sin\left(\frac{n\pi x}{b}\right) \cos\left(\frac{m\pi y}{b}\right) \tag{3.135}$$

3.10 Case Study: Distribution of Temperature in a Peltier Element

A rectangular, homogeneous film ($ABCD$) has two borders (AC, BC) thermally insulated. The other two are kept at a temperature $T = T_0$. Find the stationary distribution of temperature (with respect to T_0) supposing that the film absorbs heat homogeneously at a constant rate Q (constant by unit time and unit surface).

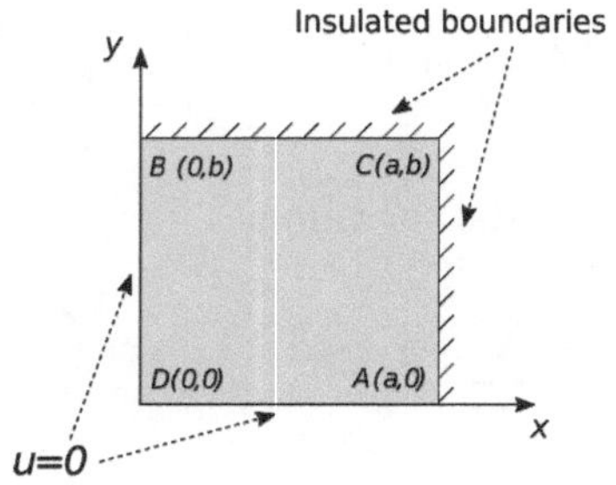

Figure 3.13

Mathematical formulation

We will formulate the solution $u(x, y, t)$ with respect to the temperature of the environment $T = T_0$.

The solution which will produce fewer Gibbs phenomena consists in expanding the solution in orthogonal functions in one of the directions with homogeneous boundaries. An alternative way is, expanding the solution in eigenfunctions of the Sturm–Liouville problem in the two homogeneous directions, replacing the solution in the non-homogeneous equation and find the coefficients of the expansion. This method is possible thanks to the presence of two pairs of opposite homogeneous boundaries.

Mathematical formulation subtracting T_0 to the solution:

$$\left\{\begin{array}{l} C\rho\dfrac{\partial u}{\partial t} - k\left(\dfrac{\partial^2 u}{\partial x^2} + \dfrac{\partial^2 u}{\partial y^2}\right) = -Q \\ \\ \text{Stationary equation : } k\left(\dfrac{\partial^2 u}{\partial x^2} + \dfrac{\partial^2 u}{\partial y^2}\right) = Q \\ \\ u(x=0) = \left.\dfrac{\partial u}{\partial x}\right|_{x=a} = 0 \\ \\ u(y=0) = \left.\dfrac{\partial u}{\partial y}\right|_{y=b} = 0 \end{array}\right\} \quad (3.136)$$

Method 1

In this method we will use an expansion in one-dimensional eigenfunctions.

Sturm–Liouville problem

Searching a solution as $\sum Y(y)X(x)$, we formulate the Sturm–Liouville problem only for the x direction:

$$\left\{ \begin{array}{l} \dfrac{d^2X}{dx^2} + \lambda X = 0 \\ \\ X(x=0) = \left.\dfrac{dX}{dx}\right|_{x=a} = 0 \end{array} \right\} \tag{3.137}$$

The eigenfunctions are:

$$X_n(x) = \sin\left(\frac{\pi(2n+1)}{2a}x\right) \tag{3.138}$$

The eigenvalues are:

$$\lambda_n = \left[\frac{\pi(2n+1)}{2a}\right]^2 \tag{3.139}$$

General solution

The general solution is:

$$u(x,y) = \sum_{n=0}^{\infty} Y_n(y)\sin\left(\frac{\pi(2n+1)}{2a}x\right) \tag{3.140}$$

Replacing this solution into the heat equation we have:

$$-\sum_{n=0}^{\infty} Y_n(y)\left[\frac{\pi(2n+1)}{2a}\right]^2 \sin\left(\frac{\pi(2n+1)}{2a}x\right)$$
$$+\sum_{n=0}^{\infty}\frac{d^2Y_n}{dy^2}\sin\left(\frac{\pi(2n+1)}{2a}x\right) = \frac{Q}{k} = \sum_{n=0}^{\infty} C_n \sin\left(\frac{\pi(2n+1)}{2a}x\right) \tag{3.141}$$

where

$$C_n = \frac{2}{a}\int_0^a \frac{Q}{k}\sin\left(\frac{\pi(2n+1)}{2a}x\right)dx = \frac{4Q}{k\pi(2n+1)} \tag{3.142}$$

In this case what we did is basically first expand the inhomogeneous part of the equation in Fourier series of orthogonal eigenfunctions in the x direction and apply the orthogonality conditions so that we have the non-homogeneous differential equation in the y variable. The equation to find $Y(y)$ then will be:

$$\frac{d^2Y_n}{dy^2} - \left[\frac{\pi(2n+1)}{2a}\right]^2 Y_n = \frac{4Q}{k\pi(2n+1)} \tag{3.143}$$

$$Y_n(y=0) = \left.\frac{dY_n}{dy}\right|_{y=b} = 0 \tag{3.144}$$

Final solution

We look for the solution as the sum of the homogeneous equation and the particular solution. The solution of the homogeneous equation $\left(\text{using the condition: } \left.\frac{dY_n}{dy}\right|_{y=b} = 0\right)$ is:

$$Y_{n,hom}(y) = A_n \cosh\left(\frac{\pi(2n+1)}{2a}(b-y)\right) \tag{3.145}$$

The particular solution is:

$$Y_p(y) = -\frac{16Qa^2}{k\pi^3(2n+1)^3} \tag{3.146}$$

This form of the solution already satisfies the boundary condition for $y = b$. Applying the last boundary condition, we have: $Y_n(0) = 0$:

$$Y_{n,hom}(0) + Y_p(0) = A_n \cosh\left(\frac{\pi(2n+1)}{2a}b\right) - \frac{16Qa^2}{k\pi^3(2n+1)^3} = 0 \tag{3.147}$$

The coefficients are:

$$A_n = \frac{16Qa^2}{k\pi^3(2n+1)^3}\frac{1}{\cosh\left(\frac{\pi(2n+1)}{2a}b\right)} \tag{3.148}$$

The final solution is:

$$u(x,y) = \frac{16Qa^2}{k\pi^3}\sum_{n=0}^{\infty}\frac{1}{(2n+1)^3}\left[\frac{\cosh\left(\frac{(2n+1)(b-y)\pi}{2a}\right)}{\cosh\left(\frac{(2n+1)b\pi}{2a}\right)} - 1\right] \times \sin\left(\frac{\pi(2n+1)}{2a}x\right) \tag{3.149}$$

In the figure we represent the numeric solution of the problem obtained with MATLAB'S PDE Toolbox module.

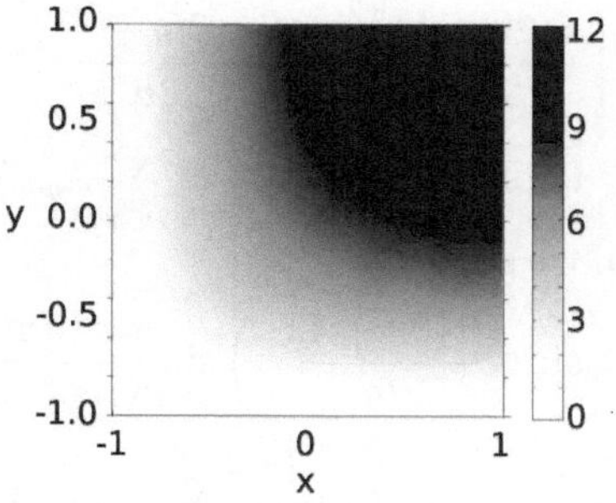

Figure 3.14

Method 2

The alternative method consists in expanding the solution in orthogonal functions in two dimensions:

$$u(x, y) = \sum_{n,m} A_{nm} X_n(x) Y_m(y) \tag{3.150}$$

Although it is mathematically possible, this method which uses the homogeneity of the boundaries, would result in a higher number of Fourier series summations and therefore more Gibbs phenomena (that is, the divergence of the solution next to the boundaries, if there is an increasing contribution of harmonics for higher eigenvalues).

Sturm–Liouville problem

To find $X_n(x)Y_m(y)$ we solve the Sturm-Liouville problem in two dimensions.

$$\left\{ \begin{array}{l} \dfrac{\partial^2 v}{\partial x^2} + \dfrac{\partial^2 v}{\partial y^2} = -\lambda_{nm} v \\ \\ v(x=0) = \left.\dfrac{\partial v}{\partial x}\right|_{x=a} = 0 \\ v(y=0) = \left.\dfrac{\partial v}{\partial y}\right|_{y=b} = 0 \end{array} \right\} \tag{3.151}$$

$$X_n(x)Y_m(y) = \sin\left(\frac{\pi(2n+1)}{2a} x\right) \sin\left(\frac{\pi(2m+1)}{2b} y\right) \tag{3.152}$$

$$\lambda_{nm} = \left[\frac{\pi(2n+1)}{2a}\right]^2 + \left[\frac{\pi(2m+1)}{2b}\right]^2 \tag{3.153}$$

General solution Replacing $u(x, y) = \sum\limits_{n,m} A_{nm}X_n(x)Y_m(y)$ in the Poisson's equation we get:

$$= -\sum_{n,m} A_{nm}\lambda_{nm} \sin\left(\frac{\pi(2n+1)}{2a}x\right) \sin\left(\frac{\pi(2m+1)}{2b}y\right) = -\frac{Q}{k} \tag{3.154}$$

Multiplying both sides by $\sin\left(\frac{\pi(2n+1)}{2a}x\right) \sin\left(\frac{\pi(2m+1)}{2b}y\right)$

and integrating between $\int\limits_0^a \int\limits_0^b dxdy$ gives:

$$A_{nm}\lambda_{nm} \left\|\sin\left(\frac{\pi(2n+1)}{2a}x\right)\right\|^2 \cdot \left\|\sin\left(\frac{\pi(2m+1)}{2b}y\right)\right\|^2 = \frac{Q}{k}\int_0^a\int_0^b \sin\left(\frac{\pi(2n+1)}{2a}x\right) \sin\left(\frac{\pi(2m+1)}{2b}y\right) dxdy \tag{3.155}$$

$$A_{nm} = \frac{4Q}{abk\lambda_{nm}}\int_0^a\int_0^b sin\left(\frac{\pi(2n+1)}{2a}x\right) \sin\left(\frac{\pi(2m+1)}{2b}y\right) dxdy \tag{3.156}$$

The Poisson problem previously solved could be part of another, more complicated, problem related to the variation of temperature in the same element: This could be the problem, and the method to solve it, which would use the previous results: the two boundaries (AC and BC) of a homogeneous rectangular film $ABCD$ are thermally insulated. The other two are kept at a temperature T_0. Find the distribution of temperature of that element as a function of time supposing that the initial temperature (before $t = 0$) is $T = T_0$, and that starting at $t = 0$ the film loses heat homogeneously at a constant pace Q=constant (by unit time and unit surface).

Note: As before, the solution $u(x, y, t)$ will be formulated with respect to the temperature of the surroundings $T = T_0$. The solution will be sought as the sum of the stationary solution, $w(x, y)$ at times $t \to \infty$ and the transient solution $v(x, y, t)$.

Idea for a solution:

$$\left\{\begin{array}{l} C\rho\dfrac{\partial u}{\partial t} - k\left(\dfrac{\partial^2 u}{\partial x^2} + \dfrac{\partial^2 u}{\partial y^2}\right) = -Q \\ u(x=0) = \left.\dfrac{\partial u}{\partial x}\right|_{x=a} = 0 \\ u(y=0) = \left.\dfrac{\partial u}{\partial y}\right|_{y=b} = 0 \end{array}\right\} \tag{3.157}$$

Replacing $u(x, y, t) = v(x, y, t) + w(x, y)$ in the previous problem we separate the problem in two:

$$\text{Problem 1:}\left\{\begin{array}{l} C\rho\dfrac{\partial v}{\partial t} - k\left(\dfrac{\partial^2 v}{\partial x^2} + \dfrac{\partial^2 v}{\partial y^2}\right) = 0 \\ v(x=0) = \left.\dfrac{\partial v}{\partial x}\right|_{x=a} = 0 \\ v(y=0) = \left.\dfrac{\partial v}{\partial y}\right|_{y=b} = 0 \\ v(x, y, t=0) = -w(x, y) \end{array}\right\} \tag{3.158}$$

$$\text{Problem 2:}\left\{\begin{array}{l} -k\left(\dfrac{\partial^2 w}{\partial x^2} + \dfrac{\partial^2 w}{\partial y^2}\right) = -Q \\ w(0, y) = \left.\dfrac{\partial w}{\partial x}\right|_{x=a} = 0 \\ w(x, 0) = \left.\dfrac{\partial w}{\partial y}\right|_{y=b} = 0 \end{array}\right\} \tag{3.159}$$

Final solution

Problem 2 has already been solved. The solution of problem 1 has the form of an expansion of orthogonal eigenfunctions in two dimensions:

$$v(x, y, t) = \sum_{n,m} A_{nm} e^{\left(-\lambda_{nm}\frac{k}{C\rho}\right)t} \sin\left(\frac{\pi(2n+1)}{2a}x\right) \sin\left(\frac{\pi(2m+1)}{2b}y\right) \tag{3.160}$$

The coefficients A_{nm} are found by applying the initial conditions.

3.11 Case Study: Charged Filament inside a Prism

A metallic tube has a square cross section $L \times L$ and infinite length. In the interior of the tube there is a charged filament, also of infinite length, with a linear density of charge Q. Find the electrostatic potential inside the tube.

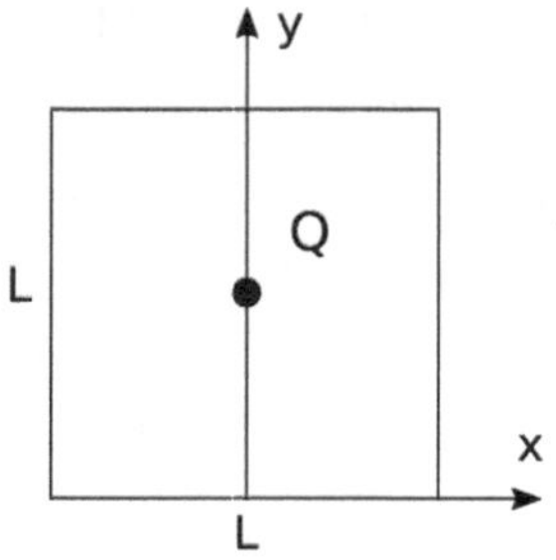

Figure 3.15

Method 1

Mathematical formulation

$$\left\{ \begin{array}{l} \dfrac{\partial^2 u}{\partial x^2} + \dfrac{\partial^2 u}{\partial y^2} = f(x, y) = -\dfrac{Q}{\epsilon_0}\delta(x)\delta(y - L/2) \\ u(-L/2, y) = u(+L/2, y) = u(x, 0) = u(x, L) = 0 \end{array} \right\} \quad (3.161)$$

being ε_0 the vacuum electric permittivity.

The presence of homogeneous boundary conditions allows to expand the solution in orthogonal eigenfunctions.

Note: since in this particular case we want to expand the solution in a summation of orthogonal functions in the vertical direction, it is more convenient to fix the position of the y axis over the base of the rectangle.

Sturm–Liouville problem

In the present case, to minimize the influence of the Gibbs phenomenon we are going to seek the solution as an expansion in eigenfunctions of the homogeneous Sturm–Liouville problem in y.

$$u(x, y) = \sum_n X_n(x) \sin\left(\frac{\pi n y}{L}\right) \tag{3.162}$$

being $X_n(x)$ the coefficients of the expansion which depend on the x coordinate.

General solution

Replacing this expression in equation (3.161) we get:

$$\sum_n \left[\frac{d^2 X_n}{dx^2} - \lambda_n X_n\right] \sin\left(\frac{\pi n y}{L}\right) = -\frac{Q}{\epsilon_0}\delta(x)\delta(y - L/2) \tag{3.163}$$

with $\lambda_n = (\frac{\pi n}{L})^2$. Using the orthogonality properties of the eigenfunctions $\sin\left(\frac{\pi n y}{L}\right)$ we arrive at the inhomogeneous equation for $X_n(x)$, with the corresponding boundary conditions:

$$\left\{\begin{array}{l} \dfrac{\partial^2 X_n}{\partial x^2} - \lambda_n X_n = f_n = \left\{\begin{array}{l} -\dfrac{2Q}{L\epsilon_0}\delta(x)\sin\left(\dfrac{\pi n}{2}\right) \to n = 2k+1 \\ 0 \to n = 2k \end{array}\right\} \\ \\ X_n(-L/2) = 0 \\ X_n(+L/2) = 0 \end{array}\right\} \tag{3.164}$$

Final solution

In the zone of the system corresponding to a homogeneous equation (that is, outside the special point describing the point charge) we get the following solutions:

$$X_n^+ = C_+ \sinh\left(\frac{\pi n}{L}\left(\frac{L}{2} - x\right)\right) \tag{3.165}$$

$$X_n^- = C_- \sinh\left(\frac{\pi n}{L}\left(\frac{L}{2} + x\right)\right) \tag{3.166}$$

From the symmetry conditions of the solution, we get: $C_+ = C_-$.

On the other hand, integrating equation (3.164) around the charge Q (in an environment of width $\epsilon \to 0$) we will get an expression for the difference of the derivatives of the solution:

$$\left.\frac{dX^{+}}{dx}\right|_{+\epsilon} - \left.\frac{dX^{-}}{dx}\right|_{-\epsilon} = \frac{2}{L}\frac{Q}{\epsilon_0}(-1)^{(2k+1)} \to \tag{3.167}$$

$$(-2C_{+})\frac{\pi(2k+1)}{L}\cosh\left(\frac{\pi(2k+1)}{2}\right) = \frac{2}{L}\frac{Q}{\epsilon_0}(-1)^{(2k+1)} \tag{3.168}$$

$$C_{+} = C_{-} = -\frac{Q}{\epsilon_0}\frac{(-1)^{(2k+1)}}{\pi(2k+1)\cosh\left(\frac{\pi(2k+1)}{2}\right)} \tag{3.169}$$

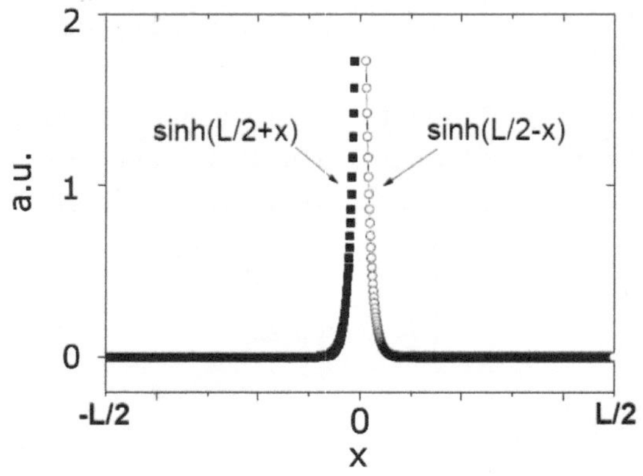

Figure 3.16

Then the solution is:

$$u(x,y) = \sum_{k=0}^{\infty}\left\{\begin{array}{ll} C_{+}\sinh\left[\left(\frac{\pi(2k+1)}{L}\right)\cdot\left(\frac{L}{2}-x\right)\right] & x>0 \\ C_{-}\sinh\left[\left(\frac{\pi(2k+1)}{L}\right)\cdot\left(\frac{L}{2}+x\right)\right] & x<0 \end{array}\right\}$$

$$\times \sin\left(\frac{\pi(2k+1)}{L}y\right) \tag{3.170}$$

The figure shows the solutions for $X_n(x)$.

Method 2

In this particular case, the idea is to develop the solution separating it in a summation of orthogonal functions both in vertical and horizontal directions. Then it will be more convenient to fix the position of the axes on the sides of the rectangle.

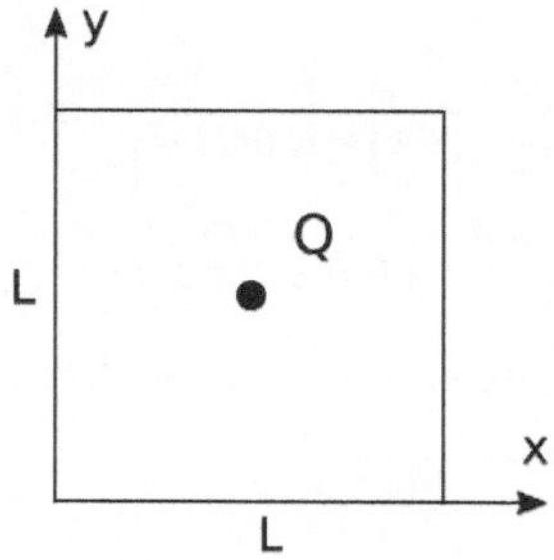

Figure 3.17

Passing the axes x, y through the sides of the rectangle, the corresponding Poisson's problem will be:

$$\left\{ \begin{array}{l} \dfrac{\partial^2 u}{\partial x^2} + \dfrac{\partial^2 u}{\partial y^2} = f(x, y) = -\dfrac{Q}{\epsilon_0}\delta(x - L/2)\delta(y - L/2) \\ u(0, y) = u(L, y) = u(x, 0) = u(x, L) = 0 \end{array} \right\} \quad (3.171)$$

General solution

We seek the solution in the form:

$$u(x, y) = \sum_n A_{nm} \sin\left(\frac{\pi n}{L}x\right) \sin\left(\frac{\pi m}{L}y\right) \quad (3.172)$$

Replacing in Poisson's equation:

$$\frac{\partial^2 u}{\partial x^2} + \frac{\partial^2 u}{\partial y^2} = -\frac{Q}{\epsilon_0}\delta(x - L/2)\delta(y - L/2) \quad (3.173)$$

$$-\sum_{nm} A_{nm} \left[\left(\frac{\pi n}{L}\right)^2 + \left(\frac{\pi m}{L}\right)^2\right] \sin\left(\frac{\pi n}{L}x\right) \sin\left(\frac{\pi n}{L}y\right)$$
$$= -\frac{Q}{\epsilon_0}\delta(x - L/2)\delta(y - L/2) \quad (3.174)$$

Final solution

Using the orthogonality of the eigenfunctions $\sin(\frac{\pi nx}{L})$; $\sin(\frac{\pi my}{L})$

$$A_{nm}\left[\left(\frac{\pi n}{L}\right)^2+\left(\frac{\pi m}{L}\right)^2\right]=\frac{4}{L^2}\frac{Q}{\epsilon_0}\sin\left(\frac{\pi n}{2}\right)\sin\left(\frac{\pi m}{2}\right) \quad (3.175)$$

Solution:

$$u(x,y)=\frac{4Q}{\epsilon_0 L^2}\sum_{n,m}\frac{1}{\left[(\frac{\pi n}{L})^2+(\frac{\pi m}{L})^2\right]}\sin\left(\frac{\pi n}{2}\right)\sin\left(\frac{\pi m}{2}\right)$$
$$\times\sin\left(\frac{\pi n}{L}x\right)\sin\left(\frac{\pi m}{L}y\right) \quad (3.176)$$

Only the odd terms of the summation, with $n=2k+1; m=2f+1$, will remain.

3.12 Case Study: Capacity in a Rectangular Tube

Two very thin parallel plates with density of charge Q_0 and $-Q_0$ are situated inside a metallic, rectangular box, infinite in the z direction. The faces at $x = 0$ and $x = a$ are grounded (zero potential), whereas the other two are metallic with a surface charge density σ. Find the distribution of electrostatic potential inside the box. Note: the solution will be independent of z due to the symmetry in that direction.

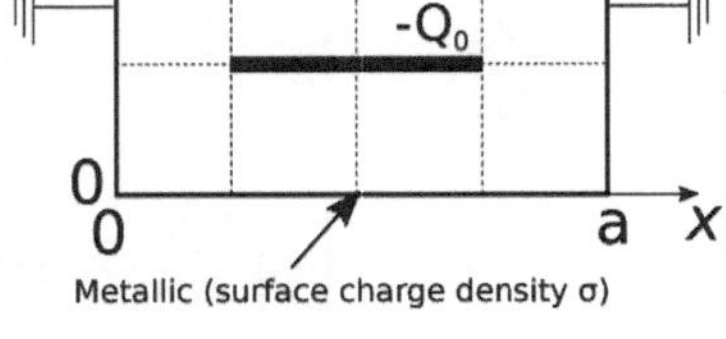

Figure 3.18

Mathematical formulation

There is no dependence on the z variable due to the symmetry.

$$\left\{\begin{array}{l} \Delta u(x, y, z) = f(x, y, z) \\ u(0, y) = u(a, y) = 0 \\ \left.\dfrac{\partial u}{\partial y}\right|_{y=0} = -\dfrac{\sigma}{\varepsilon_0} \\ \left.\dfrac{\partial u}{\partial y}\right|_{y=b} = \dfrac{\sigma}{\varepsilon_0} \end{array}\right\} \tag{3.177}$$

with the inhomogeneity of the equation expressed in the following manner:

$$\left\{\begin{array}{l} f(x, y, z) = g(x)h(y) \\ g(x) = \left\{\begin{array}{ll} 0 & (0 \leq x \leq \frac{a}{4}) \\ -\dfrac{Q_0}{\varepsilon_0} & \left(\dfrac{a}{4} \leq x \leq \dfrac{3a}{4}\right) \\ 0 & \left(\dfrac{3a}{4} \leq x \leq 1\right) \end{array}\right\} \\ h(y) = \delta\left(y - \dfrac{3b}{4}\right) - \delta\left(y - \dfrac{b}{4}\right) \end{array}\right\} \tag{3.178}$$

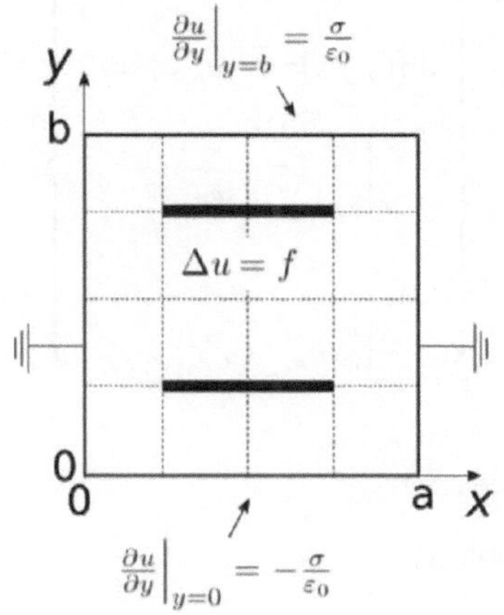

Figure 3.19

As this is a non-homogeneous problem with non-homogeneous boundary conditions, we will separate the problem in the sum of two, which consider independently the non-homogeneous equation $v(x, y)$ and the homogeneous boundary conditions $w(x, y)$.

$$u(x, y) = v(x, y) + w(x, y) \tag{3.179}$$

The first problem is non-homogeneous with homogeneous boundary conditions.

$$\left\{\begin{array}{l} \Delta v(x, y) = f(x, y) \\ v(0, y) = v(a, y) = 0 \\ \left.\dfrac{\partial v}{\partial y}\right|_{y=0} = 0 \\ \left.\dfrac{\partial v}{\partial y}\right|_{y=b} = 0 \end{array}\right\} \tag{3.180}$$

The second problem is homogeneous with non-homogeneous boundary conditions.

$$\left\{\begin{array}{l} \Delta w(x, y) = 0 \\ w(0, y) = w(a, y) = 0 \\ \left.\dfrac{\partial w}{\partial y}\right|_{y=0} = -\dfrac{\sigma}{\varepsilon_0} \\ \left.\dfrac{\partial w}{\partial y}\right|_{y=b} = \dfrac{\sigma}{\varepsilon_0} \end{array}\right\} \tag{3.181}$$

Solution of problem 1

Sturm–Liouville problem

The solution is sought by expanding into eigenfunctions of the Sturm–Liouville problem:

$$\left\{\begin{array}{l} \Delta v(x,y) + \lambda v(x,y) = 0 \\ v(0,y) = v(a,y) = 0 \\ \left.\dfrac{\partial v}{\partial y}\right|_{y=0} = 0 \\ \left.\dfrac{\partial v}{\partial y}\right|_{y=b} = 0 \end{array}\right\} \tag{3.182}$$

The eigenfunctions and eigenvalues are well known:

$$v(x,y) = \sin\left(\frac{\pi n}{a}x\right)\cos\left(\frac{\pi m}{b}y\right) \tag{3.183}$$

$$\lambda_{nm} = \left(\frac{\pi n}{a}\right)^2 + \left(\frac{\pi m}{b}\right)^2 \tag{3.184}$$

General solution

The solution of the non-homogeneous problem 3.180 will be looked for in the form:

$$v(x,y) = \sum_{n,m} A_{nm}\sin\left(\frac{\pi n}{a}x\right)\cos\left(\frac{\pi m}{b}y\right) \tag{3.185}$$

Replacing in the equation (3.180)

$$-\sum_{n,m} \lambda_{nm}A_{nm}\sin\left(\frac{\pi n}{a}x\right)\cos\left(\frac{\pi m}{b}y\right) = f(x,y) \tag{3.186}$$

Multiplying both sides by $\sin\left(\frac{\pi n}{a}x\right)\cos\left(\frac{\pi m}{b}y\right)$ and integrating in $\int_0^a dx \int_0^b dy$ we have:

$$\lambda_{nm}A_{nm}\int_0^a \sin^2\left(\frac{\pi n}{a}x\right)dx\int_0^b \cos^2\left(\frac{\pi m}{b}y\right)dy$$

$$= \int_0^a g(x)\sin\left(\frac{\pi n}{a}x\right)dx\int_0^b h(y)\cos\left(\frac{\pi m}{b}y\right)dy \tag{3.187}$$

For $m \geq 1$:

$$\int_0^a \sin^2\left(\frac{\pi n}{a}x\right)dx = \frac{a}{2} \qquad \int_0^b \cos^2\left(\frac{\pi m}{b}y\right)dy = \frac{b}{2} \tag{3.188}$$

$$A_{nm} = \frac{\int\limits_0^a g(x)dx \int\limits_0^b h(y)dy}{\lambda_{nm}(\frac{ab}{4})}$$

$$= -\frac{Q_0}{\varepsilon_0} \frac{\int\limits_{\frac{a}{4}}^{\frac{3a}{4}} \sin\left(\frac{\pi n}{a}x\right) dx \left\{\cos\left(\frac{\pi m}{b}\frac{3b}{4}\right) - \cos\left(\frac{\pi m}{b}\frac{b}{4}\right)\right\}}{\lambda_{nm}(\frac{ab}{4})}$$

$$= \frac{Q_0}{\varepsilon_0} \frac{4}{ab \cdot \lambda_{nm}} \frac{a}{\pi n} \left[\cos\left(\frac{\pi m}{4}\right) - \cos\left(\frac{3\pi m}{4}\right)\right]$$

$$\times \left[\cos\left(\frac{\pi n}{4}\right) - \cos\left(\frac{3\pi n}{4}\right)\right] \tag{3.189}$$

Take into account that for $m = 0$, the modulus and the coefficients A_{nm} will differ in a factor 2 due to:

$$\int\limits_0^b \cos^2\left(\frac{\pi m}{b}y\right) dy = b \tag{3.190}$$

Solution of problem 2

General solution

We seek the solution by expanding into orthogonal eigenfunctions in the x direction, since the problem has homogeneous boundary conditions:

$$w(x, y) = \sum_n Y_n(y) \sin\left(\frac{\pi n}{a}x\right) \tag{3.191}$$

This summation is replaced in the equation: $\Delta w(x, z) = 0$. We arrive at the following equation for $Y_n(y)$:

$$\frac{d^2 Y_n}{dy^2} - \left(\frac{\pi n}{a}\right)^2 Y_n(y) = 0 \tag{3.192}$$

With the general solution:

$$Y_n(y) = C_n \sinh\left(\frac{\pi n}{a}y\right) + D_n \cosh\left(\frac{\pi n}{a}y\right) \tag{3.193}$$

We apply the boundary conditions for $y = 0$:

$$\sum_n \left[C_n \frac{\pi n}{a}\cosh\left(\frac{\pi n}{a}0\right) - D_n \frac{\pi n}{a}\sinh\left(\frac{\pi n}{a}0\right)\right]\sin\left(\frac{\pi n}{a}x\right) = -\frac{\sigma}{\varepsilon_0} \tag{3.194}$$

$$\sum_n C_n \frac{\pi n}{a}\sin\left(\frac{\pi n}{a}x\right) = -\frac{\sigma}{\varepsilon_0} \tag{3.195}$$

Using the orthogonality of the eigenfunctions $\sin\left(\frac{\pi n}{a}x\right)$ we find one of the coefficients:

$$\begin{aligned}
C_n &= -\frac{1}{\frac{\pi n}{a}\int\limits_0^a \sin^2\left(\frac{\pi n}{a}x\right)dx}\frac{\sigma}{\varepsilon_0}\int\limits_0^a \sin\left(\frac{\pi n}{a}x\right)dx \\
&= -\frac{1}{\frac{\pi n}{a}\frac{a}{2}}\frac{\sigma}{\varepsilon_0}\int\limits_0^a \sin\left(\frac{\pi n}{a}x\right)dx \\
&= \frac{2}{\pi n}\frac{\sigma}{\varepsilon_0}\frac{a}{\pi n}[\cos(\pi n) - 1] = \frac{2a}{(\pi n)^2}\frac{\sigma}{\varepsilon_0}[(-1)^n - 1]
\end{aligned} \tag{3.196}$$

Then:

$$C_n = \frac{2a}{(\pi n)^2}\frac{\sigma}{\varepsilon_0}[(-1)^n - 1] \tag{3.197}$$

The other boundary condition for $y = b$ is applied:

$$\sum_n \left[C_n \frac{\pi n}{a}\cosh\left(\frac{\pi n}{a}b\right) - D_n \frac{\pi n}{a}\sinh\left(\frac{\pi n}{a}b\right)\right]\sin\left(\frac{\pi n}{a}x\right) = \frac{\sigma}{\varepsilon_0} \tag{3.198}$$

Using the orthogonality for the eigenfunctions $\sin\left(\frac{\pi n}{a}x\right)$

$$C_n \frac{\pi n}{a}\cosh\left(\frac{\pi n}{a}b\right) - D_n \frac{\pi n}{a}\sinh\left(\frac{\pi n}{a}b\right) = \left(\frac{2}{a}\right)\frac{\sigma}{\varepsilon_0}\int\limits_0^a \sin\left(\frac{\pi n}{a}x\right)dx \tag{3.199}$$

$$C_n \cosh\left(\frac{\pi n}{a}b\right) - D_n \sinh\left(\frac{\pi n}{a}b\right) = \frac{2a}{(\pi n)^2}\frac{\sigma}{\varepsilon_0}[1 - (-1)^n] = -C_n \tag{3.200}$$

Then:

$$D_n = C_n \frac{[1 + \cosh(\frac{\pi n}{a}b)]}{\sinh(\frac{\pi n}{a}b)} = \frac{2a}{(\pi n)^2}\frac{\sigma}{\varepsilon_0}[(-1)^n - 1]\frac{[1 + \cosh(\frac{\pi n}{a}b)]}{\sinh(\frac{\pi n}{a}b)} \tag{3.201}$$

3.13 Temperature Distribution inside a Box Heated by Two Transistors

Find the stationary distribution of temperature $T(x, y)$ inside a box with thermal conductivity k. The box is infinite in the z direction and semi-infinite in the $y > 0$ direction. The face at $x = a$ is in contact with a thermal reservoir at $T = 0$. The face at $x = 0$ is thermally insulated. The face at $y = 0$ receives a heat flux from the exterior according to what is shown in the figure (A is the heat flux density).

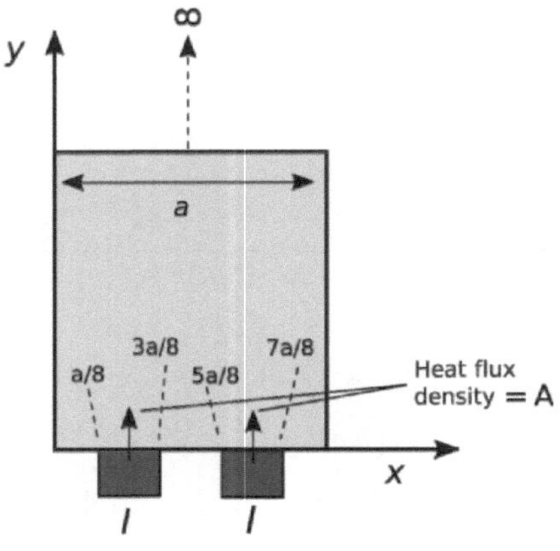

Figure 3.20

Solution

We need to solve Laplace's equation in two dimensions with the following boundary conditions:

$$\begin{cases} \left.\dfrac{\partial u}{\partial x}\right|_{x=0} = 0; \qquad u(x = a) = 0; \\ -k \left.\dfrac{\partial u}{\partial y}\right|_{y=0} = \left\{\begin{array}{ll} 0 & (0 < x < a/8) \\ A & (a/8 < x < 3a/8) \\ 0 & (3a/8 < x < 5a/8) \\ A & (5a/8 < x < 7a/8) \\ 0 & (7a/8a < x < a) \end{array}\right\} = Q(x) \\ u(y = +\infty) = 0) \end{cases} \tag{3.202}$$

Mathematical formulation:

$$\begin{cases} \dfrac{\partial^2 u}{\partial x^2} + \dfrac{\partial^2 u}{\partial y^2} = 0 \\ u(x = a) = \left.\dfrac{\partial u}{\partial x}\right|_{x=0} = 0 \\ \left.\dfrac{\partial u}{\partial y}\right|_{y=0} = -\dfrac{Q(x)}{k}; \\ u(y = +\infty) = 0 \end{cases} \tag{3.203}$$

Sturm–Liouville problem

We first seek the general solution of the problem.

$$\begin{cases} \dfrac{\partial^2 u}{\partial x^2} + \dfrac{\partial^2 u}{\partial y^2} = 0 \\ \left.\dfrac{\partial u}{\partial x}\right|_{x=0} = u(x = a) = 0 \end{cases} \tag{3.204}$$

Separating variables:

$$u(x, y) = X(x)Y(y) \tag{3.205}$$

We arrive at two second order differential equations.

We have the following Sturm–Liouville problem for the x direction:

$$\begin{cases} \dfrac{d^2 X}{dx^2} + \lambda X = 0 \\ \left.\dfrac{dX}{dx}\right|_{x=0} = X(a) = 0 \\ X_n(x) = \cos\left(\dfrac{[2n+1]\pi}{2a}x\right) \qquad n = 0, 1, 2, \ldots \end{cases} \tag{3.206}$$

For the y direction we have:

$$\frac{d^2 Y}{dy^2} - \lambda Y = 0 \tag{3.207}$$

With a solution:

$$Y_n(y) = A_n e^{\left(-\frac{[2n+1]\pi}{2a}y\right)} + B_n e^{\left(+\frac{[2n+1]\pi}{2a}y\right)} \qquad n = 0, 1, 2, \ldots \tag{3.208}$$

General solution

Then the general solution could be sought as an expansion in orthogonal function:

$$u(x,y) = \sum \left[A_n e^{\left(-\frac{[2n+1]\pi}{2a} y\right)} + B_n e^{\left(\frac{[2n+1]\pi}{2a} y\right)} \right] \times \cos\left(\frac{[2n+1]\pi}{2a} x\right) \qquad (n = 0, 1, 2, \ldots) \quad (3.209)$$

Final solution

From the condition $u(x, +\infty) = 0$

We arrive at $B_n = 0$ with $(n = 0, 1, 2, \ldots)$.

Imposing the boundary conditions:

$$\left.\frac{\partial u}{\partial y}\right|_{y=0} = \sum_n \left(-\frac{[2n+1]\pi}{2a}\right) A_n e^{\left(-\frac{[2n+1]\pi}{2a} 0\right)} \cos\left(\frac{[2n+1]\pi}{2a} x\right) = -\frac{Q(x)}{k} \quad (3.210)$$

Using the orthogonality of the $\cos\left(\frac{[2n+1]\pi}{2a} x\right)$ eigenfunctions:

We find the value of the A_n coefficients:

$$-\frac{[2n+1]\pi}{2a} A_n \int_0^a \left[\cos\left(\frac{[2n+1]\pi}{2a} x\right)\right]^2 dx = -\frac{1}{k} \int_0^a Q(x) \cos\left(\frac{[2n+1]\pi}{2a} x\right) dx \quad (3.211)$$

$$A_n = \frac{4}{k[2n+1]\pi} \int_0^a Q(x) \cos\left(\frac{[2n+1]\pi}{2a} x\right) dx \quad (3.212)$$

Finally the solution is:

$$u(x,y) = \sum A_n e^{\left(-\frac{[2n+1]\pi}{2a} y\right)} \cos\left(\frac{[2n+1]\pi}{2a} x\right) \qquad (n = 0, 1, 2, \ldots) \quad (3.213)$$

Chapter 4

Three-Dimensional Problems

This chapter expands to three dimensions the Fourier method introduced in the previous chapters. Just like before, when it is reasonable, we will use the same stages to standardize the solution process of wave, diffusion, Laplace and Poisson problems. In order to expand the solution in orthogonal functions one should choose correctly the directions in which the oppositely situated interfaces have homogeneous boundary conditions.

4.1 Stationary Temperature Distribution inside a Prism with a Thin Heater in One of Its Faces

Find the stationary distribution of temperature inside a prism (of dimensions a, b, c) with a thermal conductivity coefficient $k = 1$. The face $x = 0$ generates a power $W = I^2R$ [J/s] due to heat sources distributed as a very thin homogeneous heater (for example, a resistor R which carries an electric current I). The heater is placed inside the prism, very close to the surface (at a depth $\varepsilon \to 0$). All the heat generated is distributed towards the inside of the prism. The faces $z = 0$ and $z = c$ are kept at a temperature $T = 0$. All the remaining faces are thermally insulated.

Mathematical Methods for Physics: Problems and Solutions
Farkhad G. Aliev and Antonio Lara

ISBN 978-981-4968-71-3 (Hardcover), 978-1-003-41088-1 (eBook)
www.jennystanford.com

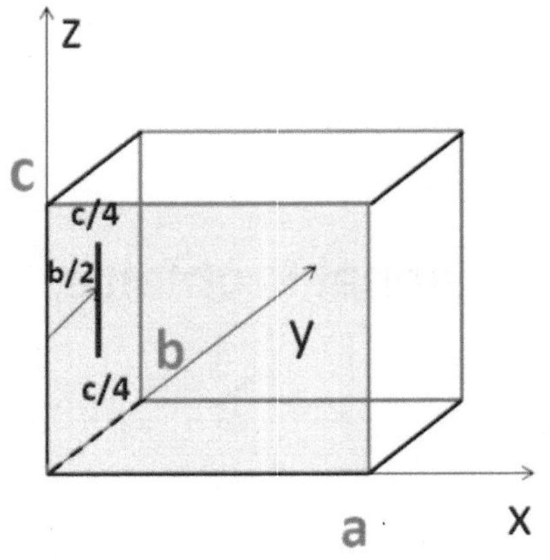

Figure 4.1

Mathematical formulation

Since the heat sources are distributed in one of the faces, we solve Laplace's equation in a prism in which five of the six faces have homogeneous boundary conditions.

$$\left\{\begin{array}{l} \Delta u(x,y,z)=0 \\ u(z=0)=u(z=c)=0 \\ \left.\dfrac{\partial u}{\partial x}\right|_{x=a}=\left.\dfrac{\partial u}{\partial y}\right|_{y=0}=\left.\dfrac{\partial u}{\partial y}\right|_{y=b}=0 \\ -k\left.\dfrac{\partial u}{\partial x}\right|_{x=0}=f(y,z)=\left\{\begin{array}{cc} 0 & 0<z<\dfrac{c}{4} \\ Q\delta\left(y-\dfrac{b}{2}\right) & \dfrac{c}{4}<z<\dfrac{3c}{4} \\ 0 & \dfrac{3c}{4}<z<c \end{array}\right\} \end{array}\right\} \tag{4.1}$$

The value of the Q constant is determined using the condition that the integral of the density of flux $f(y,z)$ through all the surface of the face $(0,y,z)$ equals the total emitted flux W. Applying this normalization condition, we have $Q=2W/c$

Sturm–Liouville problem

The solution is an expansion of orthogonal eigenfunctions, corresponding to the planes with x constant, for which the boundary conditions are homogeneous. In the $y-z$ plane this Sturm–Liouville problem will be solved:

$$\left\{\begin{array}{l} \Delta v(y,z)+\lambda v(y,z)=0 \\ v(z=0)=v(z=c)=0 \\ \left.\dfrac{\partial v}{\partial y}\right|_{y=0}=\left.\dfrac{\partial v}{\partial y}\right|_{y=b}=0 \end{array}\right\} \tag{4.2}$$

which gives the eigenfunctions:

$$v_{nm}=\cos\left(\frac{\pi n}{b}y\right)\sin\left(\frac{\pi m}{c}z\right) \tag{4.3}$$

and the eigenvalues

$$\lambda_{nm}=\left(\frac{\pi}{L}\right)^2[n^2+m^2] \tag{4.4}$$

General solution The general solution can be expanded in the base of orthogonal functions v_{nm}. The coefficients of the expansion will depend on the x coordinate:

$$u(x,y,z)=\sum_{n,m}w_{nm}(x)v_{nm}(y,z) \tag{4.5}$$

Replacing this solution in the equation $\Delta u(x,y,z)=0$ and using the orthogonality of the eigenfunctions we get the equations for $w_{nm}(x)$:

$$\frac{d^2w_{nm}}{dx^2}-\lambda_{nm}w_{nm}=0 \tag{4.6}$$

since $\lambda_{nm}>0$ the solutions are a combination of exponential functions or, more easily, since the second boundary condition for $x=a$ is: $\left.\frac{dw_{nm}}{dx}\right|_{x=a}=0$ we can present the solution in a more compact way:

$$w_{nm}(x)=A_{nm}\cosh\left(\sqrt{\lambda_{nm}}[a-x]\right) \tag{4.7}$$

Final solution

Now we can apply the first boundary condition for $x = 0$ to find the coefficients A_{nm}:

$$\left.\frac{\partial u}{\partial x}\right|_{x=0} = \sum_{n,m} \left.\frac{dw_{nm}}{dx}\right|_{x=0} v_{nm}(y, z) = \tag{4.8}$$

$$= -\sum_{n,m} A_{nm}\sqrt{\lambda_{nm}} \sinh\left(a\sqrt{\lambda_{nm}}\right) v_{nm}(y, z) = -f(y, z)/k \tag{4.9}$$

Using the orthogonality of $v_{nm}(y, z)$

$$A_{nm}\sqrt{\lambda_{nm}} \sinh\left(a\sqrt{\lambda_{nm}}\right) = \frac{\int_0^b \int_0^c f(y, z)\cos\left(\frac{\pi n}{b}y\right)\sin\left(\frac{\pi m}{c}z\right) dydz}{\int_0^b \int_0^c \cos^2\left(\frac{\pi n}{b}y\right)\sin^2\left(\frac{\pi m}{c}z\right) dydz} \tag{4.10}$$

Using the following relations:

$$\int_{\frac{c}{4}}^{\frac{3}{4}c} \sin\left(\frac{\pi m}{c}z\right) dz$$

$$= \left\{\begin{matrix} 0 & m = 2k,\ k = 0, 1, 2, \ldots \\ -\frac{c}{\pi m}\left(\cos\frac{3\pi m}{4} - \cos\frac{\pi m}{4}\right) & m = 2k+1,\ k = 0, 1, 2, \ldots \end{matrix}\right\} \tag{4.11}$$

$$\int_0^b \delta\left(y - \frac{b}{2}\right)\cos\left(\frac{\pi n}{b}y\right) dy = \cos\left(\frac{\pi n}{2}\right)$$

$$= \left\{\begin{matrix} 0 & n = 2l+1,\ l = 0, 1, 2\ldots \\ (-1)^l & n = 2l,\ l = 0, 1, 2\ldots \end{matrix}\right\} \tag{4.12}$$

We can present the coefficients A_{lk} as:

$$A_{lk} = (-1)^l \frac{Qc}{\pi(2k+1)\sqrt{\lambda_{lk}}\sinh\left(a\sqrt{\lambda_{lk}}\right)}\frac{bc}{4} \times \left[\cos\left(\frac{3\pi(2k+1)}{4}\right) - \cos\left(\frac{\pi(2k+1)}{4}\right)\right] \tag{4.13}$$

with

$$\lambda_{lk} = \left(\frac{\pi}{L}\right)^2 \left[(2l)^2 + (2k+1)^2\right] \tag{4.14}$$

The final solution is:

$$u(x, y, z) = \sum_{l,k} A_{lk} \cosh\left(\sqrt{\lambda_{lk}}[a-x]\right) \cos\left(\frac{\pi 2l}{b} y\right) \times \sin\left(\frac{\pi(2k+1)}{c} z\right) \tag{4.15}$$

4.2 Case Study: Forced Gas Oscillations in a Prism: Case of a Homogeneous Force

A prism of square cross section ($b \times b$) and length L is open in one of its ends. One of its lateral faces (massless, i.e., with no inertia) can move. A force acts on that wall, perpendicular to it, with the value $F(t) = F_0 \sin(\omega_0 t)$, which creates periodic variations in the pressure next to it. Find the pressure of the gas next to the opposite wall, assuming no resonance conditions occur.

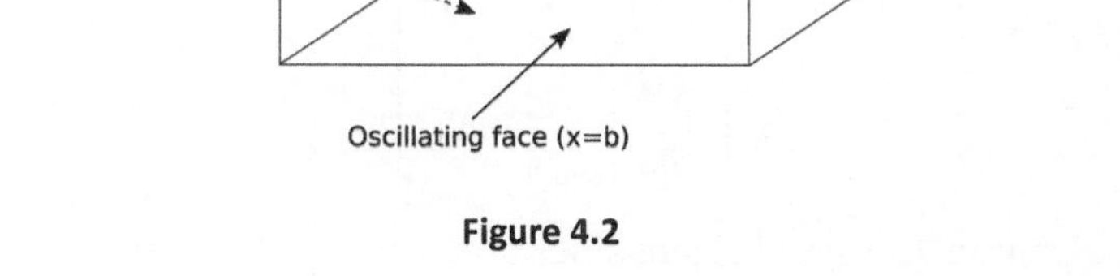

Figure 4.2

Mathematical formulation

The equation for the oscillations in a 3D gas are presented in terms of $u = P - P_0$ (pressure relative to the equilibrium pressure). The problem consists of two open boundaries, three closed and one which moves periodically:

$$\left\{\begin{array}{l} \dfrac{\partial u^2(x, y, z, t)}{\partial t^2} - a^2 \Delta u = 0 \\ \left.\dfrac{\partial u}{\partial x}\right|_{x=0} = 0 \\ u(x = b) = \dfrac{F_0}{Lb} \sin(\omega_0 t) \\ \left.\dfrac{\partial u}{\partial y}\right|_{y=0} = 0 \\ u(y = L) = 0 \\ \left.\dfrac{\partial u}{\partial z}\right|_{z=0} = \left.\dfrac{\partial u}{\partial z}\right|_{z=b} = 0 \end{array}\right\} \tag{4.16}$$

Sturm–Liouville Problem

Since the boundary conditions for (y, z) are homogeneous, the solution will consist in an expansion into orthogonal eigenfunctions. We separate variables:

$$u(x, y, z, t) = \sum Q(x, t) W(y, z) \tag{4.17}$$

Where $W(x, z)$ are the solutions of the Sturm–Liouville problem.

$$\left\{\begin{array}{l} \Delta W + \lambda W = 0 \\ \left.\dfrac{\partial W}{\partial y}\right|_{y=0} = 0 \\ W(y = L) = 0 \\ \left.\dfrac{\partial W}{\partial z}\right|_{z=0,b} = 0 \end{array}\right\} \tag{4.18}$$

The eigenfunctions and eigenvalues are:

$$W_{nm}(x, z) = \cos\left(\frac{\pi(2n+1)}{2L} y\right) \cos\left(\frac{\pi m}{b} z\right) \tag{4.19}$$

$$\lambda_{nm} = \left[\frac{\pi(2n+1)}{2L}\right]^2 + \left[\frac{\pi m}{b}\right]^2 \quad (n,\ m = 0, 1, 2, \ldots) \tag{4.20}$$

General solution

We seek a general solution in the form:

$$u(x, y, z, t) = \sum_{nm} Q_{nm}(x) \sin(\omega_0 t)\, W_{nm}(y, z) \tag{4.21}$$

This solution is replaced into the wave equation:

$$\sum_{nm} Q_{nm}(x) \frac{d^2 \sin(\omega_0 t)}{dt^2} \times W_{nm}(y, z) - a^2 \frac{d^2 Q_{nm}}{dx^2} \sin(\omega_0 t) \times W_{nm}(y, z) - a^2 Q_{nm}(x, t) \times \sin(\omega_0 t) \Delta W = 0 \tag{4.22}$$

$$\sum_{nm} \left[-\omega_0{}^2 Q_{nm}(x) - a^2 \frac{d^2 Q_{nm}}{dx^2} + a^2 \lambda_{nm} Q_{nm}(x) \right] \times W_{nm}(y, z) = 0 \tag{4.23}$$

Final solution

Using the orthogonality of $W(y, z)$ it is possible to arrive at the equation for the coefficients $Q_{nm}(x)$:

$$\frac{d^2 Q_{nm}}{dx^2} + \frac{\left[\omega_0{}^2 - a^2 \lambda_{nm}\right]}{a^2} Q_{nm}(x) = 0 \tag{4.24}$$

Applying the first boundary condition:

$$\left.\frac{\partial u}{\partial x}\right|_{x=0} \rightarrow \left.\frac{d Q_{nm}}{dx}\right|_{x=0} = 0 \tag{4.25}$$

Applying the second boundary condition:

$$u(x = b) = \sum_{nm} Q_{nm}(b) \sin(\omega_0 t) \times W_{nm}(x, z) = \frac{F_0}{Lb} \sin(\omega_0 t) \tag{4.26}$$

We use the orthogonality of $W(y, z)$ and arrive at another boundary condition for $Q_{nm}(b, t)$

$$Q_{nm}(b) = \frac{F_0}{Lb} \frac{1}{|W_{nm}(y, z)|^2} \int_0^L \int_0^b W_{nm}(y, z) dy dz = A_{nm} \tag{4.27}$$

with

$$\int_0^L\int_0^b \cos\left(\frac{\pi(2n+1)}{2L}y\right)\times\cos\left(\frac{\pi m}{b}z\right)dydz = \left\{\begin{array}{ll} 0 & (m\neq 0) \\ \dfrac{2L(-1)^n}{\pi(2n+1)} & (m=0)\end{array}\right\} \tag{4.28}$$

That is, only the coefficients $Q_{n0}(b) = A_{n0}$ have non-zero values since the profile of the eigenfunctions in the z direction has several constants due to the plain profile of the pressure on the applied to the mobile wall. Then the problem for $Q_{n0}(x,t)$ is:

$$\left\{\begin{array}{l} \dfrac{d^2Q_{n0}}{dx^2}+\dfrac{[\omega_0{}^2-a^2\lambda_{nm}]}{a^2}Q_{n0}(x)=0 \\ \left.\dfrac{dQ_{n0}}{dx}\right|_{x=0}=0 \\ Q_{n0}(b)=A_{n0}\end{array}\right\} \tag{4.29}$$

Now two different cases with two qualitatively different solutions will be considered:

Case 1

$$\omega_0^2-a^2\lambda_{n0}>0 \tag{4.30}$$

For high frequencies (depending on n) we will get oscillatory solutions:

Imposing the boundary condition $\frac{dQ_{n0}}{dx}\Big|_{x=0}=0$:

$$Q_{n0}(x)=C\cos\left(\sqrt{\frac{\omega_0^2}{a^2}-\lambda_{n0}}\,x\right) \tag{4.31}$$

Imposing the second boundary condition:

$$Q_{n0}(b)=C\cos\left(\sqrt{\frac{\omega^2}{a^2}-\lambda_{n0}}\,b\right)=A_{n0} \tag{4.32}$$

Then:

$$C=\frac{A_{n0}}{\cos\left(\sqrt{\frac{\omega^2}{a^2}-\lambda_{n0}}\,b\right)} \tag{4.33}$$

The magnitudes of the modes when $\omega_0^2>a^2\lambda_{n0}$ are:

$$Q_{n0}(x)=\frac{A_{n0}}{\cos\left(\sqrt{\frac{\omega^2}{a^2}-\lambda_{n0}}\,b\right)}\cos\left(\sqrt{\frac{\omega_0^2}{a^2}-\lambda_{n0}}\,x\right) \tag{4.34}$$

Case 2:

$$\omega_0^2 - a^2\lambda_{n0} < 0 \tag{4.35}$$

The solutions are exponentially decaying:

$$\left\{\begin{array}{l} \dfrac{d^2 Q_{n0}}{dx^2} - [\lambda_{n0} - \dfrac{\omega_0^2}{a^2}]Q_{n0}(x) = 0 \\ \left.\dfrac{dQ_{n0}}{dx}\right|_{x=0} = 0 \\ Q_{n0}(b) = A_{n0} \end{array}\right\} \tag{4.36}$$

Imposing the boundary condition $\left.\frac{dQ_{n0}}{dx}\right|_{x=0} = 0$:

$$Q_{n0}(x) = C\cosh\left(\sqrt{\lambda_{n0} - \frac{\omega_0^2}{a^2}}\,x\right) \tag{4.37}$$

Imposing the second boundary condition:

$$Q_{n0}(b) = C\cosh\left(\sqrt{\lambda_{n0} - \frac{\omega_0^2}{a^2}}\,b\right) = A_{n0} \tag{4.38}$$

Then:

$$C = \frac{A_{n0}}{\cosh\left(\sqrt{\lambda_{n0} - \frac{\omega_0^2}{a^2}}\,b\right)} \tag{4.39}$$

Solution for the amplitudes at low excitation frequencies:

$$Q_{n0}(x) = \frac{A_{n0}}{\cosh\left(\sqrt{\lambda_{n0} - \frac{\omega_0^2}{a^2}}\,b\right)}\cosh\left(\sqrt{\lambda_{n0} - \frac{\omega_0^2}{a^2}}\,x\right) \tag{4.40}$$

Finally we get to the pressure at the wall opposite to the oscillating one (this is, for $x = 0$):

$$\begin{aligned} u(0, y, z, t) &= \sin(\omega_0 t)\sum_n \frac{A_{n0}}{\cos\left(\sqrt{\frac{\omega^2}{a^2} - \lambda_{n0}}\,b\right)}\cos\left(\frac{\pi(2n+1)}{2L}y\right) \\ &[n : \omega_0^2 > a^2\lambda_{n0}] \\ &+ \sin(\omega_0 t)\sum_n \frac{A_{n0}}{\cosh\left(\sqrt{\lambda_{n0} - \frac{\omega_0^2}{a^2}}\,b\right)}\cos\left(\frac{\pi(2n+1)}{2L}y\right) \\ &[n : \omega_0^2 < a^2\lambda_{n0}] \end{aligned} \tag{4.41}$$

4.3 Case Study: Forced Gas Oscillations in a Prism: Case of an Inhomogeneous Force

A tube of square section, of side length b is closed on one of its ends. On the other there are two pistons which move with velocities $U(t)$ and $-U(t)$. Supposing that $U(t) = A\sin(\omega t)$ and that the tube length is $L >> b$, find the variation of pressure inside the tube.

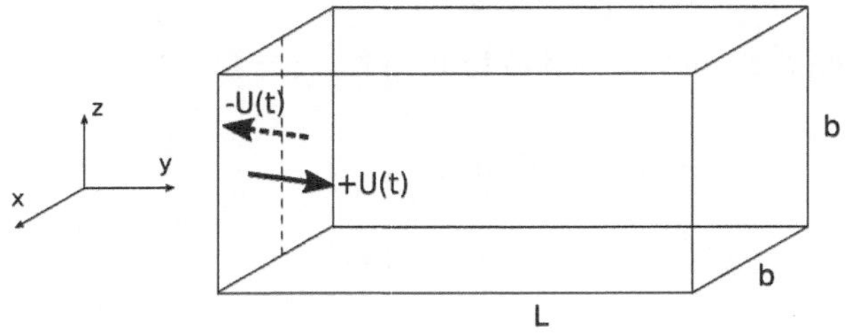

Figure 4.3

Mathematical formulation

The equation for the oscillations of a gas in 3D (in terms of the pressure $u = P - P_0$, with P_0 being the equilibrium pressure) is:

$$\frac{\partial u^2(x, y, z, t)}{\partial t^2} - a^2\Delta u = 0 \tag{4.42}$$

Formulation of the boundary conditions for the five closed boundaries:

$$\left.\frac{\partial u}{\partial x}\right|_{x=0,b} = \left.\frac{\partial u}{\partial z}\right|_{z=0,b} = \left.\frac{\partial u}{\partial y}\right|_{y=L} = 0 \tag{4.43}$$

For the forced boundary ($y = 0$) we need to use the relation between the velocity of the gas and its local pressure:

$$\rho_0\frac{\partial \mathbf{V}}{\partial t} = -\nabla P = -\nabla u(x, y, z) \tag{4.44}$$

Considering the projection in y of the equation (4.44) and the plane $y = 0$ we get the following relation:

$$\rho_0\frac{\partial V_y}{\partial t} = -\nabla_y u(x, y, z) \tag{4.45}$$

from which we get at the boundary conditions for the plane $y = 0$:

$$\left.\frac{\partial u}{\partial y}\right|_{y=0} = \left\{ \begin{array}{l} -\rho_0 A\omega \cos(\omega t)\ (0 < x < \frac{b}{2}) \\ \rho_0 A\omega \cos(\omega t)\ \ (\frac{b}{2} < x < b) \end{array} \right\} \tag{4.46}$$

It is the last boundary condition and is non-homogeneous. Furthermore, intuitively we see that the solution of the problem does not depend on z.

Sturm–Liouville problem

Since the boundary conditions for x and z are homogeneous we will expand the solution into eigenfunctions which are solutions of the Sturm–Liouville problem in those directions, using separation of variables:

$$u(x, y, z, t) = Q(y, t) \times W(x, z) \tag{4.47}$$

$$\left\{ \begin{array}{l} \Delta W + \lambda W = 0 \\ \left.\dfrac{\partial W}{\partial \vec{\mathbf{n}}}\right|_{z=0,b;x=0,b} = 0 \end{array} \right\} \tag{4.48}$$

The normalized eigenfunctions and eigenvalues are:

$$W_{nm}(x, z) = \frac{2}{b} \cos\left(\frac{\pi n}{b} x\right) \cos\left(\frac{\pi m}{b} z\right) \tag{4.49}$$

$$\lambda_{nm} = \left(\frac{\pi}{b}\right)^2 [n^2 + m^2]\ (n, m = 1, 2, \ldots) \tag{4.50}$$

Note that this type of normalization is not applicable to cases with n, $m = 0$.

General solution

$$u(x, y, z, t) = \sum_{nm} Q_{nm}(y, t) W_{nm}(x, z) \tag{4.51}$$

Replacing 4.47 into the wave equation

$$\sum_{nm} \frac{\partial^2 Q_{nm}(y, t)}{\partial t^2} W_{nm}(x, z) - a^2 \frac{\partial^2 Q_{nm}(y, t)}{\partial y^2} W_{nm}(x, z) - a^2 Q_{nm}(y, t) \Delta W = 0 \tag{4.52}$$

$$\sum_{nm} \left[\frac{\partial^2 Q_{nm}(y, t)}{\partial t^2} - a^2 \frac{\partial^2 Q_{nm}(y, t)}{\partial y^2} + a^2 \lambda_{nm} Q_{nm}(y, t) \right] W_{nm}(x, z) = 0 \tag{4.53}$$

Final solution

Using the orthogonality of $W(x, z)$ we get at the equation for the coefficients $Q_{nm}(y, t)$:

$$\frac{\partial^2 Q_{nm}(y,t)}{\partial t^2} - a^2 \frac{\partial^2 Q_{nm}(y,t)}{\partial y^2} + a^2 \lambda_{nm} Q_{nm}(y,t) = 0 \tag{4.54}$$

We need to include the boundary condition:

$$\left.\frac{\partial u}{\partial y}\right|_{y=L} = \sum_{nm} \left.\frac{Q_{nm}(y,t)}{\partial y}\right|_{y=L} W_{nm}(x,z) = 0 \tag{4.55}$$

$$\rightarrow \left.\frac{\partial Q_{nm}(y,t)}{\partial y}\right|_{y=L} = 0 \tag{4.56}$$

The boundary condition for the oscillating face is:

$$\left.\frac{\partial u}{\partial y}\right|_{y=0} = \sum_{nm} \left.\frac{\partial Q_{nm}(y,t)}{\partial y}\right|_{y=0} W_{nm}(x,z)$$
$$= \left\{ \begin{array}{ll} -\rho_0 A\omega \cos(\omega t) & \left(0 < x < \dfrac{b}{2}\right) \\ \rho_0 A\omega \cos(\omega t) & \left(\dfrac{b}{2} < x < b\right) \end{array} \right\} \tag{4.57}$$

Using the orthogonality of $W(x, z)$ we arrive at the condition:

$$\int_0^b \int_0^b \left.\frac{\partial u}{\partial y}\right|_{y=0} W_{nm}(x,z)dxdz = \left.\frac{\partial Q_{nm}(y,t)}{\partial y}\right|_{y=0} |W_{nm}(x,z)|^2$$
$$= \left.\frac{\partial Q_{nm}(y,t)}{\partial y}\right|_{y=0} \tag{4.58}$$

Then the equation for $Q_{nm}(y, t)$ is:

$$\left\{ \begin{array}{l} \dfrac{\partial^2 Q_{nm}(y,t)}{\partial t^2} - a^2 \dfrac{\partial^2 Q_{nm}(y,t)}{\partial y^2} + a^2 \lambda_{nm} Q_{nm}(y,t) = 0 \\ \left.\dfrac{\partial Q_{nm}(y,t)}{\partial y}\right|_{y=L} = 0 \\ \left.\dfrac{\partial Q_{nm}(y,t)]}{\partial y}\right|_{y=0} = \int_0^b \int_0^b \left.\dfrac{\partial u}{\partial y}\right|_{y=0} W_{nm}(x,z)dxdz \end{array} \right\} \tag{4.59}$$

Since the boundary conditions are for the derivatives, the solution would require to define a constant. We can check that when n, $m =$ 0:

$$\left.\frac{\partial Q_{00}(y,t)}{\partial y}\right|_{y=0} = \int_0^b\int_0^b \left.\frac{\partial u}{\partial y}\right|_{y=0} W_{00}(x,z)dxdz$$

$$= \frac{1}{b}\int_0^b\int_0^b \left.\frac{\partial u}{\partial y}\right|_{y=0} dxdz = 0 \qquad (4.60)$$

$\rightarrow Q_{00}(y,t) =$ const=(if the initial conditions are null) $= 0$

When n, $m \neq 0$:

$$\left.\frac{\partial Q_{nm}(y,t)}{\partial y}\right|_{y=0} = \int_0^b\int_0^b \left.\frac{\partial u}{\partial y}\right|_{y=0} W_{nm}(x,z)dxdz$$

$$= \int_0^b \left.\frac{\partial u}{\partial y}\right|_{y=0} \frac{2}{b}\cos\left(\frac{\pi n}{b}x\right)dx \int_0^b \cos\left(\frac{\pi m}{b}z\right)dz = \qquad (4.61)$$

$$= \int_0^b \left.\frac{\partial u}{\partial y}\right|_{y=0} \frac{2}{b}\cos\left(\frac{\pi n}{b}x\right)dx \times \left|\frac{b}{\pi m}\sin\left(\frac{\pi m}{b}z\right)\right|_0^b = 0 \qquad (4.62)$$

Then the modes with n and m different from zero are not excited (due to the symmetry of the problem there are no excited modes in the z direction). Finally we need to check the case with $n \neq 0, m = 0$

$$\left.\frac{\partial Q_{n0}(y,t)}{\partial y}\right|_{y=0} = \int_0^b\int_0^b \left.\frac{\partial u}{\partial y}\right|_{y=0} W_{n0}(x,z)dxdz$$

$$= \int_0^b \sqrt{\frac{2}{b}}dz \int_0^b \left.\frac{\partial u}{\partial y}\right|_{y=0} \cos\left(\frac{\pi n}{b}x\right)dx = \qquad (4.63)$$

$$= \sqrt{2}A\omega\cos(\omega t)\left[-\left(\frac{b}{\pi n}\sin(\frac{\pi n}{b}x\right)_0^{b/2} + \left(\frac{b}{\pi n}\sin\frac{\pi n}{b}x\right)_{b/2}^{b}\right]$$

$$= \frac{2\sqrt{2}b}{\pi n}A\omega\left[\sin\left(\frac{\pi n}{2}\right)\right]\cos(\omega t) = \varphi_n\cos(\omega t)$$

Then the problem for $Q_{n0}(y, t)$ is:

$$\left\{\begin{array}{l} \dfrac{\partial^2 Q_{n0}(y,t)}{\partial t^2} - a^2 \dfrac{\partial^2 Q_{n0}(y,t)}{\partial y^2} + a^2 \lambda_{nm} Q_{n0}(y,t) = 0 \\ \left.\dfrac{\partial Q_{nm}(y,t)}{\partial y}\right|_{y=L} = 0 \\ \left.\dfrac{\partial Q_{nm}(y,t)}{\partial y}\right|_{y=0} = \varphi_n \cos(\omega t) \end{array}\right\} \tag{4.64}$$

We obtain an equation analogous to that of a string attached to an elastic plane, with a free right boundary and the left boundary moving periodically, with a derivative proportional to $\cos(\omega t)$. The solution of the last problem will be sought as $Q_{n0}(y, t) = H_n(y)\cos(\omega t)$:

$$-\omega^2 H_n(y) - a^2 \frac{\partial^2 H_n(y)}{\partial y^2} + a^2 \lambda_{n0} H_n(y) = 0 \tag{4.65}$$

or

$$\frac{d^2 H_n(y)}{dy^2} + \left(\frac{\omega^2}{a^2} - \lambda_{n0}\right) H_n(y) = 0 \tag{4.66}$$

First boundary condition:

$$\left.\frac{d Q_{nm}(y,t)}{dy}\right|_{y=L} = \left.\frac{d H_n(y)}{dy}\right|_{y=L} \cos(\omega t) = 0 \tag{4.67}$$

$$\rightarrow \left.\frac{d H_n(y)}{dy}\right|_{y=L} = 0 \tag{4.68}$$

Second boundary condition:

$$\left.\frac{\partial Q_{nm}(y,t)]}{\partial y}\right|_{y=0} = \left.\frac{d H_n(y)}{dy}\right|_{y=0} \cos(\omega t) = \varphi_n \cos(\omega t) \tag{4.69}$$

$$\rightarrow \left.\frac{d H_n(y)}{dy}\right|_{y=0} = \varphi_n \tag{4.70}$$

Equation for $H_n(y)$:

$$\left\{\begin{array}{l} \dfrac{d^2 H_n(y)}{dy^2} + \left(\dfrac{\omega^2}{a^2} - \lambda_{n0}\right) H_n(y) = 0 \\ \left.\dfrac{d H_n(y)}{dy}\right|_{y=L} = 0 \\ \left.\dfrac{d H_n(y)}{dy}\right|_{y=0} = \varphi_n \end{array}\right\} \tag{4.71}$$

First case:

$$\left(\frac{\omega^2}{a^2} - \lambda_{n0}\right) > 0 \tag{4.72}$$

We get oscillating solutions. Using the first boundary condition:

$$H_n(y) = C \cos\left(\sqrt{\frac{\omega^2}{a^2} - \lambda_{n0}}\,(L - y)\right) \tag{4.73}$$

Imposing the second boundary condition:

$$\left.\frac{\partial H_n(y)}{\partial y}\right|_{y=0} = -C\sqrt{\frac{\omega^2}{a^2} - \lambda_{n0}}\,\sin\left(\sqrt{\frac{\omega^2}{a^2} - \lambda_{n0}}\,L\right) = \varphi_n \tag{4.74}$$

Then

$$C = -\frac{\varphi_n}{\sqrt{\frac{\omega^2}{a^2} - \lambda_{n0}}\,\sin\left(\sqrt{\frac{\omega^2}{a^2} - \lambda_{n0}}\,L\right)} \tag{4.75}$$

Then the magnitudes

$$Q_{n0}(y,t) = -\frac{\varphi_n \cos(\omega t)\cos\left(\sqrt{\frac{\omega^2}{a^2} - \lambda_{n0}}\,(L - y)\right)}{\sqrt{\frac{\omega^2}{a^2} - \lambda_{n0}}\,\sin\left(\sqrt{\frac{\omega^2}{a^2} - \lambda_{n0}}L\right)} \tag{4.76}$$

The final solution in this case is:

$$u(x,y,z,t) = \sum_n Q_{n0}(y,t)W_{n0}(x,z) \tag{4.77}$$

The minimum frequency of the induced sound is:

$$\omega_{\min}^2 = a^2\lambda_{10} = a^2\left(\frac{\pi}{b}\right)^2[1^2 + 0^2] = \left(\frac{\pi}{b}a\right)^2 \tag{4.78}$$

We consider the second case: $\left(\frac{\omega^2}{a^2} - \lambda_{n0}\right) < 0$ providing exponentially decreasing solutions.

$$\left\{\begin{array}{l} \dfrac{\partial^2 H_n(y)}{\partial y^2} - \left(\lambda_{n0} - \dfrac{\omega^2}{a^2}\right)H_n(y) = 0 \\ \left.\dfrac{\partial H_n(y)}{\partial y}\right|_{y=L} = 0 \\ \left.\dfrac{\partial H_n(y)}{\partial y}\right|_{y=0} = \varphi_n \end{array}\right\} \tag{4.79}$$

From the first boundary condition:

$$H_n(y) = C\cosh\left(\sqrt{\lambda_{n0} - \frac{\omega^2}{a^2}}\right)(L - y).$$

Imposing the second boundary condition:

$$\left.\frac{dH_n(y)}{dy}\right|_{y=0} = C\sqrt{\lambda_{n0} - \frac{\omega^2}{a^2}}\sinh\left(\sqrt{\lambda_{n0} - \frac{\omega^2}{a^2}}\,L\right) = \varphi_n \qquad (4.80)$$

Then:

$$C = \frac{\varphi_n}{\sqrt{\lambda_{n0} - \frac{\omega^2}{a^2}}\sinh\left(\sqrt{\lambda_{n0} - \frac{\omega^2}{a^2}}\,L\right)} \qquad (4.81)$$

So are the amplitudes obtained:

$$Q_{n0}(y, t) = \frac{\varphi_n\cos(\omega t)\cosh\sqrt{\lambda_{n0} - \frac{\omega^2}{a^2}}\,(L - y)}{\sqrt{\lambda_{n0} - \frac{\omega^2}{a^2}}\sinh\left(\sqrt{\lambda_{n0} - \frac{\omega^2}{a^2}}\,L\right)} \qquad (4.82)$$

For $\omega < \omega_{\min} = (\frac{\pi}{b}a) \Leftrightarrow u(x, y, z, t) = \sum Q_{n0}(y, t)W_{n0}(x, z)$

Final note: in the case of coincidence of the frequency of the piston with any of the frequencies of the standing waves in the perpendicular direction of propagation (in the y direction), these transversal waves will be excited, suppressing any solution in the y direction.

Mathematically this is reflected in the fact that there are no solutions in the y direction for:

$$\left(\frac{\omega^2}{a^2} = \lambda_{n0}\right) \qquad (4.83)$$

which satisfies both boundary conditions, being it a straight line with zero derivative at $y = L$ and with finite value of the derivative for $y = 0$.

4.4 Case Study: Optimization of the Size of an Atomic Bomb: Diffusion Equation in Cartesian Coordinates

Calculate the critical size needed to control the diffusion processes in a radioactive medium in the shape of a cube of side L. Consider a medium with neutron diffusion (concentration $= C$) produced by a fission chain reaction.

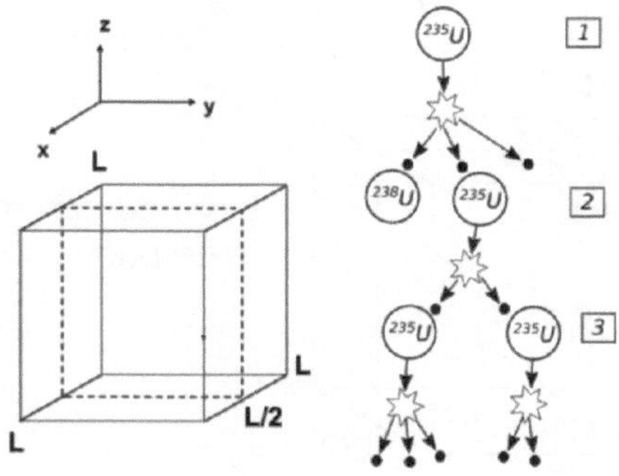

Figure 4.4

Considerations

The neutrons multiply at each location at a rate proportional to their concentration:

$$\frac{\partial C(x, y, z)}{\partial t} = \beta C(x, y, z) \tag{4.84}$$

Equation to solve:

$$\frac{\partial C(x, y, z, t)}{\partial t} - D\Delta C(x, y, z, t) = \beta C(x, y, z, t) \tag{4.85}$$

or

$$\frac{\partial C(x, y, z, t)}{\partial t} - \beta C(x, y, z, t) - D\Delta C(x, y, z, t) = 0 \tag{4.86}$$

These processes have two limit situations:

(a) If the size of the radioactive sample is small in comparison with the mean free path of the neutrons, they will escape very fast without colliding with the uranium atoms. Therefore, there won't be an effective production of new neutrons and the contribution to the diffusion ($\Delta C(x, y, z, t)$) will be much higher than the term which describes the production of neutrons due to their effective escape outside the bomb. Then in the first case of small size (with a small ratio between volume and surface):

$$\beta C(x, y, z, t) << D\Delta C(x, y, z, t) \tag{4.87}$$

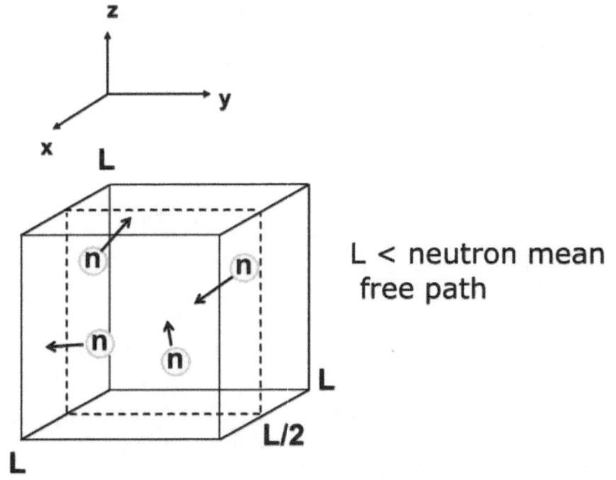

Figure 4.5

In this limit the equation to be solved would have the form of a diffusion equation:

$$\frac{\partial C(x, y, z, t)}{\partial t} - D\Delta C(x, y, z, t) = 0 \tag{4.88}$$

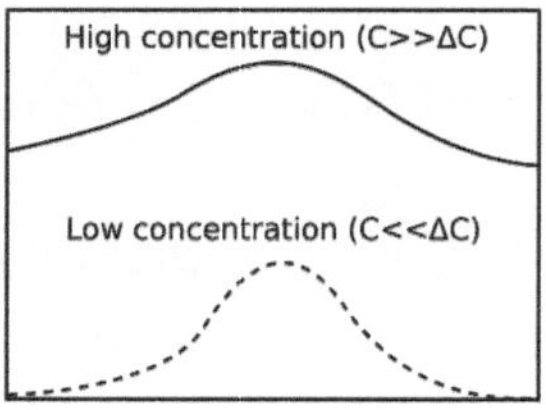

Figure 4.6

(b) If the sample is very large in comparison with the mean free path the neutrons will collide with the uranium atoms with a higher chance, which will trigger the chain reaction: The contribution of the diffusion ΔC should be lower than the production of neutrons βC.

$$\beta C(x, y, z, t) >> D\Delta C(x, y, z, t) \tag{4.89}$$

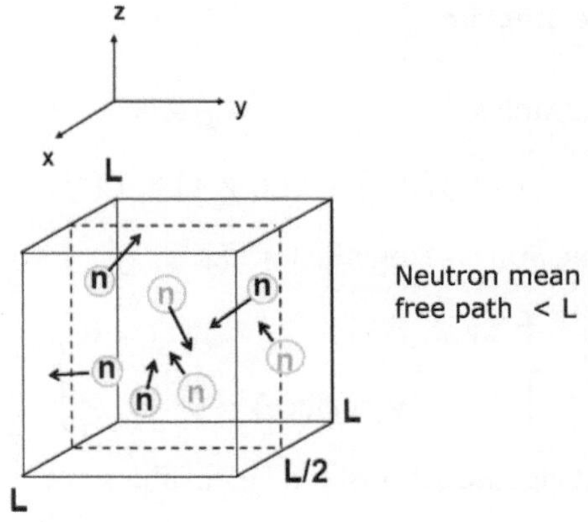

Figure 4.7

In this second limit the equation to be solved is:

$$\frac{\partial C(x, y, z, t)}{\partial t} - \beta C(x, y, z, t) = 0 \tag{4.90}$$

This equation has an exponential solution:

$$C(x, y, z, t) \sim e^{\beta t} \tag{4.91}$$

We seek the moment of the transition of the solution between the stable case and the exponential growth. For a cube of side L:

$$\left\{ \begin{array}{c} \frac{\partial C(x,y,z,t)}{\partial t} - \beta C(x, y, z, t) - D\Delta C(x, y, z, t) = 0 \\ C(\text{surface}, t) = 0 \\ C(x, y, z, 0) = \varphi(x, y, z) \end{array} \right\} \tag{4.92}$$

Notes:

(a) The null boundary conditions are used as an approximation to simplify the solution.
(b) The null boundary conditions describe the fact that the neutrons escape very fast from the surface and their concentration there is taken to be null.

Sturm–Liouville problem

We separate variables:

$$C(x, y, z, t) = v(x, y, z) \times T(t) \tag{4.93}$$

and arrive at the Sturm–Liouville for $v(x, y, z)$.

$$\begin{Bmatrix} \Delta v(x, y, z) + \lambda v(x, y, z) = 0 \\ v(\text{surface}) = 0 \end{Bmatrix} \tag{4.94}$$

the normalized eigenfunctions and eigenvalues are:

$$\begin{aligned} v_{nmk}(x, y, z) &= X_n(x) Y_m(y) Z_k(z) \\ &= \left(\frac{2}{L}\right)^{3/2} \sin\left(\frac{\pi n}{L}x\right) \sin\left(\frac{\pi m}{L}y\right) \sin\left(\frac{\pi k}{L}z\right) \end{aligned} \tag{4.95}$$

$$\lambda_{nmk} = \left(\frac{\pi}{L}\right)^2 [n^2 + m^2 + k^2] \tag{4.96}$$

being n, m, k positive integers.

General solution

We replace $C(x, y, z, t) = v_{nmk}(x, y, z) \cdot T_{nmk}(t)$ into equation (4.92). We arrive at the equation for the magnitude of the modes T_{nmk}:

$$\frac{\partial T_{nmk}(t)}{\partial t} - \beta T_{nmk}(t) + D\lambda_{nmk} T_{nmk}(t) = 0 \tag{4.97}$$

The solution is: $T_{nmk}(t) = A_{nmk} e^{(\beta - D\lambda_{nmk})t}$ The general solution of equation (4.92) is:

$$C(x, y, z, t) = \sum_{n,m,k} A_{nmk} e^{(\beta - D\lambda_{nmk})t} \times \sin\left(\frac{\pi n}{L}x\right) \sin\left(\frac{\pi m}{L}y\right) \sin\left(\frac{\pi k}{L}z\right) \tag{4.98}$$

a) Let us consider that the condition of stability of the solution: $\beta - D\lambda_{nmk} < 0$ is satisfied for every eigenvalue λ_{nmk}.

Then, the solution is totally stable if $\beta - D\lambda_{111} < 0$ is satisfied (then it will be satisfied for the rest of the λ_{nmk}).

Final solution

To summarize the first part of the solution: the condition of stability is:

$$\beta - D\lambda_{111} = \beta - 3D\left(\frac{\pi}{L}\right)^2 < 0 \tag{4.99}$$

$$L < \pi\sqrt{\frac{3D}{\beta}} = L_{cr} \tag{4.100}$$

For $L < L_{cr}$ the solution is *stable.* For $L > L_{cr}$ the solution is *unstable.*

Now we consider the stability of the half of the cube to calculate its critical size.

Eigenvalues of the solution for half of the cube:

$$\lambda_{2nmk} = \left(\frac{\pi}{L}\right)^2 [(2n)^2 + m^2 + k^2] \tag{4.101}$$

Normalized eigenfunctions:

$$v_{nmk(x,y,z)} = \left(\frac{2}{L}\right)^2 \left(\frac{4}{L}\right) \sin\left(\frac{\pi 2n}{L}x\right) \sin\left(\frac{\pi m}{L}y\right) \sin\left(\frac{\pi k}{L}z\right) \tag{4.102}$$

$$\lambda_{\min} = \lambda_{211} = \left(\frac{\pi}{L}\right)^2 [(2)^2 + 1^2 + 1^2] = 6\left(\frac{\pi}{L}\right)^2 \tag{4.103}$$

$$L_{cr} = \pi\sqrt{\frac{6D}{\beta}} \tag{4.104}$$

Therefore if we consider a cube of size: $\pi\sqrt{\frac{3D}{\beta}} < L < \pi\sqrt{\frac{6D}{\beta}}$

The bomb could explode when the two halves are brought together, but each of the halves is stable.

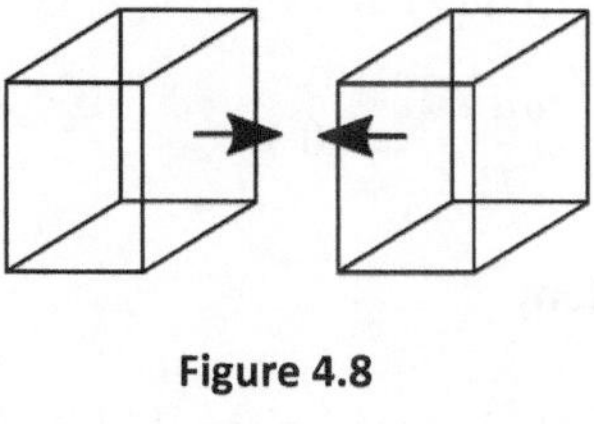

Figure 4.8

4.5 Oscillations of a Gas in a Cube

We have a gas (with speed of sound c) inside a cubic container of side L. This container is divided in two equal halves, with densities of the gas equal to $\rho_0+\rho_1$ and $\rho_0-\rho_1$ in each half (consider that $\rho_1 << \rho_0$). At the instant $t = 0$ the division of the container is removed. Find the vibrations of the gas density as a function of time.

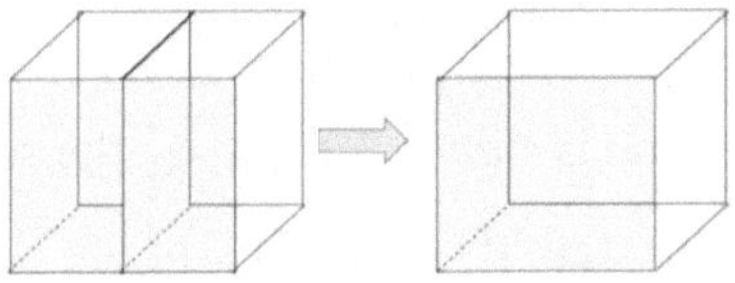

Figure 4.9

Mathematical formulation

We seek a solution for $u(x, y, z, t) = \rho - \rho_0$, this is, the relative variation of the gas density with respect to the equilibrium density ρ_0.

$$\frac{\partial^2 u(x, y, z, t)}{\partial t^2} = c^2 \Delta u(x, y, z, t) \qquad (4.105)$$

With boundary conditions $\frac{\partial u(x,y,z,t)}{\partial n} = 0$ (being n the direction normal to each surface).

The initial conditions are:

$$u(x, y, z, 0) = \left\{ \begin{array}{l} f(x) \times g(y) \times v(z) \\ f(x) = \left\{ \begin{array}{ll} \rho_1 & \left(0 < x < \frac{L}{2}\right) \\ -\rho_1 & \left(\frac{L}{2} < x < L\right) \end{array} \right\} \\ g(y) = 1 \ (0 < y < L) \\ v(z) = 1 \ (0 < z < L) \end{array} \right\}$$

$$\left.\frac{\partial u}{\partial t}\right|_{t=0} = 0 \qquad (4.106)$$

Sturm–Liouville problem

The solution is expanded in orthogonal eigenfunctions in all three spatial directions, since all boundary conditions are homogeneous.

The eigenfunctions and eigenvalues are known to be:

$$\lambda_{n,m,k} = \left(\frac{\pi n}{L}\right)^2 + \left(\frac{\pi m}{L}\right)^2 + \left(\frac{\pi k}{L}\right)^2 \quad (4.107)$$

$$X_n(x)Y_m(y)Z_k(z) = \cos\left(\frac{\pi n}{L}x\right)\cos\left(\frac{\pi m}{L}y\right)\cos\left(\frac{\pi k}{L}z\right) \quad (4.108)$$

General solution

$$u(x, y, z, t) = \sum_{n,m,k} A_{nmk}\cos\left(\sqrt{\lambda_{n,m,k}}ct\right) \times \cos\left(\frac{\pi n}{L}x\right)\cos\left(\frac{\pi m}{L}y\right)\cos\left(\frac{\pi k}{L}z\right) \quad (4.109)$$

Final solution

The A_{nmk} coefficients are obtained from the initial conditions, using the orthogonality of the eigenfunctions.

$$u(x, y, z, t = 0) = f(x) \times g(y) \times v(z) = \sum_{n,m,k} A_{nmk}\cos\left(\frac{\pi n}{L}x\right)\cos\left(\frac{\pi m}{L}y\right)\cos\left(\frac{\pi k}{L}z\right) \quad (4.110)$$

Both sides are multiplied by $X_n(x)Y_n(y)Z_m(z)$ and integrated:

$$\int_0^L\int_0^L\int_0^L dxdydz$$

Only the coefficients with n different from zero but with $m, k = 0$ remain. Then:

$$\begin{aligned} A_{n0,0} &= \left(\frac{2}{L}\right)^2 \int_0^L f(x)\cos\left(\frac{\pi n}{L}x\right)dx \\ &= \frac{4\rho_1}{L^2}\left[\int_0^{L/2}\cos\left(\frac{\pi n}{L}x\right)dx - \int_{L/2}^{L}\cos\left(\frac{\pi n}{L}x\right)dx\right] \\ &= \frac{4\rho_1}{L^2}\frac{L}{\pi n}\left(\sin\left(\frac{\pi n}{2}\right) - \sin(\pi n) + \sin\left(\frac{\pi n}{2}\right)\right) \\ &= \frac{8\rho_1}{L^2}\frac{L}{\pi n}\left(\sin\left(\frac{\pi n}{2}\right)\right) \end{aligned} \quad (4.111)$$

4.6 Stationary Temperature Distribution inside a Prism

Find the stationary distribution of temperature inside a prism of dimensions a, b, c. The temperature of the face $x = 0$ is kept as $T(x = 0, y, z, t) = Ayz$. The faces $z = 0$ and $z = c$ are kept at $T = 0$. All remaining faces are thermally insulated.

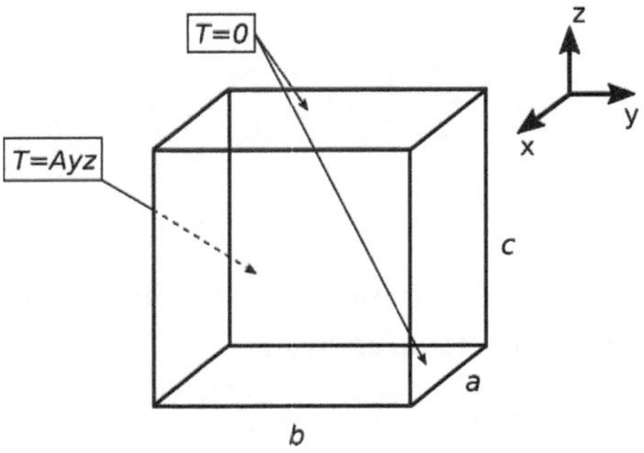

Figure 4.10

We must solve Laplace's equation in a prism in which five of the six faces have homogeneous boundary conditions.

Mathematical formulation

$$\left\{\begin{array}{l} \Delta u(x, y, z) = 0 \\ u(x = 0) = Ayz \\ u(z = 0) = u(z = c) = 0 \\ \left.\dfrac{\partial u}{\partial x}\right|_{x=a} = \left.\dfrac{\partial u}{\partial y}\right|_{y=0} = \left.\dfrac{\partial u}{\partial y}\right|_{y=b} = 0 \end{array}\right\} \qquad (4.112)$$

We first consider the faces with x constant, for which the boundary conditions are homogeneous.

Sturm–Liouville problem

In the YZ plane we can solve the following Sturm–Liouville problem:

$$\left\{\begin{array}{l} \Delta v(y,z) + \lambda v(y,z) = 0 \\ v(z=0) = v(z=c) = 0 \\ \left.\dfrac{\partial v}{\partial y}\right|_{y=0} = \left.\dfrac{\partial v}{\partial y}\right|_{y=b} = 0 \end{array}\right\} \tag{4.113}$$

with eigenfunctions:

$$v_{nm} = \cos\left(\frac{\pi n}{b}y\right)\sin\left(\frac{\pi m}{c}z\right) \tag{4.114}$$

and eigenvalues:

$$\lambda_{nm} = \left(\frac{\pi n}{b}\right)^2 + \left(\frac{\pi m}{c}\right)^2 \qquad (n = 0, 1, 2, 3...);\ (m = 1, 2, 3...) \tag{4.115}$$

General solution

The general solution can be expanded in the base of the v_{nm} and the coefficients will depend on the x coordinate:

$$u(x,y,z) = \sum_{n,m} w_{nm}(x) v_{nm}(y,z) \tag{4.116}$$

Replacing into Laplace's equation: $\Delta u(x,y,z) = 0$ we get the equations we need to solve:

$$\frac{d^2 w_{nm}(x)}{dx^2} - \lambda_{nm} w_{nm}(x) = 0 \tag{4.117}$$

with the boundary condition:

$$\left.\frac{dw_{nm}(x)}{dx}\right|_{x=a} = 0 \tag{4.118}$$

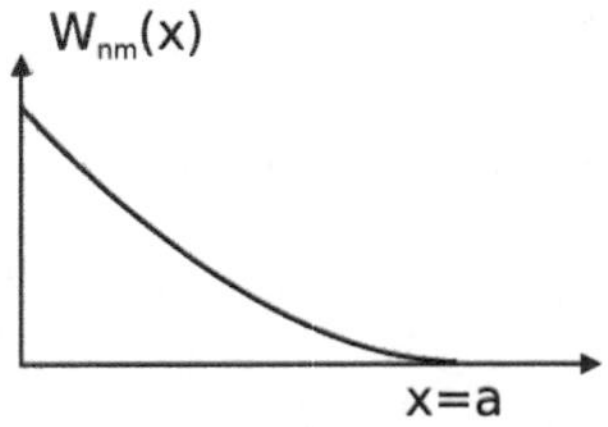

Figure 4.11

With this type of condition is more convenient to search for a solution of the form:

$$w_{nm}(x) = A_{nm} \sinh(\sqrt{\lambda_{nm}}x) + B_{nm} \cosh(\sqrt{\lambda_{nm}}(a - x)) \quad (4.119)$$

Final solution

Applying the boundary condition, we have:

$$\begin{aligned} w_{nm}(x)_x(x = a) = A_{nm}\sqrt{\lambda_{nm}} \cosh(\sqrt{\lambda_{nm}}a) \\ + B_{nm}\sqrt{\lambda_{nm}} \sinh((a - a)) = 0 \end{aligned} \quad (4.120)$$

We conclude that the coefficients $A_{nm} = 0$, and:

$$w_{nm}(x) = B_{nm} \cosh(\sqrt{\lambda_{nm}}(a - x)) \quad (4.121)$$

The general solution using the second boundary condition is:

$$u(x, y, z) = \sum_{n,m} B_{nm} \cosh(\sqrt{\lambda_{nm}}(a - x))v_{nm}(y, z) \quad (4.122)$$

We use the first boundary condition to find B_{nm}:

$$u(0, y, z) = Ayz = \sum_{n,m} B_{nm} \cosh(\sqrt{\lambda_{nm}}a)v_{nm}(y, z) \quad (4.123)$$

Using the orthogonality of $v_{nm}(y, z)$ we arrive at the expression for B_{nm}

$$B_{nm} = \frac{A}{\cosh(\sqrt{\lambda_{nm}}a)} \frac{1}{\int_0^b \int_0^c \cos^2\left(\frac{\pi n}{b}y\right) \sin^2\left(\frac{\pi m}{c}z\right) dydz} \int_0^b \int_0^c yz \cos\left(\frac{\pi n}{b}y\right) \sin\left(\frac{\pi m}{c}z\right) dydz \quad (4.124)$$

Note on the modulus of the eigenfunctions for different values of n, m.

$$\int_0^c \sin^2\left(\frac{\pi m}{c}z\right) dz = \frac{c}{2} \tag{4.125}$$

$$\int_0^b \cos^2\left(\frac{\pi n}{b}y\right) dy = \begin{Bmatrix} \frac{b}{2}\ (n \neq 0) \\ b\ (n = 0) \end{Bmatrix} \tag{4.126}$$

Furthermore:

$$\int_0^c z \sin\left(\frac{\pi m}{c}z\right) dz = \frac{c^2}{\pi m}(-1)^{m+1} \tag{4.127}$$

$$\int_0^b y \cos\left(\frac{\pi n}{c}y\right) dy = \begin{Bmatrix} \frac{b^2}{\pi^2 n^2}[(-1)^n - 1]\ (n \neq 0) \\ \frac{b^2}{2} \qquad (n = 0) \end{Bmatrix} \tag{4.128}$$

Then:

$$B_{0m} = \frac{Abc}{\pi m} \frac{1}{\cosh(\sqrt{\lambda_{nm}}a)}(-1)^{m+1} \tag{4.129}$$

$$B_{nm} = \frac{4Abc}{\cosh(\sqrt{\lambda_{nm}}a)} \frac{(-1)^m}{\pi^3 n^2 m}[(-1)^n - 1] \tag{4.130}$$

4.7 Variation of the Temperature inside a Cube: From Poisson to a Diffusion Problem

A thermal reservoir keeps the surface of a cube of size $L \times L \times L$ at a temperature T_0. In the center of this cube there is a point heat source which supplies q units of heat by unit time. This heat source is suddenly turned off at $t = 0$. Find the temperature in the inside of the cube from that moment if the thermal conductivity of the material is k.

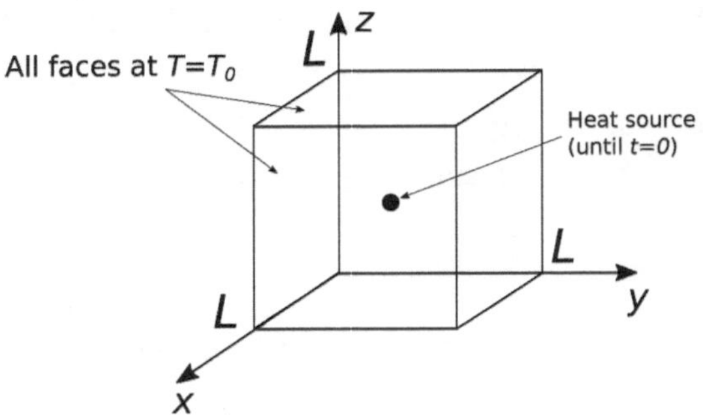

Figure 4.12

Mathematical formulation

First we need to solve Poisson's equation for $t < 0$ to get the initial conditions for the diffusion problem, which starts at $t = 0$.

Mathematical formulation of problem 1 ($t > 0$):

$$\left\{\begin{array}{l} C\rho\dfrac{\partial u(x,y,z,t)}{\partial t} - k\Delta u(x,y,z,t) = 0 \\ u(0,y,z,t) = u(L,y,z,t) = u(x,0,z,t) = u(x,L,z,t) \\ = u(x,y,0,t) = u(x,y,L,t) = T_0 \\ u(x,y,z,0) = f(x,y,z) \end{array}\right\} \tag{4.131}$$

To find $f(x, y, z)$ we need to formulate and solve the stationary problem 2:

$$\left\{\begin{array}{l} \Delta u(x,y,z) = -\dfrac{q}{k}\delta\left(x-\dfrac{L}{2}\right)\delta\left(y-\dfrac{L}{2}\right)\delta\left(z-\dfrac{L}{2}\right) \\ u(0,y,z) = u(L,y,z) = u(x,0,z) = u(x,L,z) \\ = u(x,y,0) = u(x,y,L) = T_0 \end{array}\right\} \tag{4.132}$$

We shift the solution by a value T_0 ($v(x, y, z) = u(x, y, z) - T_0$) and seek the solution by expanding $v(x, y, z)$ in a summation of orthogonal eigenfunctions, corresponding to the following Sturm–Liouville problem (homogeneous boundary conditions of the first type):

$$\left\{\begin{array}{l}\Delta v(x, y, z) + \lambda v(x, y, z) = 0 \\ v(0, y, z) = v(L, y, z) = v(x, 0, z) = v(x, L, z) \\ = v(x, y, 0) = v(x, y, L) = 0\end{array}\right\} \tag{4.133}$$

Sturm–Liouville problem

Seeking the solution with the method of separation of variables (for example in the normalized form), we get:

$$v(x, y, z) = \frac{2\sqrt{2}}{L^{3/2}} \sin\left(\frac{\pi n}{L}x\right) \sin\left(\frac{\pi m}{L}y\right) \sin\left(\frac{\pi l}{L}z\right) \tag{4.134}$$

$$\lambda_{nml} = \left(\frac{\pi n}{L}\right)^2 + \left(\frac{\pi m}{L}\right)^2 + \left(\frac{\pi l}{L}\right)^2 \tag{4.135}$$

Then we look for the solution of the non-homogeneous equation (4.132) as:

$$v(x, y, z) = \sum_{n,m,l} A_{nml} \frac{2\sqrt{2}}{L^{3/2}} \sin\left(\frac{\pi n}{L}x\right) \sin\left(\frac{\pi m}{L}y\right) \sin\left(\frac{\pi l}{L}z\right) \tag{4.136}$$

$$\sum_{n,m,j} A_{nml} \frac{2\sqrt{2}}{L^{3/2}} \lambda_{nml} \sin\left(\frac{\pi n}{L}x\right) \sin\left(\frac{\pi m}{L}y\right) \sin\left(\frac{\pi l}{L}z\right)$$
$$= -\frac{q}{k} \delta\left(x - \frac{L}{2}\right) \delta\left(y - \frac{L}{2}\right) \delta\left(z - \frac{L}{2}\right) \tag{4.137}$$

Using the orthogonality of the eigenfunctions we will get the coefficients of the sum.

$$A_{nml} = \frac{q}{k\lambda_{nml}} \frac{L^{3/2}}{2\sqrt{2}} \frac{8}{L^3} \int_0^L \int_0^L \int_0^L \delta\left(x - \frac{L}{2}\right) \delta\left(y - \frac{L}{2}\right) \delta\left(z - \frac{L}{2}\right)$$

$$\sin\left(\frac{\pi n}{L}x\right) \sin\left(\frac{\pi m}{L}y\right) \sin\left(\frac{\pi l}{L}z\right) dxdydz =$$

$$= \frac{q}{k\lambda_{nml}} \frac{2\sqrt{2}}{L^{3/2}} \sin\left(\frac{\pi n}{2}\right) \sin\left(\frac{\pi m}{2}\right) \sin\left(\frac{\pi l}{2}\right) = \frac{q}{k\lambda_{nml}} \frac{2\sqrt{2}}{L^{3/2}} (-1)^{n+m+l} \tag{4.138}$$

A_{nml} only finite for $n = 2n' + 1;\ m = 2m' + 1;\ l = 2l' + 1$ (that is, odd integers). Then

$$u(x, y, z) = f(x, y, z) = T_0 + \frac{8q}{kL^3} \sum_{n,m,l} \frac{(-1)^{n+m+l}}{\left(\frac{\pi n}{L}\right)^2 + \left(\frac{\pi m}{L}\right)^2 + \left(\frac{\pi l}{L}\right)^2}$$
$$\times \sin\left(\frac{\pi n}{L} x\right) \sin\left(\frac{\pi m}{L} y\right) \sin\left(\frac{\pi l}{L} z\right) \tag{4.139}$$

where $n = 2n' + 1;\ m = 2m' + 1;\ l = 2l' + 1$

We now move on to the solution of problem 1 ($t > 0$), again shifting the solution by T_0.

$$w(x, y, z, t) = u(x, y, z, t) - T_0 \tag{4.140}$$

$$\begin{Bmatrix} C\rho \dfrac{\partial w(x, y, z, t)}{\partial t} - k\Delta w(x, y, z, t) = 0 \rightarrow \dfrac{\partial w(x, y, z, t)}{\partial t} \\ \quad - \dfrac{1}{\chi} \Delta w(x, y, z, t) = 0 \\ w(0, y, z, t) = w(L, y, z, t) = w(x, 0, z, t) = w(x, L, z, t) \\ \quad = w(x, y, 0, t) = w(x, y, L, t) = 0 \\ w(x, y, z, 0) = f(x, y, z) - T_0 \end{Bmatrix} \tag{4.141}$$

Separating variables in the form $w(x, y, z, t) = T(t)X(x)Y(y)Z(z)$ we get to the solution in terms of the already obtained eigenfunctions.

$$w(x, y, z) = \sum_{n,m,l} C_{nml} e^{-\lambda_{nml}\chi t} \frac{2\sqrt{2}}{L^{3/2}} \sin\left(\frac{\pi n}{L} x\right) \sin\left(\frac{\pi m}{L} y\right) \sin\left(\frac{\pi l}{L} z\right) \tag{4.142}$$

Final solution

Applying the initial condition:

$$w(x, y, z, 0) = f(x, y, z) - T_0 =$$

$$= \frac{8q}{kL^3} \sum_{n,m,l} \frac{(-1)^{n+m+l}}{(\frac{\pi n}{L})^2 + (\frac{\pi m}{L})^2 + (\frac{\pi l}{L})^2} \sin\left(\frac{\pi n}{L}x\right) \sin\left(\frac{\pi m}{L}y\right) \sin\left(\frac{\pi l}{L}z\right)$$

$$= \sum_{n,m,l} C_{nml} \frac{2\sqrt{2}}{L^{3/2}} \sin\left(\frac{\pi n}{L}x\right) \sin\left(\frac{\pi m}{L}y\right) \sin\left(\frac{\pi l}{L}z\right) \qquad (4.143)$$

and using the orthogonality of the eigenfunctions $\sin\left(\frac{\pi n}{L}x\right)$, $\sin\left(\frac{\pi m}{L}y\right)$, $\sin\left(\frac{\pi l}{L}z\right)$, we get the coefficients:

$$C_{nml} = \frac{q}{k} \frac{2\sqrt{2}}{L^{3/2}} \frac{(-1)^{n+m+l}}{\left(\frac{\pi n}{L}\right)^2 + \left(\frac{\pi m}{L}\right)^2 + \left(\frac{\pi l}{L}\right)^2}$$

$$n = 2n' + 1; \; m = 2m' + 1; \; l = 2l' + 1 \qquad (4.144)$$

4.8 Variation of the Pressure inside a Rectangular Prism due to the Periodic Action of a Piston

Find the temporal variation of the relative pressure (with initial value equal to zero) inside the rectangular prism shown in the figure below if starting at $t = 0$ the pressure in one of the faces varies as $P_0 \sin(\omega t)$.

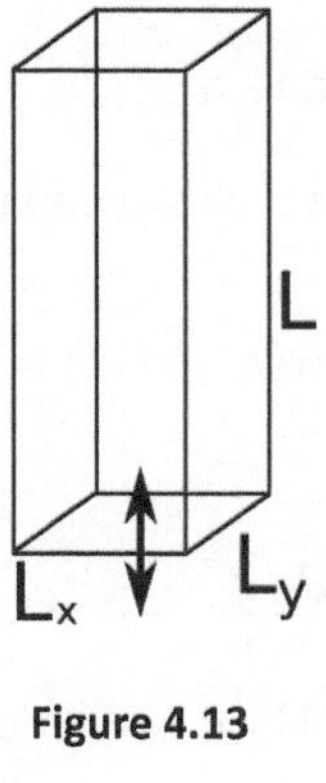

Figure 4.13

Mathematical formulation

We will look for the solution $S(x, y, z, t)$ as the sum of the stationary solution $u(x, y, z, t)$ established at $t \to \infty$ (which will be reached due to friction), and the transient solution $w(x, y, z, t)$. The equation to be solved is:

$$\left\{\begin{array}{l} \dfrac{\partial^2 S(x, y, z, t)}{\partial t^2} - c^2 \Delta S(x, y, z, t) = 0 \\ S(x, y, z, t = 0) = 0 \\ \left\{\dfrac{\partial}{\partial \vec{n}} u(x, y, z)\right\}_\Sigma = 0 \\ (\Sigma = \text{all surfaces except the base}) \\ (\vec{n} = \text{perpendicular to every surface}) \\ S(x, y, 0, t) = P_0 \sin(\omega t) \end{array}\right\} \tag{4.145}$$

Replacing the following form of the solution:

$$S(x, y, z, t) = w(x, y, z, t) + v(x, y, z) \sin(\omega t) \tag{4.146}$$

With the stationary solution (without initial conditions), considered to be $u = v(x, y, z) \sin(\omega t)$. We have:

$$\frac{\partial^2 w}{\partial t^2} - \omega^2 v \sin(\omega t) - c^2 \Delta w - c^2 \Delta v \sin(\omega t) = 0 \tag{4.147}$$

The problem is separated in two:

Formulation of the problem for the transitory part:

$$\left\{\begin{array}{l} \dfrac{\partial^2 w(x, y, z, t)}{\partial t^2} - c^2 \Delta w(x, y, z, t) = 0 \\ w(x, y, z, t = 0) = 0 \\ \text{Conditions 1-5:} \quad \dfrac{\partial w(x, y, z, t)}{\partial n} = 0 \\ \qquad\qquad \text{(all homogeneous of the second type)} \\ w(x, y, 0, t) = 0 \end{array}\right\} \tag{4.148}$$

This problem has a rather straightforward solution, as an expansion of three types of orthogonal functions. We just need to find the initial conditions by solving the problem for $v(x, y, z)$.

Formulation of problem (2) for the spatial part of the stationary solution:

$$\begin{cases} -\omega^2 v(x,y,z) - c^2 \Delta v(x,y,z) = 0 \\ \text{Conditions 1-5: } \dfrac{\partial v(x,y,z)}{\partial n} = 0 \\ \qquad\qquad\qquad \text{(all homogeneous of the second type)} \\ v(x,y,0) = P_0 \end{cases} \tag{4.149}$$

We will explain the initial details of how to find the stationary solution:

$$\Delta v(x,y,z) + \frac{\omega^2}{c^2} v(x,y,z) = 0 \tag{4.150}$$

General solution

The solution for $v(x, y, z)$ (of the Poisson type) is expanded as:

$$v(x,y,z) = \sum_{nm} C_{nm}(z) w_{nm}(x,y) \tag{4.151}$$

Where the coefficients $w_{nm}(x, y)$ are obtained by solving the Sturm–Liouville problem in 2D:

$$\begin{Bmatrix} \Delta w_{nm}(x,y) = -\lambda_{nm} w_{nm}(x,y) \\ \text{Second type boundary conditions} \end{Bmatrix} \tag{4.152}$$

We replace the summation into Poisson's equation.

$$\sum C_{nm}(z)\Delta w_{nm}(x,y) + \sum \frac{d^2 C_{nm}(z)}{dz^2} w_{nm}(x,y) + \frac{\omega^2}{c^2} \sum C_{nm}(z) w_{nm}(x,y) = 0 \tag{4.153}$$

We use the result to solve the Sturm–Liouville problem:

$$\sum C_{nm}(z)[-\lambda_{nm} w_{nm}(x,y)] + \sum \frac{d^2 C_{nm}(z)}{dz^2} w_{nm}(x,y) + \frac{\omega^2}{c^2} \sum C_{nm}(z) w_{nm}(x,y) = 0 \tag{4.154}$$

The sum is rewritten as:

$$\sum \left\{ C_{nm}(z)[-\lambda_{nm}] + \frac{d^2 C_{nm}(z)}{dz^2} + \frac{\omega^2}{c^2} C_{nm}(z) \right\} w_{nm}(x,y) = 0 \tag{4.155}$$

Final solution

Applying the orthogonality of the $w_{nm}(x, y)$ functions we arrive at a solution for the z variable:

$$\frac{d^2C_{nm}(z)}{dz^2} + \left[\frac{\omega^2}{c^2} - \lambda_{nm}\right] C_{nm}(z) \tag{4.156}$$

We have these boundary conditions:

$$\text{At } (z = L) \rightarrow \left.\frac{\partial v(x, y, x)}{\partial z}\right|_{z=L} = \sum \left.\frac{dC_{nm}(z)}{dz}\right|_{z=L} w_{nm}(x, y)$$

$$= 0 \rightarrow \left.\frac{dC_{nm}(z)}{dz}\right|_{z=L} = 0$$

$$\text{At } (z = 0) \rightarrow v(x, y, 0) = \sum C_{nm}(0) w_{nm}(x, y) = P_0$$

Using the orthogonality of the $w_{nm}(x, y)$ eigenfunctions we get:

$$C_{nm}(0) = \frac{\iint P_0 w_{nm}(x, y) dxdy}{\iint [w_{nm}(x, y)]^2 dxdy} \tag{4.157}$$

4.9 Case Study: Variation of Temperature inside a Prism: From Laplace to Poisson Problems

A prism has dimensions $a \times b \times c$. One of its faces is in contact with a thermal reservoir at a temperature T_0, another isothermally insulated (this face shares an edge with the former) and the rest are in contact with a reservoir at a temperature T_1. At the instant $t = 0$ the first face is put also at a temperature T_1, and the center of the prism starts to act as a heat point source which supplies Q [J/s].

(1) Find the variation of the temperature with time.
(2) How will the heat flux across the prism surface change with time?

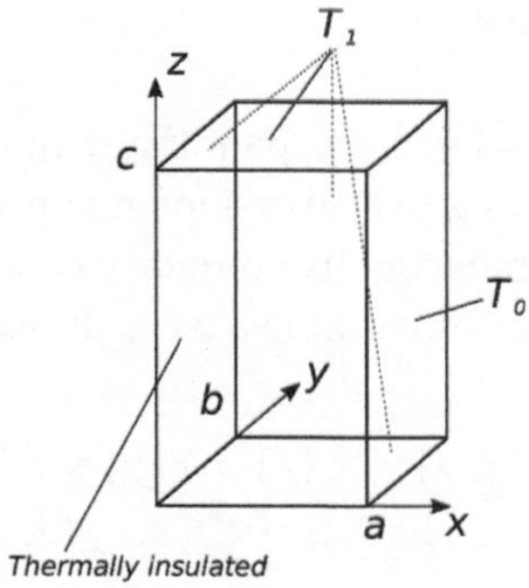

Figure 4.14

First we attempt to solve Laplace's equation in a prism to obtain the initial conditions which will be used in Poisson's problem (from $t = 0$ onwards).

Mathematical formulation

Formulation for $t < 0$:

$$\left.\begin{cases} \Delta u(x, y, z) = 0 \\ u(0, y, z) = u(x, b, z) = u(x, y, 0) = u(0, y, c) = T_1 \\ u(a, y, z) = T_0 \\ \left.\dfrac{\partial u}{\partial y}\right|_{y=0} = 0 \end{cases}\right\} \tag{4.158}$$

To create homogeneous boundary conditions, we shift the origin of temperatures: $u(x, y, z) = g(x, y, z) + T_1$. The problem for $g(x, y, z)$ is:

$$\left.\begin{cases} \Delta g(x, y, z) = 0 \\ \left.\dfrac{\partial g}{\partial y}\right|_{y=0} = 0; \quad g(a, y, z) = T_0 - T_1 \\ g(0, y, z) = g(x, b, z) = g(x, y, 0) = g(0, y, c) = 0 \end{cases}\right\} \tag{4.159}$$

Sturm–Liouville problem

We can seek the solution by expanding it into orthogonal eigenfunctions, corresponding to Sturm–Liouville problems in the y and z directions (with homogeneous boundary conditions). Seeking the solution by separation of variables, we will use this to choose the signs of the constants.

$$g(x, y, z) = X(x) \times Y(y) \times Z(z) \tag{4.160}$$

From $\Delta g(x, y, z) = 0 \rightarrow \frac{1}{X}\frac{d^2X}{dx^2} + \frac{1}{Y}\frac{d^2Y}{dy^2} + \frac{1}{Z}\frac{d^2Z}{dz^2} = 0$

Sturm–Liouville problem in the y direction:

$$\begin{aligned} &\frac{d^2Y}{dy^2} + \alpha^2 Y = 0 \\ &\text{Boundary conditions: } \left.\frac{\partial Y(y)}{\partial y}\right|_{y=0} = 0;\ Y(b) = 0 \end{aligned} \tag{4.161}$$

Provides us with the eigenfunctions $Y_m = \cos\left(\frac{\pi(2m+1)}{2b}y\right);\quad m = 0, 1, 2, 3\ldots$ and eigenvalues $\alpha_m = \frac{\pi(2m+1)}{2b}$. The Sturm–Liouville problem in the z direction is:

$$\left\{\begin{aligned} &\frac{d^2Z}{dz^2} + \beta^2 Z = 0 \\ &\text{Boundary conditions: } Z(0) = Z(c) = 0 \end{aligned}\right\} \tag{4.162}$$

Gives us the eigenfunctions $Z_l = \sin\left(\frac{\pi l}{c}z\right); l = 1, 2, 3\ldots$ and eigenvalues $\beta_l = \frac{\pi l}{c}$

The problem in the x direction is not a Sturm–Liouville problem, since the boundary at $x = a$ is inhomogeneous of the first kind.

$$\left\{\begin{aligned} &\frac{d^2X}{dx^2} - (\alpha_m^2 + \beta_l^2)X = 0 \\ &\text{Boundary conditions: } X(0) = 0; \\ &\text{The condition on the opposite face is: } g(a, y, z) = T_0 - T_1 \end{aligned}\right\} \tag{4.163}$$

General solution

The general solution for $X(x)$ is written more conveniently as:

$$X(x) = E\sinh\left(\sqrt{\alpha_m^2 + \beta_l^2}x\right) + H\cosh\left(\sqrt{\alpha_m^2 + \beta_l^2}x\right) \tag{4.164}$$

The condition $X(0) = 0$ gives $H = 0$.

Then the general solution to the Laplace's problem will be:

$$g(x, y, z) = \sum_{l,m} A_{ml} \sinh\left(\sqrt{\alpha_m^2 + \beta_l^2}x\right) \times \cos\left(\frac{\pi(2m+1)}{2b}y\right) \times \sin\left(\frac{\pi l}{c}z\right) \quad (4.165)$$

Final solution

Imposing the condition $g(a, y, z) = T_0 - T_1$ we get the coefficients of the summation:

$$T_0 - T_1 = \sum_{l,m} A_{ml} \sinh\left(\sqrt{\alpha_m^2 + \beta_l^2}a\right) \times \cos\left(\frac{\pi(2m+1)}{2b}y\right) \times \sin\left(\frac{\pi l}{c}z\right) \quad (4.166)$$

Taking advantage of the orthogonality of the eigenfunctions $Y_m Z_l$.

$$\int_0^b \int_0^c (T_0 - T_1) \cos\left(\frac{\pi(2m+1)}{2b}y\right) \times \sin\left(\frac{\pi l}{c}z\right) dydz = A_{ml} \sinh\left(\sqrt{\alpha_m^2 + \beta_l^2}a\right) \left\|\cos\left(\frac{\pi(2m+1)}{2b}y\right)\right\|^2 \times \left\|\sin\left(\frac{\pi l}{c}z\right)\right\|^2 \quad (4.167)$$

$$A_{ml} = \frac{4}{bc}\frac{(T_0 - T_1)}{\sinh[\sqrt{(\alpha_m^2 + \beta_l^2)}a]} \int_0^b \int_0^c \cos\left(\frac{\pi(2m+1)}{2b}y\right) \times \sin\left(\frac{\pi l}{c}z\right) dydz \quad (4.168)$$

$$A_{ml} = \frac{4}{bc}\frac{(T_0 - T_1)}{\sinh\left(\sqrt{(\alpha_m^2 + \beta_l^2)}a\right)} \frac{2bc}{\pi^2 l(2m+1)}(-1)^m[1 - (-1)^l] \rightarrow \quad (4.169)$$

$$A_{ml} = \left\{\begin{array}{ll} (-1)^m \dfrac{16}{bc}\dfrac{bc}{\pi^2 l(2m+1)} \dfrac{(T_0 - T_1)}{\sinh\left(\sqrt{\left[\frac{\pi(2m+1)}{2b}\right]^2 + \left(\frac{\pi n}{c}\right)^2}a\right)} & (l\text{=odd}) \\ 0 & (l\text{=even}) \end{array}\right\} \quad (4.170)$$

Then the solution of Laplace's problem (which gives the initial condition at $t = 0$) is:

$$u(x, y, z, 0) = T_1 + 16(T_0 - T_1)\times \tag{4.171}$$

$$\times \sum_{m,l=odd} (-1)^m \frac{1}{\pi^2 l(2m+1)} \frac{1}{\sinh\left(\sqrt{[\frac{\pi(2m+1)}{2b}]^2 + (\frac{\pi n}{c})^2 a}\right)} \times$$

$$\times \sinh\left(\sqrt{\left[\frac{\pi(2m+1)}{2b}\right]^2 + \left(\frac{\pi n}{c}\right)^2 x}\right) \times \cos\left(\frac{\pi(2m+1)}{2b} y\right)$$

$$\times \sin\left(\frac{\pi l}{c} z\right)$$

Mathematical formulation of Poisson's problem ($t > 0$)

$$\left\{\begin{array}{l} \frac{\partial u}{\partial t}(x, y, z, t) - \chi \Delta u(x, y, z, t) = \frac{q(x, y, z)}{C\rho} \\ q(x, y, z) = Q\delta\left(x - \frac{a}{2}\right)\delta\left(y - \frac{b}{2}\right)\delta\left(z - \frac{c}{2}\right) \\ u(0, y, z) = u(a, y, z) = u(x, b, z) = u(x, y, 0) = u(0, y, c) = T_1 \\ \left.\frac{\partial u}{\partial t}\right|_{y=0} = 0 \\ u(x, y, z, 0) = T_1 + g(x, y, z) \end{array}\right\} \tag{4.172}$$

This problem can be solved by separating it into two: one for $w(x, y, z)$, non-homogeneous (Poisson), which corresponds to the solution at infinite times ($t = +\infty$) and transient one $v(x, y, z, t)$ as a function of time, which is a solution of the homogeneous problem:

$$u(x, y, z, t) = v(x, y, z, t) + w(x, y, z) \tag{4.173}$$

Replacing into the formulation of Poisson's problem we have the equation:

$$\frac{\partial v(x, y, z, t)}{\partial t} - \chi \Delta v(x, y, z, t) - \chi \Delta w(x, y, z) = \frac{q(x, y, z)}{C\rho} \tag{4.174}$$

$$\left\{\begin{array}{l} \frac{\partial v(x, y, z, t)}{\partial t} - \chi \Delta v(x, y, z, t) = 0 \\ \chi \Delta w(x, y, z) = -\frac{Q}{C\rho}\delta\left(x - \frac{a}{2}\right)\delta\left(y - \frac{b}{2}\right)\delta\left(z - \frac{c}{2}\right) \end{array}\right\} \tag{4.175}$$

Distributing the boundary conditions between the two problems:

$$\left\{\begin{array}{l} w(0,y,z)=w(a,y,z)=w(x,b,z)=w(x,y,0) \\ \qquad =w(0,y,c)=T_1 \\ \left.\dfrac{\partial w}{\partial y}\right|_{y=0}=0 \end{array}\right\} \tag{4.176}$$

$$\left\{\begin{array}{l} v(0,y,z)=v(a,y,z)=v(x,b,z)=v(x,y,0)=v(0,y,c)=0 \\ \left.\dfrac{\partial v}{\partial y}\right|_{y=0}=0 \end{array}\right\} \tag{4.177}$$

We also can obtain the initial condition for the temporal problem using the relation:

$$u(x,y,z,0)=v(x,y,z,0)+w(x,y,z) \tag{4.178}$$

Then $v(x,y,z,0) = u(x,y,z,0) - w(x,y,z)$. The function $u(x,y,z,0)$ has been obtained from the solution of the Laplace's problem. Then we solve first the stationary problem:

$$\left\{\begin{array}{l} \Delta w(x,y,z)=-\dfrac{Q}{\chi C\rho}\delta\left(x-\dfrac{a}{2}\right)\delta\left(y-\dfrac{b}{2}\right)\delta\left(z-\dfrac{c}{2}\right) \\ w(0,y,z)=w(a,y,z)=w(x,b,z)=w(x,y,0) \\ \qquad =w(0,y,c)=T_1 \\ \left.\dfrac{\partial w}{\partial y}\right|_{y=0}=0 \end{array}\right\} \tag{4.179}$$

If we seek the solution as $w(x,y,z)=T_1+f(x,y,z)$ we will get for $f(x,y,z)$ the problem with homogeneous boundary conditions.

$$\left\{\begin{array}{l} \Delta f(x,y,z)=-\dfrac{Q}{\chi C\rho}\delta\left(x-\dfrac{a}{2}\right)\delta\left(y-\dfrac{b}{2}\right)\delta\left(z-\dfrac{c}{2}\right) \\ f(0,y,z)=f(a,y,z)=f(x,b,z)=f(x,y,0)=f(0,y,c)=0 \\ \left.\dfrac{\partial f}{\partial y}\right|_{y=0}=0 \end{array}\right\} \tag{4.180}$$

Sturm–Liouville problem

The solution will be expanded into eigenfunctions of the following Sturm–Liouville problem:

$$\begin{Bmatrix} \Delta s(x, y, z) + \lambda^2 s(x, y, z) = 0 \\ s(0, y, z) = s(a, y, z) = s(x, b, z) = s(x, y, 0) = s(0, y, c) = 0 \\ \left.\dfrac{\partial s}{\partial y}\right|_{y=0} = 0 \end{Bmatrix} \tag{4.181}$$

They are known:

$$s_{nml}(x, y, z) = \sin\left(\frac{\pi n}{a}x\right) \times \cos\left(\frac{\pi(2m+1)}{2b}y\right) \times \sin\left(\frac{\pi l}{c}z\right) \tag{4.182}$$

$$\lambda_{nml}^2 = \left(\frac{\pi n}{a}\right)^2 + \left(\frac{\pi(2m+1)}{2b}\right)^2 + \left(\frac{\pi l}{c}\right)^2 \tag{4.183}$$

General solution

The solution will be sought in the form:

$$\begin{aligned} f(x, y, z) &= \sum_{n,m,l} C_{nml} s_{nml}(x, y, z) \\ &= \sum_{n,m,l} C_{nml} \sin\left(\frac{\pi n}{a}x\right) \cos\left(\frac{\pi(2m+1)}{2b}y\right) \sin\left(\frac{\pi l}{c}z\right) \end{aligned} \tag{4.184}$$

Replacing into Poisson's equation:

$$\sum_{n,m,l} -C_{nml}\lambda_{nml}^2 s_{nml}(x, y, z) = -\frac{Q}{\chi C\rho}\delta\left(x - \frac{a}{2}\right)\delta\left(y - \frac{b}{2}\right)\delta\left(z - \frac{c}{2}\right) \tag{4.185}$$

Using the orthogonality:

$$C_{nml} = \frac{8}{\lambda_{nml}^2 abc}\frac{Q}{\chi C\rho}\int_0^a\int_0^b\int_0^c \delta\left(x - \frac{a}{2}\right)\delta\left(y - \frac{b}{2}\right)\delta\left(z - \frac{c}{2}\right) \times$$

$$\sin\left(\frac{\pi n}{a}x\right)\cos\left(\frac{\pi(2m+1)}{2b}y\right)\sin\left(\frac{\pi l}{c}z\right) dxdydz =$$

$$= \frac{8}{\lambda_{nml}^2 abc} \frac{Q}{\chi C\rho} \sin\left(\frac{\pi n}{2}\right) \cos\left(\frac{\pi(2m+1)}{4}\right) \sin\left(\frac{\pi l}{2}\right) \quad (4.186)$$

It is obvious that $C_{nml} = 0$ for even values of n, l. Then the solution for the stationary part is:

$$w(x, y, z) = T_1 + \frac{8Q}{\chi C\rho abc} \sum_{n,m,l} \frac{1}{\lambda_{nml}^2} \sin\left(\frac{\pi n}{2}\right) \cos\left(\frac{\pi(2m+1)}{4}\right) \times \sin\left(\frac{\pi l}{2}\right) \sin\left(\frac{\pi n}{a}x\right) \cos\left(\frac{\pi(2m+1)}{2b}y\right) \sin\left(\frac{\pi l}{c}z\right) \quad (4.187)$$

Now we solve the transient part.

Mathematical formulation

$$\begin{cases} \dfrac{\partial v(x, y, z, t)}{\partial t} - \chi \Delta v(x, y, z, t) = 0 \\ v(0, y, z) = v(a, y, z) = v(x, b, z) = v(x, y, 0) = v(0, y, c) = 0 \\ \left.\dfrac{\partial v}{\partial y}\right|_{y=0} = 0 \\ v(x, y, z, t = 0) = u(x, y, z, t = 0) - w(x, y, z) \end{cases} \quad (4.188)$$

The solution is sought by separating the temporal and spatial parts as:

$$v(x, y, z, t) = \sum_{n,m,l} A_{nml} T_{nml}(t) s_{nml}(x, y, z) = \sum_{n,m,l} A_{nml} T_{nml}(t) \sin\left(\frac{\pi n}{a}x\right) \cos\left(\frac{\pi(2m+1)}{2b}y\right) \sin\left(\frac{\pi l}{c}z\right) \quad (4.189)$$

Replacing in the equation to be solved, we will obtain an equation for the temporal part:

$$\sum_{n,m,l} A_{nml} \left\{ \frac{dT_{nml}(t)}{dt} + \chi \lambda_{nml}^2 T_{nml}(t) \right\} s_{nml}(x, y, z) = 0 \quad (4.190)$$

From here $T_{nml}(t) = Const \cdot e^{(-\chi \lambda_{nml}^2 t)}$

$$v(x, y, z, t) = \sum_{n,m,l} A_{nml} e^{(-\chi \lambda_{nml}^2 t)} s_{nml}(x, y, z) \quad (4.191)$$

Final solution

The constants A_{nml} are obtained from the initial conditions.

$$v(x,y,z,0)=u(x,y,z,0)-w(x,y,z)=\sum_{n,m,l}A_{nml}\times s_{nml}(x,y,z) \tag{4.192}$$

In particular, we can write:

$$\sum_{n,m,l}A_{nml}\times s_{nml}(x,y,z)=T_1+16(T_0-T_1)$$
$$\times\sum_{m,l=odd}(-1)^m\frac{1}{\pi^2 l(2m+1)}\frac{1}{\sinh\left(\sqrt{\left[\frac{\pi(2m+1)}{2b}\right]^2+\left(\frac{\pi n}{c}\right)^2}a\right)}$$
$$\times\sinh\left(\sqrt{\left[\frac{\pi(2m+1)}{2b}\right]^2+\left(\frac{\pi n}{c}\right)^2}x\right)\times\cos\left(\frac{\pi(2m+1)}{2b}y\right)$$
$$\times\sin\left(\frac{\pi l}{c}z\right)-T_1-\frac{8Q}{\chi C\rho abc}\sum_{n,m,l}\frac{1}{\lambda_{nml}^2}\sin\left(\frac{\pi n}{2}\right)\cos\left(\frac{\pi(2m+1)}{4}\right)$$
$$\times\sin\left(\frac{\pi l}{2}\right)\sin\left(\frac{\pi n}{a}x\right)\cos\left(\frac{\pi(2m+1)}{2b}y\right)\sin\left(\frac{\pi l}{c}z\right) \tag{4.193}$$

Using the orthogonality properties:

$$A_{nml}\left(\frac{abc}{8}\right)=\frac{bc}{4}(-1)^m\frac{16(T_0-T_1)}{\pi^2 l(2m+1)}\frac{1}{\sinh\left(\sqrt{\left[\frac{\pi(2m+1)}{2b}\right]^2+\left(\frac{\pi n}{c}\right)^2}a\right)}$$
$$\times\int_0^a\sin\left(\frac{\pi n}{a}x\right)\times\sinh\left(\sqrt{\left[\frac{\pi(2m+1)}{2b}\right]^2+\left(\frac{\pi n}{c}\right)^2}x\right)dx$$
$$-\frac{abc}{8}\frac{8Q}{\chi C\rho abc\lambda_{nml}^2}\sin\left(\frac{\pi n}{2}\right)\cos\left(\frac{\pi(2m+1)}{4}\right)$$
$$\times\sin\left(\frac{\pi l}{2}\right)-T_1\frac{abc}{8} \tag{4.194}$$

From integral tables we have:

$$\int\sin(ax)\times\sinh(bx)dx=\frac{b\sin(ax)\times\cosh(bx)-a\cos(ax)\times\sinh(bx)}{a^2+b^2}+\text{const.}$$

Then:

$$A_{nml} = (-1)^{m+n+1} \frac{(T_0 - T_1)}{\pi^2 l(2m+1)} \frac{32}{[\frac{\pi(2m+1)}{2b}]^2 + (\frac{\pi n}{c})^2 + (\frac{\pi n}{a})^2}$$
$$- \frac{Q}{\chi C\rho} \sin\left(\frac{\pi n}{2}\right) \cos\left(\frac{\pi(2m+1)}{4}\right) \sin\left(\frac{\pi l}{2}\right) \quad (4.195)$$

Then the solution for $t > 0$ is:

$$u(x, y, z, t) = \sum_{n,m,l} A_{nml} \times e^{(-\chi\lambda_{nml}^2 t)} \sin\left(\frac{\pi n}{a}x\right) \cos\left(\frac{\pi(2m+1)}{2b}y\right)$$
$$\times \sin\left(\frac{\pi l}{c}z\right) + T_1 + \frac{8Q}{\chi C\rho abc} \sum_{n,m,l} \frac{1}{\lambda_{nml}^2} \sin\left(\frac{\pi n}{2}\right)$$
$$\times \cos\left(\frac{\pi(2m+1)}{4}\right) \sin\left(\frac{\pi l}{2}\right) \sin\left(\frac{\pi n}{a}x\right)$$
$$\times \cos\left(\frac{\pi(2m+1)}{2b}y\right) \sin\left(\frac{\pi l}{c}z\right) \quad (4.196)$$

To find the heat across the surface:

$$Q_{\text{sup}} = -\int\int\int_S (\mathbf{w} \times \mathbf{n})ds = \{\mathbf{w} = -k\nabla u(x, y, z, t)\}$$
$$= -\int\int\int_V (\nabla \times \mathbf{w})dV = k\int\int\int_V (\Delta u)dV$$
$$= k\int\int\int_V (\Delta v + \Delta w)dV$$
$$= k\int\int\int_V \left[\Delta v - \frac{Q}{\chi C\rho}\delta\left(x - \frac{a}{2}\right)\delta\left(y - \frac{b}{2}\right)\delta\left(z - \frac{c}{2}\right)\right] dV$$
$$= k\int\int\int_V \Delta v dV - \frac{kQ}{\chi C\rho} \quad (4.197)$$

Since we have:

$$\Delta v = \Delta\left[\sum_{n,m,l} A_{nml} e^{(-\chi\lambda_{nml}^2 t)} \times s_{nml}(x, y, z)\right]$$
$$= -\sum_{n,m,l} \lambda_{nml}^2 A_{nml} \times e^{(-\chi\lambda_{nml}^2 t)} \times s_{nml}(x, y, z) \quad (4.198)$$

The total heat across the surface by unit time is:

$$
\begin{aligned}
Q_{\text{sup}} &= -\frac{kQ}{\chi C\rho} - \sum_{n,m,l} k\lambda_{nml}^2 A_{nml} \times e^{(-\chi\lambda_{nml}^2 t)} \int\int\int_V s_{nml}(x, y, z)]dV \\
&= -\frac{kQ}{\chi C\rho} - \sum_{n,m,l} k\lambda_{nml}^2 A_{nml} \\
&\times \frac{2abc}{\pi^3 nl(2m+1)}(-1)^m \times [1-(-1)^n] \times [1-(-1)^l]
\end{aligned}
$$

4.10 Case Study: Distribution of Temperature inside a Periodically Heated Prism

A prism of length L and square cross section $a \times a$, with thermal diffusivity coefficient χ has all of its surface thermally insulated except for one of its bases. In the middle of this base the temperature varies periodically as $T_0 \sin(\omega t)$, meanwhile the temperature in the other half is $-T_0 \sin(\omega t)$. These oscillations exist at all times $(-\infty < t < \infty)$. Find the distribution of temperature inside a prism as a function of position and time.

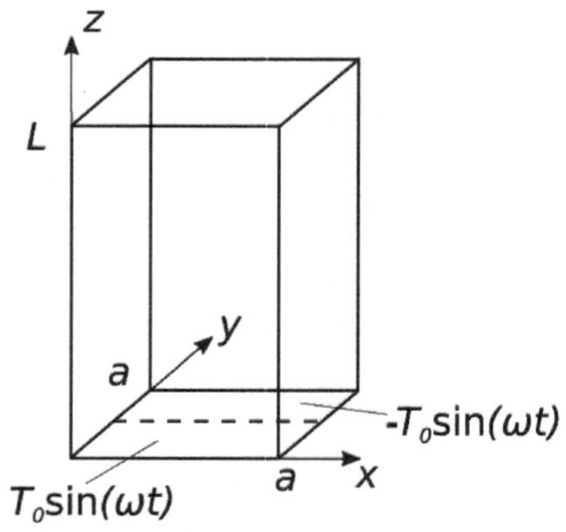

Figure 4.15

Mathematical formulation

$$\left\{\begin{array}{l}\frac{\partial}{dt}u(x,y,z,t)-\chi\,\Delta u(x,y,z,t)=0\\ \left\{\frac{\partial}{\partial\vec{\mathbf{n}}}u(x,y,z)\right\}_{\Sigma}=0\\ (\Sigma=\text{all surface except the base})\\ (\vec{\mathbf{n}}=\text{perpendicular to every surface})\\ \text{Boundary conditions at the base: } u(x,y,0,t)=F(x,y)\sin(\omega t)\\ F(x,y)=\left\{\begin{array}{l}T_0\ \left(0<y<\frac{a}{2}\right)\\ -T_0\ \left(\frac{a}{2}<y<a\right)\end{array}\right\}\end{array}\right\} \tag{4.199}$$

The solution will be sought as an expansion in orthogonal eigenfunctions in the x, y directions, corresponding to homogeneous boundary conditions.

$$u(x,y,z,t)=\sum_{nm}\Theta_{nm}(z,t)w_{nm}(x,y) \tag{4.200}$$

Sturm–Liouville problem

We formulate the Sturm–Liouville problem for $w_{nm}(x, y)$

$$\left\{\begin{array}{l}\Delta w_{nm}(x,y)+\lambda_{nm}w_{nm}(x,y)=0\\ \left[\frac{\partial}{dx}w(x,y)\right]_{x=0,a}=\left[\frac{\partial}{dy}w(x,y)\right]_{y=0,a}=0\end{array}\right\} \tag{4.201}$$

The corresponding eigenfunctions and eigenvalues are:

$$w_{nm}(x,y)=\cos\left(\frac{\pi n}{a}x\right)\cos\left(\frac{\pi m}{a}y\right) \tag{4.202}$$

$$\lambda_{nm}=\left(\frac{\pi}{a}\right)^2[n^2+m^2]\text{ with } m,n=0,1,2,3\ldots \tag{4.203}$$

General solution

Replacing $u(x, y, z, t) = \sum_{nm} \Theta_{nm}(z, t) \cos\left(\frac{\pi n}{a}x\right) \cos\left(\frac{\pi m}{a}y\right)$ into (4.199) we arrive at equations to find $\Theta_{nm}(z, t)$. The result of the substitution is:

$$\sum_{nm} \frac{\partial \Theta_{nm}(z, t)}{dt} w_{nm}(x, y) - \chi \left[\Theta_{nm}(z, t)\Delta w_{nm}(x, y) + \frac{\partial^2 \Theta_{nm}(z, t)}{dz^2} w_{nm}(x, y)\right] = 0 \quad (4.204)$$

Applying the orthogonality of $w_{nm}(x, y)$:

$$\sum_{nm} \left[\frac{\partial}{dt}\Theta_{nm}(z, t) - \chi \frac{\partial^2}{dz^2}\Theta_{nm}(z, t) + \chi \lambda_{nm}\Theta_{nm}(z, t)\right] \omega_{nm}(x, y) = 0 \quad (4.205)$$

$$\frac{\partial}{dt}\Theta_{nm}(z, t) - \chi \left[\frac{\partial^2}{dz^2}\Theta_{nm}(z, t) - \lambda_{nm}\Theta_{nm}(z, t)\right] = 0 \quad (4.206)$$

Formulation of the problem in terms of the variables (z,t):

$$\left\{\begin{array}{l} \dfrac{\partial \Theta_{nm}(z, t)}{dt} - \chi \left[\dfrac{\partial^2 \Theta_{nm}(z, t)}{dz^2} - \lambda_{nm}\Theta_{nm}(z, t)\right] = 0 \\ \text{First boundary condition: } \left.\dfrac{\partial \Theta_{nm}(z, t)}{dz}\right|_{z=L} = 0 \\ \text{Second boundary condition: ?} \end{array}\right\} \quad (4.207)$$

We use the orthogonality of $w_{nm}(x, y)$ to find the second boundary condition.

$$u(x, y, 0, t) = \sum_{nm} \Theta_{nm}(0, t) \cos\left(\frac{\pi n}{a}x\right) \cos\left(\frac{\pi m}{a}y\right) = F(x, y)\sin(\omega t) \quad (4.208)$$

Multiplying both sides by $\cos(\frac{\pi n}{a}x)\cos(\frac{\pi m}{a}y)$ and integrating:

$$\int_0^a\int_0^a \Theta_{nm}(0, t) \left[\cos\left(\frac{\pi n}{a}x\right)\right]^2 \left[\cos\left(\frac{\pi m}{a}y\right)\right]^2 dxdy = \sin(\omega t) \int_0^a\int_0^a F(x, y)\cos\left(\frac{\pi n}{a}x\right) \cos\left(\frac{\pi m}{a}y\right) dxdy \quad (4.209)$$

We get:

$$\Theta_{nm}(0,t) = \sin(\omega t)\frac{\int_0^a\int_0^a F(x,y)\cos(\frac{\pi n}{a}x)\cos(\frac{\pi m}{a}y)dxdy}{\int_0^a\int_0^a[\cos(\frac{\pi n}{a}x)]^2[\cos(\frac{\pi m}{a}y)]^2dxdy}$$
$$= \zeta_{nm}\sin(\omega t) \tag{4.210}$$

The second boundary condition is presented as an imaginary exponential function, to search the solution of (4.207) in complex form, and then take only the imaginary part. Using real functions we would not be able to cancel them out after the substitution, due to the lack of second derivatives.

This form allows to easily separate the variables z and t at the cost of having to solve an equation for the complex function $Z(z)$. We define the following function $J_{nm}(z,t) = Z_{nm}(z)e^{(i\omega t)}$

$$\Theta_{nm}(z,t) = Im\{J_{nm}(z,t)\} \tag{4.211}$$

$$\rightarrow \left\{\begin{array}{l} \dfrac{\partial J_{nm}(z,t)}{dt} - \chi\left[\dfrac{\partial^2 J_{nm}(z,t)}{dz^2} - \lambda_{nm}J_{nm}(z,t)\right] = 0 \\ \left.\dfrac{\partial J_{nm}(z,t)}{dz}\right|_{z=L} = 0 \\ J_{nm}(0,t) = \zeta_{nm}e^{(i\omega t)} \end{array}\right\} \tag{4.212}$$

Eliminating the exponentials which multiply each member of the equation:

$$\rightarrow \left\{\begin{array}{l} i\omega Z_{nm}(z) - \chi\left[\dfrac{\partial^2 Z_{nm}(z)}{dz^2} - \lambda_{nm}Z_{nm}(z)\right] = 0 \\ \left.\dfrac{dZ_{nm}(z)}{dz}\right|_{z=L} = 0 \\ Z_{nm}(0) = \zeta_{nm} \end{array}\right\} \tag{4.213}$$

or

$$\left\{\begin{array}{l} \dfrac{\partial^2 Z_{nm}(z)}{dz^2} - Z_{nm}(z)\left[\lambda_{nm} + \dfrac{i\omega}{\chi}\right] = 0 \\ \left.\frac{dZ_{nm}(z)}{dz}\right|_{z=L} = 0 \\ Z_{nm}(0) = \zeta_{nm} \end{array}\right\} \tag{4.214}$$

We can check that ζ_{nm} is finite only for values $m = 2k+1$ (this is, odd) and for $n = 0$:

$$\zeta_{0k} = \frac{\int_0^a\int_0^a F(x,y)\cos(\frac{\pi 0}{a}x)\cos(\frac{\pi m}{a}y)dxdy}{\int_0^a\int_0^a[\cos(\frac{\pi 0}{a}x)]^2[\cos(\frac{\pi m}{a}y)]^2dxdy} = \frac{8T_0}{\pi(2k+1)}(-1)^k \tag{4.215}$$

We seek the general solution in the form:

$$Z_{nm}(z) = A_{nm}e^{z/C_{nm}} + B_{nm}e^{-z/C_{nm}} \tag{4.216}$$

with $C_{nm} = \frac{1}{\sqrt{\lambda_{nm}+\frac{i\omega}{\chi}}}$

Final solution

Imposing $n = 0,\ m = 2k+1$:

From the first boundary condition:

$$\left.\frac{dZ_{0k}(z)}{dz}\right|_{z=L} = \frac{A_{0k}}{C_{0k}}e^{(L/C_{0k})} - \frac{B_{0k}}{C_{0k}}e^{(-L/C_{0k})} = 0 \tag{4.217}$$

To the second boundary condition:

$$Z_{nm}(0) = A_{0k} + B_{0k} = \frac{8T_0}{\pi(2k+1)}(-1)^k \tag{4.218}$$

Eliminating the index $n = 0$, since the rest of the coefficients with $n > 0$ are zero:

$$\left\{\begin{array}{l} A_k e^{(L/C_k)} - B_k e^{(-L/C_k)} = 0 \\ A_k + B_k = \frac{8T_0}{\pi(2k+1)}(-1)^k \end{array}\right\} \tag{4.219}$$

$$\left\{\begin{array}{l} A_k = B_k e^{(-2L/C_k)} \\ B_k[e^{(-2L/C_k)} + 1] = \frac{8T_0}{\pi(2k+1)}(-1)^k \end{array}\right\} \tag{4.220}$$

$$B_k = \frac{8T_0}{\pi(2k+1)[e^{(-2L/C_k)}+1]}(-1)^k \tag{4.221}$$

$$A_k = \frac{8T_0e^{(-2L/C_k)}}{\pi(2k+1)[e^{(-2L/C_k)}+1]}(-1)^k = \frac{8T_0}{\pi(2k+1)[e^{(2L/C_k)}+1]}(-1)^k \tag{4.222}$$

From

$$Z_k(z) \rightarrow J_k(z,t) = Z_k(z)e^{(i\omega t)} \quad (4.223)$$

$$\Theta_k(z,t) = \mathrm{Im}\{J_k(z,t)\} \quad (4.224)$$

$$u(x,y,z,t) = u(y,z,t) = \sum_k \Theta_k(z,t)\cos\left(\frac{\pi\,[2k+1]}{a}y\right) \quad (4.225)$$

Note: observing the form of the solution (which does not depend on x) we see that the propagating thermal wave (independent of x) preserves the transversal symmetry of the profile of the heating of the heated surface.

4.11 Heating Rectangular Resistor with Different Boundary Conditions

A rectangular bar ($a \times b \times c$) with an electric resistance R and thermal conductivity coefficient k is heated with a constant current I that produces a heat $Q = \frac{I^2 R}{abc}$ per unit time and volume. The heat exchange with the outer medium at zero temperature occurs through the surface of the bar, according to Newton's law. Determine the stationary distribution of temperature in the bar.

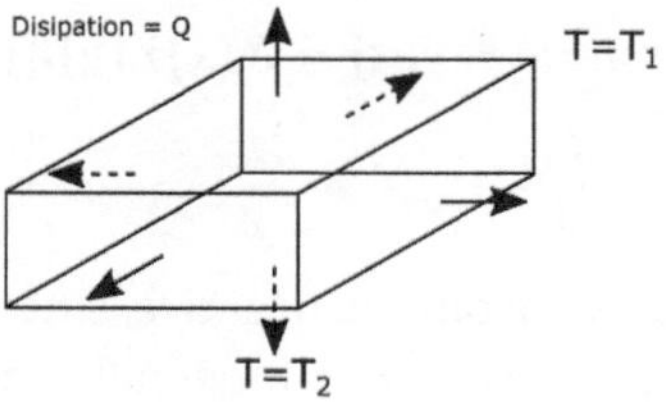

Figure 4.16

Mathematical formulation

$$-k\Delta u(x,y,z) = f(x,y,z) = \frac{I^2 R}{abc} \quad (4.226)$$

$$\Delta u(x,y,z) = -\frac{I^2 R}{k \times abc} = \frac{-Q}{k} \quad (4.227)$$

Boundary condition:

$$\frac{\partial u(x, y, z)}{\partial n} + Au(x, y, z) = 0 \qquad (4.228)$$

where n is normal to each surface and $A = \frac{\alpha}{k}$, being α the Newton's law heat exchange constant.

Boundary conditions (the subindices indicate derivatives with respect to the variable of the subindex, for example $u_x = \frac{\partial u}{\partial x}$, etc.):

$$\left\{\begin{array}{l} u_x(0, y, z) + Au(0, y, z) = 0 \\ u_x(a, y, z) - Au(a, y, z) = 0 \\ u_y(x, 0, z) + Au(x, 0, z) = 0 \\ u_y(x, b, z) - Au(x, b, z) = 0 \\ u_z(x, y, 0) + Au(x, y, 0) = 0 \\ u_z(x, y, c) - Au(x, y, c) = 0 \end{array}\right\} \qquad (4.229)$$

Sturm–Liouville problem

The idea is to seek the solution by expanding it in a series of orthogonal eigenfunctions in the three spatial directions (that is, solutions of the Sturm–Liouville problem with homogeneous boundary conditions of the third kind).

$$\Delta u(x, y, z) + \lambda u(x, y, z) = 0 \qquad (4.230)$$

Separating variables as: $u(x, y, z) = X(x)Y(y)Z(z)$, we have:

$$\frac{X_{xx}}{X} + \frac{Y_{yy}}{Y} + \frac{Z_{zz}}{Z} = -\lambda \qquad (4.231)$$

(we impose that the λ constant be negative, since we are looking for the expansion of the solution in orthogonal functions, in this case sinusoidal).

We separate the Sturm–Liouville problem in three different, independent problems, in the x, y, z directions:

$$\frac{X_{xx}}{X} = -\nu; \qquad \frac{Y_{yy}}{Y} = -\mu; \qquad \frac{Z_{zz}}{Z} = -\chi \qquad (4.232)$$

The general solution will be sought as:

$$u = \sum C_{nmk} \times X_n(x) \times Y_m(y) \times Z_k(z) \qquad (4.233)$$

We find as an example the form of the $X_n(x)$ eigenfunctions, solving the following problem (the solutions for the functions $Y_m(y)$ and $Z_k(z)$ are found in a similar manner):

$$\frac{d^2X(x)}{dx^2} + \nu X(x) = 0 \tag{4.234}$$

With the boundary conditions:

$$\left\{ \begin{array}{l} X_x(0) + AX(0) = 0 \\ X_x(a) - AX(a) = 0 \end{array} \right\} \tag{4.235}$$

General solution

The general solution for $X_n(x)$ is:

$$X_n(x) = A_n \cos(\sqrt{\nu_n}x) + B_n \sin(\sqrt{\nu_n}x) \tag{4.236}$$

We seek the values of A_n and B_n with the boundary conditions:

First condition: $X_x(0) + AX(0) = 0 \rightarrow A_n A + \sqrt{\nu_n} B_n = 0$

Second condition: $X_x(a) - AX(a) = 0 \rightarrow A_n[-\sqrt{\nu_n}\sin(\sqrt{\nu_n}a) - A\cos(\sqrt{\nu_n}a)] + B_n[\sqrt{\nu_n}\cos(\sqrt{\nu_n}a) - A\sin(\sqrt{\nu_n}a)] = 0$

The values of A and B can be obtained from this system of equations by seeking that the determinant of the system be null. With this we will have the equation to find the ν_n eigenvalues and the respective eigenfunctions for the y and z coordinates.

Equating the determinant of the matrix of the equations for the two boundary conditions to zero we end up arriving at an equation for the eigenvalues:

$$\frac{A^2 - \nu_n}{2A\sqrt{\nu_n}} = \tan^{-1}(\sqrt{\nu_n}a) \tag{4.237}$$

We would obtain the eigenfunctions as (replacing the ratio between the A_n and B_n coefficients):

$$X_n(x) = -\frac{\sqrt{\nu_n}}{A} B_n \cos(\sqrt{\nu_n}x) + B_n \sin(\sqrt{\nu_n}x) \tag{4.238}$$

A similar procedure should give us $Y(y)$ and $Z(z)$.

To find the solution of the non-homogeneous equation (4.255), we expand the solution as a sum of the obtained orthogonal functions:

$$\Delta u_{nmk}(x, y, z) + \lambda_{nmk} u_{nmk}(x, y, z) = 0 \tag{4.239}$$

We replace $u = \sum_{nmk} C_{nmk} \cdot u_{nmk}(x, y, z)$ into the equation:

$$\Delta u(x, y, z) = \frac{-Q}{k} \tag{4.240}$$

And arrive at the relation:

$$\sum_{nmk} C_{nmk} \Delta u_{nmk}(x, y, z) = \frac{-Q}{k} \tag{4.241}$$

Using the solutions of the Sturm–Liouville problem, we have:

$$\sum_{nmk} -\lambda_{n,m,k} C_{nmk} u_{nmk}(x, y, z) = \frac{-Q}{k} \tag{4.242}$$

with $\lambda_{n,m,k} = \nu_n + \mu_m + \chi_k$

We will now use the orthogonality of the $u_{nmk}(x, y, z) = X_n(x) Y_m(y) Z_k(z)$ eigenfunctions to find the C_{nmk} coefficients.

We arrive at:

$$-\lambda_{n,m,k} C_{nmk} \|u_{nmk}(x, y, z)\|^2 = \frac{-Q}{k} \int_0^a \int_0^b \int_0^c X_n(x) Y_m(y) Z_k(z) dxdydz \tag{4.243}$$

Finally:

$$C_{nmk} = \frac{Q \int_0^a \int_0^b \int_0^c X_n(x) Y_m(y) Z_k(z) dxdydz}{k \times \lambda_{n,m,k} \|u_{nmk}(x, y, z)\|^2} \tag{4.244}$$

Note: in this case we attempt to separate the solution in two:

$$u = v + w \tag{4.245}$$

one of them being $v(x, y, z)$, the solution to the homogeneous equation with homogeneous boundary conditions of the third kind and the other being $w(x, y, z)$, the solution of the non-homogeneous equation, leads us nowhere.

It is clear that the boundary conditions applied to each of these functions, when they are all considered at the same time, must recover the boundary conditions of the initial problem.

Mathematical formulation of the problem for v

$$\Delta v(x, y, z) = 0 \tag{4.246}$$

Boundary conditions: $\frac{\partial v(x,y,z)}{\partial n} + Av(x, y, z) = 0$ (n is normal to each of the surfaces), which explicitly turns out to be:

$$\left\{\begin{array}{l} v_x(0, y, z) + Av(0, y, z) = 0 \\ v_x(a, y, z) - Av(a, y, z) = 0 \\ v_y(x, 0, z) + Av(x, 0, z) = 0 \\ v_y(x, b, z) - Av(x, b, z) = 0 \\ v_z(x, y, 0) + Av(x, y, 0) = 0 \\ v_z(x, y, c) - Av(x, y, c) = 0 \end{array}\right\} \tag{4.247}$$

Mathematical formulation of the problem for w

$$\Delta w(x, y, z) = -\frac{Q}{k} \tag{4.248}$$

$$\left\{\begin{array}{l} w_x(0, y, z) + Aw(0, y, z) = 0 \\ w_x(a, y, z) - Aw(a, y, z) = 0 \\ w_y(x, 0, z) + Aw(x, 0, z) = 0 \\ w_y(x, b, z) - Aw(x, b, z) = 0 \\ w_z(x, y, 0) + Aw(x, y, 0) = 0 \\ w_z(x, y, c) - Aw(x, y, c) = 0 \end{array}\right\} \tag{4.249}$$

Solution

The solution for the Laplace's problem for v is trivial:

$$v(x, y, z) = 0 \tag{4.250}$$

since the totality of the heat crossing the surfaces is zero:

The solution of the problem for $w(x, y, z)$ is the same, as we previously showed for $u(x, y, z)$.

4.12 Heating of a Rectangular Resistor with the Same Boundary Conditions

A rectangular bar ($a \times b \times c$) with an electric resistance R and thermal conductivity coefficient k is heated with a constant current

I that produces a heat $Q = \frac{I^2R}{abc}$ per unit time and volume. The heat exchange with the outer medium at temperature T_1 occurs through the surface of the bar, according to Newton's law, except at the base ($z = 0$) where the exchange takes place with the outer medium at a temperature T_2.

Determine the stationary distribution of temperature in the bar.

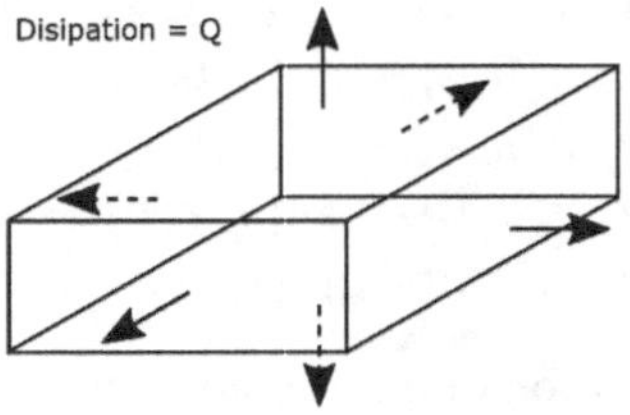

Figure 4.17

Mathematical formulation

$$-k\Delta u(x, y, z) = f(x, y, z) = \frac{I^2R}{abc} \tag{4.251}$$

$$\Delta u(x, y, z) = -\frac{I^2R}{k \times abc} = \frac{-Q}{k} \tag{4.252}$$

Boundary condition:

$$\frac{\partial u(x, y, z)}{\partial n} + Au(x, y, z) = 0 \tag{4.253}$$

where n is the normal direction to each surface and $A = \frac{\alpha}{k}$, being α the Newton's law heat exchange constant.

One we subtract the T_1 value from the solution, the boundary conditions will be as follows (the subindices indicate derivatives with respect to the variable of the subindex, for example $u_x = \frac{\partial u}{\partial x}$, etc.):

$$\left\{\begin{array}{l} u_x(0, y, z) + Au(0, y, z) = 0 \\ u_x(a, y, z) - Au(a, y, z) = 0 \\ u_y(x, 0, z) + Au(x, 0, z) = 0 \\ u_y(x, b, z) - Au(x, b, z) = 0 \\ u_z(x, y, 0) + Au(x, y, 0) = T_2 - T_1 \\ u_z(x, y, c) - Au(x, y, c) = 0 \end{array}\right\} \tag{4.254}$$

Next we split problem in two: $u = v + w$

We have an inhomogeneous problem w with all the boundary conditions being homogeneous of the third type (this problem has been already solved previously) and a Laplace equation for v with all boundary conditions homogeneous of the third type, except the base, which is inhomogeneous of the third type.

$$\Delta v(x, y, z) = 0 \tag{4.255}$$

Boundary conditions:

$$\left\{\begin{array}{l} v_x(0, y, z) + Av(0, y, z) = 0 \\ v_x(a, y, z) - Av(a, y, z) = 0 \\ v_y(x, 0, z) + Av(x, 0, z) = 0 \\ v_y(x, b, z) - Av(x, b, z) = 0 \\ v_z(x, y, 0) + Av(x, y, 0) = T_2 - T_1 \\ v_z(x, y, c) - Au(x, y, c) = 0 \end{array}\right\} \tag{4.256}$$

Sturm–Liouville problem

The idea is to seek the solution by expanding it in a series of orthogonal eigenfunctions in the two spatial directions x, y (that is, solutions of the Sturm–Liouville problem with homogeneous boundary conditions of the third kind) and finding the solution for the differential equation in the z direction with two boundary conditions.

Separating variables as: $u(x, y, z) = X(x)Y(y)Z(z)$ and substituting into the Laplace equation, we have:

$$\frac{X_{xx}}{X} + \frac{Y_{yy}}{Y} + \frac{Z_{zz}}{Z} = 0 \tag{4.257}$$

$$\frac{X_{xx}}{X} + \frac{Y_{yy}}{Y} = -\lambda \tag{4.258}$$

(we impose that the λ constant be negative, since we are looking for the expansion of the solution in orthogonal functions, in this case sinusoidal).

We separate the Sturm–Liouville problem in two different, independent problems, in the x, y directions:

$$\frac{X_{xx}}{X} = -\nu; \qquad \frac{Y_{yy}}{Y} = -\mu; \tag{4.259}$$

The general solution will be sought as:

$$v = \sum X_n(x) \times Y_m(y) \times Z(z) \tag{4.260}$$

We find as an example the form of the $X_n(x)$ eigenfunctions, solving the following problem (the solutions for the functions $Y_m(y)$ are found in a similar manner):

$$\frac{d^2X(x)}{dx^2} + \nu X(x) = 0 \tag{4.261}$$

With the boundary conditions:

$$\left\{ \begin{matrix} X_x(0) + AX(0) = 0 \\ X_x(a) - AX(a) = 0 \end{matrix} \right\} \tag{4.262}$$

General solution

The general solution for $X_n(x)$ is:

$$X_n(x) = A_n \cos(\sqrt{\nu_n}x) + B_n \sin(\sqrt{\nu_n}x) \tag{4.263}$$

We seek the values of A_n and B_n with the boundary conditions:

First condition: $X_x(0) + AX(0) = 0 \rightarrow A_nA + \sqrt{\nu_n}B_n = 0$

Second condition: $X_x(a) - AX(a) = 0 \rightarrow A_n[-\sqrt{\nu_n}\sin(\sqrt{\nu_n}a) - A\cos(\sqrt{\nu_n}a)] + B_n[\sqrt{\nu_n}\cos(\sqrt{\nu_n}a) - A\sin(\sqrt{\nu_n}a)] = 0$

The values of A and B can be obtained from this system of equations by seeking that the determinant of the system be null. With this we will have the equation to find the ν_n eigenvalues and the respective eigenfunctions for the y and z coordinates.

Equating the determinant of the matrix of the equations for the two boundary conditions to zero we end up arriving at an equation for the eigenvalues:

$$\frac{A^2 - \nu_n}{2A\sqrt{\nu_n}} = \tan^{-1}(\sqrt{\nu_n}a) \tag{4.264}$$

We would obtain the eigenfunctions as (replacing the ratio between the A_n and B_n coefficients):

$$X_n(x) = -\frac{\sqrt{\nu_n}}{A}B_n\cos(\sqrt{\nu_n}x) + B_n\sin(\sqrt{\nu_n}x) \tag{4.265}$$

A similar procedure should give us $Y(y)$.

We therefore come to formulate the equation for $Z(z)$:

$$\frac{d^2 Z(z)}{dx^2} - \nu Z(z) = 0 \tag{4.266}$$

which has following general solution:

$$Z(z) = C_{mn} \cosh(\sqrt{(\lambda z)}) + B_{mn} \cosh(\sqrt{(\lambda z)})$$

The general solution therefore will be

$$v = \sum C_{mn} \cosh(\sqrt{(\lambda z)}) + B_{mn} \cosh(\sqrt{(\lambda z)}) \times X_n(x) \times Y_m(y) \tag{4.267}$$

Applying the boundary conditions and the orthogonality conditions of $X_n(x)$ and $Y_m(y)$ we arrive at a system of two equations and find the values of C_{mn} and B_{mn}.

Note: An alternative way to solve the formulated 3D Poisson problem with one of the boundaries being inhomogeneous of the 3rd type is to search directly the solution in form:

$$u = \sum Z(z) \times X_n(x) \times Y_m(y) \tag{4.268}$$

Once we substitute the above expression into the 3D Poisson equation and use the fact that $X_n(x)$ and $Y_n(x)$ are set of orthogonal functions, as they are solutions of 2D Sturm–Liouville problem, we arrive to inhomogeneous equation for the $Z(z)$ function with one of two boundaries (z=0) being inhomogeneous of the 3rd kind.

The general solution for $Z(z)$ will be represented as sum previously found solution of homogeneous equation and particular solution Z_p (which is a constant A already defined by the form of 1D second order differential equation).

$$Z(z) = Z_h(z) + Z_p = C_{mn} \cosh(\sqrt{(\lambda z)}) + D_{mn} \cosh(\sqrt{(\lambda z)}) + A$$

The solution for the C_{mn} and D_{mn} in $Z(z)$ will be found after application of two (homogeneous and inhomogeneous) boundary conditions of the 3rd type to the general solution $u = \sum Z(z) \cdot X_n(x) \cdot Y_m(y)$.

4.13 Case Study: Distribution of Photocarriers Induced by a Laser

Suppose that a metallic bar (dimensions a, b, c, see Figure) with diffusion coefficient equal to 1 is in contact (its base) with a charge-neutral semiconductor. Until $t = 0$ a laser had been shone over a very small area (normally called “spot”) while the remaining part of the contact allows carrier leakage through semi-transparent to diffusion interface (with leakage flux proportional with factor of B to the difference in metal and semiconductor surface concentrations). The distribution of generated photocarriers near the interface providing local flux injection can be approximated by a Dirac’s Delta function (with flux rate proportional to A factor). The carriers were injected from the semiconductor to the metal in a stationary manner (until $t = 0$) through a slit situated in the proximities of the central point of contact of the bar with the semiconductor.

Determine the distribution of carriers inside the bar as a function of time if at $t = 0$ the laser is turned off and never used again.

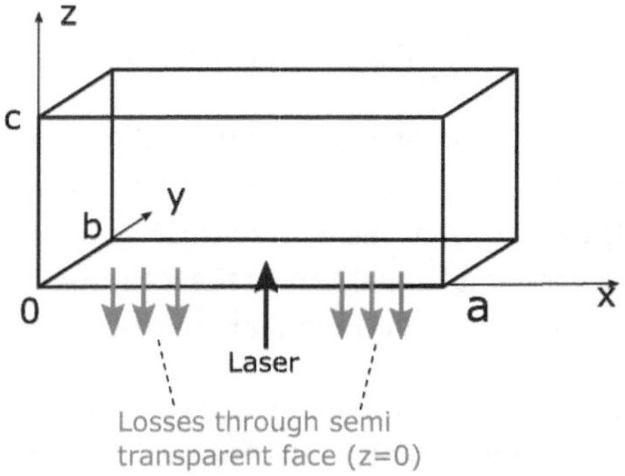

Figure 4.18

Solution:

We first seek the stationary distribution of carriers up until $t = 0$.

If we consider that the sources of photoelectrons are at the interface, we need to solve Laplace’s equation in a prism in which five out of the six faces have homogeneous boundary conditions. The central part of the face at ($z = 0$) injects photoelectrons through a small square slit ($\varepsilon << a, b, c$), corresponding to the laser spot). The rest of the parts of the $z = 0$ face are semi-transparent to the electrons. Then the boundary conditions are not homogeneous for the face at $z = 0$.

Mathematical formulation

$$\left\{ \begin{array}{l} \Delta u(x, y, z) = 0 \\ \left.\dfrac{\partial u}{\partial z}\right|_{z=0} + Bu(z=0) = A \times \delta(x - a/2) \times \delta(y - b/2) \\ \left.\dfrac{\partial u}{\partial x}\right|_{x=0} = \left.\dfrac{\partial u}{\partial x}\right|_{x=a} = \left.\dfrac{\partial u}{\partial y}\right|_{y=0} = \left.\dfrac{\partial u}{\partial y}\right|_{y=b} = \left.\dfrac{\partial u}{\partial z}\right|_{z=c} = 0 \end{array} \right\} \tag{4.269}$$

To express the non-homogeneous boundary condition, we have used that $z = 0$ is semitransparent for the electrons (condition described using boundary conditions of the third type) and at the application point of the laser appears an excess of carriers described by a Dirac's Delta function.

Sturm–Liouville problem

We expand the solution in orthogonal eigenfunction, corresponding to four homogeneous surfaces. We can arrive at the same result by separating variables, replacing the solution of Laplace's equation and using the dimensions that have homogeneous boundary conditions to lower the dimensionality of the derivatives thanks to the solution of the Sturm–Liouville problem.

In the $x - y$ plane (with homogeneous boundary conditions), we solve the problem:

$$\left\{ \begin{array}{l} \Delta v(x, y) + \lambda v(x, y) = 0 \\ \left.\dfrac{\partial v}{\partial x}\right|_{x=0} = \left.\dfrac{\partial v}{\partial x}\right|_{x=a} = \left.\dfrac{\partial v}{\partial y}\right|_{y=0} = \left.\dfrac{\partial v}{\partial y}\right|_{y=b} = 0 \end{array} \right. \tag{4.270}$$

with eigenfunctions $v_{nm} = \cos(\frac{\pi n}{a}x)\cos(\frac{\pi m}{b}y)$

and eigenvalues $\lambda_{nm} = \pi^2[(\frac{n}{a})^2 + (\frac{m}{b})^2]$.

General solution

The general solution can be expanded in the base of the v_{nm} functions and the coefficients will depend on the x coordinate:

$$u(x, y, z) = \sum_{n,m} w_{nm}(z) v_{nm}(x, y) \tag{4.271}$$

Replacing in the Laplace's equation: $\Delta u(x, y, z) = 0$

We arrive at the equations we need to solve:

$$\frac{d^2 w_{nm}(z)}{dz^2} - \lambda_{nm} w_{nm}(z) = 0 \tag{4.272}$$

with the boundary condition: $\frac{dw_{nm}}{dz}\Big|_{z=c} = 0$

$$w_{nm}(z) = F_{nm} \sinh(\sqrt{\lambda_{nm}} x) + C_{nm} \cosh[\sqrt{\lambda_{nm}}(c - z)] \tag{4.273}$$

Applying the boundary condition:

$$\frac{dw_{nm}}{dz}\Big|_{z=c} = 0 = F_{nm}\sqrt{\lambda_{nm}} \cosh\left(\sqrt{\lambda_{nm}} c\right) + C_{nm}\sqrt{\lambda_{nm}} \sinh(c - c) = 0 \tag{4.274}$$

Then we get the $F_{nm} = 0$ and $w_{nm}(z) = C_{nm} \cosh[\sqrt{\lambda_{nm}}(c - z)]$ coefficients.

The general solution will be:

$$\begin{aligned} u(x, y, z) &= \sum_{n,m} C_{nm} \cosh\left[\sqrt{\lambda_{nm}}(c - z)\right] v_{nm}(x, y) \\ &= u(x, y, z, 0) = f(x, y, z) \end{aligned} \tag{4.275}$$

We use the second boundary condition to find C_{nm}:

$$\begin{aligned} &\frac{\partial u}{\partial z}\Big|_{z=0} + Bu(z = 0) \\ &\quad = -\sum_{n,m} C_{nm}\sqrt{\lambda_{nm}} \sinh[\sqrt{\lambda_{nm}}(c)] v_{nm}(x, y) \\ &\qquad + B\sum_{n,m} C_{nm} \cosh[\sqrt{\lambda_{nm}}(c)] v_{nm}(x, y) \\ &\quad = A \times \delta(x - a/2) \times \delta(y - b/2) \end{aligned} \tag{4.276}$$

Final solution

Using the orthogonality of $v_{nm}(y, z)$ we arrive at the expression for B_{nm}:

$$C_{nm}\left\{-\sqrt{\lambda_{nm}}\sinh\left[\sqrt{\lambda_{nm}}(c)\right]+B\cosh\left[\sqrt{\lambda_{nm}}(c)\right]\right\}\times$$
$$\int_0^a\int_0^b \cos^2\left(\frac{\pi n}{a}x\right)\cos^2\left(\frac{\pi m}{b}y\right)dxdy$$
$$= A\times\int_0^a\int_0^b \delta(x-a/2)\delta(y-b/2)\cos\left(\frac{\pi n}{a}x\right)\cos\left(\frac{\pi m}{b}y\right)dxdy$$
$$= A\times\cos\left(\frac{\pi n}{2}\right)\cos\left(\frac{\pi m}{2}\right) \tag{4.277}$$

Then

$$C_{nm}=\frac{A}{B\cosh[\sqrt{\lambda_{nm}}(c)]-\sqrt{\lambda_{nm}}\sinh[\sqrt{\lambda_{nm}}(c)]}\times\frac{1}{\int_0^a\int_0^b\cos^2\left(\frac{\pi n}{a}x\right)\cos^2\left(\frac{\pi m}{b}y\right)dxdy} \tag{4.278}$$

Note about the modulus of the eigenfunctions for different values of n (or m):

$$\int_0^b \cos^2\left(\frac{\pi n}{b}y\right)dy=\left\{\begin{matrix}\frac{b}{2}\ (n\neq 0)\\ \\ b\ (n=0)\end{matrix}\right\} \tag{4.279}$$

Next we will consider the temporal variation of the concentration of photoelectrons after turning the laser off.

Mathematical formulation

$$\left\{\begin{array}{l}\frac{\partial u}{\partial t}-\chi\Delta u=0\\ \left.\frac{\partial u}{\partial z}\right|_{z=0}+Bu(z=0)=0\\ \left.\frac{\partial u}{\partial x}\right|_{x=0}=\left.\frac{\partial u}{\partial x}\right|_{x=a}=\left.\frac{\partial u}{\partial y}\right|_{y=0}=\left.\frac{\partial u}{\partial y}\right|_{y=b}=\left.\frac{\partial u}{\partial z}\right|_{z=c}=0\\ u(x,y,x,t=0)=f(x,y,z)\end{array}\right\} \tag{4.280}$$

We seek the solution by separating spatial and temporal variables.

$$u=Q(t)\times v(x,y,z) \tag{4.281}$$

Sturm–Liouville problem

$v(x, y, z)$ are eigenfunctions of the Sturm-Liouville problem.

$$\left\{ \begin{array}{l} \Delta v + \lambda v = 0 \\ \left.\dfrac{\partial v}{\partial z}\right|_{z=0} + Bv(z=0) = 0 \\ \left.\dfrac{\partial v}{\partial x}\right|_{x=0} = \left.\dfrac{\partial v}{\partial x}\right|_{x=a} = \left.\dfrac{\partial v}{\partial y}\right|_{y=0} = \left.\dfrac{\partial v}{\partial y}\right|_{y=b} = \left.\dfrac{\partial v}{\partial z}\right|_{z=c} = 0 \end{array} \right\} \tag{4.282}$$

We separate variables: $v(x, y, z)) = X(x) \cdot Y(y) \cdot Z(z)$

$$\Delta v + \lambda v = 0 \rightarrow \frac{1}{X(x)} \frac{d^2 X(x)}{dx^2} + \frac{1}{Y(y)} \frac{d^2 Y(y)}{dy^2} + \frac{1}{Z(z)} \frac{d^2 Z(z)}{dz^2} = -\lambda \tag{4.283}$$

$$-\left(\frac{\pi n}{a}\right)^2 - \left(\frac{\pi m}{b}\right)^2 - \nu_k = -\lambda_{nmk} \tag{4.284}$$

To satisfy the boundary conditions we then use a solution of the type:

$$v(x, y, z) = \cos\left(\frac{\pi n}{a} x\right) \cos\left(\frac{\pi m}{b} y\right) \cos[\sqrt{\nu_k}(z - c)] \tag{4.285}$$

The equation to find the ν_k eigenvalues is:

$$\left.\frac{dv}{dz}\right|_{z=0} + Bv(z = 0) = 0 \rightarrow -\sqrt{\nu_k} \sin[\sqrt{\nu_k} c] + B \cos[\sqrt{\nu_k} c] = 0 \tag{4.286}$$

The eigenvalues are solutions of the equation:

$$\frac{\sin(\sqrt{\nu_k} c)}{\cos(\sqrt{\nu_k} c)} = \frac{\sqrt{\nu_k}}{B} \tag{4.287}$$

Equation and solution for the temporal part:

$$\frac{dQ}{dt} + \lambda_{nmk} Q = 0 \tag{4.288}$$

$$Q(t) = A_{nmk} e^{(-\lambda_{nmk} t)} \tag{4.289}$$

$$u = \sum Q(t) \times v_{nmk}(x, y, z) \tag{4.290}$$

$$= \sum A_{nmk} e^{(-\lambda_{nmk} t)} \cos\left(\frac{\pi n}{a} x\right) \cos\left(\frac{\pi m}{b} y\right) \cos[\sqrt{\nu_k}(z-c)] \tag{4.291}$$

Final solution

From the initial condition:

$$u(x, y, z, t = 0) = \sum A_{nmk} e^{(-\lambda_{nmk} 0)} \cos\left(\frac{\pi n}{a} x\right) \times \cos\left(\frac{\pi m}{b} y\right) \cos[\sqrt{\nu_k}(z - c)] = f(x, y, z) \tag{4.292}$$

we will get the coefficients of the expansion using the properties of orthogonality of the eigenfunctions.

4.14 Heater inside a Prism

A rectangular prism of dimensions a, b, c, has a heat capacity C, density of mass ρ and thermal conductivity coefficient k. Until $t = 0$ it's in thermal equilibrium at a temperature $T = T_0$. From the instant $t > 0$ onwards an electrical current I starts circulating through a wire with electrical resistance R, of length $c/2$, centered at $\left(\frac{a}{2}, \frac{b}{2}, \frac{c}{4} < z < \frac{3c}{4}\right)$ in the vertical direction. Consider that the upper and lower boundaries are thermally insulated, while the other four are in contact with a thermal reservoir at a temperature $T = T_0$.

(i) Find the variation of temperature if the applied current $I = I_0$ from $t = 0$ onwards is continuous current.
(ii) Solve the same problem if the current is alternating $I(t) = I_0 \cos(\omega t)$
(iii) Find the stationary solution (when all transient solutions damp out) if the alternating current is applied since $t = -\infty$.

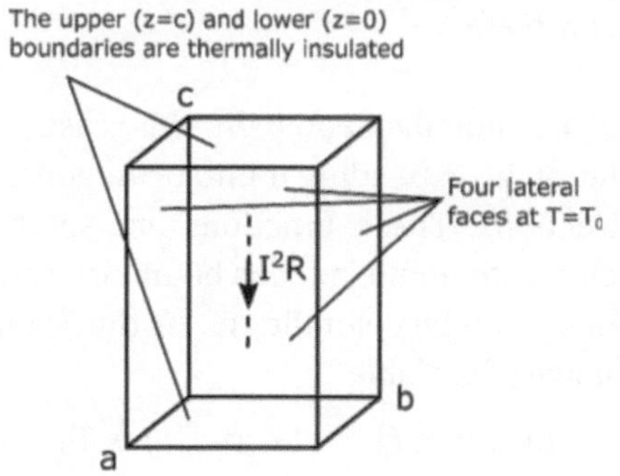

Figure 4.19

Mathematical formulation

$$C(x,y,z)\rho(x,y,z)\frac{\partial u(x,y,z,t)}{\partial t} - \frac{\partial}{\partial x}\left[k\frac{\partial u(x,y,z,t)}{\partial x}\right] - \frac{\partial}{\partial y}\left[k\frac{\partial u(x,y,z,t)}{\partial y}\right] + \frac{\partial}{\partial z}\left[k\frac{\partial u(x,y,z,t)}{\partial z}\right] = f(x,y,z,t) \tag{4.293}$$

There is a transient process until equilibrium is reached. In our case we will assume that k, C, ρ=constants.

The resulting equation with boundary conditions is:

$$\left\{\begin{array}{l} C\rho\dfrac{\partial u(x,y,z,t)}{\partial t} - k\Delta u(x,y,z) = f(x,y,z,t) \\ f(x,y,z) = \left\{\begin{array}{ll} 0 & \left(z<\frac{c}{4}\right) \\ \dfrac{2I^2R}{c}\delta\left(x-\frac{a}{2}\right)\delta\left(y-\frac{b}{2}\right) & \left(\frac{c}{4}<z<\frac{3c}{4}\right) \\ 0 & \left(z>\frac{3c}{4}\right) \end{array}\right\} \\ u(x,0,z) = u(x,b,z) = u(0,y,z) = u(a,y,z) = T_0 \\ \left.\dfrac{\partial u}{\partial z}\right|_{z=0} = \left.\dfrac{\partial u}{\partial z}\right|_{z=c} = 0 \end{array}\right\} \tag{4.294}$$

$$f(x,y,z) = \frac{2I^2R}{c}\delta\left(x-\frac{a}{2}\right)\delta\left(y-\frac{b}{2}\right)H\left[-\left(z-\frac{c}{2}\right)^2+\left(\frac{c}{4}\right)^2\right] \tag{4.295}$$

Sturm–Liouville problem

All boundaries are homogeneous. In this case it's possible to search for a solution by expanding it into orthogonal eigenfunctions in all three directions. These functions are solutions of Sturm–Liouville problems with homogeneous boundary conditions. To find the homogeneous boundary conditions in the y direction we can perform this change of variable:

$$u(x,y,z,t) = v(x,y,z,t) + T_0 \tag{4.296}$$

$$\left\{\begin{array}{ll} C\rho\dfrac{\partial v(x,y,z,t)}{\partial t} - k\Delta v(x,y,z) = f(x,y,z,t) & (t>0) \\ v(x,0,z) = v(x,b,z) = v(0,y,z) = v(a,y,z) = 0 & \\ \left.\dfrac{\partial v}{\partial z}\right|_{z=0} = \left.\dfrac{\partial v}{\partial z}\right|_{z=c} = 0 & \end{array}\right\} \tag{4.297}$$

We seek the solution by expanding in orthogonal eigenfunctions in the y, z directions, where we have homogeneous problems.

$$v(x,y,z,t) = \sum T(t)Q(x,y,z) \tag{4.298}$$

Eigenfunctions and eigenvalues of the Sturm–Liouville problem for $Q(x,y,z)$:

$$\left\{\begin{array}{l} \Delta Q(x,y,z) + \lambda Q(x,y,z) = 0 \\ Q(x,0,z) = Q(x,b,z) = Q(0,y,z) = Q(a,y,z) = 0 \\ \left.\dfrac{\partial Q}{\partial z}\right|_{z=0} = \left.\dfrac{\partial Q}{\partial z}\right|_{z=c} = 0 \end{array}\right\} \tag{4.299}$$

Eigenfunction

$$Q(x,y,z) = \sin\left(\frac{\pi n}{a}x\right)\sin\left(\frac{\pi m}{b}y\right)\cos\left(\frac{\pi l}{c}z\right) \tag{4.300}$$

$$\lambda = \left(\frac{\pi n}{a}\right)^2 + \left(\frac{\pi m}{b}\right)^2 + \left(\frac{\pi l}{c}\right)^2 \tag{4.301}$$

$n = 1, 2, 3\ldots;\ m = 1, 2, 3, \ldots;\ l = 0, 1, 2, 3\ldots$

General solution

We replace the solution $v(x,y,z,t) = \sum_{n,m,l} T_{nml}(t)Q_{nml}(x,y,z)$ into the equation:

$$C\rho\frac{\partial v(x,y,z,t)}{\partial t} - k\Delta v(x,y,z) = f(x,y,z) \tag{4.302}$$

and use the orthogonality of $Q_{nml}(x,y,z)$ we arrive at the non-homogeneous equation for $T_{nml}(t)$:

$$C\rho \sum_{n,m,l} \frac{\partial T_{nml}}{\partial t} Q_{nml}(x, y, z) + k \sum_{n,m,l} T_{nml}(t)\lambda_{nml} Q_{nml}(x, y, z) = f(x, y, z) \tag{4.303}$$

$$\sum_{n,m,l} \left[C\rho \frac{\partial T_{nml}}{\partial t} + k\lambda_{nml} T_{nml}(t) \right] Q_{nml}(x, y, z) = f(x, y, z) \tag{4.304}$$

Multiplying both sides by $Q_{nml}(x, y, z)$ and integrating we get to the non-homogeneous equation to find $T_{nml}(t)$:

$$\frac{\partial T_{nml}}{\partial t} + \frac{k}{C\rho} \lambda_{nml} T_{nml}(t) = \frac{1}{C\rho} \frac{\int_V f(x, y, z) Q_{nml}(x, y, z) dxdydz}{\int_V \left| Q_{nml}(x, y, z)^2 \right| dxdydz} = f_{nml} \tag{4.305}$$

Final solution

Seeking the solution in the form of a summation of solutions of the homogeneous equation and the particular solution, and applying the initial condition T_{nml} (0)=0 we find T_{nml} (t), and in this way the solution.

$$\int_V \left| Q_{nml}(x, y, z)^2 \right| dxdydz = \frac{abc}{8}$$

$$\int_V f(x, y, z) Q_{nml}(x, y, z) dxdydz =$$

$$\frac{2I^2 R}{l} \int_0^a \delta\left(x - \frac{a}{2}\right) \sin\left(\frac{\pi n}{a} x\right) dx \int_0^b \sin\left(\frac{\pi m}{b} y\right) \delta\left(y - \frac{b}{2}\right) dy$$
$$\int_{\frac{1}{4}c}^{\frac{3}{4}c} \cos\left(\frac{\pi l}{c} z\right) dz \tag{4.306}$$

We have:

$$\int_0^a \delta\left(x - \frac{a}{2}\right) \sin\left(\frac{\pi n}{a} x\right) dx = \sin\left(\frac{\pi n}{2}\right) \text{ (non-zero only for odd } n\text{)} \tag{4.307}$$

$$\int_0^b \sin\left(\frac{\pi m}{b}y\right)\delta\left(y-\frac{b}{2}\right)dy = \sin\left(\frac{\pi m}{2}\right) \text{ (non-zero only for odd } m) \tag{4.308}$$

$$\int_{\frac{1}{4}c}^{\frac{3}{4}c} \cos\left(\frac{\pi l}{c}z\right)dz = \frac{c}{\pi l}\left[\sin\left(\frac{3\pi l}{4}\right) - \sin\left(\frac{\pi l}{4}\right)\right] \tag{4.309}$$

The transient solution, which is composed by a particular solution and the solution of the non-homogeneous equations, and that satisfies the initial conditions, is:

$$T_{nml}(t) = -C\rho\frac{f_{nml}}{k\lambda_{nml}}[1 - e^{(-\frac{k}{C\rho}\lambda_{nml}t)}] \tag{4.310}$$

Finally the solution will be: $u(x, y, z, t) = T_0 - \sum_{n,m,l} C\rho\frac{f_{nml}}{k\lambda_{nml}}[1 - e^{(-\frac{k}{C\rho}\lambda_{nml}t)}] \sin\left(\frac{\pi n}{a}x\right)\sin\left(\frac{\pi m}{b}y\right)\cos\left(\frac{\pi l}{c}z\right)$

(ii) We now consider the solution when the applied current changes periodically in time $I = I_0\cos(\omega t)$.

The alternating current changes the non-homogeneous term of the equation and the dissipated power will be:

$$I^2R = R(I_0)^2[\cos(\omega t)]^2 = \frac{(I_0)^2R}{2} + \frac{(I_0)^2R}{2}\cos(2\omega t).$$

Then the inhomogeneous part changes to:

$$f(x,y,z) = \begin{cases} 0 & \left(z < \frac{c}{4}\right) \\ \dfrac{2\left[\frac{(I_0)^2R}{2} + \frac{(I_0)^2R}{2}\cos(2\omega t)\right]}{c}\delta\left(x-\frac{a}{2}\right)\delta\left(y-\frac{b}{2}\right) & \left(\frac{c}{4} < z < \frac{3c}{4}\right) \\ 0 & \left(z > \frac{3c}{4}\right) \end{cases} \tag{4.311}$$

In the same manner as we did in (i), replacing:

$$v(x, y, z, t) = \sum T(t)Q(x, y, z) \tag{4.312}$$

into the equation and applying the orthogonality of the $Q(x, y, z)$ functions we get to a differential equation for $T(t)$:

$$\frac{C\rho\partial T_{nml}}{\partial t} + k\lambda_{nml}T_{nml}(t) = \frac{\int_V f(x,y,z)Q_{nml}(x,y,z)dxdydz}{\int_V \left|Q_{nml}(x,y,z)^2\right| dxdydz}$$
$$= F_{nml}\frac{[1+\cos(2\omega t)]}{2} \quad (4.313)$$

The solution is made of a particular solution and a solution of the homogeneous equation.

The solution of the homogeneous equation is:

$$T_{nml}(part)(t) = A \times e^{[-\chi\lambda_{nml}t]} \quad (4.314)$$

Where $\chi = \frac{k}{C\rho}$ is the thermal diffusivity coefficient.

The particular solution of the non-homogeneous equation is sought as:

$$T_{nml}(nh)(t) = a \times \cos(2\omega t) + b \times \sin(2\omega t) + \text{Const} \quad (4.315)$$

Replacing this form into the non-homogeneous equation and applying the initial condition (trivial) we find the coefficients:

$$T_{nml}(t) = A \times e^{[-\chi\lambda_{nml}t]} + F_{nml}\left[\frac{1}{2k\lambda_{nml}} + \frac{1}{(2\omega\rho C)^2 + (k\lambda_{nml})^2}\right.$$
$$\left.\times\left[\frac{k\lambda_{nml}}{2}\cos(2\omega t) + C\rho\omega\sin(2\omega t)\right]\right] \quad (4.316)$$

with

$$A = -F_{nml}\left[\frac{1}{2k\lambda_{nml}} + \frac{k\lambda_{nml}}{2[(2\omega\rho C)^2 + (k\lambda_{nml})^2]}\right] \quad (4.317)$$

4.15 Cube with a Heater

A cube of side L has all of its surfaces thermally insulated. At its central point $(L/2, L/2, L/2)$ there is a point like heat source that supplies a heat density in the form $B \cdot \sin(\omega t)$ since $t = -\infty$. Find the stationary distribution of temperature. The thermal conductivity is k, its specific heat is C and its density is ρ.

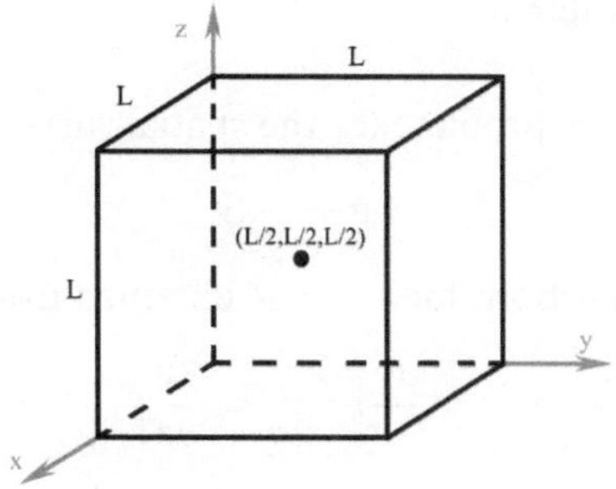

Figure 4.20

Mathematical formulation

The heat equation describes the distribution of temperature $u = u(x, y, z, t)$:

$$C\rho\frac{\partial u}{\partial t} - k\Delta u = B\sin(\omega t)\delta\left(x - \frac{L}{2}\right)\delta\left(y - \frac{L}{2}\right)\delta\left(z - \frac{L}{2}\right) \tag{4.318}$$

The boundary conditions at the insulated boundaries are:

$$\left.\frac{\partial u}{\partial x}\right|_{x=0} = \left.\frac{\partial u}{\partial x}\right|_{x=L} = \left.\frac{\partial u}{\partial y}\right|_{y=0} = \left.\frac{\partial u}{\partial y}\right|_{y=L} = \left.\frac{\partial u}{\partial z}\right|_{z=0} = \left.\frac{\partial u}{\partial z}\right|_{z=L} = 0 \tag{4.319}$$

Since the heat source has been acting for a long time, we can assume that the transient variations of temperature will have died out by now and u will oscillate with the same frequency ω as that of the heat source. In the heat equation there is a first order derivative with respect to time, so we will consider that the perturbation is of the form $Be^{i\omega t}\delta(\vec{r} - \vec{r}_0)$, with $\vec{r} = (x, y, z)$ and $\vec{r}_0 = L/2(1, 1, 1)$. Separating variables we will search for a solution of the form $\tilde{u}(x, y, z, t) = e^{i\omega t}v(x, y, z)$ and take only the imaginary part to obtain the final solution: $u = \Im\tilde{u}$. Replacing in the heat equation:

$$C\rho i\omega e^{i\omega t}v - ke^{i\omega t}\Delta v = Be^{i\omega t}\delta(\vec{r} - \vec{r}_0) \implies C\rho i\omega v - k\Delta v = B\delta(\vec{r} - \vec{r}_0) \tag{4.320}$$

Sturm–Liouville problem

The Sturm–Liouville problem for the spatial variables is:

$$\Delta v = -\lambda v \tag{4.321}$$

The boundary conditions for v can be obtained from those of u:

$$\left.\frac{\partial v}{\partial x}\right|_{x=0,\,L} = \left.\frac{\partial v}{\partial y}\right|_{y=0,\,L} = \left.\frac{\partial v}{\partial z}\right|_{z=0,\,L} = 0 \tag{4.322}$$

Separating variables: $v(x, y, z) \propto X(x)Y(y)Z(z)$, with boundary conditions $X'(0) = X'(L) = Y'(0) = Y'(L) = Z'(0) = Z'(L) = 0$.

$$\frac{\Delta v}{v} = \frac{1}{X(x)}\frac{d^2X}{dx^2} + \frac{1}{Y(y)}\frac{d^2Y}{dy^2} + \frac{1}{Z(z)}\frac{d^2Z}{dz^2} = -\lambda \tag{4.323}$$

Since each term is independent, all must be constant:

$$\begin{cases} X''(x) = -\nu X(x); & X'(0) = X'(L) = 0 \\ Y''(y) = -\mu Y(y); & Y'(0) = Y'(L) = 0 \\ Z''(z) = -\eta Z(z); & Z'(0) = Z'(L) = 0 \end{cases} \tag{4.324}$$

and the eigenvalue of the global problem will be the sum of those of the problems for each variable: $\lambda = \nu + \mu + \eta$. For the $X(x)$ function the general solution will be:

$$X(x) = A\cos(\sqrt{\nu}x) + B\sin(\sqrt{\nu}x)$$

Since $X'(0) = 0 \implies B = 0$. From the other boundary condition, to obtain a non-trivial solution: $X'(L) = 0 \implies \sqrt{\nu}L = n\pi$. For the y and z variables we have identical solutions:

$$\begin{cases} X_n(x) = \cos\left(\dfrac{n\pi x}{L}\right); & \nu_n = \left(\dfrac{n\pi}{L}\right)^2; & n = 0, 1, 2\ldots \\ Y_m(y) = \cos\left(\dfrac{m\pi y}{L}\right); & \mu_m = \left(\dfrac{m\pi}{L}\right)^2; & m = 0, 1, 2\ldots \\ Z_l(z) = \cos\left(\dfrac{l\pi z}{L}\right); & \eta_l = \left(\dfrac{l\pi}{L}\right)^2; & l = 0, 1, 2\ldots \end{cases} \tag{4.325}$$

The solution of the Sturm–Liouville problem is:

$$v(x, y, z) = A_{nml}\cos\left(\frac{n\pi x}{L}\right)\cos\left(\frac{m\pi y}{L}\right)\cos\left(\frac{l\pi z}{L}\right) \tag{4.326}$$

whose coefficients A_{nml} must be determined, with eigenvalues:

$$\lambda_{nml} = \nu_n + \mu_m + \eta_l = \frac{\pi^2}{L^2}(n^2 + m^2 + l^2) \tag{4.327}$$

Replacing into 4.320 the solution of the Sturm–Liouville problem:

$$\sum_{n,m,l} [C\rho i\omega + k\lambda_{nml}] A_{nml} X_n(x) Y_m(y) Z_l(z) = B\delta\left(x - \frac{L}{2}\right)\delta\left(y - \frac{L}{2}\right)\delta\left(z - \frac{L}{2}\right) \tag{4.328}$$

Final solution

We now apply the orthogonality of the eigenfunctions by multiplying both sides by $X_n(x)Y_m(y)Z_l(z)$ and integrate over the intervals $x \in [0, L]$, $y \in [0, L]$ and $z \in [0, L]$.

$$A_{nml}[C\rho i\omega + k\lambda_{nml}] \int_0^L\int_0^L\int_0^L \cos^2\left(\frac{n\pi x}{L}\right)\cos^2\left(\frac{m\pi y}{L}\right)\cos^2\left(\frac{l\pi z}{L}\right) dxdydz = B\int_0^L\int_0^L\int_0^L \cos\left(\frac{n\pi x}{L}\right)\cos\left(\frac{m\pi y}{L}\right)\cos\left(\frac{l\pi z}{L}\right)$$

$$\times\delta\left(x - \frac{L}{2}\right)\delta\left(y - \frac{L}{2}\right)\delta\left(z - \frac{L}{2}\right) dxdydz$$

$$= B\cos\left(\frac{n\pi}{2}\right)\cos\left(\frac{m\pi}{2}\right)\cos\left(\frac{l\pi}{2}\right)$$

$$\Longrightarrow A_{nml} = \frac{B\cos\left(\frac{n\pi}{2}\right)\cos\left(\frac{m\pi}{2}\right)\cos\left(\frac{l\pi}{2}\right)}{[C\rho i\omega + k\lambda_{nml}]\left|\cos\left(\frac{n\pi x}{L}\right)\right|^2\left|\cos\left(\frac{m\pi y}{L}\right)\right|^2\left|\cos\left(\frac{l\pi z}{L}\right)\right|^2} \tag{4.329}$$

where the square modulus of the $X_n(x)$ eigenfunction is:

$$\left|\cos\left(\frac{n\pi x}{L}\right)\right|^2 = \int_0^L \cos^2\left(\frac{n\pi x}{L}\right) dx = \begin{cases} L, & n = 0 \\ \frac{L}{2}, & n \neq 0 \end{cases}$$

with equivalent expressions for $Y_m(y)$ and $Z_l(z)$. We mention that as the average (background) temperature in the problem is not known, we should not consider the evaluation of the A_{000} term. We see that if at least one of the three indices (n, m, or l) is odd we have $A_{nml} = 0$. And if the three are even, $\cos\left(\frac{n\pi}{2}\right)\cos\left(\frac{m\pi}{2}\right)\cos\left(\frac{l\pi}{2}\right) = (-1)^{(n+m+l)/2}$.

With all this:

$$\tilde{u}(x, y, z, t) = \sum_{\substack{n,m,l \\ \text{even}}} \frac{e^{i\omega t}}{C\rho i\omega + k\lambda_{nml}} \times \frac{B(-1)^{\frac{n+m+l}{2}} \cos\left(\frac{n\pi x}{L}\right) \cos\left(\frac{m\pi y}{L}\right) \cos\left(\frac{l\pi z}{L}\right)}{\left|\cos\left(\frac{n\pi x}{L}\right)\right|^2 \left|\cos\left(\frac{m\pi y}{L}\right)\right|^2 \left|\cos\left(\frac{l\pi z}{L}\right)\right|^2} \tag{4.330}$$

To obtain the final solution $u(x, y, z, t)$ we will take the imaginary part of the last expression. We can rewrite the complex denominator:

$$C\rho i\omega + k\lambda_{nml} = \sqrt{(C\rho\omega)^2 + (k\lambda_{nml})^2}\, e^{i\,\arctan(C\rho\omega/(k\lambda_{nml}))} \tag{4.331}$$

$$\Im\left(\frac{e^{i\omega t}}{C\rho i\omega + k\lambda_{nml}}\right) = \Im\left(\frac{e^{i(\omega t - \arctan(C\rho\omega/(k\lambda_{nml})))}}{\sqrt{(C\rho\omega)^2 + (k\lambda_{nml})^2}}\right) = \frac{\sin\left(\omega t - \arctan\left(\frac{C\rho\omega}{k\lambda_{nml}}\right)\right)}{\sqrt{(C\rho\omega)^2 + (k\lambda_{nml})^2}} \tag{4.332}$$

The final solution is:

$$u(x, y, z, t) = \sum_{\substack{n,m,l \\ \text{even}}} \frac{\sin\left(\omega t - \arctan\left(\frac{C\rho\omega}{k\lambda_{nml}}\right)\right)}{\sqrt{(C\rho\omega)^2 + (k\lambda_{nml})^2}} \times \frac{B(-1)^{\frac{n+m+l}{2}} \cos\left(\frac{n\pi x}{L}\right) \cos\left(\frac{m\pi y}{L}\right) \cos\left(\frac{l\pi z}{L}\right)}{\left|\cos\left(\frac{n\pi x}{L}\right)\right|^2 \left|\cos\left(\frac{m\pi y}{L}\right)\right|^2 \left|\cos\left(\frac{l\pi z}{L}\right)\right|^2} \tag{4.333}$$

Chapter 5

Problems in Polar Coordinates

The previous chapters consider examples of problems in homogeneous spaces (using Cartesian coordinates). One of the simplest ways of breaking the symmetry is the presence of a symmetry point in a bi-dimensional space or of a symmetry axis in three-dimensional space.

The corresponding problems will be solved using cylindrical coordinates, which implies using new variables (angle, radius), keeping only a Cartesian variable to describe the space along the symmetry axis. As a consequence, the form of the Laplacian operator will change and the solutions of the Sturm–Liouville problem in the radial variable will be Bessel and Neumann functions. Also, the way to describe some features such as points, circles or thin cylinder changes, using the Dirac's Delta function in cylindrical coordinates. This chapter is limited to problems in cylindrical coordinates in two dimensions (i.e polar coordinates), or in three when the problem is infinite along the cylinder axis.

Mathematical Methods for Physics: Problems and Solutions
Farkhad G. Aliev and Antonio Lara

ISBN 978-981-4968-71-3 (Hardcover), 978-1-003-41088-1 (eBook)
www.jennystanford.com

5.1 Separation of Variables in a Circular Membrane

Find the eigenfunctions and eigenvalues of the Sturm–Liouville problem corresponding to a circular membrane of radius R with its border fixed.

Solution:

The problem is:

$$\left\{ \begin{array}{l} \Delta u + \lambda u = 0 \\ u(\rho = R) = 0 \end{array} \right\} \tag{5.1}$$

with $\lambda > 0$.

$$\frac{1}{\rho}\frac{\partial}{\partial \rho}\left[\rho \frac{\partial u}{\partial \rho}\right] + \frac{1}{\rho^2}\frac{\partial^2 u}{\partial \varphi^2} + \lambda u = 0 \tag{5.2}$$

We apply the method of separation of variables.

$$u = R(\rho) \cdot \Phi(\varphi) \tag{5.3}$$

$$\frac{1}{\rho}\frac{d}{d\rho}\left[\rho \frac{dR}{d\rho}\right]\Phi + \frac{R}{\rho^2}\frac{d^2\Phi}{d\varphi^2} + \lambda R(\rho) \cdot \Phi(\varphi) = 0 \tag{5.4}$$

We divide both sides by $R(\rho)\Phi(\varphi)$ and multiply by ρ^2

Then:

$$\frac{\rho \frac{d}{d\rho}\left[\rho \frac{dR}{d\rho}\right]}{R} + \lambda \rho^2 = -\frac{1}{\Phi}\frac{d^2\Phi}{d\varphi^2} = \mu = m^2 \tag{5.5}$$

Sturm–Liouville for the angular variable

$$\left\{ \begin{array}{l} \dfrac{d^2\Phi}{d\varphi^2} + \mu\Phi = 0 \\ \text{periodicity of } 2\pi \end{array} \right\} \tag{5.6}$$

Radial problem:

$$\frac{1}{\rho}\frac{d}{d\rho}\left[\rho \frac{dR}{d\rho}\right] + \left[\lambda - \frac{m^2}{\rho^2}\right] R = 0 \rightarrow \tag{5.7}$$

$$\frac{d^2R}{d\rho^2} + \frac{1}{\rho}\frac{dR}{d\rho} + \left[\lambda - \frac{m^2}{\rho^2}\right] R = 0 \tag{5.8}$$

This is Bessel's equation, which gives us the radial orthogonal eigenfunctions. Applying the boundary conditions will give us the eigenvalues of the problem.

5.2 Electric Potential in a Circular Sector: Case 1

A region is limited by three conductors: two perpendicular planes (electrically grounded) and a quarter of an infinite cylinder of radius R. The potential of the curved surface is V_0. Find the electrostatic potential at any point inside this region.

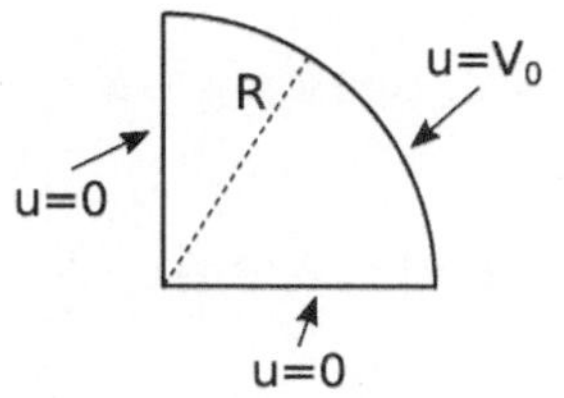

Figure 5.1

Mathematical formulation

Note: as the problem is infinite in the z direction the solution will not depend on this variable, due to symmetry reasons.

$$\left\{\begin{array}{l} \dfrac{1}{\rho}\dfrac{\partial}{\partial\rho}\left[\rho\dfrac{\partial u(\rho,\varphi)}{\partial\rho}\right]+\dfrac{1}{\rho^2}\dfrac{\partial u^2(\rho,\varphi)}{\partial\varphi^2}=0 \\ u(\rho,0)=u\left(\rho,\dfrac{\pi}{2}\right)=0 \\ u(R,\varphi)=V_0 \end{array}\right\} \tag{5.9}$$

Sturm–Liouville problem

Separating variables $u(\rho,\varphi) = Q(\rho)\Phi(\varphi)$ we arrive at two equations:

$$\frac{1}{Q}\left(\rho\frac{\partial}{\partial\rho}\left[\rho\frac{\partial Q(\rho,\varphi)}{\partial\rho}\right]\right)=-\frac{1}{\Phi}\left(\frac{\partial^2\Phi}{\partial\varphi^2}\right)=+\lambda \tag{5.10}$$

We impose the positive sign for λ so that we can expand in trigonometric functions of the angular variable, since in these variables the boundary conditions are homogeneous. The Sturm–

Liouville problem is:

$$\left\{\begin{array}{l} \dfrac{d^2\Phi}{d\varphi^2}+\lambda\Phi=0 \\ \Phi(0)=\Phi\left(\dfrac{\pi}{2}\right)=0 \end{array}\right\} \tag{5.11}$$

Eigenvalues and eigenfunctions:

$$\Phi_n(\varphi)=A_n\cos(\sqrt{\lambda}\varphi)+B_n\sin(\sqrt{\lambda}\varphi) \tag{5.12}$$

$$\Phi(0)=0\rightarrow A_n=0 \tag{5.13}$$

$$\Phi\left(\frac{\pi}{2}\right)=0\rightarrow B_n\sin\left(\sqrt{\lambda}\frac{\pi}{2}\right)=0 \tag{5.14}$$

Then $\sqrt{\lambda}=2n$ Eigenfunctions $\Phi_n(\varphi)=B_n\sin(2n\varphi)$;

Eigenvalues $\lambda=(2n)^2$; $n=1, 2, 3\ldots$

General solution

Now we solve the equation for the radial part.

$$\rho\frac{d}{d\rho}\left[\rho\frac{dQ(\rho,\varphi)}{d\rho}\right]-\lambda Q=0 \tag{5.15}$$

$$\rho^2\frac{d^2Q(\rho)}{d\rho^2}+\rho\left[\frac{dQ(\rho)}{d\rho}\right]-4n^2Q=0 \tag{5.16}$$

We look for the solution as $Q(\rho)=\rho^{\alpha}$ since all components of the equation are of the same order in the ρ variable.

Replacing, we get: $\alpha(\alpha-1)+\alpha-4n^2=0$

Then:

$$\alpha=\pm 2n \tag{5.17}$$

$$Q(\rho)=C_1\rho^{2n}+C_2\rho^{-2n} \tag{5.18}$$

Final solution

Since the solution is finite at $\rho = 0$ it is necessary that $C_2 = 0$. The general solution is:

$$u(\rho, \varphi) = \sum_n B_n \rho^{2n} \sin(2n\varphi) \tag{5.19}$$

We impose the boundary condition ($\rho = R$) to find the coefficients.

$$u(R, \varphi) = V_0 = \sum_n B_n R^{2n} \sin(2n\varphi) \tag{5.20}$$

Due to the orthogonality of the angular eigenfunctions, using the integrals:

$$\int_0^{\frac{\pi}{2}} \sin(2n\varphi) = \frac{\sin^2\left(\frac{n\pi}{2}\right)}{n} = \begin{Bmatrix} 0 \;\; n = 2k \\ \frac{1}{n} \;\; n = 2k+1 \end{Bmatrix} \tag{5.21}$$

$$\int_0^{\frac{\pi}{2}} \sin^2(2n\varphi) = \frac{\pi}{4} \tag{5.22}$$

We get the coefficients: $B_n = \frac{4V_0}{n\pi R^{2n}}$

Then:

$$u(\rho, \varphi) = \sum_{k=0}^{\infty} \frac{4V_0}{(2k+1)\pi} \left(\frac{\rho}{R}\right)^{2(2k+1)} \sin[2(2k+1)]\varphi \tag{5.23}$$

5.3 Electric Potential in a Circular Sector: Case 2

Find the distribution of electric potential inside a circular sector without charges, which spans an angle ($0 < \varphi < \alpha$), if the electric potential at the boundaries is as specified in the figure.

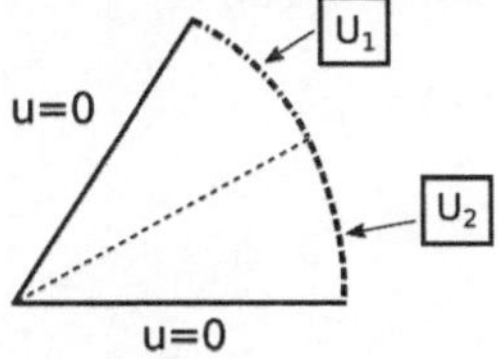

Figure 5.2

Mathematical formulation

$$\begin{Bmatrix} \Delta u = 0 \\ u(R) = f(\varphi) = \begin{Bmatrix} U_2 \left(0 < \varphi < \dfrac{\alpha}{2}\right) \\ U_1 \left(\dfrac{\alpha}{2} < \varphi < \alpha\right) \end{Bmatrix} \\ u(\rho, 0) = u(\rho, \alpha) = 0 \end{Bmatrix} \tag{5.24}$$

Sturm–Liouville problem

We separate variables to get to the eigenfunctions of the problem:

$$u(\rho, \varphi) = Q(\rho)\Phi(\varphi) \tag{5.25}$$

The Sturm–Liouville problem for $\Phi(\varphi)$ is:

$$\begin{Bmatrix} \dfrac{d^2\Phi}{d\varphi^2} + \lambda\Phi(\varphi) = 0 \\ \Phi(0) = \Phi(\alpha) = 0 \end{Bmatrix} \tag{5.26}$$

Angular eigenfunctions and eigenvalues:

$$\Phi(\varphi) = \sin\left(\frac{\pi n}{\alpha}\varphi\right) \tag{5.27}$$

$$\lambda_n = \left(\frac{\pi n}{\alpha}\right)^2 \tag{5.28}$$

General solution

The equation for the radial part is:

$$\rho^2 \frac{d^2 Q}{d\rho^2} + \rho \frac{dQ}{d\rho} - \lambda Q = 0 \tag{5.29}$$

With radial solutions: $Q(r) = \rho^{\frac{\pi n}{\alpha}}$

The general solution is:

$$u(\rho, \varphi) = \sum_{n=0}^{\infty} A_n \rho^{\frac{\pi n}{\alpha}} \sin\left(\frac{\pi n}{\alpha}\varphi\right) \tag{5.30}$$

Final solution

Imposing the boundary conditions and using the orthogonality of the eigenfunctions: $\sin\left(\frac{\pi n}{\alpha}\varphi\right)$

$$u(R,\varphi) = f(\varphi) = \sum_{n=1}^{\infty} A_n R^{\frac{\pi n}{\alpha}} \sin\left(\frac{\pi n}{\alpha}\varphi\right) \tag{5.31}$$

$$\int_0^{\alpha} f(\varphi)\sin\left(\frac{\pi n}{\alpha}\varphi\right) d\varphi = A_n R^{\frac{\pi n}{\alpha}} \int_0^{\alpha} \sin^2\left(\frac{\pi n}{\alpha}\varphi\right) d\varphi \tag{5.32}$$

We get the coefficients of the Fourier series:

$$A_n = \frac{2}{\alpha}\frac{\int_0^{\alpha} f(\varphi)\sin\left(\frac{\pi n}{\alpha}\varphi\right) d\varphi}{R^{\frac{\pi n}{\alpha}}} \tag{5.33}$$

$$\begin{aligned} A_n &= \frac{2}{\alpha R^{\frac{\pi n}{\alpha}}}\left[\int_0^{\alpha/2} U_2 \sin\left(\frac{\pi n}{\alpha}\varphi\right) d\varphi + \int_{\alpha/2}^{\alpha} U_1 \sin\left(\frac{\pi n}{\alpha}\varphi\right) d\varphi\right] = \\ &= \frac{2}{\alpha R^{\frac{\pi n}{\alpha}}}\left[U_2 \frac{\alpha}{\pi n}\left[1-\cos\left(\frac{\pi n}{2}\right)\right] + U_1 \frac{\alpha}{\pi n}\left[\cos\left(\frac{\pi n}{2}\right) - \cos(\pi n)\right]\right] \end{aligned} \tag{5.34}$$

5.4 Stationary Distribution of the Concentration of Particles in a Sector of an Infinite Cylinder

Consider a sector of a cylinder of infinite length, with radius $\rho = R$, with a diffusivity coefficient D. The angle of aperture is $0 < \varphi < \alpha$. One of the flat faces is in contact with the outer medium, which has a concentration of particles $n = 0$. The other flat face does not allow particles to go through it. The curved face exchanges particles with the outer medium with a flux density: $= -n(R,\varphi) - A\varphi$. Find the distribution of the concentration of particles inside the cylinder.

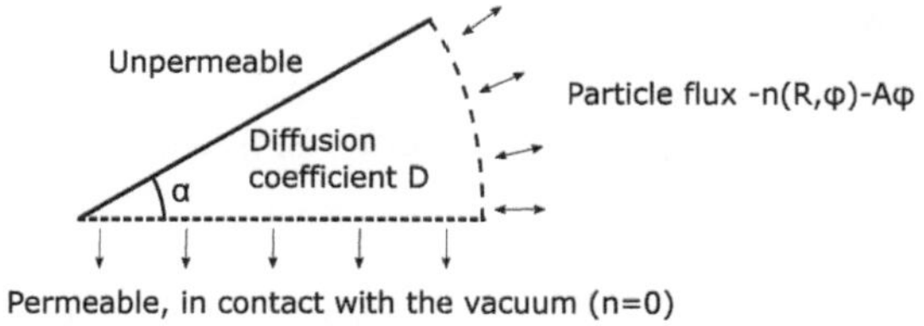

Figure 5.3

Mathematical formulation

We need to solve Laplace's problem (the diffusion equation is reduced to Laplace's one in the stationary case) in the angular sector $0 < \varphi < \alpha$, with boundary conditions:

$$\left\{\begin{array}{l} \Delta u = 0 \\ u(\rho, 0) = \left.\dfrac{\partial u}{\partial \varphi}\right|_{\varphi=\alpha} = 0 \\ -D \left.\dfrac{\partial u}{\partial \rho}\right|_{\rho=R} = -u(R, \varphi) - A\varphi \\ u(0, \varphi) < \infty \end{array}\right\} \tag{5.35}$$

Sturm–Liouville problem

Separating variables $u = \Phi(\varphi) \times Q(\rho)$ we arrive at a Sturm–Liouville problem for the angular variable, whose solution, once replaced into the Laplace's equation, gives an equation for ρ.

$$\left\{\begin{array}{l} \dfrac{d^2\Phi(\varphi)}{d\varphi^2} + \lambda\Phi(\varphi) = 0 \\ \left.\dfrac{\partial \Phi}{\partial \varphi}\right|_{\varphi=\alpha} = \Phi(0) = 0 \end{array}\right\} \tag{5.36}$$

Solution:

$$\left\{\begin{array}{l} \Phi(\varphi) = \sin\left(\pi \dfrac{(2n+1)}{2\alpha}\varphi\right) \\ \lambda_n = \left(\dfrac{\pi(2n+1)}{2\alpha}\right)^2 \qquad (n = 0, 1, 2 \ldots) \end{array}\right\} \tag{5.37}$$

General solution

In the radial direction the equation has a solution in the form of a power expansion:

$$\left\{\begin{array}{l} \rho^2 \dfrac{d^2 Q(\rho)}{d\rho^2} + \rho \dfrac{dQ(\rho)}{d\rho} - \left[\pi \dfrac{(2n+1)}{2\alpha}\right]^2 Q(\rho) = 0 \\ Q_n(\rho) = \rho^{\pm\left[\pi \frac{(2n+1)}{2\alpha}\right]} \end{array}\right\} \tag{5.38}$$

Applying the condition $u(0, \varphi) < \infty$ we obtain the form of the radial solutions:

$$Q_n(\rho) = \rho^{+\left[\pi \frac{(2n+1)}{2\alpha}\right]} \tag{5.39}$$

The general solution is:

$$u = \sum C_n \rho^{+\left[\pi \frac{(2n+1)}{2\alpha}\right]} \sin\left(\pi \frac{(2n+1)}{2\alpha} \varphi\right) \tag{5.40}$$

Final solution

We find the derivative of $u(\rho, \varphi)$ for $\rho = R$:

$$\left.\frac{\partial u}{\partial \rho}\right|_{\rho=R} = \sum C_n \left[\pi \frac{(2n+1)}{2\alpha}\right] R^{+\left[\pi \frac{(2n+1)}{2\alpha}\right]-1} \sin\left(\pi \frac{(2n+1)}{2\alpha} \varphi\right) \tag{5.41}$$

Applying the boundary condition for $\rho = R$

$$-D \sum C_n \left[\pi \frac{(2n+1)}{2\alpha}\right] R^{+\left[\pi \frac{(2n+1)}{2\alpha}\right]-1} \sin\left(\pi \frac{(2n+1)}{2\alpha} \varphi\right) = -\sum C_n R^{+\left[\pi \frac{(2n+1)}{2\alpha}\right]} \sin\left(\pi \frac{(2n+1)}{2\alpha} \varphi\right) - A\varphi \tag{5.42}$$

$$-A\varphi = \sum C_n R^{+\left[\pi \frac{(2n+1)}{2\alpha}\right]} \left(1 - D\left[\pi \frac{(2n+1)}{2\alpha}\right] \frac{1}{R}\right) \times \sin\left(\pi \frac{(2n+1)}{2\alpha} \varphi\right)$$

Finally, using the orthogonality of the angular eigenfunctions we get the coefficients C_n:

$$C_n R^{+\left[\pi \frac{(2n+1)}{2\alpha}\right]} \left(1 - D\left[\pi \frac{(2n+1)}{2\alpha}\right] \frac{1}{R}\right) = -\frac{2A}{\alpha} \int_0^{\alpha} \varphi \sin\left(\pi \frac{(2n+1)}{2\alpha} \varphi\right) d\varphi = -\frac{2A}{\alpha} \frac{(-1)^n}{\left[\pi \frac{(2n+1)}{2\alpha}\right]^2} \tag{5.43}$$

The final solution is:

$$u(\rho,\varphi) = \frac{2A}{\alpha}\sum_{n=0}^{\infty}\frac{(-1)^{n+1}}{\left[\pi\frac{(2n+1)}{2\alpha}\right]^2 R^{+\left[\pi\frac{(2n+1)}{2\alpha}\right]}\left(1-\frac{D}{R}\left[\pi\frac{(2n+1)}{2\alpha}\right]\right)}$$
$$\rho^{+\left[\pi\frac{(2n+1)}{2\alpha}\right]}\sin\left(\pi\frac{(2n+1)}{2\alpha}\varphi\right) \tag{5.44}$$

5.5 Instantaneous Hit on a Membrane with Circular Sector Form

Find the oscillations of a membrane with the form of a circular sector, of an angular aperture α and with radii r and R. The membrane is fixed in the curved part, free on one of the sides, and half-free on the other with a constant A presenting the relation between the spring constant and the membrane tension. The membrane, initially at rest, receives a point hit at the instant $t = 0$ at the location ($\varphi = 0,\ \rho = R/2$).

Note 1: Consider that the hit satisfies this condition for the initial velocity $\left.\frac{\partial u}{\partial t}\right|_{t=0}$: $\lim_{\epsilon\to 0}\int_{\Omega\epsilon}\left.\frac{\partial u}{\partial t}\right|_{t=0}\rho d\rho d\varphi = V_0$ being $\Omega\epsilon$ a surface of radius ϵ around the surroundings of the point where the hit is exerted.

Note 2: Consider that at the inner radius (r) the membrane is free. Suppose that $r/R << 1$.

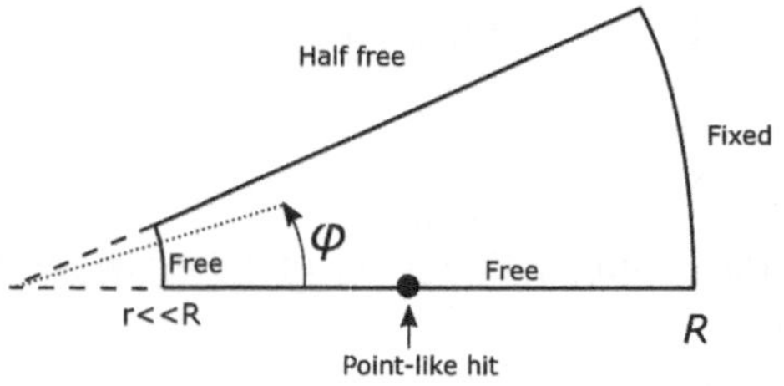

Figure 5.4

The hit is represented using the Dirac's delta function in cylindrical coordinates (see Appendix). We look for the density so that the total impulse transferred to the membrane is $\rho_0 V_0$:

$$\left.\frac{\partial u}{\partial t}\right|_{t=0} = \frac{V_0}{\rho}\delta\left(\rho - \frac{R}{2}\right)\delta(\varphi) \tag{5.45}$$

We see the relation between the total impulse transferred (I), the density of mass of the material (ρ_0) and the constant V_0.

$$I = \int_{\Omega\epsilon} \rho_0 \left.\frac{\partial u}{\partial t}\right|_{t=0} \rho d\rho d\varphi = \rho_0 V_0 \tag{5.46}$$

Mathematical formulation

$$\left\{\begin{array}{l} \dfrac{1}{a^2}\dfrac{\partial u^2(\rho,\varphi,t)}{\partial t^2} - \dfrac{1}{\rho}\dfrac{\partial}{\partial\rho}\left[\rho\dfrac{\partial u}{\partial\rho}\right] - \dfrac{1}{\rho^2}\dfrac{\partial u^2(\rho,\varphi,t)}{\partial\varphi^2} = 0 \\ u(\rho,\varphi,0) = 0; \\ \left.\dfrac{\partial u}{\partial t}\right|_{t=0} = \dfrac{V_0}{\rho}\delta\left(\rho - \dfrac{R}{2}\right)\delta(\varphi) \\ u(R,t) = 0; \left.\dfrac{\partial u}{\partial\rho}\right|_{\rho=r} = 0 \\ \left.\dfrac{\partial u}{\partial\varphi}\right|_{\varphi=\alpha} + Au(\rho,\alpha,t) = 0; \\ \left.\dfrac{\partial u}{\partial\varphi}\right|_{\varphi=0} = 0 \end{array}\right\} \tag{5.47}$$

a is the speed of sound of the membrane. The general solution for a sector of the membrane with one of the half fixed boundaries (boundary conditions of the third kind) and other of the second kind is:

$$u(\rho,\varphi,t) = \sum_{n,m}\left\{A_{nm}\cos[a\sqrt{\lambda_{nm}}t] + B_{nm}\sin[a\sqrt{\lambda_{nm}}t]\right\}$$
$$\times R_{\nu_m}(\sqrt{\lambda_{nm}}\rho)\cos(\nu_m\varphi) \tag{5.48}$$

Sturm–Liouville problem

The solution of (5.48) is obtained with the method of separation of variables:

$$u = \sum T(t)\cdot R(\rho)\cdot\Phi(\varphi) \tag{5.49}$$

When replacing this expression into the wave equation we get to the angular Sturm–Liouville problem, to lower the number of second derivatives:

$$\left\{\begin{array}{l} \dfrac{d^2\Phi(\rho,\varphi,t)}{d\varphi^2}+\nu\Phi=0 \\ \left.\dfrac{d\Phi}{d\varphi}\right|_{\varphi=\alpha}+A\Phi(\alpha)=0 \\ \left.\dfrac{d\Phi}{d\varphi}\right|_{\varphi=0}=0 \end{array}\right\} \tag{5.50}$$

To find the angular eigenfunctions first we look for the general solution of the Sturm–Liouville problem:

$$\Phi(\varphi)=C\cos(\nu\varphi)+D\sin(\nu\varphi) \tag{5.51}$$

Due to the fourth boundary condition, we have $D = 0$. The eigenvalues ν_m are sought by applying the third boundary condition:

$$\left.\frac{\partial u}{\partial\varphi}\right|_{\varphi=\alpha}+Au(\rho,\alpha,t)=0\rightarrow-\sqrt{\lambda_m}\sin(\nu_m\alpha)+A\cos(\nu_m\alpha)=0 \tag{5.52}$$

Then the eigenvalues with solutions of the equation $\tan(\nu_m\alpha)=\frac{A}{\nu_m}$ are all possible eigenvalues ν_m of the angular Sturm–Liouville (here the index m enumerates the eigenvalues).

Once the angular problem is solved, we arrive at the following equation for the radial variable (now there is no z variable and the angular eigenvalues are no longer integer numbers).

$$\frac{d^2R}{d\rho^2}+\frac{1}{\rho}\frac{dR}{d\rho}+\left(\lambda-\frac{[\nu_m]^2}{\rho^2}\right)R=0 \tag{5.53}$$

General solution

The solution of this problem gives us a set of radial solutions

$$R=C_1J_{\nu_m}(\sqrt{\lambda_{nm}}\rho)+C_2J_{-\nu_m}(\sqrt{\lambda_{nm}}\rho) \tag{5.54}$$

This is due to the fact that in general the index ν_m is not an integer.

The possible values λ_{nm} are the nth solutions of the equation obtained by imposing the condition DET=0 to the system of two equations with two unknowns at which we arrive by imposing the first and second boundary conditions:

$$\begin{Bmatrix} C_1 J_{\nu_m}(\sqrt{\lambda_{nm}}R) + C_2 J_{-\nu_m}(\sqrt{\lambda_{nm}}R) = 0 \\ C_1 J'_{\nu_m}(\sqrt{\lambda_{nm}}r) + C_2 J'_{-\nu_m}(\sqrt{\lambda_{nm}}r) = 0 \end{Bmatrix} \tag{5.55}$$

Then:

$$J_{\nu_m}(\sqrt{\lambda_{nm}}R) \cdot J'_{-\nu_m}(\sqrt{\lambda_{nm}}r) - J_{-\nu_m}(\sqrt{\lambda_{nm}}R) \cdot J'_{\nu_m}(\sqrt{\lambda_{nm}}r) = 0 \tag{5.56}$$

Also from the first or second boundary conditions we can obtain the ratio between the coefficients C_1 and C_2 and in this manner determine the form of the radial function.

General solution:

$$u(\rho, \varphi, t) = \sum_{n,m} B_{nm} \sin[a\sqrt{\lambda_{nm}}t] R_{\nu_m}(\sqrt{\lambda_{nm}}\rho) \cos(\nu_m \varphi) \tag{5.57}$$

Final solution

One the form of the general solution is known, we look for the coefficients of the expansion using the initial conditions. From the first one we have: $A_{nm} = 0$, and from the second:

$$\begin{aligned} \left.\frac{\partial u}{\partial t}\right|_{t=0} &= \frac{V_0}{\rho}\delta\left(\rho - \frac{R}{2}\right)\delta(\varphi) \\ &= \sum_{n,m} (a\sqrt{\lambda_{nm}}) B_{nm} \cos[a\sqrt{\lambda_{nm}}0] R_{\nu_m}(\sqrt{\lambda_{nm}}\rho) \cos(\nu_m \varphi) \end{aligned} \tag{5.58}$$

We use the orthogonality of the radial and angular eigenfunctions to find the coefficients B_{nm}.

Multiplying both sides of the previous relation by $R_{\nu_m}(\sqrt{\lambda_{kl}}\rho) \cos(\nu_l \varphi)$ and integrating $\int_r^R \int_0^\alpha \rho d\rho d\varphi$

$$V_0 \int_r^R \int_0^\alpha R_{\nu_l}(\sqrt{\lambda_{kl}}\rho) \cos(\nu_l \varphi) \frac{1}{\rho}\delta\left(\rho - \frac{R}{2}\right)\delta(\varphi) \rho d\rho d\varphi =$$

$$= \sum_{n,m} B_{nm}(a\sqrt{\lambda_{nm}}) \int_r^R \int_0^\alpha R_{\nu_m}(\sqrt{\lambda_{nm}}\rho) R_{\nu_l}(\sqrt{\lambda_{kl}}\rho)$$

$$\cos(\nu_l \varphi) \cos(\nu_m \varphi) \rho d\rho d\varphi \tag{5.59}$$

Due to the orthogonality of the radial and angular eigenfunctions we find the coefficients:

$$B_{nm} = \frac{V_0 \int\limits_r^R \int\limits_0^\alpha R_{\nu_l}(\sqrt{\lambda_{kl}}\rho)\cos(\nu_l\varphi)\frac{1}{\rho}\delta(\rho - \frac{R}{2})\delta(\varphi)\rho d\rho d\varphi}{(a\sqrt{\lambda_{nm}})\left\|R_{\nu_m}(\sqrt{\lambda_{nm}}\rho)\right\|^2 \|\cos(\nu_m\varphi)\|^2}$$
$$= \frac{V_0 R_{\nu_m}(\sqrt{\lambda_{nm}}\frac{R}{2})}{(a\sqrt{\lambda_{nm}})\left\|R_{\nu_m}(\sqrt{\lambda_{nm}}\rho)\right\|^2 \|\cos(\nu_m\varphi)\|^2} \tag{5.60}$$

The final solution is:

$$u(\rho, \varphi, t) = \sum_{n,m} \frac{V_0 R_{\nu_m}(\sqrt{\lambda_{nm}}\frac{R}{2})}{(a\sqrt{\lambda_{nm}})\left\|R_{\nu_m}(\sqrt{\lambda_{nm}}\rho)\right\|^2 \|\cos(\nu_m\varphi)\|^2}$$
$$\times \sin[a\sqrt{\lambda_{nm}}t] R_{\nu_m}(\sqrt{\lambda_{nm}}\rho)\cos(\nu_m\varphi) \tag{5.61}$$

5.6 Linear Heating of a Disk

Find the variation in temperature inside a disk of radius R if at $t = 0$ it is heated by uniformly distributed heat sources. The density of heat sources is equal to $AC\rho_0 t$ (being A a constant, C the heat capacity and ρ_0 the density of the material) due to the supplied power increases linearly with time. The initial temperature of the disk at $t = 0$ equals zero. The outer boundary of the disk ($\rho = R$) is kept in contact with a thermal reservoir at zero temperature. Consider that the coefficient of thermal diffusivity equals a^2.

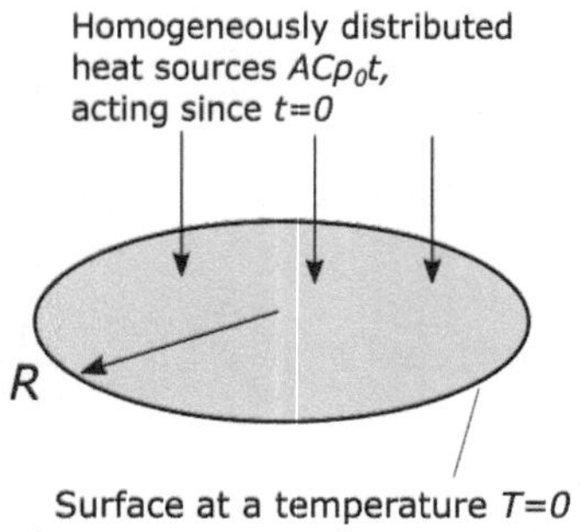

Figure 5.5

Mathematical formulation

We need to solve the non-homogeneous Fourier problem with homogeneous boundary conditions:

$$\left\{\begin{array}{l} \dfrac{\partial u}{\partial t} - a^2 \Delta u(\rho, \varphi) = \mathrm{At} \ \ (t > 0) \\ u(\rho = R, \varphi, t) = 0 \\ u(\rho, \varphi, t = 0) = 0 \end{array}\right\} \tag{5.62}$$

Sturm–Liouville problem

The solution is expanded into series of the orthogonal eigenfunctions of the Sturm–Liouville problem in a circle of radius R with homogeneous boundary conditions:

$$\left\{\begin{array}{l} \Delta v(\rho, \varphi) + \lambda v(\rho, \varphi) = 0 \\ v(R, \varphi) = 0 \end{array}\right\} \tag{5.63}$$

The eigenfunctions of the problem are known and consist of radial and angular solutions:

$$v_{nk}(\rho, \varphi) = A_{nk} J_n\left(\sqrt{\lambda_n^{(k)}}\rho\right) \cdot \cos(n\varphi) + B_{nk} J_n\left(\sqrt{\lambda_n^{(k)}}\rho\right) \cdot \sin(n\varphi) \tag{5.64}$$

being $\lambda_n^{(k)}$ solutions of the equation $J_n\left(\sqrt{\lambda_n^{(k)}}R\right) = 0$

Solution

We seek a general solution in the form:

$$u(\rho, \varphi, t) = \sum_{n,k} w_{nk}(t) v_{nk}(\rho, \varphi) \tag{5.65}$$

Since the solution must not depend on the angle it can be simplify by just keeping the eigenfunctions with angular symmetry (with $n = 0$):

$$u(\rho, \varphi, t) = \sum_{k} w_k(t) \cdot J_0\left(\sqrt{\lambda_0^{(k)}}\rho\right) \tag{5.66}$$

Replacing this general solution into equation (5.62), and using the orthogonality of the eigenfunctions:

$$J_0\left(\sqrt{\lambda_0^{(k)}}\rho\right) \tag{5.67}$$

we arrive at an equation for $w_k(t)$:

$$\left\{ \begin{array}{l} \dfrac{dw_k}{dt} + a^2\lambda_0^{(k)} w_k(t) = At \cdot f_k \\ w_k(t=0) = 0 \end{array} \right\} \tag{5.68}$$

The values of f_k are obtained from:

$$\begin{aligned} f_k &= \frac{1}{\left\| J_0\left(\sqrt{\lambda_0^{(k)}}\rho\right) \right\|^2} \int_0^R J_0\left(\sqrt{\lambda_0^{(k)}}\rho\right) \rho d\rho \\ &= \frac{2}{R \cdot \sqrt{\lambda_0^{(k)}} \cdot J_1\left(\sqrt{\lambda_0^{(k)}} R\right)} \end{aligned} \tag{5.69}$$

Here we have used: $\frac{d[x^n J_n]}{dx} = x^n J_{n-1}(x)$ and also the condition of orthogonality of the Bessel functions (the quote symbol (') indicates derivative):

$$\int_0^R J_\nu\left(\sqrt{\lambda_\nu^{(k)}}\rho\right) J_\nu\left(\sqrt{\lambda_\nu^{(l)}}\rho\right) \rho d\rho = \frac{R^2}{2}\left[J_\nu'\left(\sqrt{\lambda_\nu^{(l)}} R\right)\right]^2 \delta_{kl} \tag{5.70}$$

The solution of the problem 5.68 is expressed as the sum of a particular solution $s_p(t)$ (which subtracts the non-homogeneous part of equation 5.68) plus a homogeneous solution $s(t)$. Finally we will apply the initial condition to find the coefficients: Then the solution of 5.68 is:

$$w_k(t) = s(t) + s_p(t) \tag{5.71}$$

We propose the particular solution as:

$$s_p(t) = Bt + C \tag{5.72}$$

Replacing this expression in 5.68:

$$B = \frac{Af_k}{a^2\lambda_0^{(k)}} \tag{5.73}$$

$$C = -\frac{Af_k}{[a^2\lambda_0^{(k)}]^2} \tag{5.74}$$

Equation for the homogeneous solution $s(t)$:

$$\frac{ds}{dt} + a^2\lambda_0^{(k)} s(t) = 0 \tag{5.75}$$

$$s(0) = -s_p(0) = \frac{Af_k}{[a^2\lambda_0^{(k)}]^2} \tag{5.76}$$

The solution is an exponential function. The solution of the non-homogeneous equation which satisfies the initial conditions is:

$$\begin{aligned} w_k(t) &= \frac{Af_k}{a^2\lambda_0^{(k)}}\left[t - \frac{1}{a^2\lambda_0^{(k)}}\right] + \frac{Af_k}{[a^2\lambda_0^{(k)}]^2} e^{\left(-a^2\lambda_0^{(k)}t\right)} \\ &= \frac{Af_k}{a^2\lambda_0^{(k)}}\left[t - \frac{1}{a^2\lambda_0^{(k)}}\left(1 - e^{\left(-a^2\lambda_0^{(k)}t\right)}\right)\right] \end{aligned} \tag{5.77}$$

The final solution is:

$$u(\rho, \varphi, t) = \sum_k w_k(t) \cdot J_0\left(\sqrt{\lambda_0^{(k)}}\rho\right) \tag{5.78}$$

5.7 Case Study: Laplace's Problem in a Sector with Non-Homogeneous Boundary Conditions

The straight sides of a circular sector (with aperture angle α and radius R) are ate temperatures T_1, T_2, while the temperature of the curved part is $f(\varphi)$ (φ is the angular variable).

Find the stationary distribution of temperature inside the sector.

Mathematical formulation

We apply the principle of linearity on the Laplace's equation to solve the problem in the sector with non-homogeneous boundary conditions (see figure).

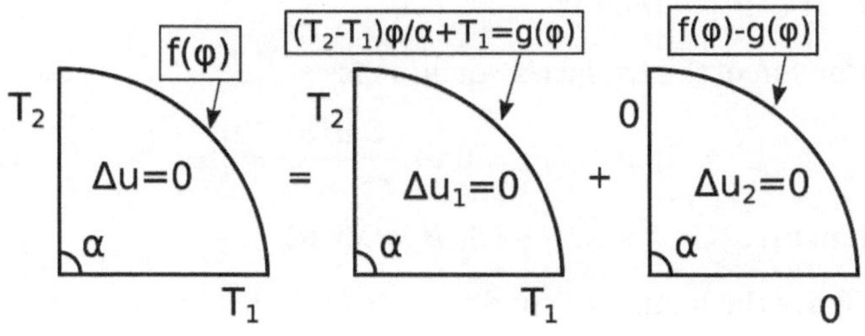

Figure 5.6

We note that the solution cannot be decomposed in any other way, since only in this manner can we have homogeneous boundary conditions in the angular variable. In this fashion we have the possibility to expand the solutions in orthogonal eigenfunctions.

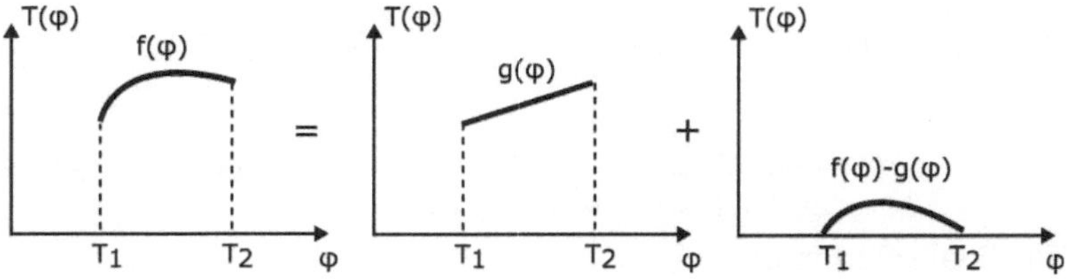

Figure 5.7

If $u(r, \varphi)$ is the solution of the initial problem $u_1(r, \varphi)$ and $u_2(r, \varphi)$ are solutions of problems (b) and (c).

Separating the solution in two we have:

$$\left\{\begin{array}{l} \Delta u(r, \varphi) = \Delta u_1(r, \varphi) + \Delta u_2(r, \varphi) = 0 \\ u(r, 0) = u_1(r, 0) + u_2(r, 0) = T_1 + 0 = T_1 \\ u(r, \alpha) = u_1(r, \alpha) + u_2(r, \alpha) = T_2 + 0 = T_2 \\ u(R, \varphi) = u_1(R, \varphi) + u_2(R, \varphi) = g(\varphi) + f(\varphi) - g(\varphi) = f(\varphi) \end{array}\right\} \tag{5.79}$$

To separate the problem in two helps to separate the solution of the initial problem (a) in two solutions, one of which only depends on the angle $u_1(r, \varphi) = u_1(\varphi)$, while the other $u_2(r, \varphi)$ has homogeneous boundary conditions, which allow to expand the solution in Fourier series for the angular variable:

If $u_1(r, \varphi) = u_1(\varphi)$ then $\frac{\partial u}{\partial r} = \frac{\partial^2 u}{\partial r^2} = 0$

The problem of the Laplace's equation is:

$$\Delta u_1(r, \varphi) = 0 \rightarrow \frac{1}{r^2} \frac{d^2 u_1}{d\varphi^2} = 0 \tag{5.80}$$

Solution: $u_1(\varphi) = A \times \varphi + B$ $(A, B = \text{Const})$

We will use the boundary conditions to find the function u_1:

First condition: $u_1(\varphi = 0) = T_1 \rightarrow B = T_1$

Second condition: $u_1(\varphi = \alpha) = T_2 \to A = \frac{T_2 - T_1}{\alpha}$

Then $u_1(\varphi) = \frac{T_2 - T_1}{\alpha} \times \varphi + T_1$ is the solution of problem (1) and does not depend on the radial variable.

General solution

The general solution of the problem has been obtained before (problem of the electric potential inside a circular sector):

$$u_2(r, \varphi) = \sum_{n=1}^{\infty} A_n r^{\frac{\pi n}{\alpha}} \sin\left(\frac{\pi n}{\alpha}\varphi\right) \tag{5.81}$$

We just need to apply the boundary condition at $r = R$ to find the coefficients by applying the orthogonality of the angular eigenfunctions

$$f(\varphi) - g(\varphi) = \sum_{n=1}^{\infty} A_n R^{\frac{\pi n}{\alpha}} \sin\left(\frac{\pi n}{\alpha}\varphi\right) \tag{5.82}$$

We will now find the solution u_2 for a specific case of a sector with radius $R = 1$. Let us consider the following boundary conditions:

$$T_1 = 0;\ T_2 = 1;\ \alpha = \frac{\pi}{4};\ f(1, \varphi) = 3\sin(4\varphi) \tag{5.83}$$

The problem to find the function $u_2(r, \varphi)$ is:

$$\left\{\begin{array}{l} \Delta u_2(r, \varphi) = 0 \\ u_2(r, 0) = 0 \\ u_2(r, \alpha) = 0 \\ u_2(R, \varphi) = f(\varphi) - g(\varphi) = 3\sin(4\varphi) - \dfrac{4\varphi}{\pi} \end{array}\right\} \tag{5.84}$$

The general solution in the specific case will be:

$$u_2(r, \varphi) = \sum (r)^{4n}[A_n \sin(4n\varphi)] \tag{5.85}$$

Final solution

The final solution will be obtained by imposing the boundary conditions:

$$u_2(R, \varphi) = 3\sin(4\varphi) - \frac{4\varphi}{\pi} = \sum (1)^{4n}[A_n \sin(4n\varphi)] \tag{5.86}$$

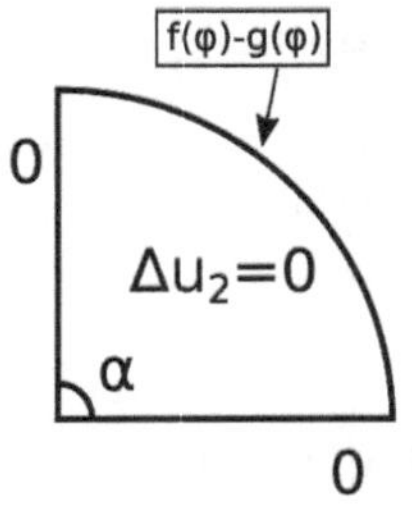

Figure 5.8

and using the orthogonality of the eigenfunctions $\sin(4n\varphi)$ in the interval $(0 < \varphi < \frac{\pi}{4})$ we arrive at the final solution:

$$A_n = \frac{2}{(\pi/4)} \int_0^{\pi/4} \left[3\sin(4\varphi) - \frac{4\varphi}{\pi}\right] \sin(4n\varphi) d\varphi$$

$$= \left\{ \begin{array}{ll} A_1 = 3 + \dfrac{2}{\pi} & n = 1) \\ A_n = 2\dfrac{(-1)^{n+1}}{\pi n} & (n \geq 2) \end{array} \right\} \tag{5.87}$$

Then:

$$u_2(r, \varphi) = \left(3 + \frac{2}{\pi}\right) r^4 \sin(4\varphi) + 2\sum_{n=2}^{\infty} \frac{(-1)^{n+1}}{\pi n} (r)^{4n} \sin(4n\varphi) \tag{5.88}$$

5.8 Case Study: Temperature Distribution in a Disk with Heaters

A disk of radius R with heat capacity $C = 1$, density $\rho_0 = 1$, thermal conductivity $k = 1$ and thermal diffusivity $\chi = 1$, has its outer surface in contact with a thermal reservoir at zero temperature . Find the stationary distribution of temperature inside the disk if in its inside there are heat sources with density $f = F \cdot xy$.

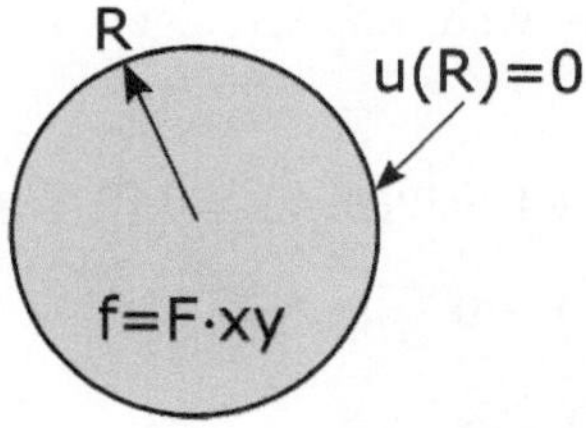

Figure 5.9

Mathematical formulation

$$\begin{cases} C\rho_0 \dfrac{\partial u}{\partial t} - k\Delta u(\rho, \varphi) = Fxy & (0 \le \rho \le R) \\ \dfrac{\partial u}{\partial t} - \chi \Delta u(\rho, \varphi) = \dfrac{F}{C\rho_0} xy = \dfrac{F}{C\rho_0}\rho^2 \sin(\varphi)\cos(\varphi) = \dfrac{F}{C\rho_0}\rho^2[\frac{1}{2}\sin(2\varphi)] & (0 \le \rho \le R) \\ \Delta u(\rho, \varphi) = -\dfrac{F}{2k}\rho^2 \sin(2\varphi) = -\dfrac{F}{2}\rho^2 \sin(2\varphi) & (0 \le \rho \le R) \\ u(R, \varphi, t) = 0 \\ u(0, \varphi, t) < \infty \end{cases} \tag{5.89}$$

Due to the symmetry of the heat sources, the problem can be solved in several ways:

Method 1

Sturm–Liouville problem

As it is a Poisson's problem in a circle without a hole at its center and with homogeneous boundary conditions, we can seek a solution in the form of a sum of orthogonal functions:

$$u(\rho, \varphi) = \sum_{n,m} [J_m(\sqrt{\lambda_{nm}}\rho) + DN_m(\sqrt{\lambda_{nm}}\rho)][A_{nm}\sin(m\varphi) + B_{nm}\cos(m\varphi)] \tag{5.90}$$

And to satisfy the second boundary condition:

$$u(\rho, \varphi) = \sum_{n,m} J_m(\sqrt{\lambda_{nm}}\rho)[A_{nm}\sin(m\varphi) + B_{nm}\cos(m\varphi)] = \sum_{n,m} C_{nm} v_{nm}(\rho, \varphi) \tag{5.91}$$

where $v_{nm}(\rho, \varphi)$ are well known eigenfunction of the Sturm–Liouville problem.

$$\left\{\begin{array}{l} \Delta v(\rho, \varphi) + \lambda v(\rho, \varphi) = 0 \ (0 \le \rho \le R) \\ v(R, \varphi) = 0 \\ v(0, \varphi) < \infty \end{array}\right\} \tag{5.92}$$

Solution

We replace this in the equation:

$$\sum_{n,m} \lambda_{nm} J_m(\sqrt{\lambda_{nm}}\rho)[A_{nm}\sin(m\varphi) + B_{nm}\cos(m\varphi)] = \frac{F}{2}\rho^2 \sin(2\varphi) \tag{5.93}$$

We use the orthogonality of the angular functions $\cos(m\varphi)$, multiplying both sides of (5.93) and integrating from 0 to 2π, with which we get at $B_{nm} = 0$.

The orthogonality of the Bessel functions and of $\sin(m\varphi)$ is used, multiplying by them both sides of (5.93) and integrating from 0 to 2π and we check that only the coefficients $A_{n2} \neq 0$.

$$A_{nm}\lambda_{n2} \int_0^R [J_m(\sqrt{\lambda_{nm}}\rho)]^2 \rho d\rho \int_0^{2\pi} |\sin(2\varphi)|^2 \, d\varphi = \frac{F}{2} \int_0^R J_m(\sqrt{\lambda_{nm}}\rho)\rho^3 d\rho \int_0^{2\pi} |\sin(2\varphi)|^2 \, d\varphi \tag{5.94}$$

Finally:

$$A_{n2} = \frac{F}{2} \frac{\int_0^R J_m(\sqrt{\lambda_{nm}}\rho)\rho^3 d\rho}{\lambda_{n2} \int_0^R [J_m(\sqrt{\lambda_{nm}}\rho)]^2 \rho d\rho} \tag{5.95}$$

Solution as an expansion of Bessel functions:

$$u(\rho, \varphi) = \sin(2\varphi) \sum_n A_{n2} J_2(\sqrt{\lambda_{n2}}\rho) \tag{5.96}$$

Method 2

We will separate the problem in two: an inhomogeneous one, where we will not care about the boundaries, with a particular solution s and another one for the function v, with a homogeneous equation, with inhomogeneous boundary conditions to satisfy the homogeneous boundary conditions of the total solution.

$$u(\rho, \varphi) = s(\rho, \varphi) + v(\rho, \varphi) \tag{5.97}$$

The heat equation is multiplied by ρ^2. The inhomogeneous problem with a particular solution must satisfy the equation:

$$\rho^2 \frac{\partial^2 u}{\partial \rho^2} + \rho \frac{\partial u}{\partial \rho} + \frac{\partial^2 u}{\partial \varphi^2} = -\frac{F}{2}\rho^4 \sin(2\varphi) \tag{5.98}$$

$$u(R, \varphi) = 0 \tag{5.99}$$

Due to the symmetry of the inhomogeneous part, the particular solution can be sought in the form:

$$s(\rho, \varphi) = w(\rho)\sin(2\varphi) \tag{5.100}$$

Replacing in (5.98) we get an equation for $w(\rho)$:

$$\rho^2 \frac{d^2 w}{d\rho^2} + \rho \frac{dw}{d\rho} - 4w = -\frac{F}{2}\rho^4 \tag{5.101}$$

We can seek the solution in the form $w(\rho) = C\rho^\alpha$, since all terms to the left side are raised to the same power:

$$\alpha(\alpha - 1)\rho^\alpha + \alpha\rho^\alpha - 4\rho^\alpha = -\frac{F}{2}\rho^4 \tag{5.102}$$

$$C[\alpha(\alpha - 1) + \alpha - 4]\rho^\alpha = -\frac{F}{2}\rho^4 \tag{5.103}$$

From where: $\alpha = 4; C = -\frac{F}{24}$. Then:

$$s(\rho, \varphi) = -\frac{F}{24}\rho^4 \sin(2\varphi) \tag{5.104}$$

We formulate the Laplace's problem for $v(\rho, \varphi)$

$$\begin{Bmatrix} \Delta v(\rho, \varphi) = 0 \\ v(R, \varphi) = -s(R, \varphi) \end{Bmatrix} \tag{5.105}$$

General solution

The general solution of the Laplace's problem in the (whole) disk is:

$$v(\rho, \varphi) = \sum_m \rho^m [A_m \sin(m\varphi) + B_m \cos(m\varphi)] \tag{5.106}$$

Final solution

Applying the boundary conditions and using the orthogonality of the angular function we find the coefficients:

$$\sum_m R^m [A_m \sin(m\varphi) + B_m \cos(m\varphi)] = \frac{F}{24} R^4 \sin(2\varphi) \tag{5.107}$$

$$B_m = 0 \tag{5.108}$$

Due to the angular asymmetry, only the A_2 coefficient remains.

$$R^2 A_2 = \frac{F}{24} R^4 \tag{5.109}$$

$$A_2 = \frac{F}{24} R^2 \tag{5.110}$$

Then: $v(\rho, \varphi) = \frac{F}{24} R^2 \rho^2 \sin(2\varphi)$. Finally the solution of the problem (as a power expansion) is:

$$u(\rho, \varphi) = s(\rho, \varphi) + v(\rho, \varphi) = \frac{F}{24} \rho^2 [R^2 - \rho^2] \sin(2\varphi) \tag{5.111}$$

Method 3

This is a modified version of Method 2. Instead of searching for a solution as an expansion in 2D orthogonal functions (radial and angular in Method 1) we will only use the orthogonality of the angular functions.

$$u(\rho, \varphi) = \sum_m R(\rho) \cdot \Phi_m(\varphi) \tag{5.112}$$

with

$$\frac{d^2 \Phi_m(\varphi)}{d\varphi^2} = -m^2 \Phi_m(\varphi) \tag{5.113}$$

Replacing the solution into the heat equation, rewritten as:

$$\rho^2 \frac{\partial^2 u}{\partial \rho^2} + \rho \frac{\partial u}{\partial \rho} + \frac{\partial^2 u}{\partial \varphi^2} = -\frac{F}{2}\rho^4 \sin(2\varphi) \quad (5.114)$$

$$u(R, \varphi) = 0 \quad (5.115)$$

$$u(0, \varphi) < \infty \quad (5.116)$$

We get

$$\sum_m \Phi_m(\varphi) \left[\rho^2 \frac{d^2 R}{d\rho^2} + \rho \frac{dR}{d\rho} - m^2 R\right] = -\frac{F}{2}\rho^4 \sin(2\varphi) \quad (5.117)$$

or

$$\sum_m [A_m \sin(m\varphi) + B_m \cos(m\varphi)] \quad \left[\rho^2 \frac{d^2 R}{d\rho^2} + \rho \frac{dR}{d\rho} - m^2 R\right]$$
$$= -\frac{F}{2}\rho^4 \sin(2\varphi) \quad (5.118)$$

Using the orthogonality of the eigenfunctions $\Phi_m(\varphi)$ we have $B_m = 0$. In a similar fashion, multiplying by $\sin(m\varphi)$ and integrating from 0 to 2π we see that only the term with $m = 2$ remains. With this we can formulate an equation for the radial function (without having normalized angular eigenfunctions we can suppose $A_2 = 1$).

$$\rho^2 \frac{d^2 R}{d\rho^2} + \rho \frac{dR}{d\rho} - 4R = -\frac{F}{2}\rho^4 \quad (5.119)$$

Similar to Method 2, we find the particular solution and sum it to the solution of the homogeneous equation, we use the boundary condition to find $R(\rho)$.

5.9 Diffusion in an Infinite Cylinder with Heat Sources

Find the distribution of temperature in an infinite cylinder of radius R if, starting at $t = 0$, inside a cylinder a heat source with density $f = -xy$ starts acting. Consider that the cylinder surface is always in contact with a heat reservoir at T_0. The heat capacity is C, the density is ρ_0, the coefficient of thermal conductivity is k.

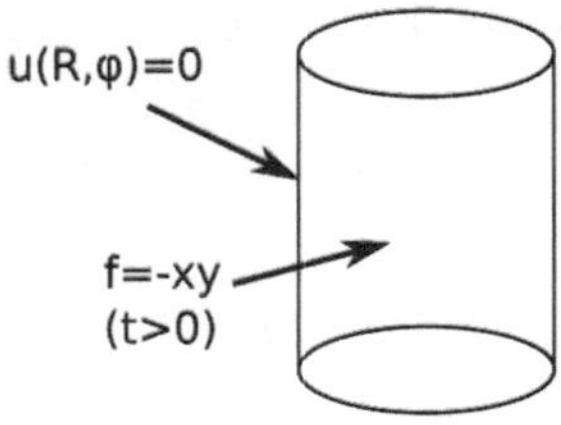

Figure 5.10

Mathematical formulation

The origin of temperatures is shifted by T_0. Due to the cylindrical symmetry, we use these coordinates:

$$\left\{ \begin{array}{l} \dfrac{\partial u}{\partial t} - \dfrac{k}{C\rho_0}\Delta u = \dfrac{\partial u}{\partial t} - \chi\Delta u = -\dfrac{1}{2C\rho_0}\rho^2 \sin(2\varphi) \\ u(R, \varphi, t) = 0 \end{array} \right\} \tag{5.120}$$

We now look for a solution by separating it into two: one which corresponds to the stationary distribution of temperature $w(\rho, \varphi)$, and another corresponding to the transient part.

$$u(\rho, \varphi, t) = w(\rho, \varphi) + v(\rho, \varphi, t) \tag{5.121}$$

$$\left\{ \begin{array}{l} \Delta w = -\dfrac{1}{2k}\rho^2 \sin(2\varphi) \\ u(R, \varphi, t) = 0 \\ \text{With no initial condition} \end{array} \right\} \tag{5.122}$$

The previous problem (5.8) contains a solution of the stationary problem for $w(\rho, \varphi)$.

The problem for $v(\rho, \varphi, t)$ is:

$$\left\{ \begin{array}{l} \dfrac{\partial v}{\partial t} - \chi\Delta v = 0 \\ v(R, \varphi, t) = 0 \\ v(\rho, \varphi, 0) = -w(\rho, \varphi) \end{array} \right\} \tag{5.123}$$

Sturm–Liouville problem

We will separate the temporal variable from the spatial ones:

$$v = \sum Q_{nm}(t) s_{nm}(\rho, \varphi) \tag{5.124}$$

This last problem will be solved by expanding the solution into orthogonal eigenfunctions of the Sturm–Liouville problem, corresponding to the homogeneous boundary of the first kind:

$$\left\{ \begin{array}{l} \Delta s(\rho, \varphi) + \lambda s(\rho, \varphi) = 0 \\ s(R, \varphi) = 0 \end{array} \right\} \tag{5.125}$$

The corresponding eigenfunctions and eigenvalues are well known.

$$[s(\rho, \varphi)]_{nm} = J_m(\sqrt{\lambda_{mn}}\rho)[A_{nm} \cos(m\varphi) + B_{nm} \sin(m\varphi)] \tag{5.126}$$

The eigenvalues correspond to the zeros of the Bessel function $J_m(\sqrt{\lambda_{mn}} R) = 0$. Then, replacing the expression $v(\rho, \varphi, t) = \sum Q_{nm}(t) s_{nm}(\rho, \varphi)$ into equation (5.123) we arrive at the equation for the temporal part, which has an exponential solution:

$$Q_{nm}(t) = e^{-\chi \lambda_{mn} t} \tag{5.127}$$

General solution

$$v(\rho, \varphi, t) = \sum e^{-\chi \lambda_{mn} t} J_m(\sqrt{\lambda_{nm}}\rho)[A_{nm} \cos(m\varphi) + B_{nm} \sin(m\varphi)] \tag{5.128}$$

Final solution

The initial condition is applied:

$$-w(\rho, \varphi) = \sum J_m(\sqrt{\lambda_{nm}}\rho)[A_{nm} \cos(m\varphi) + B_{nm} \sin(m\varphi)] \tag{5.129}$$

We use the orthogonality of the angular and radial functions to obtain the coefficients of the sum A_{nm} and B_{nm}.

5.10 Variation of the Temperature in a Quarter of a Disk

Find the variation of temperature in a membrane in the form of a quarter of a disk with radii a and b. The two straight boundaries are in contact with a thermal reservoir at zero temperature and the curved ones exchange heat with the outer medium at zero temperature according to Newton's law, with a constant h. Starting at $t = 0$ in the central part (indicated in the figure) acts a local heat source of value F [J/(cm^2· s)]:

$$f(r, \varphi, t) = F\frac{\delta[r - \frac{(a+b)}{2}] \times \delta(\varphi - \frac{\pi}{4})}{r} \tag{5.130}$$

The density of the material is ρ, the heat capacity is C and the thermal conduction coefficient is k

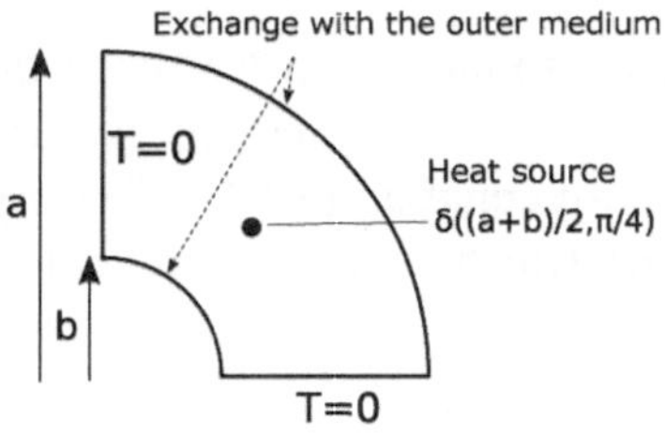

Figure 5.11

Mathematical formulation

$$\left\{\begin{array}{l} C\rho\dfrac{\partial u}{\partial t} - k\Delta u = f \rightarrow \dfrac{\partial u}{\partial t} - \chi\Delta u = \dfrac{f}{C\rho} \quad \left(\chi = \dfrac{k}{C\rho}\right) \\ -k\left.\dfrac{\partial u}{\partial r}\right|_{r=a} = +hu(a, \varphi, t) \\ -k\left.\dfrac{\partial u}{\partial r}\right|_{r=b} = -hu(b, \varphi, t) \\ u(r, 0, t) = 0 \\ u(r, \frac{\pi}{2}, t) = 0 \\ u(r, \varphi, 0) = 0 \end{array}\right\} \tag{5.131}$$

The next figure shows a schematic representation of the heat fluxes along the radial direction:

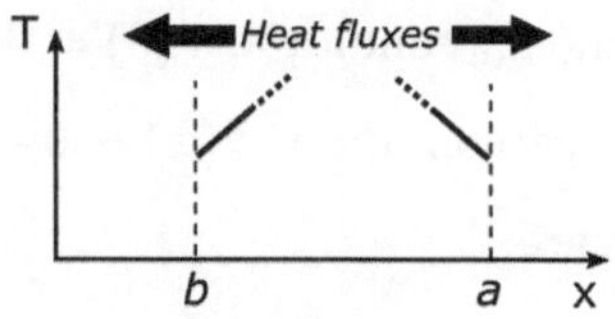

Figure 5.12

Sturm–Liouville problem

The solution is presented as a sum of eigenfunctions in two dimensions:

$$u(r, \varphi, t) = \sum Q(t)V(r, \varphi) \tag{5.132}$$

Where $V(r, \varphi)$ is the solution of the Sturm–Liouville problem:

$$\left\{\begin{array}{l} \Delta V(r, \varphi) + \lambda V(r, \varphi) = 0 \\ \left.\dfrac{\partial V}{\partial r}\right|_{r=a} + \dfrac{h}{k}V(a, \varphi) = 0 \\ \left.\dfrac{\partial V}{\partial r}\right|_{r=b} - \dfrac{h}{k}V(b, \varphi) = 0 \\ V(r, 0) = 0 \\ V(r, \dfrac{\pi}{2}) = 0 \end{array}\right\} \tag{5.133}$$

We separate variables to get to the solution. We will obtain the eigenfunctions for the radial and angular part. We take ν as a positive value to guarantee that we will have orthogonal angular eigenfunctions.

$$V(r, \varphi) = R(r) \cdot \Phi(\varphi) \tag{5.134}$$

$$\frac{r\frac{d}{dr}\left(r\frac{dR(r)}{dr}\right) + \lambda r^2 R}{R} = -\frac{1}{\Phi(\varphi)}\frac{d^2\Phi(\varphi)}{d\varphi^2} = +\nu \tag{5.135}$$

(with $\nu > 0$)

The general solution of the angular part is:

$$\Phi(\varphi) = A\cos(\sqrt{\nu}\varphi) + B\sin(\sqrt{\nu}\varphi) \tag{5.136}$$

From the second boundary condition: $\Phi(0) = 0 \rightarrow A = 0$

From the third boundary condition: $\Phi(\frac{\pi}{2}) = 0 \rightarrow \sqrt{\nu}\frac{\pi}{2} = m\pi$

Then the eigenvalues are:

$$\nu = (2m)^2 \tag{5.137}$$

The eigenfunction are:

$$\Phi(\varphi) = \sin(2m\varphi) \qquad (m = 1, 2, \ldots) \tag{5.138}$$

Equation for the radial part:

$$r\frac{d}{dr}\left(r\frac{dR(r)}{dr}\right) + [\lambda r^2 - (2m)^2]R(r) = 0 \tag{5.139}$$

General solution for the radial part:

$$R(r) = C \times J_{2m}(\sqrt{\lambda}r) + D \times N_{2m}(\sqrt{\lambda}r) \tag{5.140}$$

From the first and second boundary conditions:

$$\left\{\begin{array}{l} \frac{d}{dr}\left[C \times J_{2m}\left(\sqrt{\lambda_k^{(2m)}}r\right) + D \times N_{2m}\left(\sqrt{\lambda_k^{(2m)}}r\right)\right]_{r=a} + \\ +\frac{h}{k}\left[C \times J_{2m}(\sqrt{\lambda_k^{(2m)}}a) + D \times N_{2m}\left(\sqrt{\lambda_k^{(2m)}}a\right)\right] = 0 \\ \frac{d}{dr}\left[C \times J_{2m}\left(\sqrt{\lambda_k^{(2m)}}r\right) + D \times N_{2m}\left(\sqrt{\lambda_k^{(2m)}}r\right)\right]_{r=b} - \\ -\frac{h}{k}\left[C \times J_{2m}\left(\sqrt{\lambda_k^{(2m)}}b\right) + D \times N_{2m}\left(\sqrt{\lambda_k^{(2m)}}b\right)\right] = 0 \end{array}\right\} \tag{5.141}$$

or

$$\left\{\begin{array}{l} C\left[\frac{d}{dr}\left[J_{2m}\left(\sqrt{\lambda_k^{(2m)}}r\right)\right]_{r=a} + \frac{h}{k}J_{2m}\left(\sqrt{\lambda_k^{(2m)}}a\right)\right] + \\ +D\left[\frac{d}{dr}\left[N_{2m}\left(\sqrt{\lambda_k^{(2m)}}r\right)\right]_{r=a} + \frac{h}{k}N_{2m}\left(\sqrt{\lambda_k^{(2m)}}a\right)\right] = 0 \\ C\left[\frac{d}{dr}\left[J_{2m}\left(\sqrt{\lambda_k^{(2m)}}r\right)\right]_{r=b} - \frac{h}{k}J_{2m}\left(\sqrt{\lambda_k^{(2m)}}b\right)\right] + \\ +D\left[\frac{d}{dr}\left[N_{2m}\left(\sqrt{\lambda_k^{(2m)}}r\right)\right]_{r=b} - \frac{h}{k}N_{2m}\left(\sqrt{\lambda_k^{(2m)}}b\right)\right] = 0 \end{array}\right\} \tag{5.142}$$

By equating to zero the determinant of the previous system of equations we get the eigenvalues λ_{km}. From one of the boundary conditions, we get the ratio between the coefficients C and D:

$$C = -D\frac{\frac{d}{dr}\left[N_{2m}\left(\sqrt{\lambda_k^{(2m)}}r\right)\right]_{r=a} + \frac{h}{k}N_{2m}(\sqrt{\lambda_k^{(2m)}}a)}{\frac{d}{dr}\left[J_{2m}\left(\sqrt{\lambda_k^{(2m)}}r\right)\right]_{r=a} + \frac{h}{k}J_{2m}\left(\sqrt{\lambda_k^{(2m)}}a\right)} \tag{5.143}$$

Spatial eigenfunctions to expand the solution:

$$V_{km}(r, \varphi) = [R(r)]_{mk} \times \sin(2m\varphi) \tag{5.144}$$

Solution

Replacing the solution $u(r, \varphi, t) = \sum Q(t)V_{km}(r, \varphi)$ into the heat equation $\frac{\partial u}{\partial t} - \chi\Delta u = \frac{f}{C\rho}$ we get:

$$\sum_{km}\frac{dQ_{km}(t)}{dt}V_{km}(r, \varphi) - \chi Q_{km}(t)\sum_{km}\Delta V_{km}(r, \varphi) = \frac{f}{C\rho} \tag{5.145}$$

$$\sum_{km}\left[\frac{dQ_{km}}{dt} + \chi Q_{km}(t)\lambda_{km}\right]V_{km}(r, \varphi) = \frac{F}{C\rho}\frac{\delta\left(r - \frac{(a+b)}{2}\right) \times \delta(\varphi - \frac{\pi}{4})}{r} \tag{5.146}$$

We use the orthogonality by multiplying both sides by $V_{km}(r, \varphi)$; and integrating $\int_b^a\int_0^{\frac{\pi}{2}} r dr d\varphi$:

$$\sum_{km}\left[\frac{dQ_{km}}{dt} + \chi Q_{km}(t)\lambda_{km}\right]\int_b^a\int_0^{\frac{\pi}{2}}[V_{km}(r, \varphi)]^2 r dr d\varphi =$$

$$= \frac{F}{C\rho}\int_b^a\int_0^{\frac{\pi}{2}} V_{km}(r, \varphi)\delta\left(r - \frac{(a+b)}{2}\right) \times \delta\left(\varphi - \frac{\pi}{4}\right)dr d\varphi \tag{5.147}$$

$$\frac{dQ_{km}}{dt} + \chi Q_{km}(t)\lambda_{km}$$

$$= \frac{F}{C\rho}\frac{\int_b^a\int_0^{\frac{\pi}{2}} V_{km}(r, \varphi)\delta\left(r - \frac{(a+b)}{2}\right) \times \delta\left(\varphi - \frac{\pi}{4}\right)dr d\varphi}{\int_b^a\int_0^{\frac{\pi}{2}}[V_{km}(r, \varphi)]^2 r dr d\varphi} = \alpha_{km} \tag{5.148}$$

The solution of the equation:

$$\frac{dQ_{km}(t)}{dt} + \chi\lambda_{km}Q_{km}(t) = \alpha_{km} \tag{5.149}$$

is obtained as the sum of a particular solution:

$$Q_{km,\,part}(t) = \frac{\alpha_{km}}{\chi\lambda_{km}} \tag{5.150}$$

and the solution of the homogeneous equation:

$$Q_{km,hom}(t) = Ae^{(-\chi\lambda_{km}t)} \tag{5.151}$$

Then:

$$Q_{km}(t) = Ae^{(-\chi\lambda_{km}t)} + \frac{\alpha_{km}}{\chi\lambda_{km}} \tag{5.152}$$

Then, imposing the initial conditions to the whole solution:

$$u(r,\varphi,0) = \sum_{km}\left[Ae^{(-\chi\lambda_{km}0)} + \frac{\alpha_{km}}{\chi\lambda_{km}}\right]V_{km}(r,\varphi) = 0 \tag{5.153}$$

We find the coefficient $A = -\frac{\alpha_{km}}{\chi\lambda_{km}}$. Finally:

$$u(r,\varphi,t) = \sum_{km}\left[\frac{\alpha_{km}}{\chi\lambda_{km}}\right]\left(1 - e^{(-\chi\lambda_{km}t)}\right)[R(r)]_{mk} \times \sin(2m\varphi) \tag{5.154}$$

5.11 Oscillations of a Quarter of a Membrane

Find the amplitude of the main tone (the lowest frequency) of a membrane with fixed boundaries and the shape of 1/4 of a circle, with tension T and surface mass density ρ. A point-like hit hits the membrane at rest at $t = 0$ in the indicated location, with a total transfer of momentum K:

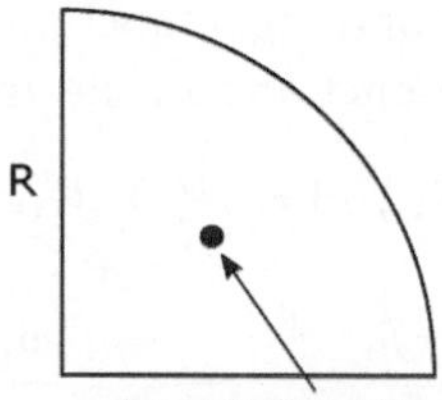

Figure 5.13

Mathematical formulation

$$\left\{\begin{array}{l} \dfrac{\partial^2 u}{\partial t^2} - a^2 \Delta u = 0 \\ a^2 = \dfrac{T}{\rho} \\ u(R, \varphi, t) = 0 \\ u(r, 0, t) = 0 \\ u\left(r, \dfrac{\pi}{2}, t\right) = 0 \\ u(r, \varphi, 0) = 0 \\ \left.\dfrac{\partial u}{\partial t}\right|_{t=0} = \dfrac{K}{\rho} \dfrac{\delta\left(r - \dfrac{R}{2}\right) \times \delta\left(\varphi - \dfrac{\pi}{4}\right)}{r} \end{array}\right\} \tag{5.155}$$

The last relation is found from the following integral of the initial velocity in the surface (S) of the disk:

$$\iint_S \rho \left.\frac{\partial u}{\partial t}\right|_{t=0} r\, dr\, d\varphi = K \tag{5.156}$$

Sturm–Liouville problem

We use separation of variables to formulate the Sturm–Liouville problem, which turns out to be:

$$\left\{\begin{array}{l} \Delta V(r, \varphi) + \lambda V(r, \varphi) = 0 \\ V(R, \varphi) = 0 \\ V(r, 0) = 0 \\ V\left(r, \dfrac{\pi}{2}\right) = 0 \end{array}\right\} \tag{5.157}$$

To find the eigenvalues and eigenfunctions we separate variables once again. We get eigenfunctions for the angular and radial parts.

$$V(r, \varphi) = \mathcal{R}(r) \cdot \Phi(\varphi) \tag{5.158}$$

$$\frac{r\frac{d}{dr}(r\frac{d\mathcal{R}(r)}{dr}) + \lambda r^2 \mathcal{R}}{\mathcal{R}} = -\frac{1}{\Phi(\varphi)}\frac{d^2\Phi(\varphi)}{d\varphi^2} = +\nu \tag{5.159}$$

(with $\nu > 0$, since the sign of the constant from the separation of variables must be greater than zero to have orthogonal angular solutions). The general solution of the angular part is:

$$\Phi(\varphi) = A\cos(\sqrt{\nu}\varphi) + B\sin(\sqrt{\nu}\varphi) \tag{5.160}$$

From the second boundary condition: $\Phi(0) = 0 \rightarrow A = 0$

From the third boundary condition: $\Phi(\frac{\pi}{2}) = 0 \rightarrow \sqrt{\nu}\frac{\pi}{2} = m\pi$

Then the angular eigenvalues are:

$$\nu = (2m)^2 \tag{5.161}$$

The angular eigenfunctions are:

$$\Phi(\varphi) = \sin(2m\varphi) \qquad (m = 1, 2, \ldots) \tag{5.162}$$

Equation for the radial part:

$$r\frac{d}{dr}\left(r\frac{d\mathcal{R}(r)}{dr}\right) + [\lambda r^2 - (2m)^2]\mathcal{R}(r) = 0 \tag{5.163}$$

General solution for the radial part:

$$\mathcal{R}(r) = C \times J_{2m}(\sqrt{\lambda}r) + D \times N_{2m}(\sqrt{\lambda}r) \tag{5.164}$$

Since $\mathcal{R}(0) = 0 \rightarrow D = 0$

From the first boundary condition, imposing $\mathcal{R}(0) = 0$ we will get the eigenvalues $\lambda_k^{(2m)}$:

$$\mathcal{R}(R) = C \times J_{2m}\left(\sqrt{\lambda_k^{(2m)}}R\right) = 0 \tag{5.165}$$

The spatial eigenfunctions to expand the solution are:

$$V_{km}(r, \varphi) = J_{2m}\left(\sqrt{\lambda_k^{(2m)}}r\right) \times \sin(2m\varphi) \tag{5.166}$$

General solution

We have the following general solution:

$$u(r,\varphi,t)=\sum Q_{km}(t)V_{km}(r,\varphi) \tag{5.167}$$

where $Q_{km}(t)$ are solutions of the equation:

$$\frac{d^2Q}{dt^2}+a^2\lambda_k^m Q_{km}(t)=0 \tag{5.168}$$

Then:

$$u_{km}(r,\varphi,t)=\sum\left[C_{km}^{(1)}\cos(a\sqrt{\lambda_k^m}t)+C_{km}^{(2)}\sin\left(a\sqrt{\lambda_k^m}t\right)\right]$$
$$\times J_{2m}\left(\sqrt{\lambda_k^m}r\right)\times\sin(2m\varphi) \tag{5.169}$$

Frequencies of the excited modes: $\omega_{km}^2=a^2\lambda_{km}$

The minimum value of ω_{km} corresponds to the minimum value of λ_{km}, which happens for $k=1$, $m=1$. Profile of the lowest mode:

$$u_{11}(r,\varphi,t)=\left[C_{11}^{(1)}\cos\left(a\sqrt{\lambda_1^1}t\right)+C_{11}^{(2)}\sin\left(a\sqrt{\lambda_1^1}t\right)\right]$$
$$\times J_2\left(\sqrt{\lambda_1^1}r\right)\times\sin(2\varphi) \tag{5.170}$$

Final solution

We search the coefficients $C_{11}^{(1)}$, $C_{11}^{(2)}$ as a result of the hit, using the initial conditions:

$$u_{11}(r,\varphi,0)=\left[C_{11}^{(1)}\cos\left(a\sqrt{\lambda_1^1}0\right)+C_{11}^{(2)}\sin\left(a\sqrt{\lambda_1^1}0\right)\right]$$
$$\times J_2\left(\sqrt{\lambda_1^1}r\right)\times\sin(2\varphi)=0 \tag{5.171}$$

We get $C_{11}^{(1)}=0$

We search $C_{11}^{(2)}$ using the initial condition for the velocity:

$$\left.\frac{\partial u}{\partial t}\right|_{t=0}=\sum\left[C_{km}^{(2)}a\sqrt{\lambda_k^m}\cos\left(a\sqrt{\lambda_k^m}0\right)\right]J_{2m}\left(\sqrt{\lambda_k^m}r\right)\times\sin(2m\varphi)$$
$$=\frac{K}{\rho}\frac{\delta(r-\frac{R}{2})\times\delta(\varphi-\frac{\pi}{4})}{r} \tag{5.172}$$

Since we are only interested in the frequency of the main mode ($k, m = 1$), we multiply the former expression by the corresponding eigenfunction:

$$J_2\left(\sqrt{\lambda_1^1}r\right) \times \sin(2\varphi) \tag{5.173}$$

and using the orthogonality of the other eigenfunction we find the amplitude of the main tone:

$$\sum[C_{km}^{(2)}a\sqrt{\lambda_k^m}] \int\int J_{2m}(\sqrt{\lambda_k^m}r)J_2\left(\sqrt{\lambda_1^1}r\right) rdr \times \sin(2m\varphi)\sin(2\varphi)$$
$$= \int\int \frac{K}{\rho}\frac{\delta(r-\frac{R}{2})}{r}J_2\left(\sqrt{\lambda_1^1}r\right)\sin(2\varphi)\delta\left(\varphi-\frac{\pi}{4}\right)rdrd\varphi \tag{5.174}$$

Only the terms with indices $k, m = 1$ remain.

$$C_{1,1}^{(2)}a\sqrt{\lambda_1^1}\left|J_2\left(\sqrt{\lambda_1^1}r\right)\right|^2 \times \frac{|\sin(2\varphi)|^2}{4} = \frac{K}{\rho}J_2\left(\sqrt{\lambda_1^1}\frac{R}{2}\right)\sin\left(2\frac{\pi}{4}\right) \tag{5.175}$$

The final result for the amplitude of the main tone is:

$$u_{1,1}(r,\varphi,t) = \frac{\frac{K}{\rho}J_2(\sqrt{\lambda_1^1}\frac{R}{2})\sin(2\frac{\pi}{4})}{a\sqrt{\lambda_1^1}\left|J_2\left(\sqrt{\lambda_1^1}r\right)\right|^2 \times \frac{|\sin(2\varphi)|^2}{4}}$$
$$\times J_2\left(\sqrt{\lambda_1^1}r\right)\sin(2\varphi)\sin\left(a\sqrt{\lambda_1^1}t\right) \tag{5.176}$$

5.12 Case Study: Variation of the Temperature in a Cylinder with a Thin Heater

A disk has a hole in its center. The outer surface (radius R_1) is thermally insulated. The outer surface (radius R_2) is put into contact with a thermal reservoir at zero temperature. Find the variations of temperature as a function of time if the disk was in thermal equilibrium at zero temperature and, starting at $t = 0$ a thin heater of power P with the shape of half a ring, centered and with radius $R = (R_1 + R_2)/2$ between the inner and outer radii of the disk. The thermal conductivity of the material is k, the heat capacity is C and the density is ρ_0.

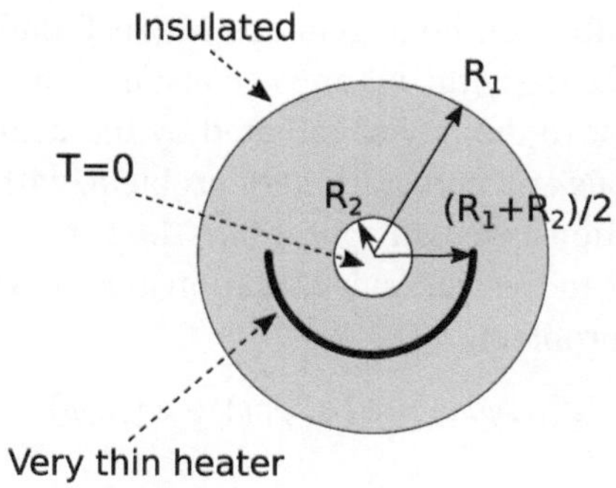

Figure 5.14

Solution:

We seek the solution as the sum of a transient function and a stationary one.

Mathematical formulation

There is no z variable.

$$\left\{\begin{array}{l} C\rho_0\dfrac{\partial u}{\partial t} - k\Delta u(\rho,\varphi,t) = f(\rho,\varphi)\ (R_1 \le \rho \le R_2); (t>0) \\ \dfrac{\partial u}{\partial t} - \frac{k}{C\rho_0}\Delta u(\rho,\varphi,t) = \frac{f(\rho,\varphi)}{C\rho_0} \\ u(\rho,\varphi,t=0)=0 \\ u(R_2,\varphi,t)=0 \\ \left.\dfrac{\partial u}{\partial\rho}\right|_{\rho=R_1} = 0 \end{array}\right\} \tag{5.177}$$

We define mathematically the non-homogeneous part of the equation:

$$\left\{\begin{array}{l} f(\rho,\varphi) = g(\rho)\cdot h(\varphi) \\ g(\rho) = \dfrac{P}{\pi\rho}\delta\left(\rho - \dfrac{(R_1+R_2)}{2}\right) \\ h(\varphi) = \left\{\begin{array}{l} 0\ \ (0\le\varphi\le\pi) \\ 1\ (\pi<\varphi<2\pi) \end{array}\right\} \end{array}\right\} \tag{5.178}$$

The constant of the non-homogeneous part is found by integrating the density of heat (with unknown constant) in the whole area and integrating it to the power radiated by the heater. Since it is a non-homogeneous and non-stationary problem, with homogeneous boundary conditions, we will split it into the sum of two problems: one corresponds to the stationary solution at $t \to \infty$, and the other is the transient problem.

$$u(\rho, \varphi, t) = v(\rho, \varphi, t) + w(\rho, \varphi) \tag{5.179}$$

$$\left\{\begin{array}{l} \dfrac{\partial v}{\partial t} - \dfrac{k}{C\rho_0}\Delta v(\rho, \varphi, t) = 0 \quad (R_1 \le \rho \le R_2); (t > 0) \\ v(\rho, \varphi, t = 0) = -w\ (\rho, \varphi) \\ v(R_2, \varphi, t) = 0 \\ \left.\dfrac{\partial v}{\partial \rho}\right|_{\rho=R_1} = 0 \end{array}\right\} \tag{5.180}$$

The second problem is homogeneous with non-homogeneous boundary conditions:

$$\left\{\begin{array}{l} \Delta w(\rho, \varphi) = -\dfrac{f(\rho, \varphi)}{k} \quad (R_1 \le \rho \le R_2) \\ w(R_2, \varphi) = 0 \\ \left.\dfrac{\partial w}{\partial \rho}\right|_{\rho=R_1} = 0 \end{array}\right\} \tag{5.181}$$

We start with the stationary problem.

Sturm–Liouville problem

We seek the radial solution by expanding by eigenfunctions of the Sturm–Liouville problem:

$$\left\{\begin{array}{l} \Delta w(\rho, \varphi) + \lambda w(\rho, \varphi) = 0 \\ w(R_2, \varphi) = 0 \\ \left.\dfrac{\partial w}{\partial \rho}\right|_{\rho=R_1} = 0 \end{array}\right\} \tag{5.182}$$

The orthogonal eigenfunctions and the eigenvalues are well known:

$$w_{nm}(\rho, \varphi) = [A_{nm} J_m(\sqrt{\lambda_{nm}}\rho) + B_{nm} N_m(\sqrt{\lambda_{nm}}\rho)]e^{(im\varphi)} \tag{5.183}$$

For the time being, this form only considers the periodicity of the solution (with period 2π and the general form of the radial part: a combination of two linearly independent solutions (Bessel and Neumann).

Applying the boundary conditions, we will obtain the eigenvalues λ_{nm}:

$$\left\{\begin{array}{l} w(R_2,\varphi)=0\rightarrow[A_{nm}J_m(\sqrt{\lambda_{nm}}R_2)+B_{nm}N_m(\sqrt{\lambda_{nm}}R_2)]=0 \\ \left.\dfrac{\partial w}{\partial\rho}\right|_{\rho=R_1}=0\rightarrow[\sqrt{\lambda_{nm}}A_{nm}[J_m(\sqrt{\lambda_{nm}}r)]_{\rho=R_1} \\ \qquad +B_{nm}\sqrt{\lambda_{nm}}[N_m(\sqrt{\lambda_{nm}}r)]_{\rho=R_1}=0 \end{array}\right\} \tag{5.184}$$

Solution

The solution of the determinant of the previous system of equations gives us the eigenvalues λ_{nm}. On the other hand, using the equation of the first boundary condition:

$$B_{nm}=-A_{nm}\frac{J_m(\sqrt{\lambda_{nm}}R_2)}{N_m(\sqrt{\lambda_{nm}}R_2)} \tag{5.185}$$

We can write the eigenfunctions (not normalized) as:

$$w_{nm}(\rho,\varphi)=\left[J_m(\sqrt{\lambda_{nm}}\rho)-\frac{J_m(\sqrt{\lambda_{nm}}R_2)}{N_m(\sqrt{\lambda_{nm}}R_2)}N_m(\sqrt{\lambda_{nm}}\rho)\right] \times[C_{nm}\sin(m\varphi)+D_{nm}\cos(m\varphi)] \tag{5.186}$$

The solution of the stationary problem will be sought by replacing its form:

$$w(\rho,\varphi)=\sum_{n,m}[J_m(\sqrt{\lambda_{nm}}\rho)-\frac{J_m(\sqrt{\lambda_{nm}}R_2)}{N_m(\sqrt{\lambda_{nm}}R_2)}N_m(\sqrt{\lambda_{nm}}\rho)] \times[C_{nm}\sin(m\varphi)+D_{nm}\cos(m\varphi)] \tag{5.187}$$

into equation (5.181):

$$\sum_{n,m}\lambda_{nm}\left[J_m(\sqrt{\lambda_{nm}}\rho)-\frac{J_m(\sqrt{\lambda_{nm}}R_2)}{N_m(\sqrt{\lambda_{nm}}R_2)}N_m(\sqrt{\lambda_{nm}}\rho)\right] \times[C_{nm}\sin(m\varphi)+D_{nm}\cos(m\varphi)]=\frac{g(\rho)\cdot h(\varphi)}{k} \tag{5.188}$$

Multiplying both sides by orthogonal eigenfunctions and integrating

$$\int_{R_1}^{R_2} \rho d\rho \int_0^{2\pi} \sin(m\varphi)d\varphi \tag{5.189}$$

we will get the coefficients C_{nm} and D_{nm}:

$$C_{nm}\left\| J_m(\sqrt{\lambda_{nm}}\rho) - \frac{J_m(\sqrt{\lambda_{nm}}R_2)}{N_m(\sqrt{\lambda_{nm}}R_2)} N_m(\sqrt{\lambda_{nm}}\rho)\right\|^2 \|\sin(m\varphi)\|^2 =$$

$$= \frac{1}{k\lambda_{nm}} \int_{R_1}^{R_2} J_m(\sqrt{\lambda_{nm}}\rho) - \frac{J_m(\sqrt{\lambda_{nm}}R_2)}{N_m(\sqrt{\lambda_{nm}}R_2)} N_m(\sqrt{\lambda_{nm}}\rho) g(\rho)\rho d\rho \int_0^{2\pi} h(\varphi)\sin(m\varphi)d\varphi \tag{5.190}$$

$$C_{nm} = \frac{2}{k\lambda_{nm}\left\| J_m(\sqrt{\lambda_{nm}}\rho) - \frac{J_m(\sqrt{\lambda_{nm}}R_2)}{N_m(\sqrt{\lambda_{nm}}R_2)} N_m(\sqrt{\lambda_{nm}}\rho)\right\|^2} \times$$

$$\times \frac{P}{\pi^2} \int_{R_1}^{R_2} \left[J_m(\sqrt{\lambda_{nm}}\rho) - \frac{J_m(\sqrt{\lambda_{nm}}R_2)}{N_m(\sqrt{\lambda_{nm}}R_2)} N_m(\sqrt{\lambda_{nm}}\rho)\right] \delta(\rho - R)d\rho \times$$

$$\times \int_{\pi}^{2\pi} \sin(m\varphi)d\varphi = \frac{2P}{\pi^2 k\lambda_{nm}\left\| J_m(\sqrt{\lambda_{nm}}\rho) - \frac{J_m(\sqrt{\lambda_{nm}}R_2)}{N_m(\sqrt{\lambda_{nm}}R_2)} N_m(\sqrt{\lambda_{nm}}\rho)\right\|^2} \times$$

$$\times \left[J_m(\sqrt{\lambda_{nm}}\rho_0) - \frac{J_m(\sqrt{\lambda_{nm}}R_2)}{N_m(\sqrt{\lambda_{nm}}R_2)} N_m(\sqrt{\lambda_{nm}}R)\right]$$

$$\times \left[\frac{\cos(\pi m) - \cos(2\pi m)}{m} \right] \tag{5.191}$$

In a similar way:

$$D_{nm} = \frac{2}{k\lambda_{nm}\left\| J_m(\sqrt{\lambda_{nm}}\rho) - \frac{J_m(\sqrt{\lambda_{nm}}R_2)}{N_m(\sqrt{\lambda_{nm}}R_2)} N_m(\sqrt{\lambda_{nm}}\rho)\right\|^2} \frac{P}{\pi^2} \times$$

$$\times \int_{R_1}^{R_2} [J_m(\sqrt{\lambda_{nm}}\rho) - \frac{J_m(\sqrt{\lambda_{nm}}R_2)}{N_m(\sqrt{\lambda_{nm}}R_2)} N_m(\sqrt{\lambda_{nm}}\rho)]\delta(\rho - R)d\rho \times$$

$$\times \int_{\pi}^{2\pi} \cos(m\varphi)d\varphi = \frac{2P}{\pi^2 k\lambda_{nm}\left\| J_m(\sqrt{\lambda_{nm}}\rho) - \frac{J_m(\sqrt{\lambda_{nm}}R_2)}{N_m(\sqrt{\lambda_{nm}}R_2)} N_m(\sqrt{\lambda_{nm}}\rho)\right\|^2} \times$$

$$\times \left[J_m(\sqrt{\lambda_{nm}}R) - \frac{J_m(\sqrt{\lambda_{nm}}R_2)}{N_m(\sqrt{\lambda_{nm}}R_2)} N_m(\sqrt{\lambda_{nm}}R)\right]$$

$$\times \left[\frac{\sin(2\pi m) - \sin(\pi m)}{m} \right] = 0 \ (m \geq 1) \tag{5.192}$$

The coefficients that correspond to the case $m = 0$ are:

$$D_{n0} = \frac{P}{\pi k \lambda_{n0} \left\| J_0(\sqrt{\lambda_{n0}}\rho) - \frac{J_0(\sqrt{\lambda_{n0}}R_2)}{N_0(\sqrt{\lambda_{n0}}R_2)} N_0(\sqrt{\lambda_{n0}}\rho) \right\|^2} \times$$
$$\times \left[J_m(\sqrt{\lambda_{n0}}R - \frac{J_0(\sqrt{\lambda_{n0}}R_2)}{N_0(\sqrt{\lambda_{n0}}R_2)} N_0(\sqrt{\lambda_{n0}}R) \right] \quad (5.193)$$

Note that the modulus of the angular eigenfunction in the case $m = 0$ is equal to 2π.

Solution of the problem 5.180

We seek the solution by separating the temporal and spatial variables, and expanding by the previously obtained orthogonal eigenfunctions (the solutions of the Sturm–Liouville problem with homogeneous boundary conditions):

$$\left\{ \begin{array}{l} \dfrac{\partial v}{\partial t} - \dfrac{k}{C\rho_0} \Delta v(\rho, \varphi, t) = 0 \; (R_2 \le \rho \le R_1); \; (t > 0) \\ v(\rho, \varphi, 0) = -w(\rho, \varphi) \\ v(R_2, \varphi, t) = 0 \\ \left. \dfrac{\partial v}{\partial \rho} \right|_{\rho = R_1} = 0 \end{array} \right\} \quad (5.194)$$

We seek $v(\rho, \varphi, t) = \sum s(\rho, \varphi) \cdot T(t)$

with the same orthogonal functions as in the previous part:

$$\left\{ \begin{array}{l} \Delta s(\rho, \varphi) + \lambda s(\rho, \varphi) = 0 \\ \dfrac{\partial T}{\partial t} + (\frac{k}{C\rho_0}\lambda) T(t) = 0 \end{array} \right\} \quad (5.195)$$

The temporal solutions give $\frac{dT}{dt} = e^{(-\frac{k}{C\rho_0}\lambda t)}$. The general solution is:

$$v(\rho, \varphi, t) = \sum_{n,m} e^{(-\frac{k}{C\rho_0}\lambda t)} \times \left[J_m(\sqrt{\lambda_{nm}}\rho) - \frac{J_m(\sqrt{\lambda_{nm}}R_2)}{N_m(\sqrt{\lambda_{nm}}R_2)} N_m(\sqrt{\lambda_{nm}}\rho) \right]$$
$$\times \left[K_{nm} \sin(m\varphi) + M_{nm} \cos(m\varphi) \right] \quad (5.196)$$

Applying the initial condition $v(\rho, \varphi, 0) = -w(\rho, \varphi)$ and using the orthogonality of the spatial and angular eigenfunctions and we find the coefficients $K_{nm} = -C_{nm}$; $M_{nm} = -D_{nm}$.

Note: the alternative method (less physically transparent) which will provide the same result consists of seeking the solution as $v(\rho, \varphi, t) = \sum s(\rho, \varphi) \cdot T(t)$ being $s(\rho, \varphi)$ orthogonal functions which are solutions of the previous Sturm–Liouville problem. Replacing this form into the initial non-homogeneous equation and applying the orthogonality of the functions will give us a non-homogeneous first order differential equation for the $T(t)$ function. Its solution will be sought as the sum of the solutions of the homogeneous problem and a particular solution (a constant). Applying the initial condition (trivial) we will find the coefficient of the solution of the homogeneous equation.

5.13 Forced Oscillations in a Circular Membrane

Find the oscillations of a membrane of radius R fixed at its boundary if, starting at $t = 0$ it is subjected to a force homogeneously distributed, with density $f(t) = P_0 \sin(\omega t)$. For $t < 0$ the membrane is at rest.

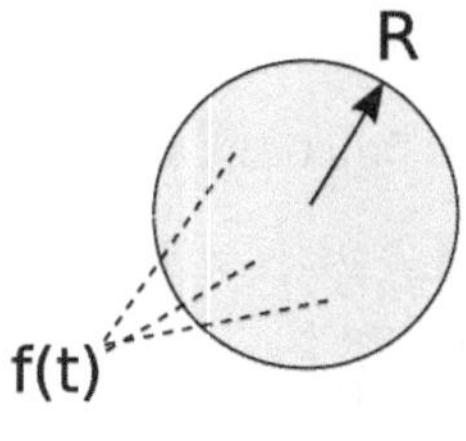

Figure 5.15

Mathematical formulation

Since the problem is symmetric in the angular coordinate the equation we need to solve is:

$$\left\{\begin{array}{l}\dfrac{1}{a^2}\dfrac{\partial u^2(\rho,t)}{\partial t^2}-\dfrac{1}{\rho}\dfrac{\partial}{\partial\rho}\left[\rho\dfrac{\partial u}{\partial\rho}\right]=(P_0/T)\sin(\omega t)\quad(t>0)\\ u(\rho,0)=\left.\dfrac{\partial u}{\partial t}\right|_{t=0}=0\\ u(R,t)=0\\ u(0,t)<\infty\end{array}\right\}\tag{5.197}$$

with $a^2 = T/\rho_0$. Using the principle of superposition we will seek the solution as the sum of:

(i) A solution of the inhomogeneous problem, $w(\rho, t)$, stationary (independent of the initial conditions).
(ii) A solution of the homogeneous problem $v(\rho, t)$ but with new initial conditions to compensate the stationary solution, so that the overall initial conditions are null.

In our case the null initial conditions correspond to the total solution.

$$u(\rho,t)=w(\rho,t)+v(\rho,t)\tag{5.198}$$

In this way the original problem is obtained as superposition, i.e., sum, of these simpler subproblems.

Problem 1:

$$\left\{\begin{array}{l}\dfrac{1}{a^2}\dfrac{\partial w^2(\rho,t)}{\partial t^2}-\dfrac{1}{\rho}\dfrac{\partial}{\partial\rho}\left[\rho\dfrac{\partial w}{\partial\rho}\right]=(P_0/T)\sin(\omega t)\\ w(R,t)=0\\ w(0,t)<\infty\end{array}\right\}\tag{5.199}$$

Problem 2:

$$\left\{\begin{array}{l}\dfrac{1}{a^2}\dfrac{\partial v^2(\rho,t)}{\partial t^2}-\dfrac{1}{\rho}\dfrac{\partial}{\partial\rho}\left[\rho\dfrac{\partial v}{\partial\rho}\right]=0\\ u(\rho,0)=w(\rho,0)+v(\rho,0)=0\rightarrow v(\rho,0)=-w(\rho,0)\\ \left.\dfrac{\partial v}{\partial t}\right|_{t=0}=-\left.\dfrac{\partial w}{\partial t}\right|_{t=0}\\ v(R,t)=0\\ v(0,t)<\infty\end{array}\right\}\tag{5.200}$$

We seek the solution of the inhomogeneous problem $w(\rho, t)$ as:

$$w(\rho, t) = A(\rho)\sin(\omega t) \tag{5.201}$$

We replace this expression into problem (1) and divide the result by a factor $\sin(\omega t)$

$$\left\{ \begin{array}{l} \rho^2 \dfrac{d^2 A}{d\rho^2} + \rho \dfrac{dA}{d\rho} + \dfrac{\rho^2\omega^2}{a^2} A = -\dfrac{P_0}{T}\rho^2 \\ A(R) = 0 \end{array} \right\} \tag{5.202}$$

The solution is the sum of the solution of the homogeneous equation (A_1) and the particular solution (A_2).

$$A(\rho) = A_1(\rho) + A_2(\rho) \tag{5.203}$$

Equation to find $A_1(\rho)$:

$$\rho^2 \frac{d^2 A_1}{d\rho^2} + \rho \frac{dA_1}{d\rho} + \frac{\rho^2\omega^2}{a^2} A_1 = 0 \tag{5.204}$$

General solution

With the change of variable $x = \frac{\rho\omega}{a}$ this equation turns into the equation for the zeroth order Bessel's equation:

$$\frac{dA}{d\rho} = \frac{dA}{dx}\frac{x}{\rho} = \frac{dA}{dx}\left(\frac{\omega}{a}\right) \tag{5.205}$$

$$\frac{d^2 A}{d\rho^2} = \frac{d^2 A}{dx^2}\left(\frac{dx}{d\rho}\right)^2 = \frac{d^2 A}{dx^2}\left(\frac{\omega}{a}\right)^2 \tag{5.206}$$

(due to $\frac{d^2x}{d\rho^2} = 0$). With this we obtain an equation for the new variable x:

$$x^2 \frac{d^2 A_1}{dx^2} + x \frac{dA_1}{dx} + x^2 A_1 = 0 \tag{5.207}$$

or

$$\frac{d^2 A_1}{dx^2} + \frac{1}{x}\frac{dA_1}{dx} + A_1 = 0 \tag{5.208}$$

With the following solution:

$$A_1 = C \cdot J_0(x) = C \cdot J_0\left(\frac{\rho\omega}{a}\right) \tag{5.209}$$

Note: The radial solution cannot include the Neumann function. We seek the particular solution $A_2(\rho)$ to satisfy the boundary conditions:

$$A(R) = A_1(R) + A_2(R) = 0 \tag{5.210}$$

The particular solution $A_2(\rho)$ is a constant:

$$A_2(\rho) = -\frac{P_0 a^2}{T\omega^2} \tag{5.211}$$

We impose the boundary conditions:

$$A(R) = A_1(R) + A_2(R) = C \cdot J_0\left(\frac{R\omega}{a}\right) - \frac{P_0 a^2}{T\omega^2} = 0 \tag{5.212}$$

And obtain the constant:

$$C = \frac{P_0 a^2}{T\omega^2} \frac{1}{J_0(\frac{R\omega}{a})} \tag{5.213}$$

Then:

$$\begin{aligned} w(\rho, t) &= A(\rho)\sin(\omega t) = [A_1(\rho) + A_2(\rho)]\sin(\omega t) \\ &= \frac{P_0 a^2}{T\omega^2}\left(\frac{J_0(\frac{\rho\omega}{a})}{J_0(\frac{R\omega}{a})} - 1\right)\sin(\omega t) \end{aligned} \tag{5.214}$$

As we know the form of $w(\rho, t)$, we now seek the solution of the homogeneous equation (which depends on the initial conditions).

$$\left\{\begin{array}{l} \dfrac{1}{a^2}\dfrac{\partial v^2(\rho, t)}{\partial t^2} - \dfrac{1}{\rho}\dfrac{\partial}{\partial \rho}\left[\rho\dfrac{\partial v}{\partial \rho}\right] = 0 \\ u(\rho, 0) = w(\rho, 0) + v(\rho, 0) = 0 \rightarrow v(\rho, 0) = -w(\rho, 0) \\ \left.\dfrac{\partial v}{\partial t}\right|_{t=0}(\rho, 0) = -\left.\dfrac{\partial w}{\partial t}\right|_{t=0}(\rho, 0) \\ v(R, t) = 0 \\ v(0, t) < \infty \end{array}\right\} \tag{5.215}$$

First initial condition:

$$v(\rho, 0) = -w(\rho, 0) = -\frac{P_0 a^2}{T\omega^2}\left(\frac{J_0(\frac{\rho\omega}{a})}{J_0\left(\frac{R\omega}{a}\right)} - 1\right)\sin(\omega 0) = 0 \tag{5.216}$$

Second initial condition:

$$\left.\frac{\partial v}{\partial t}\right|_{t=0} = -\left.\frac{\partial w}{\partial t}\right|_{t=0} = -\frac{\omega P_0 a^2}{T\omega^2}\left(\frac{J_0(\frac{\rho\omega}{a})}{J_0(\frac{R\omega}{a})} - 1\right)\cos(\omega 0)$$
$$= -\frac{P_0 a^2}{T\omega}\left(\frac{J_0(\frac{\rho\omega}{a})}{J_0(\frac{R\omega}{a})} - 1\right) \quad (5.217)$$

Sturm–Liouville problem

The homogeneous problem (5.215) is solved by separating variables:

$$v(\rho, t) = D(\rho)\cdot T(t) \quad (5.218)$$

We will obtain two differential equations independent of the angular variable:

$$\begin{Bmatrix} \rho^2\dfrac{d^2D}{d\rho^2} + \rho\dfrac{dD}{d\rho} + \lambda\rho^2 D = 0 \\ \dfrac{d^2T}{dt^2} + \lambda a^2 T = 0 \end{Bmatrix} \quad (5.219)$$

The solutions for the radial part are eigenfunctions of the following Sturm–Liouville problem:

$$\begin{Bmatrix} \dfrac{d^2D}{d\rho^2} + \dfrac{1}{\rho}\dfrac{dD}{d\rho} + \lambda D = 0 \\ D(R) = 0;\ D(0) < \infty \end{Bmatrix} \quad (5.220)$$

$$D(\rho) = J_0(\sqrt{\lambda_n}\rho); \qquad \sqrt{\lambda_n} = \frac{x_{n0}}{R} \quad (5.221)$$

where x_{n0} are the n-th zeros of the Bessel's function of argument zero $J_0(x_{n0}) = 0$. Solution of the temporal part:

$$T_n(t) = C_1\cos\left(a\frac{x_{n0}}{R}t\right) + C_2\sin\left(a\frac{x_{n0}}{R}t\right) \quad (5.222)$$

General solution for the homogeneous equation.

$$v(\rho, t) = \sum_n J_0(\sqrt{\lambda_n}\rho)\left[C_{1n}\cos\left(a\frac{x_{n0}}{R}t\right) + C_{2n}\sin\left(a\frac{x_{n0}}{R}t\right)\right] \quad (5.223)$$

Final solution

We find the coefficients of the sum using the initial conditions:

Applying the first condition: $v(\rho, 0) = 0 \rightarrow C_{1n} = 0$

Applying the second condition:

$$\left.\frac{\partial v}{\partial t}\right|_{t=0} = -\frac{P_0 a^2}{T\omega}\left(\frac{J_0(\frac{\rho\omega}{a})}{J_0(\frac{R\omega}{a})} - 1\right) = \sum_n C_{2n}(a\frac{x_{n0}}{R})J_0(\sqrt{\lambda_n}\rho)$$

From here, to find the C_{2n} coefficients, we multiply both sides of the previous relation by the orthogonal eigenfunctions $J_0(\sqrt{\lambda_k}\rho)$ and integrate $\int_0^R \rho d\rho$. We use the property of orthogonality of Bessel's functions.

$$\int_0^R J_0(\sqrt{\lambda_n}\rho)J_0(\sqrt{\lambda_k}\rho)\rho d\rho = \left\{\begin{array}{l} 0 \text{ (if n} \neq k) \\ \frac{R^2}{2}[J_0'(x_n)]^2 \text{ (if n} = k) \end{array}\right\} \quad (5.224)$$

being $J_0'(x_n)$ the derivative of the Bessel's function at the zeros $x = x_n$. Furthermore, we will use the following relation:

$$\int_0^x zJ_0(z)dz = xJ_1(x) \quad (5.225)$$

and the integral:

$$\int_0^R J_0(\sqrt{\lambda_n}\rho)J_0(k\rho)\rho d\rho = \frac{x_n J_0'(x_n)J_0(kR)}{k^2 - \lambda_n} \qquad (k \neq \lambda_n) \quad (5.226)$$

And arrive at:

$$C_{2n} = -\frac{2P_0 a\omega R^3}{Tx_n J_0'(x_n)}\left(\frac{1}{\omega^2 R^2 - x_n a^2}\right) \quad (5.227)$$

The final solution is:

$$u(\rho, t) = w(\rho, t) + v(\rho, t) = \frac{P_0 a^2}{T\omega^2}\left(\frac{J_0(\frac{\rho\omega}{a})}{J_0(\frac{R\omega}{a})} - 1\right)\sin(\omega t)$$
$$-\frac{2P_0 a\omega R^3}{T}\sum_n \frac{1}{x_n J_0'(x_n)}\left(\frac{1}{\omega^2 R^2 - x_n a^2}\right)$$
$$\times J_0(\sqrt{\lambda_n}\rho)\sin\left(a\frac{x_{n0}}{R}t\right) \quad (5.228)$$

Note: The obtained solution is valid supposing that the frequency of the outer force does not coincide with any of the resonant frequencies of the membrane. If that were the case, in the resonance condition the amplitude would increase non-stop (due to the lack of friction) and the stationary situation would never be reached the membrane would break first. In any case we would be beyond the limit of small oscillations withing which the wave equation is valid.

5.14 Case Study: Stationary Distribution of Temperature Inside the Sector of a Disk

Find the stationary distribution of temperature in a fourth of a circular ring (outer radius a, inner radius b and spans an angle $\pi/2$. The membrane (with thermal conductivity k) is insulated on three boundaries and semi-insulated in the inner curved boundary, where there is a heat exchange with the outer medium, which stays at $T = 0$ according to Newton's law with a constant h. Inside the membrane there are heat sources localized along a curved line inside the sector ($\pi/8 < \varphi < 3\pi/8$) at a distance R from the center. The total dissipated power is such that: $\lim_{\epsilon\to 0} \iint\limits_{\Omega\epsilon} f(\rho, \varphi)\rho d\rho d\varphi = F_0$ being $\Omega\epsilon$ the infinitesimal surface around the heat sources.

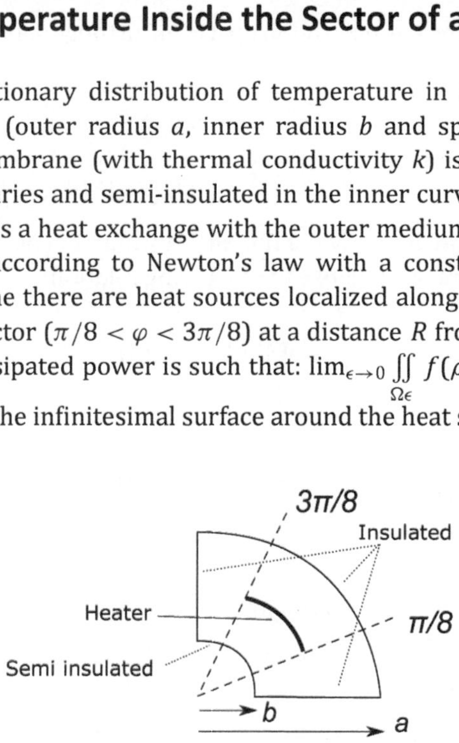

Figure 5.16

We know that:

$$f(\rho,\varphi)=\frac{A}{\rho}\delta(\rho-R)G(\varphi)=\frac{A}{\rho}\delta(\rho-R)\begin{Bmatrix}0 & \left(0<\varphi<\frac{\pi}{8}\right)\\ 1 & \left(\frac{\pi}{8}<\varphi<\frac{3\pi}{8}\right)\\ 0 & \left(\frac{3\pi}{8}<\varphi<\frac{\pi}{2}\right)\end{Bmatrix} \tag{5.229}$$

Applying this condition: $F_0=\iint\limits_{\Omega\epsilon}\frac{A}{\rho}\delta(\rho-R)G(\varphi)\rho d\rho d\varphi=\frac{\pi}{4}A$

From what we get: $A=\frac{4F_0}{\pi}$

Mathematical formulation

$$\begin{Bmatrix}-k\left[\frac{1}{\rho}\frac{\partial}{\partial\rho}\left[\rho\frac{\partial u}{\partial\rho}\right]+\frac{1}{\rho^2}\frac{\partial u^2(\rho,\varphi,t)}{\partial\varphi^2}\right]=\frac{4F_0}{\pi}\frac{1}{\rho}\delta(\rho-R)G(\varphi)\\ \left.\frac{\partial u}{\partial\varphi}\right|_{\varphi=0}=0\\ \left.\frac{\partial u}{\partial\varphi}\right|_{\varphi=\frac{\pi}{2}}=0\\ \left.\frac{\partial u}{\partial\rho}\right|_{\rho=a}=0\\ -k\left.\frac{\partial u}{\partial\rho}\right|_{\rho=b}=-hu(b,\varphi)\qquad (u>0\rightarrow \text{outwards flux})\end{Bmatrix} \tag{5.230}$$

The following figure shows a schematic representation of the heat fluxes at the boundaries along the radial direction:

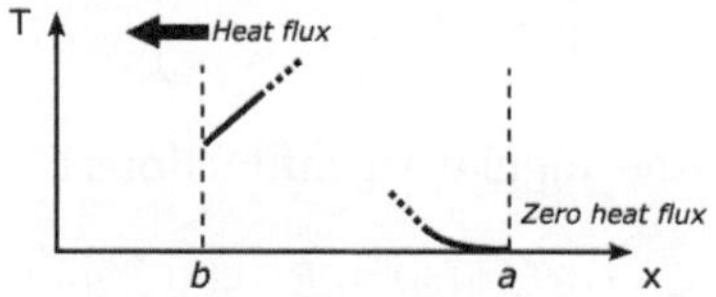

Figure 5.17

Sturm–Liouville problem

Since all boundaries are homogeneous, we can seek the solution by expanding it in eigenfunctions of the Sturm–Liouville problem.

$$u(\rho, \varphi) = \sum D_{nm} v_{nm}(\rho, \varphi) \tag{5.231}$$

Where the $v_{nm}(\rho, \varphi)$ satisfy:

$$\left\{\begin{array}{l} \Delta v(\rho, \varphi) + \lambda v(\rho, \varphi) = 0 \\ \left.\dfrac{\partial v}{\partial \varphi}\right|_{\varphi=0} = 0 \\ \left.\dfrac{\partial v}{\partial \varphi}\right|_{\varphi} = \frac{\pi}{2} = 0 \\ \left.\dfrac{\partial v}{\partial \rho}\right|_{\rho=a} = 0 \\ \left.\dfrac{\partial v}{\partial \rho}\right|_{\rho=b} - Hv(b, \varphi) = 0 \qquad (H = -\frac{h}{k}) \end{array}\right\} \tag{5.232}$$

We seek the solution using the separation of variables method:

$$v(\rho, \varphi) = R(\rho) \cdot \Phi(\varphi) \tag{5.233}$$

We get the following Sturm–Liouville problem for the angular variable φ:

$$\left\{\begin{array}{l} \dfrac{\partial \Phi^2}{\partial \varphi^2} + \mu \Phi = 0 \\ \left.\dfrac{d\Phi}{d\varphi}\right|_{\varphi=0} = 0 \\ \left.\dfrac{d\Phi}{d\varphi}\right|_{\varphi=\frac{\pi}{2}} = 0 \end{array}\right\} \tag{5.234}$$

We general form of the angular eigenfunctions is:

$$\Phi(\varphi) = C \cdot \cos(\sqrt{\mu}\varphi) + E \cdot \sin(\sqrt{\mu}\varphi) \tag{5.235}$$

Due to the first initial condition, we get $E = 0$. The eigenvalues μ_m are sought by applying the second boundary condition:

$$\left.\frac{d\Phi}{d\varphi}\right|_{\varphi=\frac{\pi}{2}} = 0 = \sin\left(\sqrt{\mu}\frac{\pi}{2}\right) \tag{5.236}$$

$$\sqrt{\mu}\frac{\pi}{2} = \pi m \tag{5.237}$$

$$\mu_m = (2m)^2 \qquad (m = 0, 1, 2, 3...) \tag{5.238}$$

By replacing $v(\rho, \varphi) = R(\rho)\cos(2m\varphi)$ into equation (5.339) we arrive at the following equation for the radial variable:

$$\frac{\partial^2 R}{\partial \rho^2} + \frac{1}{\rho}\frac{\partial R}{\partial \rho} + \left(\lambda - \frac{[2m]^2}{\rho^2}\right) R = 0 \tag{5.239}$$

The solution of this problem gives us the set of radial functions:

$$R_{nm}(\rho) = A_{nm} J_{2m}(\sqrt{\lambda_{nm}}\rho) + B_{nm} N_{2m}(\sqrt{\lambda_{nm}}\rho) \tag{5.240}$$

General solution

The general solution is:

$$u(\rho, \varphi, t) = \sum_{n,m} \left[A_{nm} J_{2m}(\sqrt{\lambda_{nm}}\rho) + B_{nm} N_{2m}(\sqrt{\lambda_{nm}}\rho)\right] \cos(2m\varphi) \tag{5.241}$$

The possible values λ_{nm} are the n-th zero of the equation, obtained by equating to zero the determinant of the system of two equations with two unknowns formed by the first and second boundary conditions:

$$\left\{\begin{array}{l} \left.\dfrac{dR}{d\rho}\right|_{\rho=a} = 0 \\ \left.\dfrac{dR}{d\rho}\right|_{\rho=b} - H\,R(b) = 0 \end{array}\right\} \tag{5.242}$$

$$\left\{\begin{array}{l} A_{nm}\sqrt{\lambda_{nm}} \left.\dfrac{dJ_{2m}(\sqrt{\lambda_{nm}}\rho)}{d\rho}\right|_{\rho=a} + B_{nm}\sqrt{\lambda_{nm}} \left.\dfrac{dN_{2m}(\sqrt{\lambda_{nm}}\rho)}{d\rho}\right|_{\rho=a} = 0 \\ A_{nm}\sqrt{\lambda_{nm}} \left.\dfrac{dJ_{2m}(\sqrt{\lambda_{nm}}\rho)}{d\rho}\right|_{\rho=b} - HJ_{2m}(\sqrt{\lambda_{nm}}b) + \\ +B_{nm}\sqrt{\lambda_{nm}} \left.\dfrac{dN_{2m}(\sqrt{\lambda_{nm}}\rho)}{d\rho}\right|_{\rho=b} - HN_{2m}(\sqrt{\lambda_{nm}}b) = 0 \end{array}\right\} \tag{5.243}$$

Also from the first or second boundary conditions can we obtain the relation between the coefficients A_{nm} and B_{nm}, and in this way be able to determine the form of the final solution.

$$B_{nm} = -A_{nm}\frac{\left.\frac{dJ_{2m}(\sqrt{\lambda_{nm}}\rho)}{d\rho}\right|_{\rho=a}}{\left.\frac{dN_{2m}(\sqrt{\lambda_{nm}}\rho)}{d\rho}\right|_{\rho=a}} \tag{5.244}$$

So that:

$$R_{nm}(\rho) = J_{2m}(\sqrt{\lambda_{nm}}\rho) - \frac{\left.\frac{dJ_{2m}(\sqrt{\lambda_{nm}}\rho)}{d\rho}\right|_{\rho=a}}{\left.\frac{dN_{2m}(\sqrt{\lambda_{nm}}\rho)}{d\rho}\right|_{\rho=a}} N_{2m}(\sqrt{\lambda_{nm}}\rho)] \tag{5.245}$$

Final solution

Once the form of the general solution has been clarified we find the coefficients of the expansion, replacing $u(\rho, \varphi, t) = \sum\limits_{n,m} A_{nm}R_{nm}(\rho)\cos(2m\varphi)$ into (5.337):

$$-\Delta u(\rho, \varphi, t) = \sum_{n,m} A_{nm}\lambda_{nm}R_{nm}(\rho)\cos(2m\varphi) = \tag{5.246}$$

$$= \sum_{n,m} A_{nm}\lambda_{nm}\left[J_{2m}(\sqrt{\lambda_{nm}}\rho) - \frac{\left.\frac{dJ_{2m}(\sqrt{\lambda_{nm}}\rho)}{d\rho}\right|_{\rho=a}}{\left.\frac{dN_{2m}(\sqrt{\lambda_{nm}}\rho)}{d\rho}\right|_{\rho=a}} N_{2m}(\sqrt{\lambda_{nm}}\rho)\right]\cos(2m\varphi)$$

$$= \frac{4F_0}{k\pi}\frac{1}{\rho}\delta(\rho - R)G(\varphi) \tag{5.247}$$

We use the orthogonality of the radial and angular eigenfunctions to find the coefficients A_{nm}. Both sides of the previous relation are multiplied by $R_{nm}(\rho)\cos(2m\varphi)$ and integrated $\int\limits_b^a\int\limits_0^{\frac{\pi}{2}} \rho d\rho d\varphi$. Due to the orthogonality the radial $R_{nm}(\rho)$ and angular $\cos(2m\varphi)$ eigenfunctions, we get the coefficients A_{nm}:

$$A_{nm}\lambda_{nm}\,\|R_{nm}(\rho)\|^2\,\|\cos(2m\varphi)\|^2$$

$$= \frac{4F_0}{k\pi}\int\limits_b^a\int\limits_0^{\frac{\pi}{2}} R_{nm}(\rho)\cos(2m\varphi)\delta(\rho - R)G(\varphi)d\rho d\varphi \tag{5.248}$$

$$A_{nm} = \frac{4F_0}{k\pi} \frac{R_{nm}(R)\frac{1}{2m}[\sin(3\pi m/4) - \sin(\pi m/4)]}{\lambda_{nm} \left\| R_{nm}(\rho) \right\|^2 \left\| \cos(2m\varphi) \right\|^2} \tag{5.249}$$

We must consider that:

$$\left\| \cos(2m\varphi) \right\|^2 = \begin{Bmatrix} \frac{\pi}{4} \ (m \neq 0) \\ \frac{\pi}{2} \ (m = 0) \end{Bmatrix} \tag{5.250}$$

$$A_{n0} = \frac{4F_0}{k\pi} \frac{R_{n0}(R)\frac{\pi}{4}}{\lambda_{n0} \left\| R_{n0}(\rho) \right\|^2 \frac{\pi}{2}} \tag{5.251}$$

Alternative method

Sturm–Liouville problem

Another way to solve the problem is shown next, expressing the solution only in a set of orthogonal eigenfunctions in the angular direction:

$$u(\rho, \varphi) = \sum R_m(\rho) \cdot \Phi_m(\varphi) \tag{5.252}$$

Where the $\Phi_m(\varphi)$ come from solving the same angular Sturm–Liouville problem:

$$\begin{Bmatrix} \frac{\partial \Phi^2(\rho, \varphi, t)}{\partial \varphi^2} + \mu \Phi = 0 \\ \left. \frac{d\Phi}{d\varphi} \right|_{\varphi=0} = 0 \\ \left. \frac{d\Phi}{d\varphi} \right|_{\varphi=\frac{\pi}{2}} = 0 \end{Bmatrix} \tag{5.253}$$

$$\Phi_m(\varphi) = \cos(2m\varphi) \tag{5.254}$$

$$\mu_m = (2m)^2 \qquad (m = 0, 1, 2, 3 \ldots) \tag{5.255}$$

Replacing $u(\rho, \varphi) = \sum_m R_m(\rho)\Phi_m(\varphi)$ in (5.337) we get to the following non-homogeneous equation for the radial variable:

$$\sum_m \Phi_m(\varphi) \frac{1}{\rho} \frac{d}{d\rho} \left(\rho \frac{d}{d\rho} R_m(\rho) \right) + R_m(\rho) \frac{1}{\rho} \frac{d^2 \Phi_m(\varphi)}{d\varphi^2} = -\frac{f(\rho, \varphi)}{k} \tag{5.256}$$

$$\sum_m \left[\frac{1}{\rho}\frac{d}{d\rho}\left(\rho\frac{d}{d\rho}R_m(\rho)\right) - R_m(\rho)\frac{4m^2}{\rho^2}\right]\cos(2m\varphi) = -\frac{f(\rho,\varphi)}{k} \tag{5.257}$$

$$\sum_m \left[\frac{d^2R_m}{d\rho^2} + \frac{1}{\rho}\frac{dR_m}{d\rho} - \frac{4m^2}{\rho^2}R_m\right]\cos(2m\varphi) = -\frac{f(\rho,\varphi)}{k} \tag{5.258}$$

Applying the orthogonality:

$$\int_0^{\frac{\pi}{2}} \left[\frac{d^2R_m}{d\rho^2} + \frac{1}{\rho}\frac{dR_m}{d\rho} - \frac{4m^2}{\rho^2}R_m\right]\|\cos(2m\varphi)\|^2$$

$$= -\frac{1}{k}\int_0^{\frac{\pi}{2}} f(\rho,\varphi)\cos(2m\varphi)d\varphi \tag{5.259}$$

Considering the case for $m = 0$:

$$\left[\frac{d^2R_m}{d\rho^2} + \frac{1}{\rho}\frac{dR_m}{d\rho} - \frac{4m^2}{\rho^2}R_m\right]\frac{\pi}{2} = -\frac{4F_0}{k\pi}\frac{1}{\rho}\delta(\rho-R)\int_0^{\frac{\pi}{2}} G(\varphi)\cos(0)d\varphi \tag{5.260}$$

$$\left[\frac{d^2R_m}{d\rho^2} + \frac{1}{\rho}\frac{dR_m}{d\rho} - \frac{4m^2}{\rho^2}R_m\right]\frac{\pi}{2} = -\frac{4F_0}{k\pi}\frac{1}{\rho}\delta(\rho-R)\frac{\pi}{4} \tag{5.261}$$

Equation to solve for $m = 0$:

$$\frac{d^2R_0}{d\rho^2} + \frac{1}{\rho}\frac{dR_0}{d\rho} = -\frac{2F_0}{k\pi}\frac{1}{\rho}\delta(\rho-R) \tag{5.262}$$

Boundary conditions:

In the case $m \neq 0$

$$\left[\frac{d^2R_m}{d\rho^2} + \frac{1}{\rho}\frac{dR_m}{d\rho} - \frac{4m^2}{\rho^2}R_m\right]\frac{\pi}{4} = -\frac{4F_0}{k\pi}\frac{1}{\rho}\delta(\rho-R)\int_0^{\frac{\pi}{2}} G(\varphi)\cos(2m\varphi)d\varphi \tag{5.263}$$

The equation to be solved is:

$$\left[\frac{d^2R_m}{d\rho^2} + \frac{1}{\rho}\frac{dR_m}{d\rho} - \frac{4m^2}{\rho^2}R_m\right] = -\frac{8F_0}{k\pi^2}\frac{1}{\rho}\delta(\rho-R)F_m \tag{5.264}$$

Where $F_m = \sin(\frac{3m\pi}{4}) - \sin(\frac{m\pi}{4})$

We solve first the case $m \neq 0$. Except at $\rho = R$ we have the following homogeneous equation:

$$\left\{ \begin{array}{l} \frac{d^2 R_m}{d\rho^2} + \frac{1}{\rho}\frac{dR_m}{d\rho} - \frac{4m^2}{\rho^2} R_m = 0 \\ \left.\frac{dR}{d\rho}\right|_{\rho=a} = 0 \\ -k\left.\frac{dR}{d\rho}\right|_{\rho=b} = -hR(b) = 0 \end{array} \right\} \tag{5.265}$$

General solution

The general solutions are:

$$\left\{ \begin{array}{l} R_m^-(\rho) = A\rho^{2m} + B\rho^{-2m} \ (\rho < R) \\ R_m^+(\rho) = C\rho^{2m} + D\rho^{-2m} \ (\rho > R) \end{array} \right\} \tag{5.266}$$

Applying the third boundary conditions:

$$2mCa^{2m-1} - 2mDa^{-2m-1} = 0 \tag{5.267}$$

Then: $C = Da^{4m}$

Applying the fourth boundary condition:

$$k(2mAb^{2m-1} - 2mBb^{-2m-1}) = h(Ab^{2m} + Bb^{-2m}) \tag{5.268}$$

$$\frac{2km}{h} = \frac{Ab^{2m} + Bb^{-2m}}{(Ab^{2m-1} - Bb^{-2m-1})} = b - \frac{2Bb}{B - Ab^{4m}} \tag{5.269}$$

$$b - \frac{2km}{h} = \frac{2Bb}{B - Ab^{4m}} \tag{5.270}$$

$$2Bb = \left(b - \frac{2km}{h}\right)(B - Ab^{4m}) = B\left(b - \frac{2km}{h}\right) - Ab^{4m}\left(b - \frac{2km}{h}\right) \tag{5.271}$$

$$B\left(2b - b + \frac{2km}{h}\right) = \left(b - \frac{2km}{h}\right)(-Ab^{4m}) \tag{5.272}$$

$$B = -Ab^{4m}\frac{(b - \frac{2km}{h})}{(b + \frac{2km}{h})} = -A\Upsilon \tag{5.273}$$

Imposing the continuity at $\rho = R$:

$$R_m^-(R) = R_m^+(R) \tag{5.274}$$

$$AR^{2m} - A\Upsilon R^{-2m} = Da^{4m}R^{2m} + DR^{-2m} = D[a^{4m}R^{2m} + R^{-2m}] \tag{5.275}$$

$$A = D\frac{[a^{4m}R^{2m} + R^{-2m}]}{R^{2m} - \Upsilon R^{-2m}} \tag{5.276}$$

We set the condition for the change in derivatives at $\rho = R$:

$$\int_{R-\varepsilon}^{R+\varepsilon} \left[\frac{d^2R_m}{d\rho^2} + \frac{1}{\rho}\frac{dR_m}{d\rho} - \frac{4m^2}{\rho^2}R_m\right]\rho d\rho = -\frac{8F_0F_m}{k\pi^2}\int_{R-\varepsilon}^{R+\varepsilon}\frac{1}{\rho}\delta(\rho - R)\rho d\rho \tag{5.277}$$

$$\int_{R-\varepsilon}^{R+\varepsilon}\frac{d^2R_m}{d\rho^2}\rho d\rho = \left(\rho\frac{dR}{d\rho} - R\right)\Bigg|_{R-\varepsilon}^{R+\varepsilon} = \left(\rho\frac{dR}{d\rho}\right)\Bigg|_{R-\varepsilon}^{R+\varepsilon} \tag{5.278}$$

Since $R\,|_{R-\varepsilon}^{R+\varepsilon} = 0$ for $\varepsilon \to 0$

$$\int_{R-\varepsilon}^{R+\varepsilon}\frac{dR_m}{d\rho}d\rho = R\,|_{R-\varepsilon}^{R+\varepsilon} = 0 \qquad (\varepsilon \to 0) \tag{5.279}$$

$$\int_{R-\varepsilon}^{R+\varepsilon}\frac{R_m}{\rho}d\rho = 0 \tag{5.280}$$

(for $\varepsilon \to 0$ as both functions are continuous and finite at $\rho = R$)

Then:

$$\rho\left.\frac{dR}{d\rho}\right|_{R-\varepsilon}^{R+\varepsilon} = R\left(\frac{dR^+}{d\rho} - \frac{dR^-}{d\rho}\right)$$

$$= -\frac{8F_0F_m}{k\pi^2}\int_{R-\varepsilon}^{R+\varepsilon}\delta(\rho - R)d\rho = -\frac{8F_0F_m}{k\pi^2} = Q_m \tag{5.281}$$

Replacing R^+, R^-

$$2mCR^{2m-1} - 2mDR^{-2m-1} - 2mAR^{2m-1} + 2mBR^{-2m-1} = Q_m \tag{5.282}$$

Renaming (to simplify the equations):

$$\sigma = a^{4m} \tag{5.283}$$

$$\Upsilon = b^{4m}\frac{(b - \frac{2km}{h})}{(b + \frac{2km}{h})} \tag{5.284}$$

$$\delta = \frac{[a^{4m} R^{2m} + R^{-2m}]}{R^{2m} - \Upsilon R^{-2m}} \tag{5.285}$$

We obtain:

$$C = \sigma D;\ B = -\Upsilon A;\ A = \delta D \tag{5.286}$$

Equation (5.282) is left as:

$$\sigma D R^{2m-1} - D R^{-2m-1} - \delta D R^{2m-1} - \Upsilon \delta D R^{-2m-1} = \frac{Q_m}{2m} \tag{5.287}$$

So we arrive at the value of the last coefficient of the expansion:

$$\begin{aligned} D = D_m &= \frac{Q_m}{2m[(\sigma - \delta) R^{2m-1} - (1 - \Upsilon\delta) R^{-2m-1}]} \\ &= -\frac{4F_0 F_m}{mk\pi^2} \frac{1}{[(\sigma - \delta) R^{2m-1} - (1 - \Upsilon\delta) R^{-2m-1}]} \end{aligned} \tag{5.288}$$

$$A_m = \delta D_m \tag{5.289}$$

$$B_m = -\Upsilon \delta D_m \tag{5.290}$$

$$C_m = \sigma D_m \tag{5.291}$$

Finally we consider the case $m = 0$:

$$\frac{d^2 R_m}{d\rho^2} + \frac{1}{\rho}\frac{dR_m}{d\rho} = -\frac{2F_0}{k\pi}\frac{1}{\rho}\delta(\rho - R) \tag{5.292}$$

The general solution of the homogeneous equation in this case is:

$$R_m(\rho) = \begin{Bmatrix} R_0^-(\rho) = A_0 + B_0 \ln \rho (\rho < R) \\ R_0^+(\rho) = C_0 + D_0 \ln \rho (\rho > R) \end{Bmatrix} \tag{5.293}$$

$$\left.\frac{dR}{d\rho}\right|_{\rho = a} = 0 = \frac{D_0}{a} = 0 \rightarrow D_0 = 0 \tag{5.294}$$

$$k\frac{B_0}{b} = h(A_0 + B_0 \ln b) \tag{5.295}$$

$$A_0 = B_0 \left(\frac{k}{hb} - \ln b\right) \tag{5.296}$$

Final solution

From the continuity condition:

$$R_0^-(R) = R_0^+(R) \tag{5.297}$$

$$A_0 + B_0 \ln R = C_0 \tag{5.298}$$

Finally the condition of the change of derivatives is used:

$$\int_{R-\varepsilon}^{R+\varepsilon} \left[\frac{d^2 R_0}{d\rho^2} + \frac{1}{\rho}\frac{dR_0}{d\rho}\right] \rho d\rho = -\frac{2F_0}{k\pi} \int_{R-\varepsilon}^{R+\varepsilon} \frac{1}{\rho}\delta(\rho - R)\rho d\rho \tag{5.299}$$

Using the same arguments as in the previous discussion:

$$R\left(\left.\frac{dR^+}{d\rho}\right|_R - \left.\frac{dR^-}{d\rho}\right|_R\right) = -\frac{2F_0}{k\pi} \tag{5.300}$$

$$R\left(0 - \frac{B_0}{R}\right) = -\frac{2F_0}{k\pi} \tag{5.301}$$

Then:

$$B_0 = \frac{2F_0}{k\pi};\ A_0 = \frac{2F_0}{k\pi}\left(\frac{k}{hb} - \ln b\right);\ C_0 = \frac{2F_0}{k\pi}\left(\frac{k}{hb} - \ln b + \ln R\right) \tag{5.302}$$

From the final solution:

$$u(\rho,\varphi) = \begin{Bmatrix} A_0 + B_0 \ln \rho + \sum_{m=1}^{\infty} [A_m \rho^{2m} + B_m \rho^{-2m}] \cos(2m\varphi) \ \ (\text{for } \rho < R) \\ C_0 + \sum_{m=1}^{\infty} [C_m \rho^{2m} + D_m \rho^{-2m}] \cos(2m\varphi) \ \ (\text{for } \rho > R) \end{Bmatrix} \tag{5.303}$$

Note: the second method is more complex from the mathematical point of view but usually describes better the solutions in the proximities of the anomalous points without using expansions in two or three dimensions in terms of trigonometric functions, minimizing the so called "Gibbs phenomena".

5.15 Variation of the Temperature in Two Semi-Cylinders

Find the temporal variation of the temperature in a cylinder composed of two infinite semi-cylinders, with all the curved surfaces thermally insulated, except for the flat surfaces in contact. The inner radius is R_1 while the outer radius is R_2. Initially ($t < 0$), before being thermally connected, the temperatures of the semi-cylinders were $T_0 - T_1$ and $T_0 + T_1$.

Starting at $t = 0$ the cylinders are united thermally to form a whole cylinder, completely insulated from the outside. Consider that the thermal contact is perfect between the two halves, so that the thermal conductivity k is homogeneous across the cylinder. The heat capacity is C and the density is ρ. The thermal diffusivity coefficient is $a^2 = k/\rho C$.

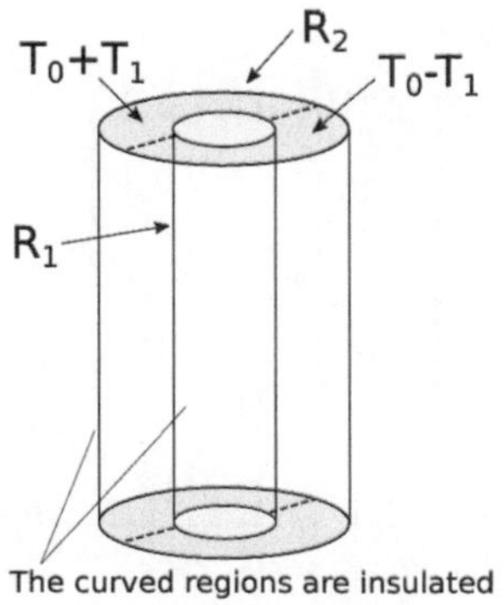

Figure 5.18

Mathematical formulation

We seek the solution as the relative variation of temperature with respect to the thermal equilibrium temperature ($T = T_0$), which will happen at infinite times.

$$u(\rho, \varphi, t) = T'(\rho, \varphi, t) - T_0 \tag{5.304}$$

$$u(\rho, \varphi, t \to \infty) = 0 \tag{5.305}$$

The problem to be solved is (there is no z variable, since the cylinder is infinite):

$$\left\{\begin{array}{l} \frac{\partial u(\rho,\varphi,t)}{\partial t} - a^2\Delta u(\rho,\varphi,t) = 0 \\ u(\rho,\varphi,t=0) = \left\{\begin{array}{ll} T_1 & (0<\varphi<\pi) \\ -T_1 & (\pi<\varphi<2\pi) \end{array}\right\} \\ \left.\dfrac{\partial u}{\partial \rho}\right|_{\rho=R_1} = 0 \\ \left.\dfrac{\partial u}{\partial \rho}\right|_{\rho=R_2} = 0 \end{array}\right\} \tag{5.306}$$

Sturm–Liouville problem

We separate variables:

$$u = W(\rho,\varphi)\cdot T(t) \tag{5.307}$$

We start with the eigenfunctions of the Sturm–Liouville problems for the (ρ, φ) variables:

$$\left\{\begin{array}{l} \Delta W + \lambda W = 0 \\ \left.\dfrac{\partial W}{\partial \rho}\right|_{\rho=R_1} = 0 \\ \left.\dfrac{\partial W}{\partial \rho}\right|_{\rho=R_2} = 0 \end{array}\right\} \tag{5.308}$$

$$\frac{1}{\rho}\frac{\partial}{\partial\rho}\left[\rho\frac{\partial W}{\partial\rho}\right] + \frac{1}{\rho^2}\frac{\partial^2 W}{\partial\varphi^2} + \lambda W = 0 \tag{5.309}$$

Separating variables once again $W = R(\rho)\cdot\Phi(\varphi)$

$$\frac{\rho\frac{d}{d\rho}[\rho\frac{dR}{d\rho}] + \lambda\rho^2 R}{R} = -\frac{1}{\Phi}\frac{d^2\Phi}{d\varphi^2} = \mu \tag{5.310}$$

Angular Sturm–Liouville problem:

$$\left\{\begin{array}{l} \dfrac{d^2\Phi}{d\varphi^2} + \mu\Phi = 0 \\ \Phi(\varphi) = \Phi(\varphi + 2\pi) \end{array}\right\} \tag{5.311}$$

Eigenfunctions and eigenvalues:

$$\Phi(\varphi) = A\cos(m\varphi) + B\sin(m\varphi) \tag{5.312}$$

The radial problem is:

$$\rho\frac{d}{d\rho}\left[\rho\frac{dR}{d\rho}\right]+\lambda\rho^2R-m^2R=0 \tag{5.313}$$

$$\rho\frac{d}{d\rho}\left[\rho\frac{dR}{d\rho}\right]+[\lambda\rho^2-m^2]R=0 \tag{5.314}$$

General solution for the radial equation:

$$R(\rho)=CJ_m(\sqrt{\lambda}\rho)+DN_m(\sqrt{\lambda}\rho) \tag{5.315}$$

Due to the solution not being necessarily finite at $\rho=0$, in general $C, D\neq 0$.

Let us calculate now the eigenvalues of the Sturm–Liouville problem. We know that the general solution is:

$$u=\sum R(\rho)\Phi(\varphi)Q(t) \tag{5.316}$$

From the boundary conditions: $\left.\dfrac{\partial u}{\partial\rho}\right|_{\rho=R_{1,2}}=0$ we deduce:

$$\sum\left.\frac{dR}{d\rho}\right|_{\rho=R_{1,2}}\Phi(\varphi)Q(t)=0$$

Since $\Phi(\varphi)$; $Z(z)$ can have any value, to satisfy the boundary conditions it is necessary that:

$$\left.\frac{dR}{d\rho}\right|_{\rho=R_{1,2}}=0$$

We have two equations for finding the eigenvalues of the radial problem:

$$\frac{d}{d\rho}[CJ_m(\sqrt{\lambda}\rho)+DN_m(\sqrt{\lambda}\rho)]=0$$

Applying the boundary conditions:

$$\left\{\begin{aligned}\frac{d}{d\rho}\left[CJ_m(\sqrt{\lambda}\rho)+DN_m(\sqrt{\lambda}\rho)\right]_{(\rho=R_1)}=0\\ \frac{d}{d\rho}\left[CJ_m(\sqrt{\lambda}\rho)+DN_m(\sqrt{\lambda}\rho)\right]_{(\rho=R_2)}=0\end{aligned}\right\} \tag{5.317}$$

Eigenvalue equation to calculate λ_{nm} (comes from equating to zero the determinant of the previous system of equations):

$$J'_m(\sqrt{\lambda}R_1)\cdot N'_m(\sqrt{\lambda}R_2)=J'_m(\sqrt{\lambda}R_2)\cdot N'_m(\sqrt{\lambda}R_1) \tag{5.318}$$

The radial orthogonal eigenfunctions (not normalized) are obtained by using, for example, the first equation:

$$C_{nm} J'_m(\sqrt{\lambda_{nm}} R_1) = -D_{nm} N'_m(\sqrt{\lambda_{nm}} R_1) \tag{5.319}$$

Then:

$$D_{nm} = -C_{nm} \frac{J'_m(\sqrt{\lambda_{nm}} R_1)}{N'_m(\sqrt{\lambda_{nm}} R_1)} \tag{5.320}$$

Finally, the radial eigenfunctions will be:

$$R_{nm}(\rho) = J_m(\sqrt{\lambda_{nm}} \rho) - \frac{J'_m(\sqrt{\lambda_{nm}} R_1)}{N'_m(\sqrt{\lambda_{nm}} R_1)} N_m(\sqrt{\lambda_{nm}} \rho) \tag{5.321}$$

General solution

The general solution is:

$$\begin{aligned} u &= \sum_{n=1;m=0}^{\infty} W_{nm}(\rho, \varphi) Q_{nm}(t) \\ &= \sum_{n=1;m=0}^{\infty} \left[J_m(\sqrt{\lambda_{nm}} \rho) - \frac{J'_m(\sqrt{\lambda_{nm}} R_1)}{N'_m(\sqrt{\lambda_{nm}} R_1)} N_m(\sqrt{\lambda_{nm}} \rho) \right] \\ &\quad \times [A_{nm} \cos(m\varphi) + B_{nm} \sin(m\varphi)] Q_{nm}(t) \end{aligned} \tag{5.322}$$

Replacing the solution in the heat equation $\frac{\partial u(\rho,\varphi,t)}{\partial t} - a^2 \Delta u(\rho, \varphi, t) = 0$ we get:

$$\sum_{n=1;m=0}^{\infty} \left[\frac{dQ_{nm}(t)}{dt} + a^2 \lambda_{nm} Q_{nm}(t) \right] \cdot W_{nm}(\rho, \varphi) = 0 \tag{5.323}$$

Problem for the temporal coefficients:

$$\left\{ \begin{array}{l} \dfrac{dQ_{nm}(t)}{dt} + a^2 \lambda_{n,m} Q_{nm}(t) = 0 \\ Q_{nmk}(t) = e^{(-a^2 \lambda_{n,m} t)} \end{array} \right\} \tag{5.324}$$

The general solution is:

$$\begin{aligned} u = &\sum_{n=1;m=0}^{\infty} e^{(-a^2 \lambda_{n,m} t)} \left[J_m(\sqrt{\lambda_{nm}} \rho) - \frac{J'_m(\sqrt{\lambda_{nm}} R_1)}{N'_m(\sqrt{\lambda_{nm}} R_1)} N_m(\sqrt{\lambda_{nm}} \rho) \right] \\ &\times [A_{nm} \cos(m\varphi) + B_{nm} \sin(m\varphi)] \end{aligned} \tag{5.325}$$

Final solution

To find the coefficients we impose the initial conditions:

$$u(\rho,\varphi,0)=\left\{\begin{matrix} T_1 & (0<\varphi<\pi) \\ -T_1 & (\pi<\varphi<2\pi) \end{matrix}\right\}=$$
$$=\sum_{n=1;m=0}^{\infty}\left[J_m(\sqrt{\lambda_{nm}}\rho)-\frac{J'_m(\sqrt{\lambda_{nm}}R_1)}{N'_m(\sqrt{\lambda_{nm}}R_1)}N_m(\sqrt{\lambda_{nm}}\rho)\right]$$
$$\times[A_{nm}\cos(m\varphi)+B_{nm}\sin(m\varphi)] \qquad (5.326)$$

The initial condition is antisymmetric with respect to the angle (φ). This implies that all coefficients corresponding to terms which are symmetric in the angular variable must be null (it can be checked mathematically by multiplying by $\cos(m\varphi)$, since the function on the left part is antisymmetric in the angular variable, just like $\sin(m\varphi)$).

Then we will have the relation:

$$0=A_{nm}\left\|\cos(m\varphi)\right\|^2\rightarrow A_{nm}=0 \qquad (5.327)$$

To find B_{nmk} we multiply both sides by $\sin(m'\varphi)R_{nm}(\rho)$ and integrate between the limits $\int_{R_1}^{R_2}\int_{0}^{2\pi}\rho d\rho d\varphi$.

We get the result:

$$\int_{R_1}^{R_2}\left[J_m(\sqrt{\lambda_{nm}}\rho)-\frac{J'_m(\sqrt{\lambda_{nm}}R_1)}{N'_m(\sqrt{\lambda_{nm}}R_1)}N_m(\sqrt{\lambda_{nm}}\rho)\right]\rho d\rho$$
$$\left[T_1\int_0^{\pi}\sin(m\varphi)d\varphi-T_1\int_{\pi}^{2\pi}\sin(m\varphi)d\varphi)\right]=$$
$$2T_1\int_{R_1}^{R_2}\left[J_m(\sqrt{\lambda_{nm}}\rho)-\frac{J'_m(\sqrt{\lambda_{nm}}R_1)}{N'_m(\sqrt{\lambda_{nm}}R_1)}N_m(\sqrt{\lambda_{nm}}\rho)\right]\rho d\rho$$
$$\left[\int_0^{\pi}\sin(m\varphi)d\varphi\right]=$$

$$= B_{nm} \left\| J_m(\sqrt{\lambda_{nm}}\rho) - \frac{J'_m(\sqrt{\lambda}R_1)}{N'_m(\sqrt{\lambda}R_1)} N_m(\sqrt{\lambda_{nm}}\rho)] \right\|^2 \|\sin(m\varphi)\|^2$$

$$B_{nm} = 2T_1 \left[\frac{(1-(-1)^m}{m} \right]$$

$$\times \frac{\int_{R_1}^{R_2} [J_m(\sqrt{\lambda_{nm}}\rho) - \frac{J'_m(\sqrt{\lambda_{nm}}R_1)}{N'_m(\sqrt{\lambda_{nm}}R_1)} N_m(\sqrt{\lambda_{nm}}\rho)]\rho d\rho}{\left\| [J_m(\sqrt{\lambda_{nm}}\rho) - \frac{J'_m(\sqrt{\lambda}R_1)}{N'_m(\sqrt{\lambda}R_1)} N_m(\sqrt{\lambda_{nm}}\rho)] \right\|^2 \|\sin(m\varphi)]\|^2} \tag{5.328}$$

The final solution is:

$$u(\rho, \varphi, t) = T_0 + \sum_{n=1;m=1}^{\infty} B_{nm} e^{(-a^2 \lambda_{n,m} t)}$$
$$\times \left[J_m(\sqrt{\lambda_{nm}}\rho) - \frac{J'_m(\sqrt{\lambda_{nm}}R_1)}{N'_m(\sqrt{\lambda_{nm}}R_1)} N_m(\sqrt{\lambda_{nm}}\rho) \right] \sin(m\varphi) \tag{5.329}$$

5.16 Stationary Temperature inside an Infinite Cylindrical Tube

Find the stationary distribution of temperature $u(\rho, \varphi)$ in an infinite cylindrical tube with radii $\rho_1 = 1$ and $\rho_2 = 2$ if the temperature in on the inner surface is $u(1) = \sin^2(\varphi)$, whereas the temperature on the outer surface is $u(2) = 0$.

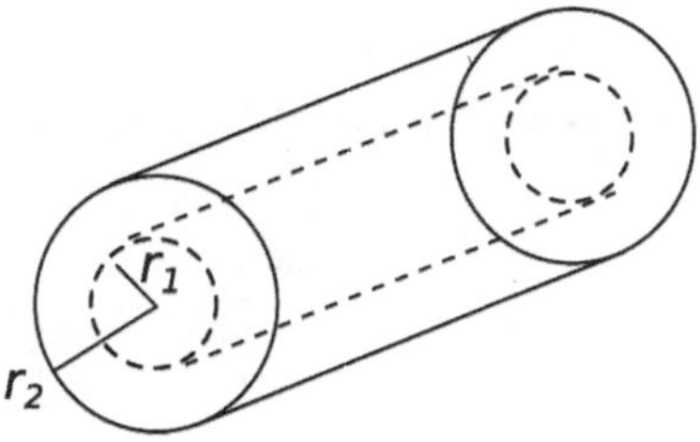

Figure 5.19

Mathematical formulation

Due to the symmetry we will solve the problem in cylindrical coordinates. Since the cylinder is infinite there is no dependence on the z coordinate. The mathematical formulation is that of Laplace's problem in a disk with the specified boundary conditions.

General solution

In these conditions the general solution for the Laplace's equation $\Delta u = 0$ is:

$$u(\rho,\varphi) = C_1 \ln(\rho) + C_2 + \sum_{n\geq 1}(A_n\rho^n + B_n\rho^{-n})\cos(n\varphi) + (D_n\rho^n + E_n\rho^{-n})\sin(n\varphi) \tag{5.330}$$

We impose the boundary conditions:

$$u(1,\varphi) = C_2 + \sum_{n\geq 1}(A_n + B_n)\cos(n\varphi) + (D_n + E_n)\sin(n\varphi) = \sin^2(\varphi) = \frac{1}{2} - \frac{1}{2}\cos(2\varphi) \tag{5.331}$$

$$u(2,\varphi) = C_1 \ln(2) + C_2 + \sum_{n\geq 1}(2^n A_n + 2^{-n} B_n)\cos(n\varphi) + (2^n D_n + 2^{-n} E_n)\sin(n\varphi) = 0 \tag{5.332}$$

Here we have attempted to use the boundary conditions at the inner surface in a form corresponding to the Fourier series expansion.

We can use the orthogonality of the angular eigenfunctions or use the easiest method to equate the Fourier coefficients so that the relations are valid for all the φ angles.

$$\begin{Bmatrix} C_2 = \dfrac{1}{2} \\ C_1 \ln 2 + C_2 = 0 \\ A_2 + B_2 = -\dfrac{1}{2} \\ 4A_2 + \dfrac{1}{4}B_2 = 0 \end{Bmatrix} \tag{5.333}$$

Final solution

The solutions are:

$$\begin{Bmatrix} C_1 = -\dfrac{1}{2\ln 2} \\ C_2 = \dfrac{1}{2} \\ A_2 = \dfrac{1}{30} \\ B_2 = -\dfrac{8}{15} \end{Bmatrix} \tag{5.334}$$

The final solution is:

$$u(\rho,\varphi) = \frac{1}{2}\left(1 - \frac{\ln\rho}{\ln 2}\right) + \left(\frac{\rho^2}{30} - \frac{8}{15\rho^2}\right)\cos(2\varphi) \tag{5.335}$$

5.17 Case Study: Time Variation of the Density of Viruses Emitted by a Thin Filament Placed in a Sector of a Disk

Find the distribution of density of viruses (diffusion coefficient D) as a function of time in a two-dimensional space with the shape of a fourth of a circular ring (spanning an angle $\pi/2$, with outer radius a and inner radius b). The membrane is impermeable in three of its boundaries and semi-permeable in the inner curved boundary, due to the exchange of viruses with the outer medium, with a constant concentration $n = n_0$. This exchange happens according to Newton's law with a constant factor h.

Inside the membrane there is a source of viruses in the form of a thin curved line inside the angular sector ($\pi/8 < \varphi < 3\pi/8$) at a distance R from the center. It starts to release viruses since $t = 0$. The total flux (number of viruses per unit time) is such that: $\lim_{\epsilon\to 0} \iint\limits_{\Omega\epsilon} f(\rho,\varphi)\rho d\rho d\varphi = F_0$ being $\Omega\epsilon$ the infinitesimal surface around the source of virus.

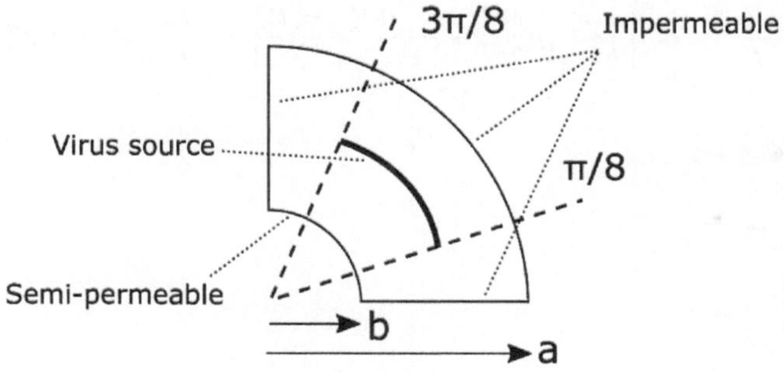

Figure 5.20

Deduction of the inhomogeneous part of the equation

It is clear that

$$f(\rho,\varphi)=\frac{A}{\rho}\delta(\rho-R)G(\varphi)=\frac{A}{\rho}\delta(\rho-R)\left\{\begin{matrix}0\left(0<\varphi<\frac{\pi}{8}\right)\\1\left(\frac{\pi}{8}<\varphi<\frac{3\pi}{8}\right)\\0\left(\frac{3\pi}{8}<\varphi<\frac{\pi}{2}\right)\end{matrix}\right\} \tag{5.336}$$

Applying this condition: $F_0=\iint\limits_{\Omega\epsilon}\frac{A}{\rho}\delta(\rho-R)G(\varphi)\rho d\rho d\varphi=\frac{\pi}{4}A$

We get: $A=\frac{4F_0}{\pi}$

Mathematical formulation

Subtracting the virus concentration of the outer medium n_0 from the solution we transform the semi-permeable boundary in a homogeneous boundary of the third kind without affecting the type of the other boundaries. The formulation of the temporal variation of concentration $u(\rho,\varphi,t)$ with respect to n_0 will be the following:

$$\left\{\begin{array}{l}\dfrac{\partial u(\rho,\varphi,t)}{\partial t}-D\left[\dfrac{1}{\rho}\dfrac{\partial}{\partial\rho}\left[\rho\dfrac{\partial u}{\partial\rho}\right]+\dfrac{1}{\rho^2}\dfrac{\partial u^2(\rho,\varphi,t)}{\partial\varphi^2}\right]=\dfrac{4F_0}{\pi}\dfrac{1}{\rho}\delta(\rho-R)G(\varphi)\\ \left.\dfrac{\partial u}{\partial\varphi}\right|_{\varphi=0}=0\\ \left.\dfrac{\partial u}{\partial\varphi}\right|_{\varphi=\frac{\pi}{2}}=0\\ \left.\dfrac{\partial u}{\partial\rho}\right|_{\rho=a}=0\\ -D\left.\dfrac{\partial u}{\partial\rho}\right|_{\rho=b}=-hu(b,\varphi)\qquad(u>0\rightarrow\text{Outward flux})\end{array}\right\} \tag{5.337}$$

Sturm–Liouville problem

As all the boundaries are homogeneous, we can seek the solution by expanding it in orthogonal eigenfunctions that solve the Sturm–Liouville problem:

$$u(\rho,\varphi)=\sum T_{nm}(t)v_{nm}(\rho,\varphi) \tag{5.338}$$

Where the $v_{nm}(\rho,\varphi)$ satisfy:

$$\left\{\begin{array}{l}\Delta v(\rho,\varphi)+\lambda v(\rho,\varphi)=0\\ \left.\dfrac{\partial v}{\partial\varphi}\right|_{\varphi=0}=0\\ \left.\dfrac{\partial v}{\partial\varphi}\right|_{\varphi}=\dfrac{\pi}{2}=0\\ \left.\dfrac{\partial v}{\partial\rho}\right|_{\rho=a}=0\\ \left.\dfrac{\partial v}{\partial\rho}\right|_{\rho=b}-Hv(b,\varphi)=0\qquad(H=-\frac{h}{D})\end{array}\right\} \tag{5.339}$$

We seek the solution using the method of separation of variables:

$$v(\rho,\varphi)=R(\rho)\cdot\Phi(\varphi) \tag{5.340}$$

We arrive at the Sturm–Liouville problem for the angular variable:

$$\left\{\begin{array}{l}\dfrac{\partial \Phi^2(\rho,\varphi,t)}{\partial \varphi^2}+\mu\Phi=0\\ \left.\dfrac{d\Phi}{d\varphi}\right|_{\varphi=0}=0\\ \left.\dfrac{d\Phi}{d\varphi}\right|_{\varphi=\dfrac{\pi}{2}}=0\end{array}\right\} \tag{5.341}$$

The angular eigenfunctions in their general form are:

$$\Phi(\varphi)=C\cdot\cos(\sqrt{\mu}\varphi)+E\cdot\sin(\sqrt{\mu}\varphi) \tag{5.342}$$

Due to the first boundary condition, we have $E=0$. The eigenvalues μ_m are sought applying the second boundary condition:

$$\left.\frac{d\Phi}{d\varphi}\right|_{\varphi=\frac{\pi}{2}}=0=\sin\left(\sqrt{\mu}\frac{\pi}{2}\right) \tag{5.343}$$

$$\sqrt{\mu}\frac{\pi}{2}=\pi m \tag{5.344}$$

$$\mu_m=(2m)^2 \qquad (m=0,1,2,3\ldots) \tag{5.345}$$

When we replace $v(\rho,\varphi)=R(\rho)\cos(2m\varphi)$ into equation (5.339) we get at the following equation for the radial variable:

$$\frac{\partial^2 R}{\partial\rho^2}+\frac{1}{\rho}\frac{\partial R}{\partial\rho}+\left(\lambda-\frac{[2m]^2}{\rho^2}\right)R=0 \tag{5.346}$$

The solution of this problem gives us the set of radial solutions:

$$R_{nm}(\rho)=A_{nm}J_{2m}(\sqrt{\lambda_{nm}}\rho)+B_{nm}N_{2m}(\sqrt{\lambda_{nm}}\rho) \tag{5.347}$$

The possible values of λ_{nm} are the n-th zero of the equation obtained when we equal to zero the determinant of the system of equations of the first and second boundary conditions:

$$\left\{\begin{array}{l}\left.\dfrac{dR}{d\rho}\right|_{\rho=a}=0\\ \left.\dfrac{dR}{d\rho}\right|_{\rho=b}-H\,R(b)=0\end{array}\right\} \tag{5.348}$$

$$\left\{\begin{array}{l} A_{nm}\sqrt{\lambda_{nm}}\left.\dfrac{dJ_{2m}(\sqrt{\lambda_{nm}}\rho)}{d\rho}\right|_{\rho=a} + B_{nm}\sqrt{\lambda_{nm}}\left.\dfrac{dN_{2m}(\sqrt{\lambda_{nm}}\rho)}{d\rho}\right|_{\rho=a} = 0 \\ A_{nm}\sqrt{\lambda_{nm}}\left.\dfrac{dJ_{2m}(\sqrt{\lambda_{nm}}\rho)}{d\rho}\right|_{\rho=b} - H J_{2m}(\sqrt{\lambda_{nm}}b) + \\ +B_{nm}\sqrt{\lambda_{nm}}\left.\dfrac{dN_{2m}(\sqrt{\lambda_{nm}}\rho)}{d\rho}\right|_{\rho=b} - H N_{2m}(\sqrt{\lambda_{nm}}b) = 0 \end{array}\right\} \tag{5.349}$$

We can also find the ration of the A_{nm} and B_{nm} coefficients from the first or second equation and, in this way, determine the form of the radial solution.

$$B_{nm} = -A_{nm}\frac{\left.\dfrac{dJ_{2m}(\sqrt{\lambda_{nm}}\rho)}{d\rho}\right|_{\rho=a}}{\left.\dfrac{dN_{2m}(\sqrt{\lambda_{nm}}\rho)}{d\rho}\right|_{\rho=a}} \tag{5.350}$$

So that:

$$R_{nm}(\rho) = J_{2m}(\sqrt{\lambda_{nm}}\rho) - \frac{\left.\dfrac{dJ_{2m}(\sqrt{\lambda_{nm}}\rho)}{d\rho}\right|_{\rho=a}}{\left.\dfrac{dN_{2m}(\sqrt{\lambda_{nm}}\rho)}{d\rho}\right|_{\rho=a}} N_{2m}(\sqrt{\lambda_{nm}}\rho)] \tag{5.351}$$

General solution

The general solution will be:

$$u(\rho, \varphi, t) = \sum_{n,m} T_{nm}(t) R_{nm}(\rho)\cos(2m\varphi) \tag{5.352}$$

Final solution

Once we clarify the form of the general solution we will search the coefficients of the expansion, replacing $u(\rho, \varphi, t) = \sum_{n,m} T_{nm}(t) R_{nm}(\rho)\cos(2m\varphi)$ in (5.337):

$$\sum_{n,m}\left[\frac{\partial T_{nm}(t)}{\partial t} + D\lambda_{nm}T_{nm}(t)\right] v_{nm}(\rho, \varphi) = \frac{4F_0}{\pi}\frac{1}{\rho}\delta(\rho - R)G(\varphi) \tag{5.353}$$

Using the orthogonality in the radial and angular eigenfunctions to arrive at the equation for the $T_{nm}(t)$ coefficients, we can multiply both sides of the previous relation by $R_{nm}(\rho)\cos(2m\varphi)$ and integrate $\int_b^a \int_0^{\frac{\pi}{2}} \rho d\rho d\varphi$. Due to the orthogonality of the radial $R_{nm}(\rho)$ and angular $\cos(2m\varphi)$ eigenfunctions we get the following equation for the coefficients $T_{nm}(t)$:

$$\frac{\partial T_{nm}(t)}{\partial t}+D\lambda_{nm}T_{nm}(t)=\frac{4F_0}{\pi}\int_b^a\int_0^{\frac{\pi}{2}} R_{nm}(\rho)\cos(2m\varphi)\delta(\rho-R)G(\varphi)d\rho d\varphi \tag{5.354}$$

or

$$\frac{\partial T_{nm}(t)}{\partial t}+D\lambda_{nm}T_{nm}(t)=\frac{4F_0}{\pi}\frac{R_{nm}(R)\frac{1}{2m}[\sin(3m\pi/4)-\sin(m\pi/4)]}{\|R_{nm}(\rho)\|^2\,\|\cos(2m\varphi)\|^2} = F_{nm} \tag{5.355}$$

We need to consider separately the terms corresponding to $m = 0$ since:

$$\|\cos(2m\varphi)\|^2 = \left\{\begin{matrix} \frac{\pi}{4}\ (m\neq 0) \\ \frac{\pi}{2}\ (m=0) \end{matrix}\right\} \tag{5.356}$$

$$F_{n0} = \frac{4F_0}{\pi}\frac{R_{n0}(R)\frac{\pi}{4}}{\|R_{n0}(\rho)\|^2\,\frac{\pi}{2}} \tag{5.357}$$

All that is left to do is to solve the equation for $T_{nm}(t)$ by searching the solution as the sum of the homogeneous solution and a particular solution and impose the initial conditions $T_{nm}(0) = 0$.

$$T_{nm}(t) = T_{nm,h}(t) + T_{nm,p}(t) = A_{nm}\,e^{(-D\lambda_{nm}t)} + \frac{F_{nm}}{D\lambda_{nm}} \tag{5.358}$$

Imposing the initial conditions we get:

$$A_{nm} = -\frac{F_{nm}}{D\lambda_{nm}} \tag{5.359}$$

Finally

$$u(\rho,\varphi,t) = \sum_{n,m}\frac{F_{nm}}{D\lambda_{nm}}[1-e^{(-D\lambda_{nm}t)}]R_{nm}(\rho)\cos(2m\varphi) \tag{5.360}$$

Chapter 6

Problems in Cylindrical Coordinates

We now start solving problems in cylindrical coordinates in three spatial dimensions. The new aspect with respect to the 2D case is the need to consider carefully the direction along which we set the orthogonality conditions. This analysis is especially relevant when we solve Laplace problems since, depending on the type of non-homogeneous boundary conditions the radial solution will change substantially, from Bessel functions of first or second type to Bessel functions of imaginary argument (modified Bessel functions).

Just like in the previous chapters in the case of having to solve problems with non-homogeneous boundary conditions in the azimuthal angle, it is important to ensure (inserting a constant or a compensatory function) that both boundaries are homogeneous.

6.1 General Solution of the Heat Equation in a Finite Cylinder with a Hole

Find the general solution for the temporal variation of temperature in a finite cylinder of height h with inner and outer radii R_1 and R_2. Both curved surfaces are thermally insulated. The flat surfaces are

Mathematical Methods for Physics: Problems and Solutions
Farkhad G. Aliev and Antonio Lara

ISBN 978-981-4968-71-3 (Hardcover), 978-1-003-41088-1 (eBook)
www.jennystanford.com

in contact with a thermal reservoir at $T = 0$. At the initial moment the temperature is given by $f(\rho, \varphi, z)$.

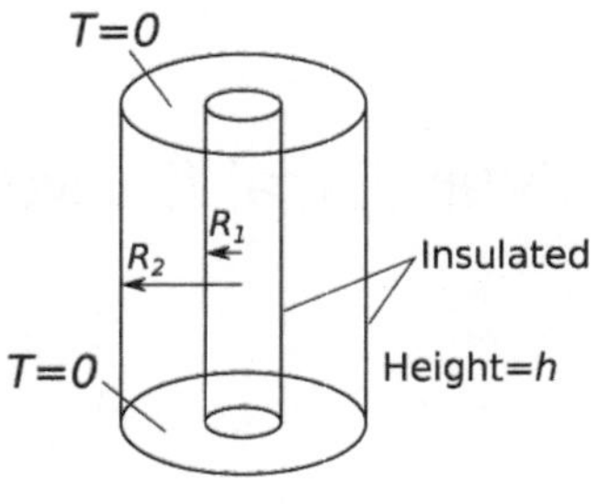

Figure 6.1

Mathematical formulation

$$\left\{\begin{array}{l} \dfrac{\partial u(\rho, \varphi, t)}{\partial t} - a^2 \Delta u(\rho, \varphi, z, t) = 0 \\ u(\rho, \varphi, z, 0) = f(\rho, \varphi, z) \\ \left.\dfrac{\partial u}{\partial \rho}\right|_{\rho=R_1} = 0 \\ \left.\dfrac{\partial u}{\partial \rho}\right|_{\rho=R_2} = 0 \\ u(\rho, \varphi, 0, t) = 0 \\ u(\rho, \varphi, h, t) = 0 \end{array}\right\} \tag{6.1}$$

Sturm–Liouville problem

We separate variables:

$$u = W(\rho, \varphi, z) \cdot T(t) \tag{6.2}$$

We need to find the eigenfunctions of the Sturm–Liouville problems in the variables (ρ, φ, z) to eliminate the second derivatives of the Laplacian.

Auxiliary problem:

$$\begin{Bmatrix} \Delta W + \lambda W = 0 \\ \left.\dfrac{\partial W}{\partial \rho}\right|_{\rho=R_1} = 0 \\ \left.\dfrac{\partial W}{\partial \rho}\right|_{\rho=R_2} = 0 \\ W(\rho, \varphi, 0) = 0 \\ W(\rho, \varphi, h) = 0 \end{Bmatrix} \tag{6.3}$$

$$\frac{1}{\rho}\frac{\partial}{\partial \rho}\left[\rho\frac{\partial W}{\partial \rho}\right] + \frac{1}{\rho^2}\frac{\partial^2 W}{\partial \varphi^2} + \frac{\partial^2 W}{\partial z^2} + \lambda W = 0 \tag{6.4}$$

Separating variables once again: $W = \mathcal{R}(\rho) \cdot \Phi(\varphi) \cdot Z(z)$ we get to the angular and vertical Sturm–Liouville problems:

$$\begin{Bmatrix} \dfrac{d^2\Phi}{d\varphi^2} + \mu\Phi = 0 \\ \Phi(\varphi) = \Phi(\varphi + 2\pi) \end{Bmatrix} \tag{6.5}$$

Eigenfunctions and eigenvalues:

$$\Phi(\varphi) = A\cos(m\varphi) + B\sin(m\varphi) \qquad \mu = m^2 \quad (m \text{ integers}) \tag{6.6}$$

Sturm–Liouville problem in the vertical direction:

$$\begin{Bmatrix} \dfrac{d^2 Z}{dz^2} + \nu Z = 0 \\ Z(0) = Z(h) = 0 \end{Bmatrix} \tag{6.7}$$

Eigenfunctions and eigenvalues

$$Z(z) = \sin\left(\frac{\pi n}{h} z\right) \tag{6.8}$$

$$\nu = \left(\frac{\pi n}{h}\right)^2 \tag{6.9}$$

(with n integers greater than zero).

The radial problem is:

$$\rho\frac{d}{d\rho}\left[\rho\frac{d\mathcal{R}}{d\rho}\right] + \left[(\lambda - \nu)\rho^2 - m^2\right]\mathcal{R} = 0 \tag{6.10}$$

or

$$\frac{d^2\mathcal{R}}{d\rho^2} + \frac{1}{\rho}\left[\frac{d\mathcal{R}}{d\rho}\right] + \left[(\lambda - \nu) - \frac{m^2}{\rho^2}\right]\mathcal{R} = 0 \tag{6.11}$$

General solution of the radial function:

$$\mathcal{R}(\rho) = CJ_m(\sqrt{[\lambda - \nu]} \cdot \rho) + DN_m(\sqrt{[\lambda - \nu]} \cdot \rho) \tag{6.12}$$

General solution

We seek the general solution: $u = \sum \mathcal{R}(\rho)\Phi(\varphi)Z(z)Q(t)$

From the first boundary condition:

$$\left.\frac{\partial u}{\partial \rho}\right|_{\rho=R_1,R_2} = 0 \rightarrow \sum \left.\frac{d\mathcal{R}}{d\rho}\right|_{\rho=R_1,R_2} \Phi(\varphi)Z(z)Q(t) = 0 \tag{6.13}$$

Since the $\Phi(\varphi)$; $Z(z)$; $Q(t)$ can be any value, $\left.\frac{d\mathcal{R}}{d\rho}\right|_{\rho=R_1,R_2} = 0$ to satisfy the first two boundary conditions.

We have two equations to find the eigenvalues (tagged with the index k) from the radial problem:

$$\left\{\begin{aligned} &\frac{d}{d\rho}[CJ_m(\sqrt{[\lambda - \nu]} \cdot \rho) + DN_m(\sqrt{[\lambda - \nu]} \cdot \rho)]_{\rho=R_1} = 0 \\ &\frac{d}{d\rho}[CJ_m(\sqrt{[\lambda - \nu]} \cdot \rho) + DN_m(\sqrt{[\lambda - \nu]} \cdot \rho)]_{\rho=R_2} = 0 \end{aligned}\right\} \tag{6.14}$$

Equation to find the eigenvalues λ_{nmk} :

$$\begin{aligned} &J'_m([\sqrt{\lambda - \nu}] \cdot R_1) \cdot N'_m(\sqrt{[\lambda - \nu]} \cdot R_2) \\ &= J'_m(\sqrt{[\lambda - \nu]} \cdot R_2) \cdot N'_m(\sqrt{[\lambda - \nu]} \cdot R_1) \end{aligned} \tag{6.15}$$

Being J' and N' the derivatives of the Bessel's and Neumann's functions. Using the relation

$$C_{nmk}J'_m(\sqrt{[\lambda_{nmk} - \nu_n]} \cdot R_1) = -D_{nm}kN'_m(\sqrt{[\lambda_{nmk} - \nu_n]} \cdot R_1) \tag{6.16}$$

We get at:

$$D_{nmk} = -C_{nmk}\frac{J'_m(\sqrt{[\lambda_{nmk}-\nu_n]}\cdot R_1)}{N'_m(\sqrt{[\lambda_{nmk}-\nu_n]}\cdot R_1)} \tag{6.17}$$

Obtaining the (not normalized) orthogonal radial eigenfunctions

$$\mathcal{R}_{nmk}(\rho) = J_m(\sqrt{[\lambda_{nmk}-\nu_n]}\cdot\rho) - \frac{J'_m(\sqrt{\lambda_{nmk}-\nu_n}R_1)}{N'_m(\sqrt{\lambda_{nmk}-\nu_n}R_1)}N_m(\sqrt{[\lambda_{nmk}-\nu_n]}\cdot\rho) \tag{6.18}$$

General solution:

$$\begin{aligned} u &= \sum_{n=1;m=0}^{\infty} W_{nmk}(\rho,\varphi,z)Q_{nmk}(t) = \\ &= \sum_{n,k=1;m=0}^{\infty} \sin\left(\frac{\pi n}{h}z\right)\Bigg[J_m(\sqrt{\lambda_{nmk}-\nu_n}\rho) \\ &\quad -\frac{J'_m(\sqrt{[\lambda_{nmk}-\nu_n]}\cdot R_1)}{N'_m(\sqrt{[\lambda_{nmk}-\nu_n]}\cdot R_1)}N_m(\sqrt{[\lambda_{nmk}-\nu_n]}\cdot\rho)\Bigg]\times \\ &\quad \times[A_{nmk}\cos(m\varphi)+B_{nmk}\sin(m\varphi)]Q_{nmk}(t) \end{aligned} \tag{6.19}$$

Final solution

To conclude, we will find the temporal coefficients:

$$\left\{\begin{array}{l} \frac{d}{dt}Q_{nmk}(t) + a^2\lambda_{nmk}Q_{nmk}(t) = 0 \\ Q_{nmk}(t) = e^{(-a^2\lambda_{nmk}t)} \end{array}\right\} \tag{6.20}$$

The general solution is:

$$\begin{aligned} u &= \sum_{n,k=1;m=0}^{\infty} \sin\left(\frac{\pi n}{h}z\right)e^{(-a^2\lambda_{nmk}t)}\Bigg[J_m(\sqrt{[\lambda_{nmk}-\nu_n]}\cdot\rho) \\ &\quad -\frac{J'_m(\sqrt{[\lambda_{nmk}-\nu_n]}\cdot R_1)}{N'_m(\sqrt{[\lambda_{nmk}-\nu_n]}\cdot R_1)}N_m(\sqrt{[\lambda_{nmk}-\nu_n]}\cdot\rho)\Bigg] \\ &\quad \times[A_{nm}\cos(m\varphi)+B_{nm}\sin(m\varphi)] \end{aligned} \tag{6.21}$$

To find the coefficients we need to impose the initial conditions and use the orthogonality of the radial, angular and vertical eigenfunctions.

6.2 Case Study: Heating of a Cylinder

Consider a cylinder (radius R and height L) with its curved surface thermally insulated. The thermal conductivity, heat capacity and density of the material are k, C, ρ respectively. Until $t = 0$ the cylinder is at thermal equilibrium and with both flat surfaces in contact with a thermal reservoir at T_0. At $t = 0$ the thermal reservoir at T_0 is removed at the upper face, and a heat flux $J(\rho)$ starts going through this boundary. Find: (i) The new stationary distribution of temperature (that is, the distribution of temperature after an infinite time and that is independent of time). (ii) The variation of the distribution of temperature in the cylinder as a function of time after $t = 0$.

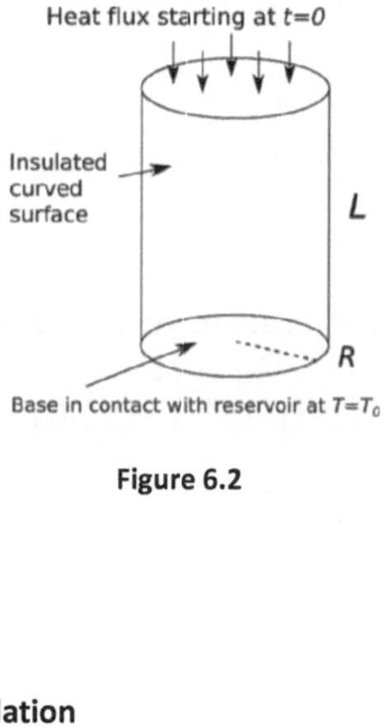

Figure 6.2

Mathematical formulation

To simplify the calculations we subtract T_0 from the solution. Furthermore, due to the angular symmetry of the supplied flux, as well as of the boundary conditions, the solution will not depend on the angular variable.

$$\left\{\begin{array}{l}\dfrac{\partial u}{\partial t} - \chi \Delta u = 0 \\ u(\rho, 0) = 0 \\ \left.\dfrac{\partial u}{\partial \rho}\right|_{\rho=R} = 0 \\ -k \left.\dfrac{\partial u}{\partial z}\right|_{z=L} = \dfrac{J(\rho)}{\pi R^2}(t > 0) \\ u(\rho, z, 0) = 0\end{array}\right\} \tag{6.22}$$

We will first formulate problem (i), stationary, for a function $w(\rho, z)$ (solution at times $t \to \infty$):

$$\left\{\begin{array}{l}\Delta w = 0 \\ w(\rho, 0) = 0 \\ \left.\dfrac{\partial w}{\partial \rho}\right|_{\rho=R} = 0 \\ -k \left.\dfrac{\partial w}{\partial z}\right|_{z=L} = \dfrac{J(\rho)}{\pi R^2}\end{array}\right\} \tag{6.23}$$

We will seek the solution as the sum of the solutions of two problems: a stationary one $w(\rho, z)$ which will be the solution of Laplace's equation with a non-homogeneous boundary and the solution of the transient problem $v(\rho, z, t)$, with all its boundaries being homogeneous, such that the total solution will be:

$$u(\rho, z, t) = w(\rho, z) + v(\rho, z, t) \tag{6.24}$$

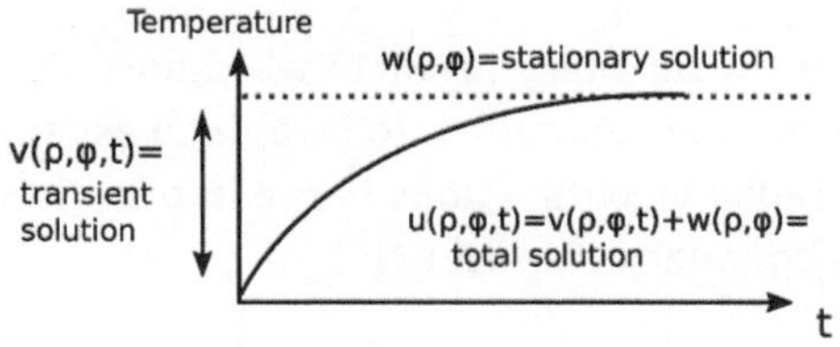

Figure 6.3

The problem for $w(\rho, z)$ does not have initial conditions, whereas the initial condition of the transient problem can be obtained from the initial condition for $u(\rho, z, t)$:

$$u(\rho, z, 0) = w(\rho, z) + v(\rho, z, 0) = 0 \tag{6.25}$$

Transient problem (ii) for $v(\rho, z, t)$, at $t > 0$ is:

$$\left\{\begin{array}{l} \dfrac{\partial v}{\partial t} - \chi \Delta v = 0 \\ v(\rho, 0) = 0 \\ \left.\dfrac{\partial v}{\partial \rho}\right|_{\rho=R} = 0 \\ \left.\dfrac{\partial v}{\partial z}\right|_{z=L} = 0 \\ v(\rho, z, 0) = -w(\rho, z) \end{array}\right\} \tag{6.26}$$

We can check that adding both problems we recover the original problem for $u(\rho, z, t)$.

Sturm–Liouville problem

We seek the solution of problem (i) by expanding it into eigenfunctions of the Sturm–Liouville for ρ to eliminate the radial Laplacian from the heat equation. In our case the boundary conditions are only homogeneous in the vertical direction (z). These facts will be important for choosing the sign of the constant of the separation of variables.

$$\frac{1}{\rho}\frac{\partial}{\partial \rho}\left[\rho \frac{\partial v}{\partial \rho}\right] + \frac{\partial^2 v}{\partial z^2} = 0 \tag{6.27}$$

$$v = \mathcal{R}(\rho) \cdot Z(z) \tag{6.28}$$

$$\frac{\frac{1}{\rho}\frac{d}{d\rho}\left[\rho \frac{d\mathcal{R}}{d\rho}\right]}{\mathcal{R}} = -\frac{1}{Z}\frac{d^2 Z}{dz^2} = -\lambda \tag{6.29}$$

With $\lambda > 0$ for the auxiliary problem we choose the *negative sign* before the constant of separation, to be able to expand the solution in orthogonal radial eigenfunctions (since in this direction we have homogeneous boundary conditions):

$$\left\{\begin{array}{l} \dfrac{d^2 Z}{dz^2} - \lambda Z = 0 \\ Z(0) = 0 \\ \left.\dfrac{dZ}{dz}\right|_{L} = \text{finite} \end{array}\right\} \tag{6.30}$$

We choose $Z(z) = A\sinh(\sqrt{\lambda}z)$ to automatically satisfy the first boundary condition. Then, after reducing the number of partial derivatives, the problem for $\mathcal{R}(\rho)$ remains like so:

$$\frac{1}{\rho}\frac{d}{d\rho}\left[\rho\frac{d\mathcal{R}}{d\rho}\right] + \lambda\mathcal{R} = 0 \tag{6.31}$$

We multiply the equation by ρ^2 and the radial problem then is:

$$\rho\frac{d}{d\rho}\left[\rho\frac{d\mathcal{R}}{d\rho}\right] + \lambda\rho^2\mathcal{R} = 0 \tag{6.32}$$

The general solution for the radial equation is a linear combination of Bessel and Neumann functions of order zero:

$$\mathcal{R}(\rho) = CJ_0(\sqrt{\lambda}\rho) + DN_0(\sqrt{\lambda}\rho) \tag{6.33}$$

Due to the solution being finite at $\rho = 0 \rightarrow D = 0$.

Calculation of the eigenvalues: general solution (for finite eigenvalues):

$$w = \sum AJ_0(\sqrt{\lambda}\rho)\sinh(\sqrt{\lambda}z) \tag{6.34}$$

From the second boundary condition we deduce: $\left.\frac{\partial u}{\partial \rho}\right|_{\rho=R} = 0 \rightarrow$ $\sum \left.\frac{d\mathcal{R}}{d\rho}\right|_{\rho=R} Z(z) = 0$. Since $Z(z)$ can have any value, the following condition must be fulfilled:

$$\left.\frac{d\mathcal{R}}{d\rho}\right|_{\rho=R} = \left.\frac{dJ_0(\sqrt{\lambda}\rho)}{d\rho}\right|_{\rho=R} \tag{6.35}$$

which gives the equation for the eigenvalues λ_n related to the null $\mu_0^{(n)}$ from the derivative of the zeroth order Bessel function:

$$\sqrt{\lambda_n}R = \mu_0^{(n)} \tag{6.36}$$

$$\lambda_n = \left[\frac{\mu_0^{(n)}}{R}\right]^2 \tag{6.37}$$

Since the first eigenvalue in this case is zero, we will consider separately its contribution to the solution.

General solution

The general solution is:

$$w(\rho, z) = \sum_{n=1}^{\infty} A_n J_0(\sqrt{\lambda_n}\rho)\sinh(\sqrt{\lambda_n}z) \tag{6.38}$$

To find the coefficients of the expansion we will apply the third boundary condition:

$$\left.\frac{\partial w}{\partial z}\right|_{z=L} = -\frac{J(\rho)}{k\pi R^2} = \sum_{n=2}^{\infty} A_n\sqrt{\lambda_n}J_0(\sqrt{\lambda_n}\rho)\cosh(\sqrt{\lambda_n}L) \tag{6.39}$$

Using the orthogonality of the radial eigenfunctions we get the coefficients. To find A_n we multiply both sides by $J_0(\sqrt{\lambda_{n'}}\rho)$ and integrate between the limits $\int_0^R \rho d\rho$:

$$A_n = -\frac{1}{k\pi R^2\sqrt{\lambda_n}} \frac{\int_0^R J(\rho)J_0(\sqrt{\lambda_n}\rho)\rho d\rho}{\left\|J_0(\frac{\mu_n}{R}\rho)\right\|^2\cosh(\sqrt{\lambda_n}L)} \tag{6.40}$$

We can simplify the results using the following relations:

$$\left\{\begin{array}{l} \int_0^a xJ_\nu\left(\frac{x_{\nu k}}{a}\right)J_\nu\left(\frac{x_{\nu l}}{a}\right)dx = \frac{a^2}{2}\left[J_\nu'\left(\frac{x_{\nu k}}{a}\right)\right]^2\delta_{kl} \\ J_0'(x) = -J_1(x) \end{array}\right\} \tag{6.41}$$

Note: due to the conditions of the curved boundary (homogeneous of the second type) the first term of the sum ($n = 1$), which corresponds to $\lambda_1 = 0$, will be treated separately since the corresponding radial equation will provide constant values and the solution of the problem in vertical direction z would be a linear function. This term describes the possible linear variation of the temperature in the z direction if the mean value of the heat flux across the surface is finite. In the case that $J(\rho) = J$ is a constant, the only term of the solution that is not null is the first one, which would give us the solution $w(z) = \frac{-Jz}{k\pi R^2}$. The rest of the terms $A_n = 0$ with $n \geq 2$ will be null due to the orthogonality of the constant (eigenfunction in the radial direction) with the rest of the eigenfunction $J_0\sqrt{\lambda_n}$.

Problem for the transient part

Due to the homogeneity of all boundaries, we expand the solution in series of orthogonal functions in two dimensions.

$$\frac{\partial v}{\partial t} - \chi \Delta v = 0 \tag{6.42}$$

$$v(\rho, z, t) = T(t) \cdot \mathcal{R}(\rho) \cdot Z(z) \tag{6.43}$$

$$\frac{dT}{dt}\mathcal{R}Z - T\chi\Delta(\mathcal{R}Z) = 0 \tag{6.44}$$

$$\frac{1}{\chi T}\frac{dT}{dt} = \frac{\Delta(\mathcal{R}Z)}{\mathcal{R}Z} = \frac{1}{\mathcal{R}\rho}\frac{d}{d\rho}\left[\rho\frac{d\mathcal{R}}{d\rho}\right] + \frac{1}{Z}\frac{d^2Z}{dz^2} = -\upsilon - \alpha = -\lambda \tag{6.45}$$

$$\left\{\begin{array}{l} \dfrac{d^2Z}{dz^2} + \alpha Z = 0 \\ Z(0) = 0 \\ \left.\dfrac{dZ}{dz}\right|_{z=L} = 0 \end{array}\right\} \tag{6.46}$$

$$Z_n(z) = \sin\left(\frac{\pi(2n+1)}{2L}z\right) \qquad (n = 0, 1, 2, 3\ldots) \tag{6.47}$$

$$\left\{\begin{array}{l} \dfrac{1}{\rho}\dfrac{d}{d\rho}\left[\rho\dfrac{d\mathcal{R}}{d\rho}\right] + \upsilon\mathcal{R} = 0 \\ \left.\dfrac{d\mathcal{R}}{d\rho}\right|_{\rho=R} = 0 \end{array}\right\} \tag{6.48}$$

Radial eigenfunctions:

$$R_k(\rho) = J_0\left(\frac{\mu_0^{(k)}}{R}\rho\right) \tag{6.49}$$

where the $\mu_0^{(k)}$ are zeros of the derivative of the zeroth order Bessel function. The eigenvalues of the Sturm–Liouville problem for the product of the radial function and the vertical one are:

$$\lambda_{nk} = \upsilon + \left(\frac{\pi(2n+1)}{2L}\right)^2 = \left[\frac{\mu_0^{(k)}}{R}\right]^2 + \left(\frac{\pi(2n+1)}{2L}\right)^2 \tag{6.50}$$

The general solution is:

$$v(\rho, z, t) = \sum_{n,k}^{\infty} T_{n,k}(t) J_0\left(\frac{\mu_0^{(k)}}{R}\rho\right) \sin\left(\frac{\pi(2n+1)}{2L}z\right) \tag{6.51}$$

Solving the equation for the temporal part $\frac{dT}{dt} + \lambda_{nk}\chi T = 0$ we get at exponential equations:

$$T_{n,k}(t) = T_{n,k}(0) e^{(-[(\frac{\mu_0^{(k)}}{R})^2 + (\frac{\pi(2n+1)}{2L})^2]\chi t)} \tag{6.52}$$

Final solutions

Finally we apply the initial conditions:

$$-w(\rho, z) = \sum_{n,k}^{\infty} T_{n,k}(0) J_0\left(\frac{\mu_0^{(k)}}{R}\rho\right) \sin\left(\frac{\pi(2n+1)}{2L}z\right) \tag{6.53}$$

Using the orthogonality of the radial eigenfunctions and the vertical ones we get the coefficients of the expansion.

$$T_{n,k}(0) = -\frac{\int_0^R \int_0^L w(\rho, z) J_0\left(\frac{\mu_0^{(k)}}{R}\rho\right) \sin\left(\frac{\pi(2n+1)}{2L}z\right) \rho d\rho dz}{\int_0^R \left|J_0\left(\frac{\mu_0^{(k)}}{R}\rho\right)\right|^2 \rho d\rho \int_0^L \left|\sin\left(\frac{\pi(2n+1)}{2L}z\right)\right|^2 dz} \tag{6.54}$$

6.3 Case Study: Stationary Distribution of Temperature inside a Semicylinder

Find the stationary distribution of temperature of a semicylinder of length L and radius ρ_0 in which three of the four surfaces are kept at different temperatures.

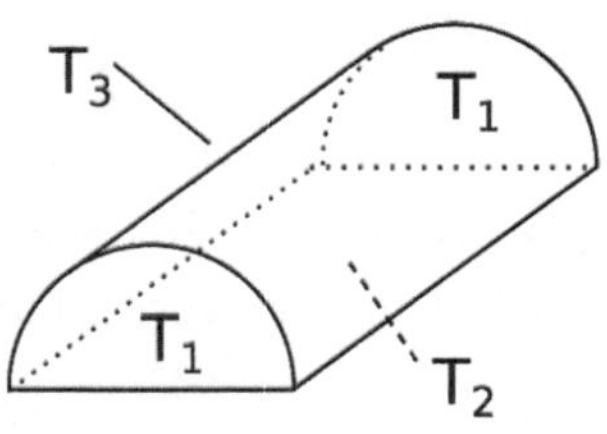

Figure 6.4

Mathematical formulation

From the theory we know we can decompose Laplace's equation, by using the linearity of the equation and the principle of superposition. To be able to expand the solution in angular eigenfunctions we subtract T_2 from the solution.

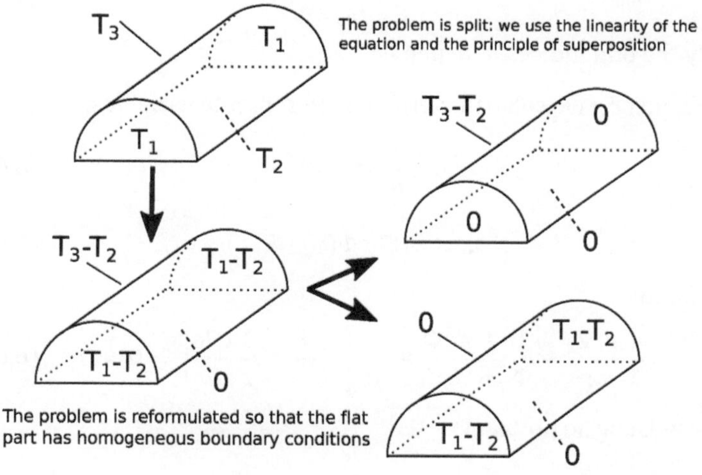

Figure 6.5

The remaining problem is split into two problems we know how to solve:

Problem (1):

$$u(\rho, \varphi, z) = \sum R_{nm}(\rho) v_{nm}(\varphi, z) \tag{6.55}$$

with:

$$v_{nm}(\varphi, z) = \Phi_n(\varphi) Z_m(z) = \sin(n\varphi) \sin\left(\frac{\pi m}{L} z\right) \tag{6.56}$$

and

$$R_{nm}(\rho) = \frac{R_{nm}(\rho_0)}{I_n(\frac{\pi m}{L}\rho_0)} I_n\left(\frac{\pi m}{L}\rho\right) \tag{6.57}$$

From the boundary conditions and orthogonality of the eigenfunctions we get the coefficients R_{nm}.

$$T_3 - T_2 = \sum R_{nm}(\rho_0) v_{nm}(\varphi, z) \tag{6.58}$$

$$R_{nm}(\rho_0)=\frac{(T_3-T_2)\int\limits_0^L \sin(\frac{\pi m}{L}z)dz\int\limits_0^\pi \sin(n\varphi)d\varphi}{\int\limits_0^L \sin^2(\frac{\pi m}{L}z)dz\int\limits_0^\pi \sin^2(n\varphi)d\varphi} \tag{6.59}$$

$$R_{nm}(\rho_0)=\frac{4(T_3-T_2)(1-(-1)^n)(1-(-1)^m)}{\pi^2 nm} \tag{6.60}$$

Only the odd indices n, m persist

Problem (2): we solve this problem, to separate variables:

$$\frac{1}{u}\Delta u=0 \tag{6.61}$$

$$u=Z(z)\cdot\Phi(\varphi)\cdot R(\rho) \tag{6.62}$$

we obtain:

$$\frac{1}{Z}\frac{d^2Z}{dz^2}+\frac{1}{\Phi}\frac{1}{\rho^2}\frac{d^2\Phi}{d\varphi^2}+\frac{1}{R}\left[\frac{d^2R}{d\rho^2}+\frac{1}{\rho}\frac{dR}{d\rho}\right]=0 \tag{6.63}$$

Sturm–Liouville problem

Solving the Sturm–Liouville problem for the angular part:

$$\frac{1}{Z}\frac{d^2Z}{dz^2}-\frac{m^2}{\rho^2}+\frac{1}{R}\left[\frac{d^2R}{d\rho^2}+\frac{1}{\rho}\frac{dR}{d\rho}\right]=0 \tag{6.64}$$

$$\Phi_m(\varphi)=\sin\left(\frac{\pi m}{\pi}\varphi\right)$$
$$=\sin(m\varphi)\ \text{(from the boundary conditions)}(m\geq 1) \tag{6.65}$$

Assigning $-\frac{m^2}{\rho^2}+\frac{1}{R}\left[\frac{d^2R}{d\rho^2}+\frac{1}{\rho}\frac{dR}{d\rho}\right]=-\lambda \quad (\lambda>0)$ we arrive at Bessel's equation for the radial part:

$$\left[\frac{d^2R}{d\rho^2}+\frac{1}{\rho}\frac{dR}{d\rho}\right]+\left(\lambda-\frac{m^2}{\rho^2}\right)R=0 \tag{6.66}$$

Applying the boundary conditions:

$$R_{km}(\rho_0)=J_m\left(\sqrt{\lambda_m^k}\rho_0\right)=0 \tag{6.67}$$

we obtain the eigenvalues λ_m^k. Finally the equation for z is:

$$\frac{d^2Z}{dz^2} - \lambda_m^k Z = 0 \quad (6.68)$$

$$Z_{km}(z) = A_{km} \sinh\left(\sqrt{\lambda_m^k} z\right) + B_{km} \cosh\left(\sqrt{\lambda_m^k} z\right) \quad (6.69)$$

General solution

The general solution of problem (2) is:

$$u(\rho, \varphi, z) = \sum \left[A_{km} \sinh\left(\sqrt{\lambda_m^k} z\right) + B_{km} \cosh\left(\sqrt{\lambda_m^k} z\right)\right] J_m\left(\sqrt{\lambda_m^k} \rho\right) \sin(m\varphi) \quad (6.70)$$

Final solution

We just need to apply the boundary conditions and to use the orthogonality of the radial and angular eigenfunctions to obtain the coefficients A_{km}, B_{km} :

$$u(\rho, \varphi, 0) = \sum B_{km} J_m\left(\sqrt{\lambda_m^k} \rho\right) \sin(m\varphi) = (T_1 - T_2) \quad (6.71)$$

$$B_{km} = \frac{(T_1 - T_2) \int\limits_0^{\rho_0} J_m\left(\sqrt{\lambda_m^k} \rho\right) \rho d\rho \int\limits_0^{\pi} \sin(m\varphi) d\varphi}{\int\limits_0^{\rho_0} [J_m\left(\sqrt{\lambda_m^k} \rho\right)]^2 \rho d\rho \int\limits_0^{\pi} \sin^2(m\varphi) d\varphi} \quad (6.72)$$

$$B_{km} = \frac{2(T_1 - T_2)[1 - (-1)^m] \int\limits_0^{\rho_0} J_m\left(\sqrt{\lambda_m^k} \rho\right) \rho d\rho}{\pi m \int_0^{\rho_0} [J_m\left(\sqrt{\lambda_m^k} \rho\right)]^2 \rho d\rho} \quad (6.73)$$

Only the odd m indices persist.

Applying another boundary condition we also find the A_{nm} indices:

$$u(\rho, \varphi. L) = \sum \left[A_{km} \sinh\left(\sqrt{\lambda_m^k} L\right) + B_{km} \cosh\left(\sqrt{\lambda_m^k} L\right)\right] \times J_m\left(\sqrt{\lambda_m^k} \rho\right) \sin(m\varphi) = (T_1 - T_2) \quad (6.74)$$

$$\left[A_{km}\sinh\left(\sqrt{\lambda_m^k}L\right)+B_{km}\cosh\left(\sqrt{\lambda_m^k}L\right)\right]$$

$$=\frac{2(T_1-T_2)[1-(-1)^m]\int\limits_0^{\rho_0}J_m\left(\sqrt{\lambda_m^k}\rho\right)\rho d\rho}{\pi m\int\limits_0^{\rho_0}[J_m\left(\sqrt{\lambda_m^k}\rho\right)]^2\rho d\rho}=B_{km} \quad (6.75)$$

$$A_{km}=\frac{B_{km}[1-\cosh\left(\sqrt{\lambda_m^k}L\right)]}{\sinh\left(\sqrt{\lambda_m^k}L\right)} \quad (6.76)$$

6.4 Case Study: Laplace's Equation in a Cylinder with No Homogeneous Contours

Find the stationary distribution of temperature in a cylinder with radius R and height L whose curved surface is in contact with a thermal reservoir at a temperature $T(\varphi, z)$. The upper base is traversed by a heat flux outwardly with a density equals to $f(\rho, \varphi)$. The lower base exchanges heat according to the Newton's law (with constant h) with the outer medium at zero temperature. Consider the thermal conductivity is $k = 1$.

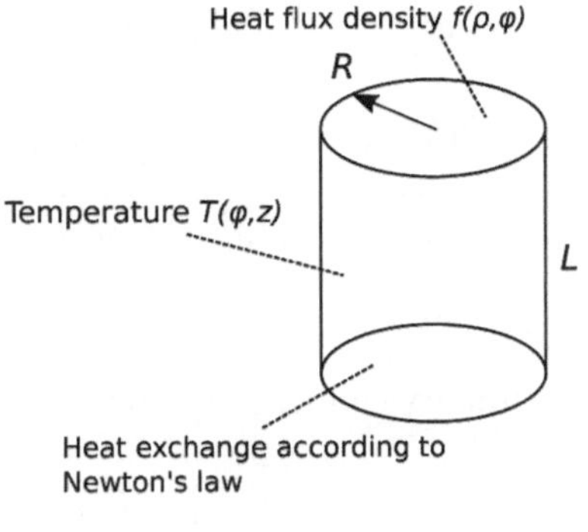

Figure 6.6

Mathematical formulation

$$\left\{\begin{array}{l} \Delta U(\rho,\varphi,z)=0 \\ U(\rho=R,\varphi,z)=T(\varphi,z) \\ -k\left.\dfrac{\partial U}{\partial z}\right|_{z=L}=f(\rho,\varphi) \\ -k\left.\dfrac{\partial U}{\partial z}\right|_{z=0}+hU=0 \end{array}\right\} \tag{6.77}$$

Using the principle of superposition we split the problem into two simpler ones:

$$U(\rho,\varphi,z)=u(\rho,\varphi,z)+v(\rho,\varphi,z) \tag{6.78}$$

In the same manner we split the boundary conditions: For the boundary at $z=L$ we have (using $k=1$):

$$-\left.\frac{\partial U}{\partial z}\right|_{z=L}=f(\rho,\varphi)\rightarrow-\left.\frac{\partial u}{\partial z}\right|_{z=L}-\left.\frac{\partial v}{\partial z}\right|_{z=L}$$
$$=f(\rho,\varphi)+0\rightarrow-\left.\frac{\partial u}{\partial z}\right|_{z=L}=0;\ -\left.\frac{\partial v}{\partial z}\right|_{z=L}=f \tag{6.79}$$

For the boundary at $z=0$ we have:

$$\left.\frac{\partial U}{\partial z}\right|_{z=0}-hU(\rho,\varphi,z=0)=0\rightarrow\left.\frac{\partial u}{\partial z}\right|_{z=0}+\left.\frac{\partial v}{\partial z}\right|_{z=0}$$
$$-hu(\rho,\varphi,z=0)-hv(\rho,\varphi,z=0)=0\rightarrow \tag{6.80}$$

$$\left.\frac{\partial u}{\partial z}\right|_{z=0}-hu(z=0)=0;\ \left.\frac{\partial v}{\partial z}\right|_{z=0}-hv(z=0)=0 \tag{6.81}$$

For the boundary at $\rho=R$ we choose:

$$u(\rho=R,\varphi,z)=T(\varphi,z);\ v(\rho=R,\varphi,z)=0 \tag{6.82}$$

The problem (1) to be solved is:

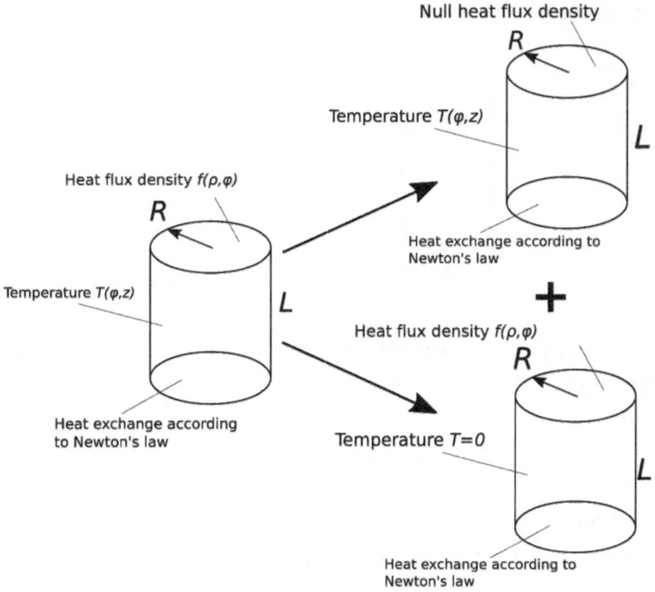

Figure 6.7

$$\left\{\begin{array}{l}\Delta u(\rho,\varphi,z)=0\\ \left.\dfrac{\partial u}{\partial z}\right|_{z=0}=hu(\rho,\varphi,z=0)\\ \left.\dfrac{\partial u}{\partial z}\right|_{z=L}=0\\ u(R,\varphi,z)=T(\varphi,z)\\ u(\rho=0,\varphi,z)<\infty\end{array}\right\} \tag{6.83}$$

Sturm–Liouville problem

Separating variables:

$$u=\mathcal{R}(\rho)\Phi(\varphi)Z(z) \tag{6.84}$$

We intend to expand the solution in eigenfunctions of the Sturm–Liouville problems for the ρ, φ, z variables to remove the corresponding second derivatives from the Laplacian. In this case the boundary conditions are homogeneous only in the vertical direction z and non-homogeneous in the radial direction ρ. These facts will influence our decision with respect to the constant of the separation of variables:

$$\frac{1}{\rho}\frac{\partial}{\partial\rho}\left[\rho\frac{\partial u}{\partial\rho}\right]+\frac{1}{\rho^2}\frac{\partial^2 u}{\partial\varphi^2}+\frac{\partial^2 u}{\partial z^2}=0 \tag{6.85}$$

Separating variables:

$$\frac{1}{\mathcal{R}}\frac{1}{\rho}\frac{d}{d\rho}\left[\rho\frac{d\mathcal{R}}{d\rho}\right]+\frac{1}{\rho^2}\frac{1}{\Phi}\frac{d^2\Phi}{d\varphi^2}=-\frac{1}{Z}\frac{d^2 Z}{dz^2}=+\lambda \tag{6.86}$$

With $\lambda > 0$ in the auxiliary problem *the positive sign is written* before the constant so that we can expand the solution in orthogonal eigenfunctions in z, since in this direction there are homogeneous boundaries of the second type.

$$\left\{\begin{array}{l}\dfrac{d^2 Z}{dz^2}+\lambda Z=0\\ \left.\dfrac{dZ}{dz}\right|_{z=0}-hZ(z=0)=0\\ \left.\dfrac{dZ}{dz}\right|_{z=L}=0\end{array}\right\} \tag{6.87}$$

We have the eigenfunctions:

$$Z(z)=A\cos\left(\sqrt{\lambda}[z-L]\right) \tag{6.88}$$

The eigenvalues λ will be solutions of the equation:

$$\tan\left(\sqrt{\lambda}L\right)=\frac{h}{\sqrt{\lambda}} \tag{6.89}$$

Then, when reducing the number of partial derivatives, the problem for $\mathcal{R}(\rho)$ is:

$$\frac{1}{\rho}\frac{d}{d\rho}\left[\rho\frac{d\mathcal{R}}{d\rho}\right]-\left(\lambda+\frac{m^2}{\rho^2}\right)\mathcal{R}=0 \tag{6.90}$$

Here we have already used the eigenvalues m^2 of the Sturm–Liouville problem for the whole cylinder. The angular eigenfunctions are: $A\cos(m\varphi)+B\sin(m\varphi)$

General solution

The general solution of the radial equation is a linear combination of modified Bessel and Neumann functions of order m:

$$\mathcal{R}(\rho) = C I_m\left(\sqrt{\lambda}\rho\right) + D K_m\left(\sqrt{\lambda}\rho\right) \tag{6.91}$$

$D = 0$ since the solution must be finite at $\rho = 0$. The general solution is:

$$u = \sum_{n=1}^{\infty}\sum_{m=0}^{\infty} I_m\left(\sqrt{\lambda}\rho\right)\left[A_{nm}\cos(m\varphi) + B_{nm}\sin(m\varphi)\right]\cos\left(\sqrt{\lambda}[z-L]\right) \tag{6.92}$$

Final solution

We will use the third boundary condition to find the coefficients:

$$u(R,\varphi,z) = T(\varphi,z) = \sum_{n=1}^{\infty}\sum_{m=0}^{\infty} I_m\left(\sqrt{\lambda}R\right)\left[A_{nm}\cos(m\varphi) + B_{nm}\sin(m\varphi)\right]\cos\left(\sqrt{\lambda}[z-L]\right) \tag{6.93}$$

We use the orthogonality of the angular and vertical eigenfunctions to obtain the coefficients. For that we multiply by the eigenfunctions in φ and z and integrate both sides of the previous equation:

$$\int_0^{2\pi}\int_0^{L} \cos(m'\varphi)\cos(\sqrt{\lambda_{n'}}[z-L])dzd\varphi$$

$$\int_0^{2\pi}\int_0^{L} T(\varphi,z)\cdot\cos(m\varphi)\cos(\sqrt{\lambda_n}[z-L])dzd\varphi =$$

$$= I_m(\sqrt{\lambda_n}R)\cdot A_{nm}\cdot\|\cos(m\varphi)\|^2\|\cos(\sqrt{\lambda_n}[z-L])\|^2 \tag{6.94}$$

$$A_{nm} = \frac{\int_0^{2\pi}\int_0^{L} T(\varphi,z)\cdot\cos(m\varphi)\cos(\sqrt{\lambda_n}[z-L])dzd\varphi}{I_m(\sqrt{\lambda_n}R)\cdot\|\cos(m\varphi)\|^2\|\cos(\sqrt{\lambda_n}[z-L])\|^2} \tag{6.95}$$

For $m \geq 0$. On the other hand, for the other coefficient, we multiply by $\sin(m\varphi)$ in the orthogonality condition and we have:

$$B_{nm} = \frac{\int_0^{2\pi}\int_0^{L} T(\varphi, z) \cdot \sin(m\varphi)\cos(\sqrt{\lambda_n}[z-L])dzd\varphi}{I_m(\sqrt{\lambda_n}R) \cdot \|\sin(m\varphi)\|^2 \|\cos(\sqrt{\lambda_n}[z-L])\|^2} \tag{6.96}$$

For $m \geq 1$.

Problem (2) is:

$$\left\{\begin{array}{c} \Delta v(\rho, \varphi, z) = 0 \\ \left.\dfrac{\partial v}{\partial z}\right|_{z=0} = hv(\rho, \varphi) \\ \left.\dfrac{\partial v}{\partial z}\right|_{z=L} = -f(\rho, \varphi) \\ v(R, \varphi, z) = 0 \\ v(\rho = 0, \varphi, z) < \infty \end{array}\right\} \tag{6.97}$$

We have here considered that the flux entering the cylinder through the upper base propagates in the negative direction.

Sturm–Liouville problem

Separating variables in a manner similar to the previous case:

$$\frac{1}{\mathcal{R}}\frac{1}{\rho}\frac{d}{d\rho}\left[\rho\frac{d\mathcal{R}}{d\rho}\right] + \frac{1}{\Phi}\frac{1}{\rho^2}\frac{d^2\Phi}{d\varphi^2} = -\frac{1}{Z}\frac{d^2 Z}{dz^2} = -\lambda \tag{6.98}$$

With $\lambda > 0$, we write the *negative sign* before the constant of separation to be able to expand the solution in radial eigenfunctions antes (since in this direction we have homogeneous boundary conditions) as well as in the angular functions:

$$\frac{d^2 Z}{dz^2} - \lambda Z = 0 \tag{6.99}$$

$$Z(z) = A\cosh(\sqrt{\lambda}z) + B\sinh(\sqrt{\lambda}z) \tag{6.100}$$

The problem for $\mathcal{R}(\rho)$ is:

$$\frac{1}{\rho}\frac{d}{d\rho}\left[\rho\frac{d\mathcal{R}}{d\rho}\right] + \left(\lambda - \frac{m^2}{\rho^2}\right)\mathcal{R} = 0 \tag{6.101}$$

Solution for the radial eigenfunctions:

$$\mathcal{R}(\rho) = CJ_m(\sqrt{\lambda}\rho) \tag{6.102}$$

The equation to find the eigenvalues by applying the boundary conditions of the radial problem is:

$$J_m(\sqrt{\lambda_{mn}}R) = 0 \tag{6.103}$$

Labelling as μ_{nm} the zeros of the Bessel function of order m:

$$\sqrt{\lambda_{mn}}R = \mu_{nm} \rightarrow \lambda_{mn} = \left[\frac{\mu_{nm}}{R}\right]^2 \tag{6.104}$$

General solution

We write the general solution of problem (2):

$$\begin{aligned} v(\rho, \varphi, z) = &\sum_{n=1}^{\infty}\sum_{m=0}^{\infty} J_m(\sqrt{\lambda_{mn}}\rho)[A_{nm}\cos(m\varphi) + B_{nm}\sin(m\varphi)] \\ &\times[\cosh(\sqrt{\lambda_{mn}}z) + C_{nm}\sinh(\sqrt{\lambda_{mn}}z)] \end{aligned} \tag{6.105}$$

To find the coefficients we will apply consecutively the first and second boundary conditions.

First condition:

$$\begin{aligned} \left.\frac{\partial v}{\partial z}\right|_{z=0} &= \sum_{n=1}^{\infty}\sum_{m=0}^{\infty} J_m(\sqrt{\lambda_{mn}}\rho)[A_{nm}\cos(m\varphi)+B_{nm}\sin(m\varphi)]C_{nm}\sqrt{\lambda_{mn}} \\ &= h\sum_{n=1,m=0}^{\infty} J_m(\sqrt{\lambda_{mn}}\rho)[A_{nm}\cos(m\varphi) + B_{nm}\sin(m\varphi)] \end{aligned} \tag{6.106}$$

From here we obtain the value of the coefficients $C_{nm} = h/\sqrt{\lambda_{mn}}$

We note that in the hypothetical case of both keeping both coefficients, for example D_{nm} and C_{nm} in the solution for the function in z, applying the boundary condition in the proper manner we get the ratio C_{nm}/D_{nm} just like we obtained it when only the C_{nm} coefficients were used (with $D_{nm} = 1$).

In the opposite case of maintaining only the D_{nm} coefficients of the $\cosh(z)$ function, taking $C_{nm} = 1$, we would obtain values of D_{nm} which are the inverse of those obtained for C_{nm} (supposing $D_{nm} = 1$). We now apply the second boundary condition:

$$-f(\rho,\varphi)=\sum_{n=1,m=0}^{\infty}J_m(\sqrt{\lambda_{mn}}\rho)[A_{nm}\cos(m\varphi)+B_{nm}\sin(m\varphi)]$$
$$\times[\lambda_{mn}\sinh(\sqrt{\lambda_{mn}}L)+h\cosh(\sqrt{\lambda_{mn}}L)] \tag{6.107}$$

Final solution

We multiply by the radial orthogonal functions (with weight ρ) and consecutively the two angular functions $\cos(m\varphi)$ and $\sin(m\varphi)$, integrating in the range of orthogonality, to find the coefficients of the expansion:

$$-\int_0^R\int_0^{2\pi} f(\rho,\varphi)J_m(\sqrt{\lambda_{mn}}\rho)\cos(m\varphi)\rho d\rho d\varphi =$$
$$A_{mn}\left[\sqrt{\lambda_{mn}}\sinh(\sqrt{\lambda_{mn}}L)+h\cosh(\sqrt{\lambda_{mn}}L)\right]$$
$$\int_0^R |J_m(\sqrt{\lambda_{mn}}\rho)|^2\rho d\rho\int_0^{2\pi}(\cos(m\varphi))^2 d\varphi \tag{6.108}$$

From where we find A_{mn}. And for the other coefficient:

$$-\int_0^R\int_0^{2\pi} f(\rho,\varphi)J_m(\sqrt{\lambda_{mn}}\rho)\sin(m\varphi)\rho d\rho d\varphi =$$
$$= B_{mn}\left[\sqrt{\lambda_{mn}}\sinh(\sqrt{\lambda_{mn}}L)+h\cosh(\sqrt{\lambda_{mn}}L)\right]\|J_{mn}\|^2\pi \tag{6.109}$$

From where we would find B_{mn}.

6.5 Heating of 1/16 of a Cylinder

Consider a cylindrical sector (radii R_2, R_1, height L) in the form of a $\pi 8$ sector, with its curved surfaces and vertical walls insulated and connected to a thermal reservoir at temperature T_0 at its base. The thermal conductivity coefficient of the material is k_0. Find the stationary distribution of temperature supposing that the total heat flux J is homogeneously distributed through its upper surface and directed towards the cylinder.

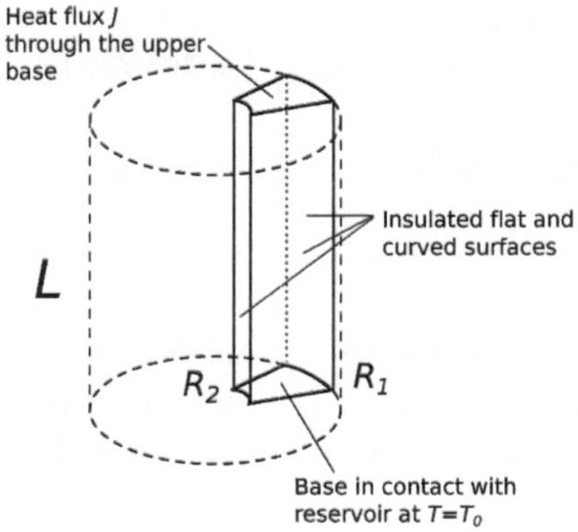

Figure 6.8

Mathematical formulation

We subtract the constant T_0 from the solution. In the present problem, the direction of the flux corresponds to the injection of heat into the cylinder.

$$\left\{\begin{array}{l} \Delta u = 0 \\ u(\rho, \varphi, z = 0) = 0 \\ -k_0 \left.\dfrac{\partial u}{\partial z}\right|_{z=L} = \dfrac{-16J}{\pi(R_1^2 - R_2^2)} \\ \left.\dfrac{\partial u}{\partial \rho}\right|_{\rho=R_1} = 0 \\ \left.\dfrac{\partial u}{\partial \rho}\right|_{\rho=R_2} = 0 \\ \left.\dfrac{\partial u}{\partial \varphi}\right|_{\varphi=0} = 0 \\ \left.\dfrac{\partial u}{\partial \varphi}\right|_{\varphi=\frac{\pi}{8}} = 0 \end{array}\right\} \tag{6.110}$$

Sturm–Liouville problem

Due to the presence of a non-homogeneous upper contour, we seek the solution by expanding it into eigenfunctions of the Sturm–

Liouville problem in the ρ and φ directions, to eliminate the angular-radial Laplacian from the heat equation.

In this case the boundary conditions are homogeneous only in the radial and azimuthal direction (of the second kind) and are non-homogeneous in the z direction. This matters when we need to decide the sign of the constant of separation.

$$\frac{1}{\rho}\frac{\partial}{\partial \rho}\left[\rho\frac{\partial u}{\partial \rho}\right]+\frac{1}{\rho^2}\frac{\partial^2 u}{\partial \varphi^2}+\frac{\partial^2 u}{\partial z^2}=0 \tag{6.111}$$

$$u=R(\rho)\cdot\Phi(\varphi)\cdot Z(z) \tag{6.112}$$

$$\frac{1}{R}\frac{1}{\rho}\frac{d}{d\rho}\left[\rho\frac{dR}{d\rho}\right]+\frac{1}{\rho^2}\frac{1}{\Phi(\varphi)}\frac{d^2\Phi(\varphi)}{d\varphi^2}=-\frac{1}{Z}\frac{d^2 Z}{dz^2}=-\lambda \tag{6.113}$$

With $\lambda > 0$, we choose the *negative sign* to be able to expand the solution in radial eigenfunctions (since in this direction there are homogeneous boundary conditions):

$$\left\{\begin{array}{l}\dfrac{d^2 Z}{dz^2}-\lambda Z=0\\ Z(0)=0\\ \left.\dfrac{dZ}{dz}\right|_{z=L}=\text{finite}\end{array}\right\} \tag{6.114}$$

We choose $Z(z) = A\sinh(\sqrt{\lambda}z)$ to automatically satisfy the first boundary condition.

$$\left\{\begin{array}{l}\dfrac{d^2\Phi(\varphi)}{d\varphi^2}+\upsilon\Phi(\varphi)=0\\ \left.\dfrac{d\phi}{d\varphi}\right|_{\varphi=0}=0\\ \left.\dfrac{d\phi}{d\varphi}\right|_{\varphi=\frac{\pi}{8}}=0\end{array}\right\} \tag{6.115}$$

$$\sqrt{\upsilon}\cdot\frac{\pi}{8}=m\pi \rightarrow \upsilon=(8m)^2 \tag{6.116}$$

$$\Phi(\varphi)=\cos(8m\varphi) \qquad (m=0,\,1,\,2,\,3...) \tag{6.117}$$

Then, by reducing the number of partial derivatives, the problem for $R(\rho)$ is:

$$\frac{1}{\rho}\frac{d}{d\rho}\left[\rho\frac{dR}{d\rho}\right]+\left[\lambda-\frac{(8m)^2}{\rho^2}\right]R=0 \tag{6.118}$$

The general solution of the radial equation is a linear combination of Bessel and Neumann functions of order $8m$:

$$R(\rho)=CJ_{8m}(\sqrt{\lambda}\rho)+DN_{8m}(\sqrt{\lambda}\rho) \tag{6.119}$$

We calculate the eigenvalues λ_{mk} by equating to zero the determinant of the system of two equations formed by radial boundary conditions. In this way we find a non-trivial combination of coefficients:

$$C\left.\frac{dJ_{8m}(\sqrt{\lambda}\rho)}{d\rho}\right|_{\rho=R_2}+D\left.\frac{dN_{8m}(\sqrt{\lambda}\rho)}{d\rho}\right|_{\rho=R_2}=0 \tag{6.120}$$

$$C\left.\frac{dJ_{8m}(\sqrt{\lambda}\rho)}{d\rho}\right|_{\rho=R_1}+D\left.\frac{dN_{8m}(\sqrt{\lambda}\rho)}{d\rho}\right|_{\rho=R_1}=0 \tag{6.121}$$

In what follows, to avoid cumbersome expressions we will use the following nomenclature for the Bessel functions:

$$\left.\frac{dJ_{8m}(\sqrt{\lambda_m}\rho)}{d\rho}\right|_{\rho=R_1}=[J_{8m}]_\rho(\sqrt{\lambda_m}R_1) \tag{6.122}$$

And an analogous expression for the Neumann functions.

General solution

The general solution is:

$$u=\sum_{k=1}^{\infty}\sum_{m=0}^{\infty}C_{mk}\left[J_{8m}(\sqrt{\lambda_{mk}}\rho)-\frac{[J_{8m}]_\rho(\sqrt{\lambda_{mk}}R_2)}{[N_{8m}]_\rho(\sqrt{\lambda_{mk}}R_2)}N_{8m}(\sqrt{\lambda_{mk}}\rho)\right]$$
$$\times\cos(8m\varphi)\sinh(\sqrt{\lambda_{mk}}z) \tag{6.123}$$

Note that the Bessel and Neumann functions are of order $8m$ and that the eigenvalues λ have both indices m and k, since we use them both for the vertical and the angular eigenfunctions. To find the coefficients of the expansion we apply the last boundary condition:

$$\left.\frac{du}{dz}\right|_L = \sum_{k=1}^{\infty}\sum_{m=0}^{\infty} C_{mk}\sqrt{\lambda_{mk}}\left[J_{8m}(\sqrt{\lambda_{mk}}\rho) - \frac{[J_{8m}]_\rho(\sqrt{\lambda_{mk}}R_2)}{[N_{8m}]_\rho(\sqrt{\lambda_{mk}}R_2)} N_{8m}(\sqrt{\lambda_{mk}}\rho)\right]\cos(8m\varphi)\cosh(\sqrt{\lambda_{mk}}L) = \frac{16J}{\pi k_0(R_1^2 - R_2^2)} \tag{6.124}$$

Final solution

Using the orthogonality of the radial and angular eigenfunctions we get the coefficients C_{mk}. We multiply both sides by:

$$\left[J_{8m}(\sqrt{\lambda_{mk}}\rho) - \frac{[J_{8m}]_\rho(\sqrt{\lambda_{mk}}R_2)}{[N_{8m}]_\rho(\sqrt{\lambda_{mk}}R_2)} N_{8m}(\sqrt{\lambda_{mk}}\rho)\right]\cos(8m\varphi) \tag{6.125}$$

and integrate between the limits $\int_{R_1}^{R_2}\int_0^{\frac{\pi}{8}} \rho d\rho d\varphi$

$$C_{mk} = \frac{16J}{\sqrt{\lambda_{mk}}\pi k_0(R_1^2 - R_2^2)} \times \frac{\int_{R_1}^{R_2}\int_0^{\frac{\pi}{8}}[J_{8m}(\sqrt{\lambda_{mk}}\rho) - \frac{[J_{8m}]_\rho(\sqrt{\lambda_{mk}}R_2)}{[N_{8m}]_\rho(\sqrt{\lambda_{mk}}R_2)} N_{8m}(\sqrt{\lambda_{mk}}\rho)]\cos(8m\varphi)\rho d\rho d\varphi}{\left\|[J_{8m}(\sqrt{\lambda_{mk}}\rho) - \frac{[J_{8m}]_\rho(\sqrt{\lambda_{mk}}R_2)}{[N_{8m}]_\rho(\sqrt{\lambda_{mk}}R_2)} N_{8m}(\sqrt{\lambda_{mk}}\rho)]\right\|^2 \|\cos(8m\varphi)\|^2\cosh(\sqrt{\lambda_{mk}}L)} \tag{6.126}$$

We have:

$$\int_0^{\frac{\pi}{8}} \cos(8m\varphi)d\varphi = \left\{\begin{array}{ll} 0 & (m \geq 1) \\ \frac{\pi}{8} & (m = 0) \end{array}\right\} \tag{6.127}$$

We see that the solution does not depend on the angular variable. Then:

$$C_{0k} = \frac{16J}{\sqrt{\lambda_{0k}}\pi k_0(R_1^2 - R_2^2)} \times \frac{\int_{R_1}^{R_2}[J_0(\sqrt{\lambda_{0k}}\rho) - \frac{[J_0]_\rho(\sqrt{\lambda_{0k}}R_2)}{[N_0]_\rho(\sqrt{\lambda_{0k}}R_2)} N_0(\sqrt{\lambda_{0k}}\rho)]\rho d\rho}{\left\|[J_0(\sqrt{\lambda_{0k}}\rho) - \frac{[J_0]_\rho(\sqrt{\lambda_{0k}}R_2)}{[N_0]_\rho(\sqrt{\lambda_{0k}}R_2)} N_0(\sqrt{\lambda_{0k}}\rho)]\right\|^2 \cosh(\sqrt{\lambda_{0k}}L)} \tag{6.128}$$

Note: we have shown a general solution path that would change little in the case of varying homogeneous boundary conditions, for example going from the second type to the first or third homogeneous type. In the case of type 2 homogeneous lateral boundary conditions, we note that due to the homogeneous distribution of injected heat flux, there are no physical reasons for this heat flux to adhere any lateral component in the radial or azimuthal directions. This implies that, for the case of contours considered, the solution in series obtained presents the development of a solution constant in angular and radial variables and which only lineally changes as a function of the vertical variable (z).

6.6 Distribution of Temperature inside a Finite Semi-Cylinder with a Centered Hole

Find the stationary distribution of temperature inside a semi-cylinder of length L with a hole (outer and inner radii a, b) if all its plane surfaces are at a temperature T_0.

Through both curved surfaces (inner and outer) the heat fluxes F_1 and F_2 are supplied (both constants have negative values). The thermal conductivity of the material is equal to k.

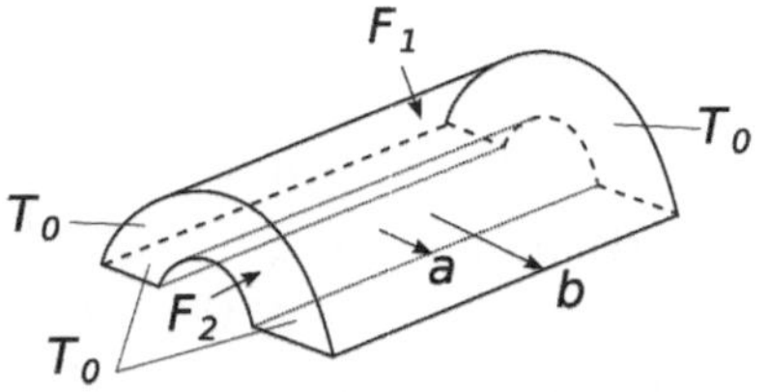

Figure 6.9

Mathematical formulation

We subtract T_0 from the solution. Due to the absence of heat sources inside the semicylinder se describe the problem with Laplace's equation:

$$\left\{\begin{array}{l} \Delta u = 0 \\ -k\left.\dfrac{\partial u}{\partial \rho}\right|_{\rho=a} = F_1 \\ -k\left.\dfrac{\partial u}{\partial \rho}\right|_{\rho=b} = -F_2 \\ u(\rho, 0, z) = u(\rho, \pi, z) = 0 \\ u(\rho, \varphi, 0) = u(\rho, \varphi, L) = 0 \end{array}\right\} \quad (6.129)$$

The F_1 and F_2 constants must be negative numbers to have a valid physical meaning.

The hole is centered in the axis and the boundary conditions of the curved surfaces are non-homogeneous of the third type. The Laplacian is expressed in cylindrical coordinates.

$$\frac{1}{\rho}\frac{\partial}{\partial \rho}\left[\rho\frac{\partial u}{\partial \rho}\right] + \frac{1}{\rho^2}\frac{\partial^2 u}{\partial \varphi^2} + \frac{\partial^2 u}{\partial z^2} = 0 \quad (6.130)$$

Sturm–Liouville problem

We separate variables to find the solution:

$$u = R(\rho)\cdot\Phi(\varphi)\cdot Z(z) \quad (6.131)$$

$$\frac{\frac{1}{\rho}\frac{d}{d\rho}\left[\rho\frac{dR}{d\rho}\right]}{R} + \frac{1}{\rho^2}\frac{\frac{d^2\Phi}{d\varphi^2}}{\Phi} + \frac{\frac{d^2 Z}{dz^2}}{Z} = 0 \quad (6.132)$$

and finding the functions $\Phi(\varphi)$; $Z(z)$ from the corresponding Sturm–Liouville problems:

$$\left\{\begin{array}{l} \dfrac{d^2 Z}{dz^2} + \lambda Z = 0 \\ Z(0) = Z(L) = 0 \end{array}\right\} \quad (6.133)$$

$$\left\{\begin{array}{l} \dfrac{d^2 \Phi}{d\varphi^2} + \mu\Phi = 0 \\ \Phi(0) = \Phi(\pi) = 0 \end{array}\right\} \quad (6.134)$$

So that the eigenfunctions are: $\Phi(\varphi)Z(z) = \sin(n\varphi)\sin(\frac{\pi m}{L}z)$ (with $n = 1, 2, 3\ldots, m = 1, 2, 3\ldots$)

The equation for the radial part is:

$$\frac{\frac{1}{\rho}\frac{d}{d\rho}\left[\rho\frac{dR}{d\rho}\right]}{R}+\frac{1}{\rho^2}(-n^2)-\left(\frac{\pi m}{L}\right)^2=0 \tag{6.135}$$

$$\frac{d^2R}{d\rho^2}+\frac{1}{\rho}\frac{dR}{d\rho}-\left[\frac{n^2}{\rho^2}+\left(\frac{\pi m}{L}\right)^2\right]R=0 \tag{6.136}$$

General solution

The general solution of the previous equation consists in Bessel functions of imaginary argument (modified Bessel function) and also McDonald functions (which are not considered for whole cylinders). The general solution is:

$$u(\rho,\varphi,z)=\sum\left[A_{nm}I_n\left(\frac{\pi m}{L}\rho\right)\right.$$
$$\left.+B_{nm}K_n\left(\frac{\pi m}{L}\rho\right)\right]\sin(n\varphi)\sin\left(\frac{\pi m}{L}z\right) \tag{6.137}$$

Final solution

To find the A_{nm} and B_{nm} coefficients we will use the first and second boundary conditions to get the general solution and apply the conditions of heat fluxes across the curved surfaces.

$$\left\{\begin{array}{l}\sum\left[A_{nm}\frac{\pi m}{L}I_n'\left(\frac{\pi m}{L}a\right)+B_{nm}\frac{\pi m}{L}K_n'\left(\frac{\pi m}{L}a\right)\right]\sin(n\varphi)\sin\left(\frac{\pi m}{L}z\right)=-\frac{F_1}{k}\\ \sum\left[A_{nm}\frac{\pi m}{L}I_n'\left(\frac{\pi m}{L}b\right)+B_{nm}\frac{\pi m}{L}K_n'\left(\frac{\pi m}{L}b\right)\right]\sin(n\varphi)\sin\left(\frac{\pi m}{L}z\right)=\frac{F_2}{k}\end{array}\right\} \tag{6.138}$$

Where I' and K' are the derivatives of the modified Bessel function and the McDonald function with respect to ρ. Using the orthogonality of the angular eigenfunctions (φ) and in the (z) direction we arrive at two equations with two unknowns to find the A_{nm} and B_{nm} coefficients.

$$\left\{\begin{array}{l}A_{nm}I_n'\left(\frac{\pi m}{L}a\right)+B_{nm}K_n'\left(\frac{\pi m}{L}a\right)=\alpha_{nm}\\ A_{nm}I_n'\left(\frac{\pi m}{L}b\right)+B_{nm}K_n'\left(\frac{\pi m}{L}b\right)=\beta_{nm}\end{array}\right\} \tag{6.139}$$

$$\alpha_{nm} = -\frac{L}{\pi m}\frac{F_1}{k\left\|\sin(n\varphi)\right\|^2\left\|\sin\left(\frac{\pi m}{L}z\right)\right\|^2}\int_0^{\pi}\int_0^{L}\sin(n\varphi)\sin\left(\frac{\pi m}{L}z\right)d\varphi dz$$

$$= -\frac{L}{\pi m}\frac{F_1}{k\frac{\pi}{2}\frac{L}{2}}\int_0^{\pi}\int_0^{L}\sin(n\varphi)\sin\left(\frac{\pi m}{L}z\right)d\varphi dz$$

$$= -\frac{4F_1}{mk}\frac{[1-\cos(n\pi)]}{n}\frac{L[1-\cos(m\pi)]}{m\pi} \quad (6.140)$$

Analogously:

$$\beta_{nm} = \frac{L}{\pi m}\frac{F_2}{k\left\|\sin(n\varphi)\right\|^2\left\|\sin\left(\frac{\pi m}{L}z\right)\right\|^2}\int_0^{\pi}\int_0^{L}\sin(n\varphi)\sin\left(\frac{\pi m}{L}z\right)d\varphi dz$$

$$= \frac{4F_2}{mk}\frac{[1-\cos(n\pi)]}{n}\frac{L[1-\cos(m\pi)]}{m\pi} \quad (6.141)$$

It is clear that both coefficients are zero for n, m = even. Finally we will get the coefficients:

$$\left\{\begin{array}{l} A_{nm} = \frac{1}{I_n'\left(\frac{\pi m}{L}a\right)}\left[\alpha_{nm} - B_{nm}K_n'\left(\frac{\pi m}{L}a\right)\right] \\ \frac{1}{I_n'\left(\frac{\pi m}{L}a\right)}\left[\alpha_{nm} - B_{nm}K_n'\left(\frac{\pi m}{L}a\right)\right]I_n'\left(\frac{\pi m}{L}b\right) + B_{nm}K_n'\left(\frac{\pi m}{L}b\right) = \beta_{nm} \\ B_{nm}\left[K_n'\left(\frac{\pi m}{L}b\right) - \frac{K_n'\left(\frac{\pi m}{L}a\right)I_n'\left(\frac{\pi m}{L}b\right)}{I_n'\left(\frac{\pi m}{L}a\right)}\right] = \beta_{nm} - \alpha_{nm}\frac{I_n'\left(\frac{\pi m}{L}b\right)}{I_n'\left(\frac{\pi m}{L}a\right)} \end{array}\right\} \quad (6.142)$$

or

$$B_{nm} = \frac{\beta_{nm}I_n'(\frac{\pi m}{L}a) - \alpha_{nm}I_n'(\frac{\pi m}{L}b)}{K_n'(\frac{\pi m}{L}b)I_n'(\frac{\pi m}{L}a) - K_n'(\frac{\pi m}{L}a)I_n'(\frac{\pi m}{L}b)} \quad (6.143)$$

$$A_{nm} = \frac{1}{I_n'(\frac{\pi m}{L}a)}\left[\alpha_{nm} - \frac{\beta_{nm}I_n'(\frac{\pi m}{L}a) - \alpha_{nm}I_n'(\frac{\pi m}{L}b)K_n'(\frac{\pi m}{L}a)]}{K_n'(\frac{\pi m}{L}b)I_n'(\frac{\pi m}{L}a) - K_n'(\frac{\pi m}{L}a)I_n'(\frac{\pi m}{L}b)}\right] \quad (6.144)$$

6.7 Distribution of Temperature inside a Hollow Cylinder with a Heater

Find the stationary distribution of temperature inside a cylinder of height L with a centered hole or radius R_1 and with an outer radius R_2. The plane surfaces are in contact with a thermal reservoir at a temperature T_0, the inner surface is at zero temperature and the outer surface exchanges heat with the outer medium at a zero temperature according to Newton's law (with constant W).

Inside the cylinder acts a heat source in the form of a thin tube or radius $\frac{R_2-R_1}{2}$ and also height L, which dissipates an energy density F.

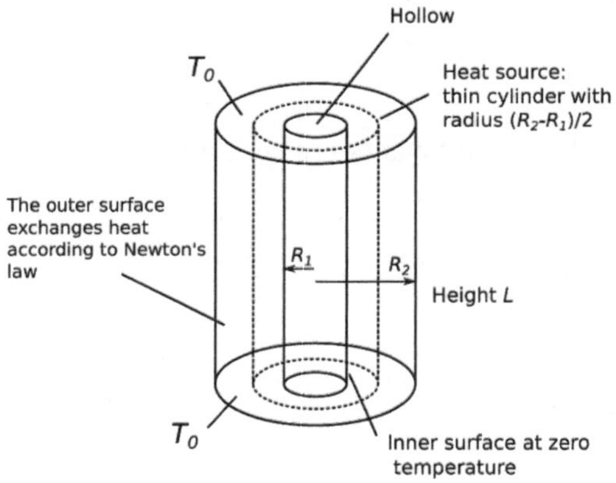

Figure 6.10

Mathematical formulation

$$\left.\begin{cases} -k\Delta u(\rho,\varphi,z) = \dfrac{F}{2\pi}\dfrac{1}{\rho}\delta\left(\rho-\left[\dfrac{R_2-R_1}{2}\right]\right) \\ u(\rho,\varphi,0) = u(\rho,\varphi,L) = T_0 \\ u(R_1,\varphi,z) = 0 \\ -k\left.\dfrac{\partial u}{\partial\rho}\right|_{\rho=R_2} = W\cdot u(R_2,\varphi,z) \end{cases}\right\} \tag{6.145}$$

Note: we will consider that the contact surface between the heater and the flat faces does not modify the boundary conditions inn z.

Sturm–Liouville problem

We will expand the solution in eigenfunctions of the Sturm–Liouville problems for the variables (ρ, φ), with the idea of eliminating the second derivatives of the Laplacian in those directions, since the boundary conditions in the vertical direction are inhomogeneous:

$$u = V(\rho, \varphi) \cdot Z(z) \tag{6.146}$$

Sturm–Liouville problem for $V(\rho, \varphi)$:

$$\left\{\begin{array}{l} \Delta_{\rho,\varphi} V(\rho, \varphi) + \lambda V(\rho, \varphi) = 0 \\ V(R_1, \varphi) = 0 \\ \left.\dfrac{\partial V}{\partial \rho}\right|_{\rho=R_2} + HV(R_2, \varphi) = 0 \\ H = \dfrac{W}{k} \end{array}\right\} \tag{6.147}$$

The eigenfunctions (using for brevity the complex form for the angular solutions) are:

$$V_{nm}(\rho, \varphi) = \left[A_{nm} J_m(\sqrt{\lambda_{nm}}\rho) + B_{nm} N_m(\sqrt{\lambda_{nm}}\rho)\right] e^{(-im\varphi)} \tag{6.148}$$

Applying the third and fourth boundary conditions:

$$\left\{\begin{array}{l} [A_{nm} J_m(\sqrt{\lambda_{nm}} R_1) + B_{nm} N_m(\sqrt{\lambda_{nm}} R_1)] = 0 \\ \sqrt{\lambda_{nm}} \left[A_{nm} J'_m(\sqrt{\lambda_{nm}} R_2) + B_{nm} N'_m(\sqrt{\lambda_{nm}} R_2)\right] + \\ +H [A_{nm} J_m(\sqrt{\lambda_{nm}} R_2) + B_{nm} N_m(\sqrt{\lambda_{nm}} R_2)] = 0 \end{array}\right\} \tag{6.149}$$

or

$$\left\{\begin{array}{l} [A_{nm} J_m(\sqrt{\lambda_{nm}} R_1) + B_{nm} N_m(\sqrt{\lambda_{nm}} R_1)] = 0 \\ A_{nm}[\sqrt{\lambda_{nm}} J'_m(\sqrt{\lambda_{nm}} R_2) + H J_m(\sqrt{\lambda_{nm}} R_2)] \\ +B_{nm}[\sqrt{\lambda_{nm}} N'_m(\sqrt{\lambda_{nm}} R_2) + N_m(\sqrt{\lambda_{nm}} R_2)] = 0 \end{array}\right\} \tag{6.150}$$

The condition DET()=0 will give us the eigenvalues λ_{nm}

Furthermore, since $B_{nm} = -A_{nm} \frac{J_m(\sqrt{\lambda_{nm}} R_1)}{N_m(\sqrt{\lambda_{nm}} R_1)}$

we get the not normalized eigenfunctions in the form:

$$V_{nm}(\rho, \varphi) = \left[J_m(\sqrt{\lambda_{nm}}\rho) - \frac{J_m(\sqrt{\lambda_{nm}} R_1)}{N_m(\sqrt{\lambda_{nm}} R_1)} N_m(\sqrt{\lambda_{nm}}\rho)\right] e^{(-im\varphi)} \tag{6.151}$$

General solution

Looking for the solution as summation of orthogonal eigenfunctions:

$$u = \sum_{nm} V_{nm}(\rho, \varphi) \cdot Z(z) \tag{6.152}$$

and replacing in the initial problem (6.145)

$$\sum_{nm} \left[\Delta_{\rho,\varphi} + \frac{d^2}{dz^2} \right] V_{nm}(\rho, \varphi) \cdot Z(z) = -\frac{F}{2\pi k}\frac{1}{\rho}\delta\left(\rho - \left[\frac{R_2 - R_1}{2}\right]\right) \tag{6.153}$$

$$\sum_{nm} \Delta_{\rho,\varphi} V_{nm}(\rho, \varphi) \cdot Z(z) + V_{nm}(\rho, \varphi)\frac{d^2}{dz^2}Z(z) = -\frac{F}{2\pi k}\frac{1}{\rho}\delta\left(\rho - \left[\frac{R_2 - R_1}{2}\right]\right) \tag{6.154}$$

$$\sum_{nm} -\lambda_{nm} V_{nm}(\rho, \varphi) \cdot Z(z) + V_{nm}(\rho, \varphi)\frac{d^2}{dz^2}Z(z) = -\frac{F}{2\pi k}\frac{1}{\rho}\delta\left(\rho - \left[\frac{R_2 - R_1}{2}\right]\right) \tag{6.155}$$

or, finally:

$$\sum_{nm} \left[\frac{d^2 Z(z)}{dz^2} - \lambda_{nm} Z(z) \right] V_{nm}(\rho, \varphi) = -\frac{F}{2\pi k}\frac{1}{\rho}\delta\left(\rho - \left[\frac{R_2 - R_1}{2}\right]\right) \tag{6.156}$$

Multiplying both sides by $V_{n'm'}(\rho, \varphi)$ and integrating $\int_{R_1}^{R_2}\int_0^{2\pi} \rho d\rho d\varphi$, we will get:

$$\left[\frac{d^2 Z(z)}{dz^2} - \lambda_{nm} Z(z) \right] \| V_{nm}(\rho, \varphi) \|^2 =$$

$$= -\frac{F}{2\pi k} \int_{R_1}^{R_2} \left[J_m(\sqrt{\lambda_{nm}}\rho) - \frac{J_m(\sqrt{\lambda_{nm}} R_1)}{N_m(\sqrt{\lambda_{nm}} R_1)} N_m(\sqrt{\lambda_{nm}}\rho) \right] \delta$$

$$\left(\rho - \left[\frac{R_2 - R_1}{2}\right]\right) d\rho \int_0^{2\pi} e^{(-im\varphi)} d\varphi =$$

$$= -\frac{F}{2\pi k} \left[J_m\left(\sqrt{\lambda_{nm}}\left[\frac{R_2 - R_1}{2}\right]\right) - \frac{J_m(\sqrt{\lambda_{nm}} R_1)}{N_m} \right.$$

$$\left. \left(\sqrt{\lambda_{nm}} R_1\right) N_m\left(\sqrt{\lambda_{nm}}\left[\frac{R_2 - R_1}{2}\right]\right)\right] \int_0^{2\pi} e^{(-im\varphi)} d\varphi \tag{6.157}$$

Since

$$\begin{Bmatrix} \int_0^{2\pi} e^{(-im\varphi)} d\varphi = 0 \ (m \neq 0) \\ 2\pi \qquad\qquad (m = 0) \end{Bmatrix} \tag{6.158}$$

Then we get at the equation for the function $Z(z)$:

$$\frac{d^2 Z(z)}{dz^2} - \lambda_{n0} Z(z) = -\frac{F}{2\pi k} \frac{J_0\left(\sqrt{\lambda_{no}}\left[\frac{R_2 - R_1}{2}\right]\right) - \frac{J_0(\sqrt{\lambda_{n0}} R_1)}{N_0(\sqrt{\lambda 0} R_1)} N_0\left(\sqrt{\lambda_{n0}}\left[\frac{R_2 - R_1}{2}\right]\right)}{\| V_{n0}(\rho, \varphi)\|^2} = \alpha_n \tag{6.159}$$

We also need to find the boundary conditions for $Z(z)$. From the first and second boundary conditions:

$$\sum_{nm} V_{nm}(\rho, \varphi) \cdot Z(0) = T_0 \tag{6.160}$$

and

$$\sum_{nm} V_{nm}(\rho, \varphi) \cdot Z(0) = T_0 \tag{6.161}$$

we get:

$$Z(0) = Z(L) = T_0 \frac{\int_{R_1}^{R_2} [J_m(\sqrt{\lambda_{nm}} \rho) - \frac{J_m(\sqrt{\lambda_{nm}} R_1)}{N_m(\sqrt{\lambda_{nm}} R_1)} N_m(\sqrt{\lambda_{nm}} \rho)] d\rho}{\| V_{n0}(\rho, \varphi)\|^2} = T_0 Z_n \tag{6.162}$$

We need to solve the non-homogeneous equation in the z direction with the corresponding boundary conditions:

$$\begin{Bmatrix} \frac{d^2 Z(z)}{dz^2} - \lambda_{n0} Z(z) = \alpha_n \\ Z(0) = Z(L) = T_0 Z_n \end{Bmatrix} \tag{6.163}$$

by searching the solution as the sum of a particular solution and the solution of the homogeneous equation. The particular solution is:

$$Z_p(z) = \frac{\alpha_n}{\lambda_{n0}} \tag{6.164}$$

The general solution of the homogeneous equation is well known:

$$Z_h(z) = C \sinh(\sqrt{\lambda_{n0}} z) + D \cosh(\sqrt{\lambda_{n0}} z) \tag{6.165}$$

Final solution

Applying the boundary conditions:

$$\left\{\begin{array}{l} C\sinh(\sqrt{\lambda_{n0}}0)+D\cosh(\sqrt{\lambda_{n0}}0)+\dfrac{\alpha_n}{\lambda_{n0}}=T_0 \\ C\sinh(\sqrt{\lambda_{n0}}L)+D\cosh(\sqrt{\lambda_{n0}}L)+\dfrac{\alpha_n}{\lambda_{n0}}=T_0 \end{array}\right\} \quad (6.166)$$

$$\left\{\begin{array}{l} D=T_0-\dfrac{\alpha_n}{\lambda_{n0}} \\ C\sinh(\sqrt{\lambda_{n0}}L)+\left(T_0-\dfrac{\alpha_n}{\lambda_{n0}}\right)\cosh(\sqrt{\lambda_{n0}}L)+\dfrac{\alpha_n}{\lambda_{n0}}=T_0 \end{array}\right\} \quad (6.167)$$

or

$$\left\{\begin{array}{l} D=T_0-\dfrac{\alpha_n}{\lambda_{n0}} \\ C=\left[T_0-\dfrac{\alpha_n}{\lambda_{n0}}\right][1-\cosh(\sqrt{\lambda_{n0}}L)]/\sinh(\sqrt{\lambda_{n0}}L) \end{array}\right\} \quad (6.168)$$

Finally:

$$\begin{aligned} u=\sum_n\left[J_0(\sqrt{\lambda_{n0}}\rho)-\frac{J_m(\sqrt{\lambda_{n0}}R_1)}{N_m(\sqrt{\lambda_{n0}}R_1)}N_0(\sqrt{\lambda_{n0}}\rho)\right]\times \\ \times\left[\left(T_0-\frac{\alpha_n}{\lambda_{n0}}\right)[1-\cosh(\sqrt{\lambda_{n0}}L)][\sinh(\sqrt{\lambda_{n0}}z)/\sinh(\sqrt{\lambda_{n0}}L)]\right. \\ \left.+\left(T_0-\frac{\alpha_n}{\lambda_{n0}}\right)\cosh(\sqrt{\lambda_{n0}}z)\right] \end{aligned} \quad (6.169)$$

6.8 Case Study: Temperature in a Cylinder with Bases Thermally Insulated

Find the stationary distribution of temperature inside a cylinder of height h with a central hole of radius R_1 and an outer radius R_2. The flat faces are thermally insulated. The inner surface is in contact with a thermal reservoir at zero temperature and the outer surface is in contact with another reservoir at a temperature $T = \cos(3\varphi)$.

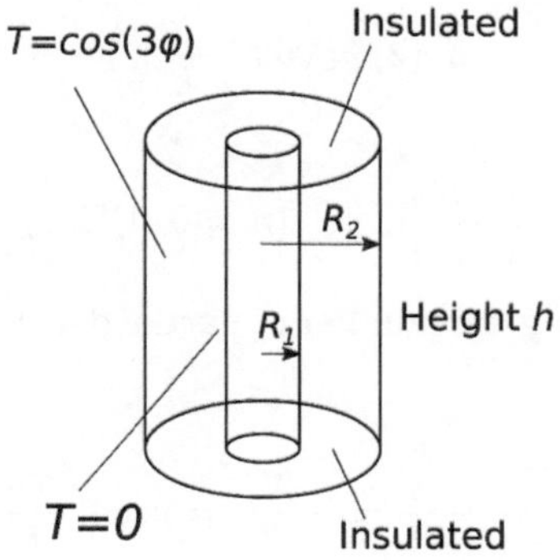

Figure 6.11

Mathematical formulation

$$\left\{\begin{array}{l}\Delta u(\rho,\varphi,z)=0\\ \left.\dfrac{\partial u}{\partial z}\right|_{z=0}=\left.\dfrac{\partial u}{\partial z}\right|_{z=h}=0\\ u(\rho=R_1,\varphi,z)=0\\ u(\rho=R_2,\varphi,z)=\cos(3\varphi)\end{array}\right\} \quad (6.170)$$

Sturm–Liouville problem

We separate variables:

$$u=V(\rho,\varphi)\cdot Z(z) \quad (6.171)$$

We seek the expansion of the solution in eigenfunctions of the Sturm–Liouville problems in the (φ, z) directions, to able to remove the second derivatives from the Laplacian, since the radial boundary condition in the radial direction is inhomogeneous.

$$\Delta[V(\rho,\varphi)\cdot Z(z)]=Z\Delta V(\rho,\varphi)+V(\rho,\varphi)\Delta Z(z)=0 \quad (6.172)$$

$$\frac{Z\Delta V+V\Delta Z}{VZ}=\frac{\Delta V}{V}+\frac{\Delta Z}{Z}=0 \quad (6.173)$$

Auxiliary problems:

$$\left\{\begin{array}{l}\dfrac{d^2Z}{dz^2}+\upsilon Z=0\\ \left.\dfrac{dZ}{dz}\right|_{z=0}=0\\ \left.\dfrac{dZ}{dz}\right|_{z=h}=0\end{array}\right\} \quad (6.174)$$

$$Z_n(z) = \cos\left(\frac{\pi n}{h} z\right) \tag{6.175}$$

$$\upsilon_n = \left(\frac{\pi n}{h}\right)^2 \qquad (n = 0, 1, 2, 3, \ldots) \tag{6.176}$$

Then, when reducing the number of partial derivatives, the problem for $V(\rho, \varphi)$ is:

$$\frac{\frac{1}{\rho}\frac{\partial}{\partial \rho}\left[\rho \frac{\partial V}{\partial \rho}\right] + \frac{1}{\rho^2}\frac{\partial^2 V}{\partial \varphi^2}}{V} = -\frac{\Delta Z}{Z} = \nu = \left(\frac{\pi n}{h}\right)^2 \tag{6.177}$$

Then:

$$\frac{1}{\rho}\frac{\partial}{\partial \rho}\left[\rho \frac{\partial V}{\partial \rho}\right] + \frac{1}{\rho^2}\frac{\partial^2 V}{\partial \varphi^2} - \left[\left(\frac{\pi n}{h}\right)^2\right] V = 0 \tag{6.178}$$

General solution

In this moment we can present the general solution as the product of angular and radial functions and eigenfunctions of the z variable.

$$u = \sum_{n=0}^{\infty} V_n(\rho, \varphi) \cdot Z_n(z) \tag{6.179}$$

We can simplify the problem by checking that the solution doesn't depend on the z variable. Precisely:

$$u(\rho = R_2) = \sum_{n=0}^{\infty} V_n(R_2, \varphi) \cdot Z_n(z) = \cos(3\varphi) \tag{6.180}$$

We multiply the equation by $\cos(\frac{\pi k}{h} z)$ and integrate it between 0 and h in the z direction. Of the whole summation, only the term with index $n = 0$ remains. Then $u = V_0(\rho, \varphi)$. The equation to find $V_0(\rho, \varphi)$ is then:

$$\frac{1}{\rho}\frac{\partial}{\partial \rho}\left[\rho \frac{\partial V_0}{\partial \rho}\right] + \frac{1}{\rho^2}\frac{\partial^2 V_0}{\partial \varphi^2} = 0 \tag{6.181}$$

We multiply the equation by ρ^2 and separate variables once again $V_0(\rho, \varphi) = \mathcal{R}(\rho) \cdot \Phi(\varphi)$

$$\frac{\rho \frac{d}{d\rho}[\rho \frac{d\mathcal{R}}{d\rho}]}{\mathcal{R}} = -\frac{1}{\Phi}\frac{d^2 \Phi}{d\varphi^2} = \eta \tag{6.182}$$

With $\eta > 0$ we arrive at a Sturm–Liouville problem to expand the solution in angular eigenfunctions:

$$\left\{\begin{array}{l} \dfrac{d^2\Phi}{d\varphi^2} + \eta\Phi = 0 \\ \Phi(\varphi) = \Phi(\varphi + 2\pi) \end{array}\right\} \tag{6.183}$$

The eigenfunctions and eigenvalues are:

$$\Phi(\varphi) = e^{(im\varphi)} \tag{6.184}$$

$$\eta = m^2 \tag{6.185}$$

The radial problem then is:

$$\rho\frac{d}{d\rho}\left[\rho\frac{d\mathcal{R}}{d\rho}\right] - m^2\mathcal{R} = 0 \tag{6.186}$$

$$\rho^2\frac{d^2\mathcal{R}}{d\rho^2} + \rho\frac{d\mathcal{R}}{d\rho} - m^2\mathcal{R} = 0 \tag{6.187}$$

The general solution is then a known function. For $m = 0$:

$$\mathcal{R}(\rho) = A_0 + B_0 \log(\rho) \tag{6.188}$$

For $m \neq 0$:

$$\mathcal{R}(\rho) = A_m\rho^m + B_m\rho^{-m} \tag{6.189}$$

We apply the fourth boundary condition:

$$u(\rho = R_2) = A_0 + B_0 \log R_2 + \sum_{m=1}^{\infty}[A_m(R_2)^m + B_m(R_2)^{-m}] \cdot \cos(m\varphi) = \cos(3\varphi) \tag{6.190}$$

Using the orthogonality of the angular eigenfunctions we see that only A_3 and B_3 remain. The terms of the expansion of the general solution which are proportional to $\sin(m\varphi)$ will be zero due to their orthogonality with the function which describes the boundary: $\cos(3\varphi)$. Furthermore, the terms A_0 and B_0 are zero due to the average of the function $\cos(3\varphi)$ in the range 2π is zero.

$$\sum_{m=1}^{\infty}[A_3(R_2)^3 + B_3(R_2)^{-3}] \cos(m\varphi) = \cos(3\varphi) \tag{6.191}$$

Then applying the orthogonality we see that the only non-trivial term of the summation is $m = 3$:

$$[A_3(R_2)^3 + B_3(R_2)^{-3}]\,|\cos(3\varphi)|^2 = |\cos(3\varphi)|^2 \tag{6.192}$$

Or alternatively:

$$A_3(R_2)^3 + B_3(R_2)^{-3} = 1 \tag{6.193}$$

We apply the third boundary condition:

$$u(\rho = R_1) = A_3(R_1)^3 + B_3(R_1)^{-3} = 0 \tag{6.194}$$

We solve

$$B_3 = \frac{R_1^6 R_2^3}{R_1^6 - R_2^6} \tag{6.195}$$

$$A_3 = -\frac{R_2^3}{R_1^6 - R_2^6} \tag{6.196}$$

Final solution

The final result is:

$$u = [A_3(\rho)^3 + B_3(\rho)^{-3}]\cos(3\varphi) = \left(\frac{R_2}{\rho}\right)^3 \frac{\rho^6 - R_1^6}{R_2^6 - R_1^6}\cos(3\varphi) \tag{6.197}$$

Additional notes: the solution could depend on the z variable due to the different conditions of the curved boundary (which would include the dependence on z) separating variables and expanding the solution in orthogonal eigenfunctions in the z direction, and as a function of the angular variable, the solution would be:

$$u(\rho,\varphi,z) = \sum_{m,n\geq 1}^{\infty} [A_{nm}\sin(m\varphi) + B_{nm}\cos(m\varphi)]\cdot \tag{6.198}$$

$$\left[I_m\left(\frac{\pi n}{h}\rho\right) - \left(\frac{I_m(\frac{\pi n}{h}R_1)}{K_m(\frac{\pi n}{h}R_1)}\right) K_m\left(\frac{\pi n}{h}\rho\right)\right]\cdot\cos\left(\frac{\pi n}{h}z\right) \tag{6.199}$$

Also the solution would depend on the z variable if the conditions on the flat boundaries had changed with the corresponding changes in the radial solution (from polynomial to modified Bessel functions) even if the curved boundary were kept as in the original formulation of the problem, that is, without a variation in the vertical direction.

Finally if we impose directly $n = 0$ in the solution we see that the McDonald function $K(x)$ diverges at $x = 0$.

If we remove the hole from the problem, we would be left with the summation formed only by radial modified Bessel functions $I_n(0) = 1$, which would mean that the solution would not depend on neither the vertical nor the radial variables, and only on the angle imposed by the boundary condition.

6.9 Case Study: Cylinder with a Heater of *xy* Symmetry

Find the stationary distribution of temperature in a cylinder of radius R and height L if inside the cylinder there are energy sources acting as a thin film in the plane $z = L/2$. The density of the heat sources is $f = xy$ and the maximum distance of the source from the center is $R/2$. Consider that the curved surface is at a constant temperature T_0. The flat faces are insulated. The thermal conductivity coefficient is k_0.

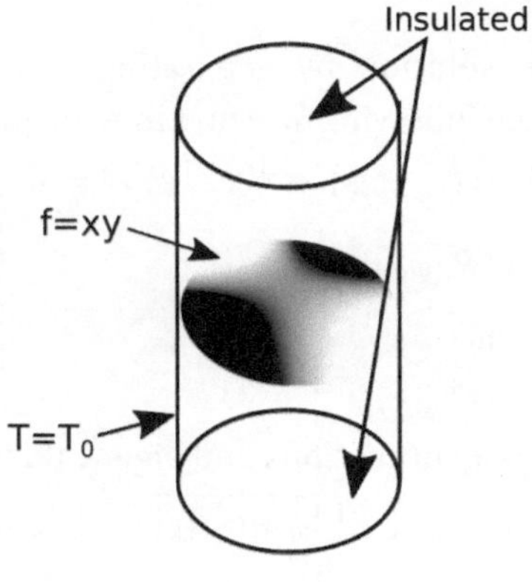

Figure 6.12

Mathematical formulation

The background temperature T_0 is subtracted from the solution.

$$\left\{\begin{array}{l} -k_0\Delta u = \rho^2\cos(\varphi)\sin(\varphi) = \frac{1}{2}\rho^2\sin(2\varphi)\cdot\delta\left(z-\frac{L}{2}\right); \quad \left(0<\rho<\frac{R}{2}\right) \\ u(R,\varphi)=0 \\ \left.\frac{\partial u}{\partial z}\right|_{z=0,L}=0 \end{array}\right\} \tag{6.200}$$

$$\left\{\begin{array}{l} \frac{1}{\rho}\frac{\partial}{\partial\rho}\left[\rho\frac{\partial u}{\partial\rho}\right]+\frac{1}{\rho^2}\frac{\partial^2 u}{\partial\varphi^2}+\frac{\partial^2 u}{\partial z^2} = \left\{\begin{array}{ll} -\frac{1}{2k_0}\rho^2\sin(2\varphi)\cdot\delta\left(z-\frac{L}{2}\right) & \left(0<\rho<\frac{R}{2}\right) \\ 0 & \left(\frac{R}{2}<\rho<R\right) \end{array}\right\} \\ u(R,\varphi)=0 \\ \left.\frac{\partial u}{\partial z}\right|_{z=0,L}=0 \end{array}\right\} \tag{6.201}$$

Note: integrating the density of heat sources can help to check that the total applied power is $W = 0$ since half of the source dissipates heat and the other half absorbs it

Sturm–Liouville problem

We now seek the solution by separating it in two orthogonal eigenfunctions of the following Sturm–Liouville problem:

$$\left\{\begin{array}{l} \Delta v(\rho,\varphi,z)+\lambda v(\rho,\varphi,z)=0 \\ v(R,\varphi)=0 \\ \left.\frac{\partial v}{\partial z}\right|_{z=0,L}=0 \end{array}\right\} \tag{6.202}$$

The corresponding eigenfunctions and eigenvalues are well known.

$$[v(\rho,\varphi)]_{nmk} = J_m(\sqrt{[\lambda_{nmk}-\mu_k]}\cdot\rho)[A_{nmk}\cos(m\varphi) + B_{nmk}\sin(m\varphi)]\cdot\cos\left(\frac{k\pi}{L}z\right) \tag{6.203}$$

with $\mu_k = (\frac{k\pi}{L})^2$; $k = 0, 1, 2, \ldots, m = 0, 1, 2, \ldots, n = 0, 1, 2, \ldots$.

The index n counts the zeros of the Bessel function, which gives the eigenvalues:

$$J_m(\sqrt{[\lambda_{nmk}-\mu_k]}\cdot R) = 0 \tag{6.204}$$

General solution

We replace the expression for the general solution:

$$u = \sum v(\rho, \varphi, z)_{nmk} \tag{6.205}$$

into equation (6.201):

$$\Delta \sum [v(\rho, \varphi, z)]_{nmk} = -\frac{1}{2k_0}\rho^2 \sin(2\varphi) \cdot \delta\left(z - \frac{L}{2}\right) \tag{6.206}$$

Note that it's one of the few cases where the solution can be sought as the sum of orthogonal eigenfunctions, since their coefficients are already present in the angular eigenfunctions.

$$\sum \lambda_{nmk}[v(\rho, \varphi, z)]_{nmk} = \frac{1}{2k_0}\rho^2 \sin(2\varphi) \cdot \delta\left(z - \frac{L}{2}\right) \tag{6.207}$$

Final solution

Using the orthogonality of the eigenfunctions we get the coefficients:

$$\begin{aligned} &\sum \lambda_{nmk} J_m(\sqrt{[\lambda_{nmk} - \mu_k]} \cdot \rho)[A_{nmk} \cos(m\varphi) \\ &+ B_{nmk} \sin(m\varphi)] \cos\left(\frac{k\pi}{L} z\right) = \\ &= \frac{1}{2k_0}\rho^2 \sin(2\varphi) \cdot \delta\left(z - \frac{L}{2}\right) \end{aligned} \tag{6.208}$$

It is clear that $A_{nmk} = 0$

$$\begin{aligned} &\sum B_{nmk}\lambda_{nmk} J_m(\sqrt{[\lambda_{nmk} - \mu_k]} \cdot \rho) \sin(m\varphi) \cos\left(\frac{k\pi}{L} z\right) \\ &= \frac{1}{2k_0}\rho^2 \sin(2\varphi) \cdot \delta\left(z - \frac{L}{2}\right) \end{aligned} \tag{6.209}$$

Multiplying both sides by $J_m(\sqrt{[\lambda_{nmk} - \mu_k]} \cdot \rho) \sin(m\varphi) \cdot \cos\left(\frac{k\pi}{L} z\right)$ and integrating between $\int_0^R \int_0^{2\pi} \int_0^L \rho d\rho d\varphi dz$ we get:

$$B_{nmk} = \frac{1}{2k_0} \frac{\int_0^{R/2} \rho^3 J_m(\sqrt{[\lambda_{nmk} - \mu_k]} \cdot \rho) d\rho \int_0^{2\pi} sin(2\varphi) \sin(m\varphi) d\varphi \int_0^L \delta(z - \frac{L}{2}) \cos\left(\frac{k\pi}{L} z\right) dz}{\lambda_{nmk} \left| J_m(\sqrt{[\lambda_{nmk} - \mu_k]} \cdot \rho)\right|^2 |\sin(m\varphi)|^2 \left|\cos(\frac{k\pi}{L} z)\right|^2} \tag{6.210}$$

Only the following coefficients remain:

$$B_{n2k} = \frac{1}{2k_0} \frac{\cos\left(\frac{k\pi}{2}\right) \int_0^{R/2} \rho^3 J_2(\sqrt{[\lambda_{n2k} - \mu_k]} \cdot \rho) d\rho}{\lambda_{n2k} \left\| J_m(\sqrt{[\lambda_{n2k} - \mu_k]} \cdot \rho) \right\|^2 \left| \cos(\frac{k\pi}{L} z) \right|^2} \quad (6.211)$$

Finally:

$$u = T_0 + \sin(2\varphi) \sum_{n,k} B_{n2k} \lambda_{n2k} J_2(\sqrt{[\lambda_{n2k} - \mu_k]} \cdot \rho) \cos\left(\frac{k\pi}{L} z\right) \quad (6.212)$$

Alternative method

Sturm–Liouville problem

Just like in the first method we convert all boundaries to homogeneous by subtracting the temperature T_0, we seek the solution as an expansion in radial and angular eigenfunctions. The vertical direction z is described by a function that we need to find:

$$\Delta \left[\sum v_{nm}(\rho, \varphi) \cdot Z_{nm}(z) \right] = \frac{1}{2k_0} \rho^2 \sin(2\varphi) \cdot \delta\left(z - \frac{L}{2}\right) \quad (6.213)$$

where $v(\rho, \varphi)_{nm}$ are the eigenfunctions of a Sturm–Liouville problem in a whole disk with first type boundary conditions.

$$v(\rho, \varphi)_{nm} = J_m(\sqrt{[\lambda_{nm}]} \cdot \rho) \sin(m\varphi) \quad (6.214)$$

Applying this condition (i.e., lowering the amount of second derivatives operators in the Laplacian from three to one) and integrating the resulting equations between the limits of the variations in the radial and angular variables $\int_0^R \int_0^{2\pi} \rho d\rho d\varphi$. We arrive at a non-homogeneous differential equation for the $Z(z)$ function with two different homogeneous boundary conditions of the second type which is solved by a solution by parts in the homogeneous ranges of the z variable. We then apply the two conditions for the insulated boundaries ($z = 0; L$), besides the condition of continuity for the functions and the condition for the difference of their derivatives.

6.10 Case Study: Semi-Cylinder with a Thin Heater

Consider a half of a cylinder (height L, radius $R = L/2$, thermal conductivity k) with its two flat horizontal sides in contact with a thermal reservoir at zero temperature and the vertical flat face is thermally insulated. The curved part is insulated except for a thin line in the form of a helix, through which a power W is uniformly supplied. Find the stationary distribution of temperature inside the cylinder, considering that all the heat generated by the resistance is directed towards the inside of the semi-cylinder.

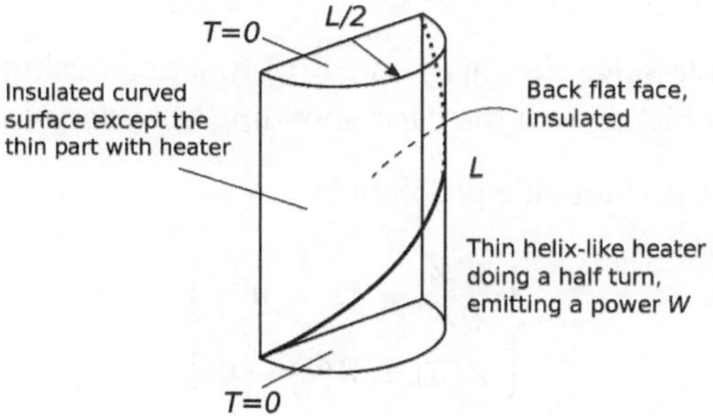

Figure 6.13

Mathematical formulation

$$\left\{\begin{array}{l} \Delta u = 0 \\ u(\rho, \varphi, 0) = 0 \\ u(\rho, \varphi, L) = 0 \\ -k\left.\dfrac{\partial u}{\partial \rho}\right|_{\rho=\frac{L}{2}} = -\dfrac{2W}{L^2\pi}\delta\left(z - \dfrac{L\varphi}{\pi}\right) \\ \left.\dfrac{\partial u}{\partial \varphi}\right|_{\varphi=0} = \left.\dfrac{\partial u}{\partial \varphi}\right|_{\varphi=\pi} = 0 \end{array}\right\} \tag{6.215}$$

Integrating in the φ angle and in the z direction the right part of the non-homogeneous boundary condition, we note that the surface power corresponds to the total power radiated by the heater, W. In cylindrical coordinates the equation to be solved is:

$$\frac{1}{\rho}\frac{\partial}{\partial\rho}\left[\rho\frac{\partial u}{\partial\rho}\right]+\frac{1}{\rho^2}\frac{\partial^2 u}{\partial\varphi^2}+\frac{\partial^2 u}{\partial z^2}=0 \tag{6.216}$$

Sturm–Liouville problem

We separate variables, taking into account that in the φ and z directions we have homogeneous boundaries:

$$u=R(\rho)\cdot\Phi(\varphi)\cdot Z(z) \tag{6.217}$$

$$\frac{1}{R}\frac{1}{\rho}\frac{d}{d\rho}\left[\rho\frac{dR}{d\rho}\right]+\frac{1}{\rho^2}\frac{1}{\Phi}\frac{d^2\Phi}{d\varphi^2}+\frac{1}{Z}\frac{d^2 Z}{dz^2}=0 \tag{6.218}$$

As always, we solve the $\Phi(\varphi)$ and $Z(z)$ functions using the Sturm–Liouville problems with the corresponding boundary conditions.

The first Sturm–Liouville problem is:

$$\left\{\begin{array}{l}\dfrac{d^2 Z}{dz^2}+\lambda Z=0\\ Z(0)=Z(L)=0\end{array}\right\} \tag{6.219}$$

The second Sturm–Liouville problem is:

$$\left\{\begin{array}{l}\dfrac{d^2\Phi}{d\varphi^2}+\mu\Phi=0\\ \left.\dfrac{d\phi}{d\varphi}\right|_{\varphi=0}=\left.\dfrac{d\Phi}{d\varphi}\right|_{\varphi=\pi}=0\end{array}\right\} \tag{6.220}$$

$$\Phi(\varphi)Z(z)=A_{nm}\cos(m\varphi)\sin\left(\frac{\pi n}{L}z\right)$$
$$(n=1,2,3\ldots,\ m=0,1,2,3\ldots) \tag{6.221}$$

The differential equation for the radial part is:

$$\frac{\frac{1}{\rho}\frac{d}{d\rho}\left[\rho\frac{dR}{d\rho}\right]}{R}+\frac{1}{\rho^2}(-m^2)-\left(\frac{\pi n}{L}\right)^2=0 \tag{6.222}$$

$$\frac{d^2 R}{d\rho^2}+\frac{1}{\rho}\frac{dR}{d\rho}-\left[\frac{m^2}{\rho^2}+\left(\frac{\pi n}{L}\right)^2\right]R=0 \tag{6.223}$$

General solution

The general solution of (6.236) is a Bessel function of imaginary argument (modified Bessel function). This time we discard the McDonald function, since it would diverge in the origin.

$$u(\rho,\varphi,z)=\sum_{n,m}A_{nm}I_m\left(\frac{\pi n}{L}\rho\right)\cos(m\varphi)\sin\left(\frac{\pi n}{L}z\right) \tag{6.224}$$

To find the A_{nm} and B_{nm} coefficients we will use the third boundary condition, differentiating the general solution and applying the condition of heat flux through the curved surface.

$$\sum_{n,m}A_{nm}\frac{\pi n}{L}\left.\frac{dI_m\left(\frac{\pi n}{L}\rho\right)}{d\rho}\right|_{\rho=\frac{L}{2}}\cos(m\varphi)\sin\left(\frac{\pi n}{L}z\right)=\frac{2W}{kL^2\pi}\delta\left(z-\frac{L\varphi}{\pi}\right) \tag{6.225}$$

Final solution

Using the orthogonality of the angular and vertical eigenfunctions we get the relation for the coefficients (with $\epsilon=\frac{\pi}{2}$ for $m\geq 1$ and $\epsilon=\pi$ for $m=0$).

$$A_{nm}=\frac{2W}{kn\pi^2L}\frac{\int_0^L\int_0^\pi\delta(z-\frac{L\varphi}{\pi}))\cos(m\varphi)\sin(\frac{\pi n}{L}z)dzd\varphi}{\left.\frac{dI_m\left(\frac{\pi n}{L}\rho\right)}{d\rho}\right|_{\rho=\frac{L}{2}}\|\cos(m\varphi)\|^2\left\|\sin(\frac{\pi m}{L}z)\right\|^2}=$$

$$=\frac{2W}{kn\epsilon\pi^2L}\frac{\int_0^\pi\sin(n\varphi)\cos(m\varphi)d\varphi}{\left.\frac{dI_m\left(\frac{\pi n}{L}\rho\right)}{d\rho}\right|_{\rho=\frac{L}{2}}\frac{L}{2}}=$$

$$=\frac{4W}{kn\epsilon\pi^2L^2}\frac{1}{\left.\frac{dI_m\left(\frac{\pi n}{L}\rho\right)}{d\rho}\right|_{\rho=\frac{L}{2}}}\int_0^\pi\sin(n\varphi)\cos(m\varphi)d\varphi=$$

$$=\frac{2W}{kn\epsilon\pi^2L^2}\frac{1}{\left.\frac{dI_m\left(\frac{\pi n}{L}\rho\right)}{d\rho}\right|_{\rho=\frac{L}{2}}}\left[\frac{1-\cos((n+m)\pi)}{(n+m)}\right.$$
$$\left.+\frac{1-\cos((n-m)\pi)}{(n-m)}\right] \tag{6.226}$$

Finally we check the units of the result. Recalling that $[k] = W/(mK)$ the obtained result will have units of temperature (K).

6.11 Half Cylinder with Two Inward Fluxes

Find the stationary distribution of temperature inside a half cylinder of radius R, height L and thermal conductivity coefficient k. The flat back surface is thermally insulated and the curved surface at the front has a temperature T_0. Each of the remaining surfaces (upper and lower) is traversed by a heat flux Q (homogeneously distributed across the boundaries) as indicated in the figure.

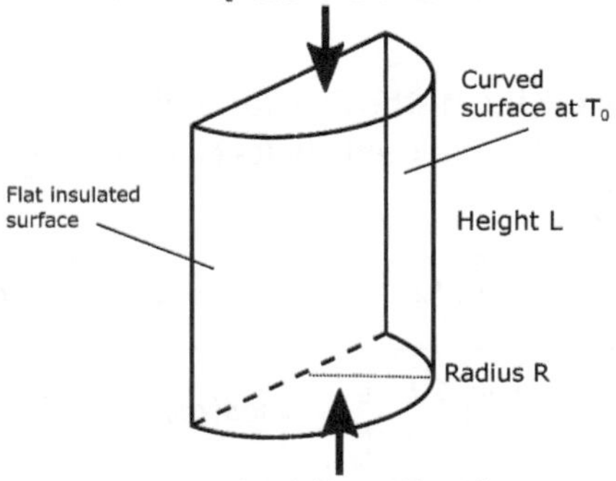

Figure 6.14

Mathematical formulation

$$\left\{\begin{array}{l} \Delta u = 0 \\ -k \left.\dfrac{\partial u}{\partial z}\right|_{z=0} = 2Q/(\pi R^2) \\ -k \left.\dfrac{\partial u}{\partial z}\right|_{z=L} = -2Q/(\pi R^2) \\ u(R, \varphi, z) = T_0 \\ \left.\dfrac{\partial u}{\partial \varphi}\right|_{\varphi=0} = \left.\dfrac{\partial u}{\partial \varphi}\right|_{\varphi=\pi} = 0 \end{array}\right\} \tag{6.227}$$

Subtracting T_0 from the solution we achieve the curved boundary with homogeneous boundary conditions, and the conditions in the rest of the boundaries don't change.

The solution can be expanded in a sum of orthogonal functions in the radial and angular variable. We will apply the method of separation of variables. In cylindrical coordinates the equation to be solved becomes:

$$\frac{1}{\rho}\frac{\partial}{\partial\rho}\left[\rho\frac{\partial u}{\partial\rho}\right]+\frac{1}{\rho^2}\frac{\partial^2 u}{\partial\varphi^2}+\frac{\partial^2 u}{\partial z^2}=0 \tag{6.228}$$

Sturm–Liouville problem

For the separation of variables method we need to consider that in the φ and z directions we have homogeneous boundaries:

$$u=R(\rho)\cdot\Phi(\varphi)\cdot Z(z) \tag{6.229}$$

$$\frac{1}{R}\frac{1}{\rho}\frac{d}{d\rho}\left[\rho\frac{dR}{d\rho}\right]+\frac{1}{\rho^2}\frac{1}{\Phi}\frac{d^2\Phi}{d\varphi^2}+\frac{1}{Z}\frac{d^2 Z}{dz^2}=0 \tag{6.230}$$

$$\frac{1}{R}\frac{1}{\rho}\frac{d}{d\rho}\left[\rho\frac{dR}{d\rho}\right]+\frac{1}{\rho^2}\frac{1}{\Phi}\frac{d^2\Phi}{d\varphi^2}=-\frac{1}{Z}\frac{d^2 Z}{dz^2}=-\lambda \tag{6.231}$$

The choice of the sign of the constant of the separation of variables is due to inability to formulate a Sturm–Liouville problem for $Z(z)$ (because of the non-homogeneous boundary conditions), whereas we can express $\Phi(\varphi)$ as a set of orthogonal eigenfunctions. The second Sturm–Liouville problem is:

$$\left\{\begin{array}{l}\dfrac{d^2\Phi}{d\varphi^2}+\mu\Phi=0\\ \left.\dfrac{d\phi}{d\varphi}\right|_{\varphi=0}=\left.\dfrac{d\Phi}{d\varphi}\right|_{\varphi=\pi}=0\end{array}\right\} \tag{6.232}$$

$$\Phi(\varphi)=\cos(m\varphi)\qquad(m=1,2,3\ldots,\ m=0,1,2,3\ldots) \tag{6.233}$$

The problem in the z direction has the form:

$$\left\{\begin{array}{l}\dfrac{d^2 Z}{dz^2}-\lambda Z=0\\ -k\left.\dfrac{\partial Z}{\partial z}\right|_{z=0}=\text{Const}\,(\rho,\phi)\\ -k\left.\dfrac{\partial u}{\partial z}\right|_{z=L}=-\text{Const}\,(\rho,\phi)\end{array}\right\} \tag{6.234}$$

$$Z(z) = A\cosh(\sqrt{\lambda}z) + B\sinh(\sqrt{\lambda}z) \quad (6.235)$$

The differential equation for the radial part is:

$$\frac{d^2R}{d\rho^2} + \frac{1}{\rho}\frac{dR}{d\rho} + \left[\lambda - \frac{m^2}{\rho^2}\right]R = 0 \quad (6.236)$$

The radial solutions are $J_m(\sqrt{\lambda}\rho)$, neglecting the Neumann functions that diverge at the origin of coordinates. The λ_{km} eigenvalues are given by the zeroes of the m-th order Bessel function: $J_m(\sqrt{\lambda}R) = 0$

General solution

$$u(\rho, \varphi, z) = \sum_{m,k} J_m(\sqrt{\lambda_{mk}}\rho)\cos(m\varphi)[A_{mk}\cosh(\sqrt{\lambda_{mk}}z) + B_{mk}\sinh(\sqrt{\lambda_{mk}}z)] \quad (6.237)$$

Final solution

To find the A_{mk} and B_{mk} coefficients we will apply two non-homogeneous boundary conditions (corresponding to the fluxes) by differentiating the general solution and applying the heat flux density condition across each of the bases.

$$-k\sum_{m,k} B_{mk}\sqrt{\lambda_{mk}}J_m(\sqrt{\lambda_{mk}}\rho)\cos(m\varphi) = 2Q/(\pi R^2) \quad (6.238)$$

$$-k\sum_{m,k}(A_{mk}\sqrt{\lambda_{mk}}\sinh(\sqrt{\lambda_{mk}}L) + B_{mk}\sqrt{\lambda_{mk}}\cosh(\sqrt{\lambda_{mk}}L))J_m (\sqrt{\lambda_{mk}}\rho)\cos(m\varphi) = -2Q/(\pi R^2) \quad (6.239)$$

Finally, using the orthogonality of the radial $J_m(\sqrt{\lambda_{mk}}\rho)$ and angular $\cos(m\varphi)$ functions we get the ration of the A_{mk} and B_{mk} coefficients:

$$B_{mk} = -\frac{2Q}{k\sqrt{\lambda_{mk}}\pi R^2}\frac{\int_0^R J_m(\sqrt{\lambda_{mk}}\rho)\rho d\rho \int_0^\pi \cos(m\varphi)d\varphi}{\int_0^R [J_m(\sqrt{\lambda_{mk}}\rho)]^2\rho d\rho \int_0^\pi [\cos(m\varphi)]^2 d\varphi} \quad (6.240)$$

Only the coefficients of index $m = 0$ remain:

$$B_{0k} = -\frac{2Q}{k\sqrt{\lambda_{0k}}\pi R^2} \frac{\int_0^R J_0(\sqrt{\lambda_{0k}\rho})\rho d\rho}{\int_0^R [J_0(\sqrt{\lambda_{0k}\rho})]^2 \rho d\rho} \tag{6.241}$$

Likewise for A_{mk} only the coefficients with $m = 0$ remain.

$$A_{0k} = \frac{2Q}{k\sqrt{\lambda_{0k}}\pi R^2} \frac{\int_0^R J_0(\sqrt{\lambda_{0k}\rho})\rho d\rho}{\sinh(\sqrt{\lambda_{0k}L}) \int_0^R [J_0(\sqrt{\lambda_{0k}\rho})]^2 \rho d\rho} - \frac{B_{0k}}{\tanh(\sqrt{\lambda_{0k}L})}$$

$$= -B_{0k}\frac{1 + \cosh(\sqrt{\lambda_{0k}L})}{\sinh(\sqrt{\lambda_{0k}L})} \tag{6.242}$$

Since for the remaining indices m we apply the orthogonality condition:

$$[A_{mk}\sqrt{\lambda_{mk}}\sinh(\sqrt{\lambda_{mk}L}) + 0\sqrt{\lambda_{mk}}\cosh(\sqrt{\lambda_{mk}L})] = 0 \tag{6.243}$$

6.12 Heat Flux through Half a Cylinder

Find the stationary distribution of temperature in a half cylinder of length L, radius ρ_0 and thermal conductivity k if a heat flux of value $J = J_0(Lz - z^2)$ enters through the curved surface and exits through the bases, being homogeneously distributed. Consider that the flat face opposite to the curved surface is thermally insulted.

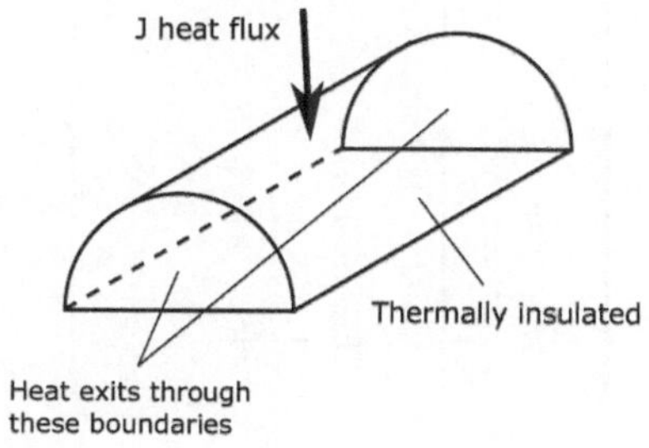

Figure 6.15

Solution:

To find all boundary conditions we must equate the flux that enters through the curved surface to the flux that exits through the lateral faces. The total flux that enters through the curved surface is

$$-\int_0^{\pi}\int_0^{L} J_0[z^2 - Lz]Rd\varphi dz = -RJ_0\pi\left[\int_0^{L} z^2dz - L\int_0^{L} zdz\right]$$
$$= -\pi RJ_0\left[\frac{L^3}{3} - \frac{L^3}{2}\right] = \frac{\pi RJ_0L^3}{6} \tag{6.244}$$

Both lateral surfaces get half of that flux:

$$\frac{W(z=0)}{S} = -k\left.\frac{\partial u}{\partial z}\right|_{z=0} = \frac{\pi RJ_0L^3}{6\left[\frac{\pi R^2}{2}\right]\cdot 2} = -\frac{J_0L^3}{6R} \quad \text{(negative flux)} \tag{6.245}$$

$$\frac{W(z=L)}{S} = -k\left.\frac{\partial u}{\partial z}\right|_{z=L} = +\frac{J_0L^3}{6R} \quad \text{(positive flux)} \tag{6.246}$$

The sign of the fluxes exiting through both sides is different since, by symmetry, they propagate in opposite directions.

Mathematical formulation

$$\left\{\begin{array}{l} \Delta u = 0 \\ -k\left.\dfrac{\partial u}{\partial \rho}\right|_{\rho=R} = J_0[Lz - z^2] \\ \left.\dfrac{\partial u}{\partial \varphi}\right|_{\varphi=0,\pi} \\ -k\left.\dfrac{\partial u}{\partial z}\right|_{z=0} = -\frac{J_0L^3}{6R} \\ -k\left.\dfrac{\partial u}{\partial z}\right|_{z=L} = +\frac{J_0L^3}{6R} \end{array}\right\} \tag{6.247}$$

We can split the problem in two:

$$u(\rho, \varphi, z) = w(\rho, \varphi, z) + v(\rho, \varphi, z)$$

Problem 1:

$$\left\{\begin{array}{l}\Delta w=0\\ -k\left.\dfrac{\partial w}{\partial \rho}\right|_{\rho=R}=J_0[Lz-z^2]\\ -k\left.\dfrac{\partial w}{\partial \varphi}\right|_{0,\pi}=0\\ -k\left.\dfrac{\partial w}{\partial z}\right|_{z=0,L}=0\end{array}\right\} \tag{6.248}$$

Problem 2:

$$\left\{\begin{array}{l}\Delta v=0\\ -k\left.\dfrac{\partial v}{\partial \rho}\right|_{\rho=R}=0\\ -k\left.\dfrac{\partial v}{\partial z}\right|_{z=0}=-\dfrac{J_0L^3}{6R}\\ -k\left.\dfrac{\partial v}{\partial z}\right|_{zL}=+\dfrac{J_0L^3}{6R}\end{array}\right\} \tag{6.249}$$

Strum Liouville problem

We separate variables for $w(\rho, \varphi, z)$:

$$w(\rho,\varphi,z)=R(\rho)\cdot\Phi(\varphi)\cdot Z(z)$$

$$\frac{1}{R}\frac{1}{\rho}\frac{\partial}{\partial \rho}\left(\rho\cdot\frac{\partial R}{\partial \rho}\right)+\frac{1}{\Phi}\frac{1}{\rho^2}\frac{\partial^2\Phi}{\partial \varphi^2}=-\frac{1}{Z}\frac{\partial^2 Z}{\partial z^2}=+\lambda$$

The Sturm–Liouville problem in the z direction is:

$$\left\{\begin{array}{l}\dfrac{d^2Z}{dz^2}+\lambda Z=0\\ \left.\dfrac{dZ}{dz}\right|_{z=0,L}=0\end{array}\right. \tag{6.250}$$

The eigenfunctions of the problem are:

$$Z(z)=\cos\left(\frac{\pi n z}{2}\right)$$

and the eigenvalues,

$$\lambda=\left(\frac{\pi n}{L}\right)^2$$

The Sturm–Liouville problem in the angular direction is:

$$\begin{cases} \dfrac{d^2\Phi}{d\varphi^2} + \lambda\Phi = 0 \\ \left.\dfrac{d\Phi}{d\varphi}\right|_{\varphi=0,\pi} = 0 \end{cases} \tag{6.251}$$

The eigenfunctions of the problem are:

$$\Phi(\varphi) = \cos(m\varphi)$$

The radial equation is:

$$\rho\frac{d}{d\rho}\left(\rho\frac{dR}{d\rho}\right) - (\lambda\rho^2 + m^2)R = 0 \tag{6.252}$$

or:

$$\frac{d^2R}{d\rho^2} + \frac{1}{\rho}\frac{dR}{d\rho} - \left[\frac{m^2}{\rho^2} + \lambda\right] = 0 \tag{6.253}$$

The radial solutions are modified Bessel functions:

$$R(\rho) = A_{nm}I_m(\sqrt{\lambda}\rho) \tag{6.254}$$

The general solution is:

$$W(\rho, \varphi, z) = \sum_{n,m} A_{nm}I_m(\sqrt{\lambda}\rho)\cos(m\varphi)\cdot\cos\left(\frac{\pi nz}{L}\right) \tag{6.255}$$

We apply the boundary conditions:

$$\begin{aligned} -k\left.\frac{\partial w}{\partial\rho}\right|_{\rho=R} &= J_0[Lz - z^2] \\ &= \sum_{n,m}\sqrt{\lambda_n}A_{nm}I'_m(\sqrt{\lambda_n}R)\cos(m\varphi)\cdot\cos\left(\frac{\pi nz}{L}\right) \end{aligned} \tag{6.256}$$

Applying the condition of orthogonality:

$$A_{nm} = -\frac{J_0\int_0^L(Lz - z^2)\cos\left(\frac{\pi n}{L}\right)dz\cdot\int_0^\pi\cos(m\varphi)d\varphi}{k\sqrt{\lambda_n}I'_m(\sqrt{\lambda_n}R)\cdot\int_0^L\int_0^\pi\cos^2(m\varphi)\cos^2\left(\frac{\pi nz}{L}\right)d\varphi dz} \tag{6.257}$$

Only the terms with $m = 0$ remain:

$$A_{n0} = -\frac{J_0\int_0^L(Lz - z^2)\cos\left(\frac{\pi n}{L}\right)dz}{k\sqrt{\lambda_n}I'_0(\sqrt{\lambda_n}R)\cdot\frac{L}{2}} \tag{6.258}$$

Now we turn to problem 2. As no boundary condition depends on φ and there are no heat fluxes at $\varphi = 0, \pi$ the solution will not depend on the angular variable.

$$\frac{1}{\rho}\frac{\partial}{\partial \rho}\left(\rho\frac{\partial V}{\partial \rho}\right) + \frac{\partial^2 V}{\partial z^2} = 0 \tag{6.259}$$

Separating variables: $V = R(\rho) \cdot Z(z)$:

$$\frac{\frac{1}{\rho}\frac{d}{d\rho}\left(\rho\frac{dR}{d\rho}\right)}{R} = -\frac{\frac{d^2 Z}{dz^2}}{Z} = -\lambda \tag{6.260}$$

The sign of λ is chosen to arrive at hyperbolic solutions in z:

$$\frac{d^2 Z}{dz^2} - \lambda Z(z) = 0 \rightarrow Z(z) = A\sinh(\sqrt{\lambda}z) + B\cosh(\sqrt{\lambda}z) \tag{6.261}$$

Radial solution:

$$\frac{1}{\rho}\frac{d}{d\rho}\left(\rho\frac{dR}{d\rho}\right) + \lambda R = 0 \rightarrow R(\rho) = J_0(\sqrt{\lambda_k}\rho) \tag{6.262}$$

The λ_k eigenvalues are found from the zeroes of the derivatives of the Bessel function:

$$\lambda_k = \left(\frac{\mu_k}{R}\right)^2 \tag{6.263}$$

General solution

The general solution is:

$$V(\rho, z) = \sum_k \left[A_k \sinh(\sqrt{\lambda_k}z) + B_k\cosh(\sqrt{\lambda_k}z)\right] J_0(\sqrt{\lambda_k}\rho) \tag{6.264}$$

Final solution

To get the constants necessary to obtain the final solution we apply the boundary conditions. Applying the boundary condition at $z = 0$:

$$-k\left.\frac{\partial V}{\partial z}\right|_{z=0} = -\frac{J_0 L^3}{6R} = -k\sum_k A_k\sqrt{\lambda_k}J_0(\sqrt{\lambda_k}\rho) \rightarrow$$

$$A_k = \frac{J_0 L^3 \int_0^R J_0(\sqrt{\lambda_k}\rho)\rho d\rho}{\sqrt{\lambda_k}\cdot k \cdot 6R\|J_0(\sqrt{\lambda_k}\rho)\|} \tag{6.265}$$

Applying now the boundary condition at $z = L$:

$$-k\left.\frac{\partial V}{\partial z}\right|_{z=L} = \sum_k \left[A_k\sqrt{\lambda_k}\cosh(\sqrt{\lambda_k}L) + B_k\sqrt{\lambda_k}\sinh(\sqrt{\lambda_k}L)\right] \times J_0(\sqrt{\lambda_k}\rho) = \frac{J_0 L^3}{6R} \quad (6.266)$$

$$B_k = \frac{1}{\sqrt{\lambda_k}\sinh(\sqrt{\lambda_k}L)}\left[\frac{J_0 L^3}{6R} - A_k\sqrt{\lambda_k}\cosh(\sqrt{\lambda_k}L)\right] \times \frac{\int_0^R J_0(\sqrt{\lambda_k}\rho)\rho d\rho}{\|J_0(\sqrt{\lambda_k}\rho)\|^2} \quad (6.267)$$

Chapter 7

Problems in Spherical Coordinates

The present chapter gives several examples of the detailed solution of problems that can be described using spherical coordinates when the symmetry break is due to a symmetry center. Spherical coordinates use three variables (azimuthal and polar angles, and radius). As a result, the form of the Laplacian operator changes, as well as the form of the solutions to the Sturm–Liouville problem in the radial variable (spherical Bessel functions) and in the polar angle (Legendre polynomials). Furthermore, also the way to describe some spatial features, such as points, circles or spherical shells changes, by using the Dirac's Delta function in spherical coordinates.

It is important to stress that in the case of problems with non-homogeneous contours in planes perpendicular to the vector of the azimuthal angle it is necessary to insert a constant or use a compensatory function in order to formulate correctly the Sturm–Liouville problem in the azimuthal angle. Only in this manner will we be able to formulate the equations as a function of the polar angle and in the appropriate cases, the radial coordinate and get solutions in the form of Legendre polynomials.

Mathematical Methods for Physics: Problems and Solutions
Farkhad G. Aliev and Antonio Lara

ISBN 978-981-4968-71-3 (Hardcover), 978-1-003-41088-1 (eBook)
www.jennystanford.com

7.1 Electric Potential between Two Spheric Shells

Find the electrostatic potential inside a sphere whose outer part ($r = b$) is at zero potential ($u(b) = 0$), whereas the inner part ($r = a$) is at a potential equal to $V_0 \sin(\theta) \sin(\varphi)$.

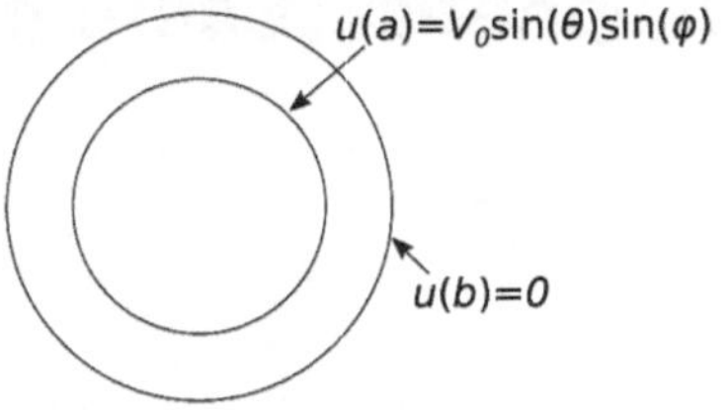

Figure 7.1

Mathematical formulation

$$\left\{\begin{array}{ll} \Delta u = 0 & (a < r < b) \\ \dfrac{1}{r^2}\dfrac{\partial}{\partial r}\left[r^2\left(\dfrac{\partial u}{\partial r}\right)\right] + \dfrac{1}{r^2}\Delta_{\theta,\varphi} u = 0 & \\ u(r=b) = f(\theta,\varphi) = V_0 \sin(\theta)\sin(\varphi) & \\ u(r=a) = 0 & \end{array}\right\} \tag{7.1}$$

We seek the solution of the equation:

$$\frac{\partial}{\partial r}\left[r^2\left(\frac{\partial u}{\partial r}\right)\right] + \Delta_{\theta,\varphi} u = 0 \tag{7.2}$$

Using the method of separation of variables:

$$u(r,\theta,\varphi) = R(r)\cdot V(\theta,\varphi) \tag{7.3}$$

$$\frac{\frac{d}{dr}[r^2(\frac{dR}{dr})]}{R} = -\frac{\Delta_{\theta,\varphi} V}{V} = \lambda > 0 \tag{7.4}$$

Sturm–Liouville problem

We use the positive sign for λ because we expect to get periodic eigenfunctions for the angular variables, while the radial solution is expected to correspond to inhomogeneous boundary conditions. We arrive at the problem:

$$\Delta_{\theta,\varphi} V + \lambda V = 0 \tag{7.5}$$

$$\frac{1}{\sin(\theta)}\frac{\partial}{\partial\theta}\left[\sin(\theta)\left(\frac{\partial V}{\partial\theta}\right)\right] + \frac{1}{\sin^2(\theta)}\frac{\partial^2 V}{\partial\varphi^2} + \lambda V = 0 \tag{7.6}$$

By separating variables once again: $V = \Theta(\theta)\cdot\Phi(\varphi)$

$$\frac{\sin(\theta)\frac{d}{d\theta}[\sin(\theta)(\frac{d\Theta}{d\theta})] + \lambda\sin^2(\theta)\cdot\Theta(\theta)}{\Theta(\theta)} = -\frac{1}{\Phi(\varphi)}\frac{d^2\Phi}{d\varphi^2} = \nu > 0 \tag{7.7}$$

We choose the positive sign for ν because we expect to obtain a periodic solution (eigenfunctions) for the angular variable (φ). The problem for (φ) is then:

$$\frac{d^2\Phi}{d\varphi^2} + \nu\Phi(\varphi) = 0 \tag{7.8}$$

$$\Phi(\varphi) = \Phi(\varphi + 2\pi) \tag{7.9}$$

The solution for the part depending on the azimuthal angle is:

$$\Phi(\varphi) = A\cos(m\varphi) + B\sin(m\varphi) \tag{7.10}$$

$$\nu = m^2 \tag{7.11}$$

Now that we have $\nu = m^2$, we can find the solution for the function $\Theta(\theta)$

$$\sin(\theta)\frac{d}{d\theta}\left[\sin(\theta)\left(\frac{d\Theta}{d\theta}\right)\right] + \lambda\sin^2(\theta)\cdot\Theta(\theta) = m^2\Theta \tag{7.12}$$

$$\sin(\theta)\frac{d}{d\theta}\left[\sin(\theta)\left(\frac{d\Theta}{d\theta}\right)\right] + [\lambda\sin^2(\theta) - m^2]\Theta(\theta) = 0 \tag{7.13}$$

The solution are Legendre polynomials:

$$\Theta(\theta) = P_n^{(m)}(\cos(\theta)) \tag{7.14}$$

With eigenvalues: $\lambda = n(n+1)$. Now, since we have $\lambda = n(n+1)$ we can find the radial function

$$\frac{d}{dr}\left[r^2\left(\frac{dR}{dr}\right)\right] - n(n+1)R = 0 \tag{7.15}$$

$$r^2\left(\frac{d^2R}{dr^2}\right) + 2r\left(\frac{dR}{dr}\right) - n(n+1)R = 0 \tag{7.16}$$

We seek the solution as:

$$R(r) = r^\alpha \tag{7.17}$$

$$r^2\alpha(\alpha-1)r^{\alpha-2} + 2r\alpha r^{\alpha-1} - n(n+1)r^\alpha = 0 \tag{7.18}$$

$$\alpha^2 + \alpha - n(n+1) = 0 \tag{7.19}$$

$$\alpha_1 = n \tag{7.20}$$

$$\alpha_2 = -n-1 \tag{7.21}$$

Therefore $R(r) = Cr^n + Dr^{-n-1}$. We cannot discard any of the two members since the problem is contained in $(a \le r \le b)$, and no divergences at the origin are considered.

General solution

We form the general solution for the problem:

$$u(r,\theta,\varphi) = \sum_{n\ge m}^{\infty}\sum_{m=0}^{\infty}[C_{nm}r^n + D_{nm}r^{-n-1}]P_n^{(m)}(\cos(\theta)) \times [A_{nm}\cos(m\varphi) + B_{nm}\sin(m\varphi)] \tag{7.22}$$

An equivalent, but more convenient way to present this summation:

$$\begin{aligned} u(r,\theta,\varphi) = &\sum_{n\ge m}^{\infty}\sum_{m=0}^{\infty}\left(\frac{b^{n+1}}{b^{2n+1}-a^{2n+1}}\right)\left(\frac{r^{2n+1}-a^{2n+1}}{r^{n+1}}\right)P_n^{(m)}(\cos(\theta)) \\ &\times [A_{nm}\cos(m\varphi) + B_{nm}\sin(m\varphi)] \\ &+ \sum_{n\ge m}^{\infty}\sum_{m=0}^{\infty}\left(\frac{a^{n+1}}{b^{2n+1}-a^{2n+1}}\right)\left(\frac{b^{2n+1}-r^{2n+1}}{r^{n+1}}\right)P_n^{(m)}(\cos(\theta)) \\ &\times [C_{nm}\cos(m\varphi) + D_{nm}\sin(m\varphi)] \end{aligned} \tag{7.23}$$

Imposing one of the boundary conditions:

$$\begin{aligned} u(b,\theta,\varphi) = &\sum_{n\ge m}^{\infty}\sum_{m=0}^{\infty}\left(\frac{b^{n+1}}{b^{2n+1}-a^{2n+1}}\right)\left(\frac{b^{2n+1}-a^{2n+1}}{b^{n+1}}\right) \\ &\times P_n^{(m)}(\cos(\theta))\cdot[A_{nm}\cos(m\varphi) + B_{nm}\sin(m\varphi)] + 0 = 0 \end{aligned} \tag{7.24}$$

Final solution

And using the orthogonality of $P_n^{(m)}(\cos(\theta))$, from $\cos(m\varphi)$ and from $\sin(m\varphi)$ we get:

$$\left\| P_n^{(m)}(\cos(\theta)) \right\|^2 \cdot A_{nm} \left\| \cos(m\varphi) \right\|^2 = 0 \rightarrow A_{nm} = 0 \tag{7.25}$$

$$\left\| P_n^{(m)}(\cos(\theta)) \right\|^2 \cdot B_{nm} \left\| \sin(m\varphi) \right\|^2 = 0 \rightarrow B_{nm} = 0 \tag{7.26}$$

Imposing the other boundary condition:

$$\sum_{n\geq m}^{\infty} \sum_{m=0}^{\infty} \left(\frac{a^{n+1}}{b^{2n+1} - a^{2n+1}} \right) \left(\frac{b^{2n+1} - a^{2n+1}}{a^{n+1}} \right) P_n^{(m)}(\cos\theta) \times [C_{nm}\cos(m\varphi) + D_{nm}\sin(m\varphi)] = V_0 \sin(\theta)\sin(\varphi) \tag{7.27}$$

Then:

$$\sum_{n\geq m}^{\infty} \sum_{m=0}^{\infty} P_n^{(m)}(\cos(\theta)) \cdot [C_{nm}\cos(m\varphi) + D_{nm}\sin(m\varphi)] = V_0 \sin(\theta)\sin(\varphi) \tag{7.28}$$

Using the orthogonality of $P_n^{(m)}(\cos\theta)$ and of $\cos(m\varphi)$ we multiply by $P_n^{(m)}(\cos\theta)$ and $\cos(m\varphi)$ and integrate $\int_0^{\pi}\int_0^{2\pi} \sin(\theta) d\theta d\varphi$. Then:

$$\left\| P_n^{(m)}(\cos(\theta)) \right\|^2 \cdot C_{nm} \left\| \cos(m\varphi) \right\|^2 = 0 \rightarrow C_{nm} = 0 \tag{7.29}$$

Finally, using the orthogonality of $P_n^{(m)}(\cos(\theta))$ and $\sin(m\varphi)$ we multiply (7.28) by $P_n^{(m)}(\cos(\theta))$ and by $\sin(m\varphi)$ and integrate $\int_0^{\pi}\int_0^{2\pi} \sin(\theta) d\theta d\varphi$. From there:

$$\sum_{n\geq m}^{\infty} \sum_{m=0}^{\infty} P_n^{(m)}(\cos\theta) \cdot D_{nm}\sin(m\varphi) = V_0 \sin(\theta)\sin(\varphi) \tag{7.30}$$

$$\left\| P_n^{(m)}(\cos(\theta)) \right\|^2 \cdot D_{nm} \left\| \sin(m\varphi) \right\|^2 = V_0 \int_0^{\pi}\int_0^{2\pi} P_n^{(m)}(\cos(\theta)) \sin^2(\theta)\sin(\varphi)\sin(m\varphi) d\theta d\varphi \tag{7.31}$$

$$\left\| P_n^{(m)}(\cos(\theta)) \right\|^2 \cdot D_{nm} \left\| \sin(m\varphi) \right\|^2$$
$$= V_0 \int_0^{\pi} P_n^{(m)}(\cos(\theta)) \sin^2(\theta) d\theta \cdot \left\| \sin(\varphi) \right\|^2 \delta_{1,m} \qquad (7.32)$$

when $m \neq 1 \rightarrow D_{nm} = 0$

when $m = 1 \rightarrow D_{n1} = \dfrac{V_0 \int_0^{\pi} P_n^{(m)}(\cos\theta) \sin^2(\theta) d\theta}{\left\| P_n^{(m)}(\cos\theta) \right\|^2}$

Since $P_1^{(1)}(\cos(\theta)) = \sin(\theta)$, of all the coefficients only $D_{11} = V_0$ remains. The final solution is:

$$u(r,\theta,\varphi) = \sum_{n\geq 1}^{\infty} \left(\frac{a^{n+1}}{b^{2n+1} - a^{2n+1}} \right) \left(\frac{b^{2n+1} - r^{2n+1}}{r^{n+1}} \right)$$
$$\times \frac{V_0 \int_0^{\pi} P_n^{(m)}(\cos(\theta)) \sin^2 \theta d\theta}{\left\| P_n^{(m)}(\cos(\theta)) \right\|^2} P_n^{(1)}(\cos(\theta)) \sin(\varphi) \qquad (7.33)$$

or more simply:

$$u(r,\theta,\varphi) = V_0 \left(\frac{a^2}{b^3 - a^3} \right) \left(\frac{b^3 - r^3}{r^2} \right) \cos(\varphi) \cdot \sin(\theta) \qquad (7.34)$$

7.2 Distribution of Temperature inside a Sphere

A sphere has radius R. Its surface is kept at a temperature that depends on the angles as: $T_0 \sin(3\theta) \cos(\varphi)$. Find the distribution of temperature inside this object.

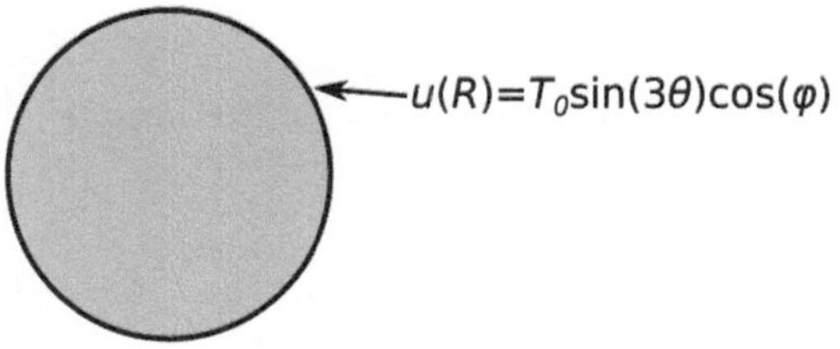

Figure 7.2

Mathematical formulation

$$\left\{\begin{array}{ll}\Delta u(r,\theta,\varphi)=0 & (0<r<R) \\ \dfrac{1}{r^2}\dfrac{\partial}{\partial r}\left[r^2\left(\dfrac{\partial u}{\partial r}\right)\right]+\dfrac{1}{r^2}\Delta_{\theta,\varphi}u=0 & \\ u(r=R)=f(\theta,\varphi)=T_0\sin(3\theta)\cos(\varphi) & \\ u(0,\theta,\varphi)<\infty & \end{array}\right\} \quad (7.35)$$

General solution

The general solution of the Laplace problem in spherical coordinates has been obtained in detail in the previous problem.

$$u(r,\theta,\varphi)=\sum_{n\geq m}^{\infty}\sum_{m=0}^{\infty}[r^n+D_{nm}r^{-n-1}]P_n^{(m)}(\cos(\theta)) \times[A_{nm}\cos(m\varphi)+B_{nm}\sin(m\varphi)] \quad (7.36)$$

Since the solution would need to be finite for $r=0\rightarrow D_{nm}=0$

$$u(r,\theta,\varphi)=\sum_{n\geq m}^{\infty}\sum_{m=0}^{\infty}[r^n]P_n^{(m)}(\cos(\theta))\cdot[A_{nm}\cos(m\varphi)+B_{nm}\sin(m\varphi)] \quad (7.37)$$

Final solution

Using the orthogonality of $\cos(m\varphi)$ and of $\sin(m\varphi)$ and applying the first boundary condition:

$$u(R,\theta,\varphi)=T_0\sin(3\theta)\cos(\varphi)=\sum_{n\geq m}^{\infty}\sum_{m=0}^{\infty}[R^n]P_n^{(m)}(\cos(\theta)) \times[A_{nm}\cos(m\varphi)+B_{nm}\sin(m\varphi)] \quad (7.38)$$

Multiplying by $\cos(m\varphi)$ and integrating from 0 to 2π:

$$T_0\sin(3\theta)=\sum_{n=1}^{\infty}A_{n1}R^nP_n^{(1)}(\cos(\theta)) \quad (7.39)$$

Multiplying by $P_n^{(1)}(\cos(\theta))$ and by $\sin(\theta)$ and integrating from 0 to π and also using the modulus of the Legendre functions we have:

$$A_{n1}=\frac{T_0}{R^n\int_0^{\pi}[P_n^{(1)}(\cos(\theta))]^2\sin(\theta)d\theta}\int_0^{\pi}\sin(3\theta)P_n^{(1)}(\cos(\theta))\sin(\theta)d\theta \quad (7.40)$$

The $\sin(3\theta)$ term can be expressed in terms of Legendre polynomials by taking into account the following relations, to simplify the calculation of the integral of the condition of orthogonality:

$$\sin(3\theta) = \sin(\theta)[4\cos^2(\theta) - 1] \tag{7.41}$$

We know that:

$$P_1^{(1)}(\cos(\theta)) = \sin(\theta) \tag{7.42}$$

$$P_3^{(1)}(\cos(\theta)) = \sin(\theta)[5\cos^2(\theta) - 1] \tag{7.43}$$

$$\begin{aligned}\sin(\theta)[4\cos^2(\theta) - 1] &= \sin(\theta)[5\cos^2(\theta) - 1]\frac{4}{5} - \frac{1}{5}\sin(\theta) \\ &= \frac{4}{5}P_3^{(1)}(\cos(\theta)) - \frac{1}{5}P_1^{(1)}(\cos(\theta))\end{aligned} \tag{7.44}$$

Only two coefficients of the summation remain:

$$A_{11} = \frac{T_0}{R^1 \int\limits_0^\pi [P_1^{(1)}(\cos\theta)]^2 \sin(\theta)d\theta} \int\limits_0^\pi \left(-\frac{1}{5}\right) P_1^{(1)}(\cos(\theta))P_1^{(1)}(\cos(\theta))\sin(\theta)d\theta = -\frac{T_0}{5R} \tag{7.45}$$

$$A_{31} = \frac{T_0}{R^3 \int\limits_0^\pi [P_3^{(1)}(\cos\theta)]^2 \sin(\theta)d\theta} \int\limits_0^\pi \frac{4}{5}P_3^{(1)}(\cos(\theta))P_3^{(1)}(\cos(\theta))\sin(\theta)d\theta = \frac{4T_0}{5R^3} \tag{7.46}$$

Finally:

$$u(r, \theta, \varphi) = \sum_{n=1,3} A_{n1}r^n P_n^{(1)}(\cos(\theta))\cos(\varphi) \tag{7.47}$$

7.3 Laplace Problem in a Sphere with a Difference of Potential

Find the electrostatic potential inside a sphere of radius R whose upper part ($0 < \theta < \pi/2$) is at a fixed potential equal to u_1 while the lower part ($\pi/2 < \theta < \pi$) is at a potential u_2.

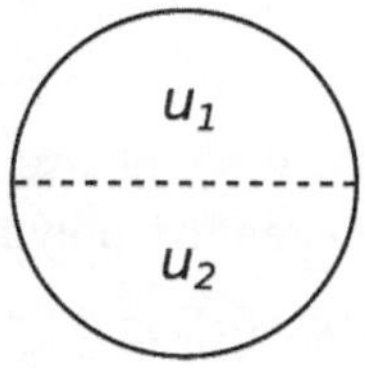

Figure 7.3

Mathematical formulation

Formulation for $v(r, \theta, \varphi) = u(r, \theta, \varphi) - u_2$

$$\left\{\begin{array}{ll} \Delta v(r, \theta, \varphi) = 0 & (0 < r < R) \\ \dfrac{1}{r^2}\dfrac{\partial}{\partial r}\left[r^2\left(\dfrac{\partial v}{\partial r}\right)\right] + \dfrac{1}{r^2}\Delta_{\theta,\varphi} v = 0 & \\ v(r = R) = f(\theta) = \left\{\begin{array}{ll} u_1 - u_2 & (0 < \theta < \pi/2) \\ 0 & (\pi/2 < \theta < \pi) \end{array}\right\} & \end{array}\right\} \tag{7.48}$$

General solution

The general solution of the Laplace problem (both internal and external) in spherical coordinates is:

$$v(r, \theta, \varphi) = \sum_{n \geq m}^{\infty}\sum_{m=0}^{\infty}\left(\frac{r}{R}\right)^n P_n^{(m)}(\cos(\theta))\cdot[A_{nm}\cos(m\varphi) + B_{nm}\sin(m\varphi)] \tag{7.49}$$

Since the solution does not depend on the angle φ we will have to expand the solution with $m = 0$:

$$v(r, \theta) = \sum_{n=0}^{\infty} A_n \left(\frac{r}{R}\right)^n P_n(\cos(\theta)) \tag{7.50}$$

Imposing the boundary condition:

$$f(\theta) = \sum_{n=0}^{\infty} A_n \left(\frac{R}{R}\right)^n P_n(\cos(\theta)) \tag{7.51}$$

Final solution

To find the coefficients A_n, we will use the orthogonality of the angular eigenfunctions (Legendre polynomials). Multiplying the previous expression by $P_{n'}(\cos(\theta))$ and integrating $\int_0^{\pi} \sin(\theta)d\theta$

$$\int_0^{\pi} f(\theta)P_{n'}(\cos(\theta))\sin(\theta)d\theta = A_n \int_0^{\pi} P_n(\cos\theta)P_{n'}(\cos(\theta))\sin(\theta)d\theta \tag{7.52}$$

$$A_n = \frac{\int_0^{\pi} f(\theta)P_n(\cos(\theta))\sin(\theta)d\theta}{\|P_n(\cos\theta)\|^2} = \frac{\int_0^{\frac{\pi}{2}} (u_1 - u_2)P_n(\cos(\theta))\sin(\theta)d\theta}{\|P_n(\cos(\theta))\|^2} \tag{7.53}$$

$$\begin{aligned} A_n &= \frac{2n+1}{2}(u_1 - u_2)\int_0^{\frac{\pi}{2}} P_n(\cos\theta)\sin(\theta)d\theta \\ &= \frac{2n+1}{2}(u_1 - u_2)\int_0^{1} P_n(x)dx \end{aligned} \tag{7.54}$$

since $\int_0^1 P_0(x)dx = 1$; $A_0 = \frac{1}{2}(u_1 - u_2)$

On the other hand, since the $P_{2k}(x)$ functions are even ($k = 1, 2, 3\ldots$):

$$\int_0^1 P_{2k}(x)dx = \frac{1}{2}\int_{-1}^{1} P_{2k}(x)dx \tag{7.55}$$

But due to the orthogonality:

$$\int_{-1}^{1} P_{2k}(x)P_0(x)dx = 0 \tag{7.56}$$

Then we will obtain: $A_{2k} = 0$; ($k = 1, 2, 3\ldots$). Finally we consider $n = 2k + 1$, with k integer numbers. In this case also:

$$\int_{-1}^{1} P_{2k+1}(x)P_0(x)dx = \int_{-1}^{1} P_{2k+1}(x)dx = 0 \tag{7.57}$$

However, since $P_{2k+1}(x)$ are now odd functions:

$$\int_0^1 P_{2k+1}(x)dx \neq 0 \tag{7.58}$$

To find this integral we use recurrence formulas:

$$(2n+1)P_n(x) = P'_{n+1}(x) - P'_{n-1}(x) \tag{7.59}$$

Then:

$$\begin{aligned}\int_0^1 P_{2k+1}(x)dx &= \frac{1}{4k+3}\int_0^1 [P'_{2k+2}(x) - P'_{2k}(x)]dx \\ &= \frac{1}{4k+3}[P_{2k+2}(0) - P_{2k}(0)]\end{aligned} \tag{7.60}$$

Since:

$$P_{2k+2}(1) = P_{2k}(1) = 1 \tag{7.61}$$

$$P_{2k+2}(1) - P_{2k}(1) = 0 \tag{7.62}$$

Finally

$$\begin{aligned}A_{2k+1} &= \frac{2(2k+1)+1}{2}(u_1 - u_2)\frac{1}{4k+3}[P_{2k+2}(0) - P_{2k}(0)] \\ &= \frac{(u_1 - u_2)}{2}[P_{2k+2}(0) - P_{2k}(0)]\end{aligned} \tag{7.63}$$

Solution:

$$\begin{aligned}u(r,\theta) &= u_2 + v(r,\theta) = u_2 + \frac{1}{2}(u_1 - u_2) + \frac{(u_1 - u_2)}{2}\sum[P_{2k+2}(0) \\ &\quad - P_{2k}(0)]\left(\frac{r}{R}\right)^{2k+1} P_{2k+1}(\cos(\theta)) \\ &= \frac{1}{2}(u_1 + u_2) + \frac{(u_1 - u_2)}{2} \\ &\quad \times \sum[P_{2k+2}(0) - P_{2k}(0)]\left(\frac{r}{R}\right)^{2k+1} P_{2k+1}(\cos(\theta))\end{aligned} \tag{7.64}$$

7.4 Electric Potential inside a Spherical Sector

Find the electrostatic potential inside a metallic spherical sector whose curved surface is grounded, whereas its flat surfaces, which comprise an angle $\pi/4$ are kept at a difference of potential V_0 with respect to ground.

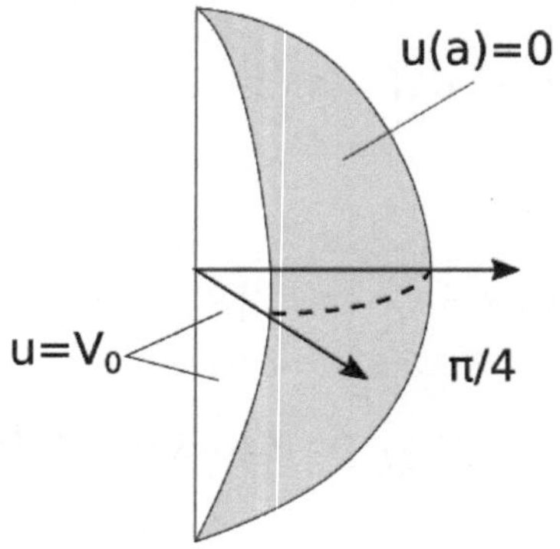

Figure 7.4

Mathematical formulation

$$\left\{\begin{array}{ll} \Delta v = 0 & (0 < r < a) \\ \dfrac{1}{r^2}\dfrac{\partial}{\partial r}\left[r^2\left(\dfrac{\partial v}{\partial r}\right)\right] + \dfrac{1}{r^2}\Delta_{\theta,\varphi} v = 0 & \\ v(r = R) = 0 & \\ v(\varphi = 0) = V_0 & \\ v\left(\varphi = \dfrac{\pi}{4}\right) = V_0 & \end{array}\right\} \tag{7.65}$$

We seek the solution to the equation in the form $u = v - V_0$ to have homogeneous boundary conditions in the azimuthal variable.

$$\left.\begin{cases} \Delta u = 0 & (0 < r < a) \\ \dfrac{1}{r^2}\dfrac{\partial}{\partial r}\left[r^2\left(\dfrac{\partial u}{\partial r}\right)\right] + \dfrac{1}{r^2}\Delta_{\theta,\varphi} u = 0 \\ u(r = a) = -V_0 \\ u(\varphi = 0) = 0 \\ u\left(\varphi = \dfrac{\pi}{4}\right) = 0 \end{cases}\right\} \tag{7.66}$$

We seek the solution of Laplace's equation by generating a solution from orthogonal functions in the angular variables.

$$\frac{\partial}{\partial r}\left[r^2\left(\frac{\partial u}{\partial r}\right)\right] + \Delta_{\theta,\varphi} u = 0 \tag{7.67}$$

Separating variables:

$$u(r, \theta, \varphi) = R(r) \cdot V(\theta, \varphi) \tag{7.68}$$

$$\frac{\frac{d}{dr}[r^2(\frac{dR}{dr})]}{R} = -\frac{\Delta_{\theta,\varphi} V}{V} = \lambda > 0 \tag{7.69}$$

Sturm–Liouville problem

We choose the positive sign for λ since we expect to obtain periodic solutions (eigenfunctions) for the angular variables.

We get to the problem:

$$\Delta_{\theta,\varphi} V + \lambda V = 0 \tag{7.70}$$

$$\frac{1}{\sin(\theta)}\frac{\partial}{\partial \theta}\left[\sin(\theta)\left(\frac{\partial V}{\partial \theta}\right)\right] + \frac{1}{\sin^2(\theta)}\frac{\partial^2 V}{\partial \varphi^2} + \lambda V = 0 \tag{7.71}$$

Once again we separate variables: $V = \Theta(\theta) \cdot \Phi(\varphi)$

$$\frac{\sin(\theta)\frac{d}{d\theta}[\sin(\theta)(\frac{d\Theta}{d\theta})] + \lambda \sin^2(\theta) \cdot \Theta(\theta)}{\Theta(\theta)} = -\frac{1}{\Phi(\varphi)}\frac{d^2\Phi}{d\varphi^2} = \nu > 0 \tag{7.72}$$

We choose the positive sign for ν because we expect to get a periodic solution for the φ angular variable.

We first find the solution for the whole sphere and then apply the boundary conditions specific to the azimuthal angle in the whole sphere.

$$\frac{d^2\Phi}{d\varphi^2} + \nu\Phi(\varphi) = 0 \tag{7.73}$$

$$\Phi(\varphi) = \Phi(\varphi + 2\pi) \tag{7.74}$$

The solution for the part that depends on the azimuthal angle is well known:

$$\Phi(\varphi) = A\cos(m\varphi) + B\sin(m\varphi) \tag{7.75}$$

$$\nu = m^2 \tag{7.76}$$

Now that we have $\nu = m^2$, we can find the solution for the $\Theta(\theta)$ function.

$$\sin(\theta)\frac{d}{d\theta}\left[\sin(\theta)\left(\frac{d\Theta}{d\theta}\right)\right] + \lambda\sin^2(\theta)\cdot\Theta(\theta) = m^2\Theta \tag{7.77}$$

$$\sin(\theta)\frac{d}{d\theta}\left[\sin(\theta)\left(\frac{d\Theta}{d\theta}\right)\right] + [\lambda\sin^2(\theta) - m^2]\Theta(\theta) = 0 \tag{7.78}$$

Its solution are Legendre polynomials.

$$\Theta(\theta) = P_n^{(m)}(\cos(\theta)) \tag{7.79}$$

With eigenvalues: $\lambda = n(n+1)$. Now, since we have $\lambda = n(n+1)$ we can find the radial function:

$$\frac{d}{dr}\left[r^2\left(\frac{dR}{dr}\right)\right] - n(n+1)R = 0 \tag{7.80}$$

$$r^2\left(\frac{d^2R}{dr^2}\right) + 2r\left(\frac{dR}{dr}\right) - n(n+1)R = 0 \tag{7.81}$$

We seek the solution as: $R(r) = r^\alpha$

$$r^2\alpha(\alpha-1)r^{\alpha-2} + 2r\alpha r^{\alpha-1} - n(n+1)r^\alpha = 0 \tag{7.82}$$

$$\alpha^2 + \alpha - n(n+1) = 0 \tag{7.83}$$

$$\alpha_1 = n \tag{7.84}$$

$$\alpha_2 = -n-1 \tag{7.85}$$

So that the radial solution in general is: $R(r) = r^n + Dr^{-n-1}$

General solution

For the general solution of the problem only the values of m will be integer for the whole sphere.

$$u(r,\theta,\varphi)=\sum_{n\geq m}^{\infty}\sum_{m=0}^{\infty}[r^n+D_{nm}r^{-n-1}]P_n^{(m)}(\cos(\theta))$$
$$\times[A_{nm}\cos(m\varphi)+B_{nm}\sin(m\varphi)] \quad (7.86)$$

Since the solution should be finite at $r=0\rightarrow D_{nm}=0$

$$u(r,\theta,\varphi)=\sum_{n\geq m}^{\infty}\sum_{m=0}^{\infty}[r^n]P_n^{(m)}(\cos(\theta))\cdot[A_{nm}\cos(m\varphi)+B_{nm}\sin(m\varphi)] \quad (7.87)$$

In the next step we can already apply the boundary conditions specific to the $\pi/4$ sector in the azimuthal angle by searching specific values of m.

$$u(\varphi=0)=0\rightarrow A_{nm}=0 \quad (7.88)$$

$$u\left(\varphi=\frac{\pi}{4}\right)=0 \quad (7.89)$$

$$\sin\left(m\frac{\pi}{4}\right)=0\rightarrow\frac{\pi}{4}=k\pi \qquad (k=0,1,2,\ldots=\text{integer});\ m=4k \quad (7.90)$$

Then:

$$u(r,\theta,\varphi)=\sum_{n\geq 4k}^{\infty}\sum_{k=1}^{\infty}B_{nk}[r^n]P_n^{(4k)}(\cos(\theta))\sin(4k\varphi) \quad (7.91)$$

Imposing the first boundary condition: $u(r=a)=-V_0$

$$-V_0=u(r,\theta,\varphi)=\sum_{n\geq 4k}^{\infty}\sum_{k=1}^{\infty}B_{nk}[a^n]P_n^{(4k)}(\cos(\theta))\sin(4k\varphi) \quad (7.92)$$

Final solution

Using now the orthogonality of $P_n^{(4k)}(\cos(\theta))$; and of $\sin(4k\varphi)$ we will get the B_{nk} coefficients. We multiply by $P_n^{(4k)}(\cos(\theta))$ and $\sin(4k\varphi)$ and integrate (7.92) between $\int_0^{\pi}\int_0^{\frac{\pi}{4}}\sin(\theta)d\theta d\varphi$:

$$-V_0\int_0^{\pi}\int_0^{\frac{\pi}{4}}P_n^{(4k)}(\cos(\theta))\sin(\theta)\sin 4k\varphi d\theta d\varphi$$
$$=B_{nk}\left\|P_n^{(4k)}(\cos(\theta))\right\|^2\cdot\|\sin(4k\varphi)\|^2 \quad (7.93)$$

$$B_{nk} = \frac{-V_0 \int\limits_0^{\frac{\pi}{4}} \sin(4k\varphi)d\varphi \int\limits_0^{\pi} P_n^{(4k)}(\cos(\theta)) \sin\theta d\theta}{\left\| P_n^{(4k)}(\cos(\theta)) \right\|^2 \cdot \|\sin(4k\varphi)\|^2}$$

$$= \frac{\frac{V_0}{4k}[(-1)^k - 1] \times \int\limits_0^{\pi} P_n^{(4k)}(\cos(\theta)) \sin\theta d\theta}{\left\| P_n^{(4k)}(\cos(\theta)) \right\|^2 \cdot \|\sin(4k\varphi)\|^2} \quad (7.94)$$

Final solution:

$$v(r, \theta, \varphi) = V_0 + \sum_{n \geq 4k}^{\infty} \sum_{k=1}^{\infty} B_{nk}[r^n] P_n^{(4k)}(\cos(\theta)) \sin(4k\varphi) \quad (7.95)$$

Additional note. It is possible to solve the problem by rotating the portion so that one of its flat surfaces lays horizontally. In this case the length of the azimuthal variable is between zero and π, whereas the polar angle will be comprised between $\pi/4$ and $\pi/2$.

7.5 Electric Potential of a Metallic Sphere inside a Homogeneous Electric Field

A uniform electric field $\vec{E}$ occupies all space, directed in the z direction, so that the corresponding electric potential is $u = -Ez$. A metallic sphere of radius R is placed in this field. Calculate the equilibrium distribution of electric potential outside the sphere supposing that the electric potential inside the surface of the sphere is zero.

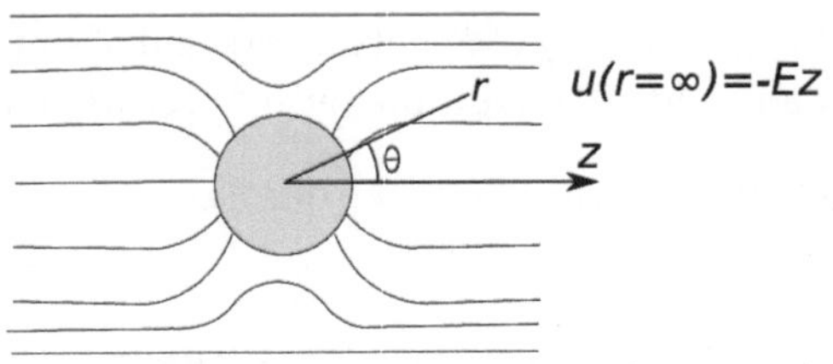

Figure 7.5

Mathematical formulation

$$\left.\begin{cases} \Delta u(r,\theta,\varphi) = 0 & (R < r < \infty) \\ u(r=\infty,\theta,\varphi) = -Ez = -Er\cos(\theta) & \\ u(r=R,\theta,\varphi) = 0 & \end{cases}\right\} \tag{7.96}$$

The general solution of the Laplace problem in spheric coordinates (problem 7.1) has the form:

$$u(r,\theta,\varphi) = \sum_{n\geq m}^{\infty}\sum_{m=0}^{\infty}[r^n + D_{nm}r^{-n-1}]P_n^{(m)}(\cos(\theta)) \times[A_{nm}\cos(m\varphi) + B_{nm}\sin(m\varphi)] \tag{7.97}$$

General solution

Both boundary conditions don't have any azimuthal dependence. In this way the solution will be presented as an expansion in Legendre functions. This can be demonstrated explicitly by integrating the boundary conditions with the orthogonal azimuthal eigenfunctions and showing that the only term that remains corresponds to the index $m = 0$.

$$u(r,\theta) = \sum_{n=0}^{\infty}[A_n r^n + B_n r^{-n-1}]P_n(\cos(\theta)) \tag{7.98}$$

We impose the first boundary condition:

$$u(r=\infty,\theta) = -Er\cos(\theta) = -ErP_1(\cos(\theta)) = \sum_{n=0}^{\infty} A_n r^n P_n(\cos\theta) \tag{7.99}$$

We impose the orthogonality (that is, multiplying by $P_k(\cos(\theta))\sin(\theta)$ and integrating from 0 to π). It can be seen that the term with $n = k = 1$ remains, and corresponds to $A_1 = -E$. Finally we apply the second boundary condition:

$$u(r=R,\theta) = 0 = B_0 + (-ER + B_1R^{-2})P_1(\cos\theta) + \sum_{n=2}^{\infty} B_n r^{-n-1}P_n(\cos\theta) \tag{7.100}$$

Final solution

Imposing the orthogonality (that is, multiplying by $P_k(\cos(\theta))\sin(\theta)$ and integrating from 0 to π) it can be shown that only the term with $n = k = 1$ remains. As a consequence we have $(-E\,R + B_1 R^{-2}) = 0$, which gives $B_1 = E\,R^3$

Then the final result will be:

$$u(r = R, \theta) = E[-r + R^3/r^2]P_1(\cos(\theta)) \tag{7.101}$$

7.6 Case Study: Variation of Temperature in a Sphere Quadrant with Non-Homogeneous Boundaries

A quadrant of a sphere is initially at thermal equilibrium at zero temperature. Find the distribution of temperature as a function of time if stating at $t = 0$ the temperature in the flat face at $\varphi = 0$ is brought into contact with a thermal reservoir at $T = T_0$ and the other flat face at $\varphi = \pi/2$ is fixed at another temperature $T = T_1$, while the temperature in the curved boundary ($r = R$) is set according to:

$$T_0 + \frac{T_1 - T_0}{\pi/2}\varphi + \frac{\delta(\theta - \pi/2)}{\sin(\theta)}\varphi \tag{7.102}$$

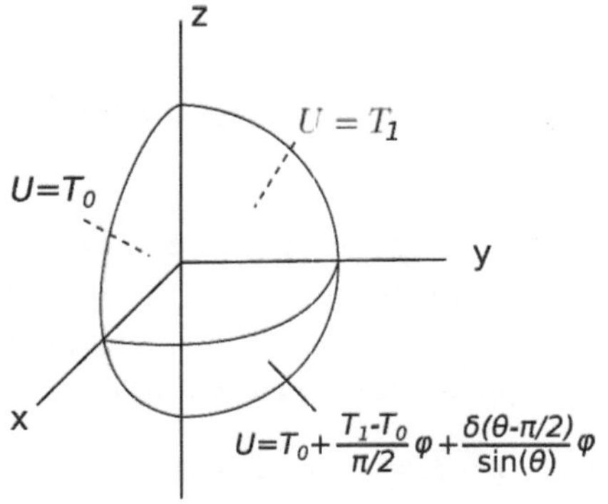

Figure 7.6

Mathematical formulation

We will find a solution as the variation of temperature with respect to the equilibrium temperature in the initial instant (U). The problem to be solved is:

$$\left\{\begin{array}{l} \dfrac{\partial U}{\partial t}-c\Delta U=0 \\ U(r,\theta,0)=T_0 \\ U(r,\theta,\pi/2)=T_1 \\ U(R,\theta,\varphi)=T_0+\dfrac{T_1-T_0}{\pi/2}\varphi+\dfrac{\delta(\theta-\pi/2)}{\sin(\theta)}\varphi \\ U(r,\theta\varphi,t=0)=0 \end{array}\right\} \qquad (7.103)$$

We will split the problem in two: one stationary with inhomogeneous boundary conditions and another with temporal evolution but with homogeneous boundary conditions:

$$U(r,\theta,\varphi,t)=T(r,\theta,\varphi,t)+E(r,\theta,\varphi), \qquad (7.104)$$

so that:

$$\left\{\begin{array}{l} \Delta E=0 \\ E(r,\theta,0)=T_0 \\ E(r,\theta,\pi/2)=T_1 \\ E(R,\theta,\varphi)=T_0+\dfrac{T_1-T_0}{\pi/2}\varphi+\dfrac{\delta(\theta-\pi/2)}{\sin(\theta)}\varphi \end{array}\right\} \qquad (7.105)$$

The boundary conditions of the previous stationary part leave the boundary conditions for the temporal part as homogeneous, but the initial condition changes:

$$\left\{\begin{array}{l} \dfrac{\partial T}{\partial t}-c\Delta T=0 \\ T(r,\theta,0)=0 \\ T(r,\theta,\pi/2)=0 \\ T(R,\theta,\varphi)=0 \\ T(r,\theta,\varphi,t=0)=-E(r,\theta,\varphi) \end{array}\right\} \qquad (7.106)$$

First we will solve the stationary part so that we can use the new initial condition in the transient part.

Solution of the stationary part

The first thing that must be done to solve the stationary part is to homogenize the boundaries in φ to be able to expand the solution of this coordinate in orthogonal eigenfunctions. To do that we will write the stationary part as:

$$E(r,\theta,\varphi) = W(r,\theta,\varphi) + T_0 + \frac{T_1 - T_0}{\pi/2}\varphi \tag{7.107}$$

In this fashion W has homogeneous boundaries in the φ coordinate and satisfies Laplace's equation:

$$\left\{\begin{array}{l} \Delta W = 0 \\ W(r,\theta,0) = 0 \\ W(r,\theta,\pi/2) = 0 \\ W(R,\theta,\varphi) = \dfrac{\delta(\theta-\pi/2)}{\sin(\theta)}\varphi \end{array}\right\} \tag{7.108}$$

It is not necessary to have homogeneous boundaries in the radial part, since that part of the solution will not be expanded in orthogonal eigenfunctions.

Sturm–Liouville problem

To solve W we separate variables:

$$W(r,\theta,\varphi) = \mathcal{R}(r)\cdot Y(\theta,\varphi) \tag{7.109}$$

Inserting this into Laplace's equation we arrive at:

$$\frac{\frac{d}{dr}\left(r^2\frac{d\mathcal{R}}{dr}\right)}{\mathcal{R}} = -\frac{\frac{1}{\sin(\theta)}\frac{\partial}{\partial\theta}\left(\sin(\theta)\frac{\partial Y}{\partial\theta}\right) + \frac{1}{\sin^2(\theta)}\frac{\partial^2 Y}{\partial\varphi^2}}{Y} = \lambda \tag{7.110}$$

With this we obtain the following two equations:

$$\left\{\begin{array}{l} r^2\dfrac{d^2\mathcal{R}}{dr^2} + 2r\dfrac{d\mathcal{R}}{dr} - \lambda\mathcal{R} = 0 \\ \dfrac{1}{\sin(\theta)}\dfrac{\partial}{\partial\theta}\left(\sin(\theta)\dfrac{\partial Y}{\partial\theta}\right) + \dfrac{1}{\sin^2(\theta)}\dfrac{\partial^2 Y}{\partial\varphi^2} + \lambda Y = 0 \end{array}\right\} \tag{7.111}$$

First we solve the equation for the angular variable Y. To do that we separate variables once again: $Y(\theta, \varphi) = \vartheta(\theta) \cdot \phi(\varphi)$. With this we get to equation:

$$\sin(\theta)\frac{\frac{d}{d\theta}\left(\sin(\theta)\frac{d\vartheta}{d\theta}\right)}{\vartheta} + \lambda\sin^2(\theta) = -\frac{\phi''}{\phi} = \mu. \tag{7.112}$$

We first solve the part for the φ coordinate, taking into account that $\phi(0) = \phi(\pi/2) = 0$ must be satisfied. With this we easily get that:

$$\phi = \sin(2m\phi) \;\; m = 1, 2, 3, \ldots \tag{7.113}$$

In this way we arrive at the value of the separation constant: $\mu = (2m)^2$. With this value we can solve the part of the θ coordinate. For that it will be necessary to perform the change of variable: $x = \cos(\theta)$. With this we get the equation:

$$(1-x^2)\frac{d^2\vartheta}{dx^2} - 2x\frac{d\vartheta}{dx} + \left(\lambda - \frac{(2m)}{1-x^2}\right)\vartheta = 0 \tag{7.114}$$

and its solution are Legendre polynomials:

$$\vartheta_n^{(2m)}(\theta) = P_n^{(2m)}(\cos(\theta)) \qquad \lambda = n(n+1) \text{ and } n \geq 2m \tag{7.115}$$

Knowing the values of λ allows to solve the equation for the radial part. Because of the form of the equation for $\mathcal{R}$ it is easy to guess that the solutions are powers of r, so that the solution of the radial part is:

$$\mathcal{R}(r) = Ar^n + Br^{-(n+1)} \tag{7.116}$$

Since the solution cannot diverge at $r = 0$ we impose $B = 0$. With all this we can now write the general solution of the stationary part:

$$W(r, \theta, \varphi) = \sum_{n=0}^{\infty}\sum_{m=1}^{n/2} A_{n,m} r^n \sin(2m\varphi) P_n^{(2m)}(\cos(\theta)) \tag{7.117}$$

Imposing the radial boundary condition:

$$W(R, \theta, \varphi) = \frac{\delta(\theta - \pi/2)}{\sin(\theta)}\varphi \tag{7.118}$$

We find the value of the coefficients $A_{n,m}$:

$$\frac{\delta(\theta - \pi/2)}{\sin(\theta)}\varphi = \sum_{n=0}^{\infty}\sum_{m=1}^{n/2} A_{n,m} R^n \sin(2m\varphi) P_n^{(2m)}(\cos(\theta)) \tag{7.119}$$

Considering that:

$$\int_{\theta=0}^{\pi} P_{n'}^{(2m)}(\cos(\theta))P_{n}^{(2m)}(\cos(\theta))\sin(\theta)d\theta = \frac{2}{2n+1}\frac{(n+2m)!}{(n-2m)!}\delta_{n.n'} \tag{7.120}$$

$$\int_{\varphi=0}^{\pi/2} \sin(2m\varphi)\sin(2m'\varphi)d\varphi = \frac{\pi}{4}\delta_{m,m'} \tag{7.121}$$

$$\int_{\varphi=0}^{\pi/2} \varphi\sin(2m\varphi)d\varphi = -\frac{\pi}{4m}\cos(m\pi) \tag{7.122}$$

Doing the respective integrals of the boundary conditions we can see that the value of the coefficients is:

$$A_{n,m} = \frac{1}{2}\frac{(-1)^{m+1}(2n+1)(n-2m)!P_{n}^{(2m)}(0)}{mR^{n}(n+2m)!} \tag{7.123}$$

Remembering that $P_{n}^{(l)}(x)$ is an even function if $n+l$ is even and it is odd if $n+l$ is odd, it is easy to realize that $A_{n,m} = 0$ if n is an odd number and $A_{n,m} \neq 0$ if n is an even number. We can now write the solution of the stationary part E. To do that we take $n = 2k$. In this way the solution is:

$$E(r,\theta,\varphi) = \sum_{k=0}^{\infty}\sum_{m=1}^{k}\frac{1}{2}\frac{(-1)^{m+1}(4k+1)(2k-2m)!P_{2k}^{(2m)}(0)}{mR^{2k}(2k+2m)!}r^{2k}$$
$$\sin(2m\varphi)P_{2k}^{(2m)}(\cos(\theta)) + T_0 + \frac{T_1-T_0}{\pi/2}\varphi \tag{7.124}$$

Solution of the temporal part

Knowing the stationary solution we can solve the temporal part, now that we explicitly know the initial condition for the transient part. We start from equation:

$$\frac{\partial T}{\partial t} - c\Delta T = 0 \tag{7.125}$$

We separate variables:

$$T(r, \theta, \varphi, t) = \tau(t) \cdot M(r, \theta, \varphi) \tag{7.126}$$

Inserting that into the equation we arrive at:

$$\frac{\tau'}{c\tau} = \frac{\Delta M}{M} = -\lambda \tag{7.127}$$

We use the negative sign in the constant of separation of variables so that the solution doesn't diverge at infinite times. From the previous equation we arrive at two equations, one for the temporal part an another for the spatial part:

$$\begin{Bmatrix} \tau' + \lambda c\tau = 0 \\ \Delta M + \lambda M = 0 \end{Bmatrix} \tag{7.128}$$

To solve the spatial part $M(r, \theta, \varphi)$ we do:

$$\Delta = \Delta_r + \frac{1}{r^2}\Delta_{\theta,\varphi} \tag{7.129}$$

Separating variables: $M = R(r) \cdot Y(\theta, \varphi)$, and imposing that the angular part satisfies a Sturm–Liouville problem, since it has homogeneous boundaries:

$$\Delta_{\theta,\varphi}Y + \nu Y = 0 \tag{7.130}$$

We separate variables once again:

$$Y(\theta, \varphi) = \vartheta(\theta) \cdot \phi(\varphi) \tag{7.131}$$

We impose that the part with ϕ satisfies a Sturm–Liouville problem, since it has homogeneous boundaries:

$$\phi'' + \mu\phi = 0 \tag{7.132}$$

With this and the boundary conditions we arrive at:

$$\phi = \sin(2m\varphi) \qquad m = 1, 2, 3, \ldots \tag{7.133}$$

Inserting this into the equation for $Y(\theta, \phi)$, just like we did previously, we get:

$$\vartheta_l^{(2m)}(\theta) = P_l^{(2m)}(\cos(\theta)) \qquad \nu = l(l+1); \quad l \geq 2m \tag{7.134}$$

Inserting this in the equation for the spatial part M and imposing that the solution doesn't diverge at $r = 0$ we find that the solution for the radial part is:

$$\mathcal{R}_l(r) = \frac{1}{\sqrt{r}}J_{l+1/2}(\sqrt{\lambda}r) \tag{7.135}$$

Solution of the temporal part $\tau(t)$

Knowing the value of the $\lambda_{i,l}$ we can solve the equation for the temporal part, whose solution ends up being:

$$\tau(t) = e^{-\lambda_{i,l}ct} \tag{7.136}$$

General solution

Knowing the solutions for the spatial and temporal part we have the general solution:

$$T(r, \theta, \varphi, t) = \sum_{i=0}^{\infty}\sum_{l=0}^{\infty}\sum_{m=1}^{l/2} A_{i,l,m}e^{-\lambda_{i,l}ct} \sin(2m\varphi)P_l^{(2m)}(\cos(\theta))\frac{1}{\sqrt{r}}J_{l+1/2}(\sqrt{\lambda}r) \tag{7.137}$$

Final solution

Finally the value of the $A_{i,l,m}$ coefficients will be obtained with the initial condition and remembering that:

$$\int_{\theta=0}^{\pi}\int_{\varphi=0}^{\pi/2}\int_{r=0}^{R} M_{i,l,m}M_{i',l',m'}r^2\sin(\theta)drd\theta d\varphi = E_{i,l,m}\delta_{i,i'}\delta_{l,l'}\delta_{m,m'} \tag{7.138}$$

with

$$M_{i,l,m} = \sin(2m\varphi)P_l^{(2m)}(\cos(\theta))\frac{1}{\sqrt{r}}J_{l+1/2}\left(\sqrt{\lambda_{i,l}}r\right) \tag{7.139}$$

Then the value of the coefficients is:

$$A_{i,l,m} = -\frac{1}{E_{i,l,m}}\int_{\theta=0}^{\pi}\int_{\varphi=0}^{\pi/2}\int_{r=0}^{R} E(r, \theta, \varphi)M_{i,l,m}r^2\sin(\theta)drd\theta d\varphi \tag{7.140}$$

Remembering that:

$$E(r, \theta, \varphi) = \sum_{k=0}^{\infty}\sum_{m=1}^{k}\frac{1}{2}\frac{(-1)^{m+1}(4k+1)(2k-2m)!P_{2k}^{(2m)}(0)}{mR^{2k}(2k+2m)!}r^{2k} \sin(2m\varphi)P_{2k}^{(2m)}(\cos(\theta)) + T_0 + \frac{T_1 - T_0}{\pi/2}\varphi \tag{7.141}$$

the numeric value of the constants $A_{i,l,m}$ is found evaluating the integrals. Finally the expression of the coefficients can be shrunken. We need to consider two cases. For odd values of l:

$$A_{i,l,m} = \frac{\pi(-1)^{m+1}}{4m}\frac{T_1 - T_o}{\pi/2}\int_{\theta=0}^{\pi}\int_{r=0}^{R} r^2$$

$$P_l^{(2m)}(\cos(\theta))\frac{1}{\sqrt{r}}J_{l+1/2}\left(\sqrt{\lambda_{i,l}}r\right)\sin(\theta)drd\theta \quad (7.142)$$

For even values ($l = 2k, k = 1, 2, 3\ldots$):

$$A_{i,l,m} = \frac{\pi(-1)^{m+1}P_{2k}^{(2m)}(0)}{4mR^{2k}}\int_{r=0}^{R} r^2\frac{1}{\sqrt{r}}J_{l+1/2}\left(\sqrt{\lambda_{i,l}}r\right)dr +$$

$$+\frac{\pi(-1)^{m+1}}{4m}\frac{T_1 - T_o}{\pi/2}\int_{\theta=0}^{\pi}\int_{r=0}^{R} r^2 P_l^{(2m)}(\cos(\theta))$$

$$\frac{1}{\sqrt{r}}J_{l+1/2}\left(\sqrt{\lambda_{i,l}}r\right)\sin(\theta)drd\theta \quad (7.143)$$

7.7 Case Study: Stationary Distribution of Temperature in a Sphere with Heat Sources

Find the stationary distribution of temperature inside a sphere or radius R with internal heat sources distributed with a density equal to $f(r, \theta, \varphi)$. The surface is in contact with a thermal reservoir at temperature $T = g(\theta, \varphi)$.

Apply the obtained result to the following particular case: Consider a sphere in contact with two thermal reservoirs at different temperatures T_1 $(0 < \varphi < \pi)$ and T_2 $(\pi < \varphi < 2\pi)$. Inside this sphere (with thermal conductivity k, heat capacity C and density ρ) along its equator (symmetrically with respect to the center) a very thin, ring-like heater of radius $R/2$ and which radiates a power W is placed.

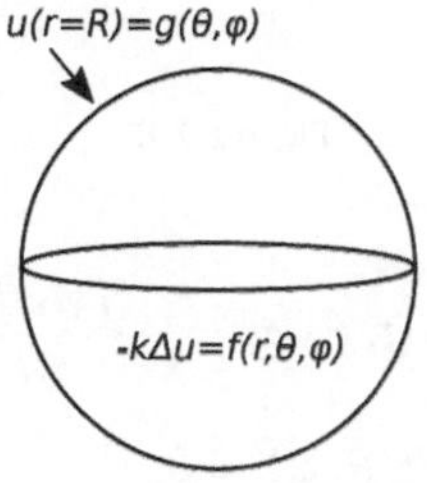

Figure 7.7

Mathematical formulation

$$\left\{\begin{array}{ll} \dfrac{\partial u}{\partial t} - a^2 \Delta u = \dfrac{f(r,\theta,\varphi)}{\rho C} & \\ u(R,\theta,\varphi) = 0 & (t<0) \\ u(R,\theta,\varphi) = g(\theta,\varphi) & (t>0) \\ a^2 = \dfrac{k}{\rho C} & \end{array}\right\} \tag{7.144}$$

General mathematical formulation for the case of a stationary process in thermal equilibrium with boundary conditions:

$$\left\{\begin{array}{l} \Delta u(r,\theta,\varphi) = -\dfrac{f(r,\theta,\varphi)}{k} \\ u(R,\theta,\varphi) = g(\theta,\varphi) \end{array}\right\} \tag{7.145}$$

To solve the problem we will decompose it in two:

$$u = u_1 + u_2 \tag{7.146}$$

where u_1 is the solution of the inhomogeneous problem and u_2 that of the homogeneous problem with inhomogeneous boundary condition (see description of the decomposition in the next figure) Problem (1):

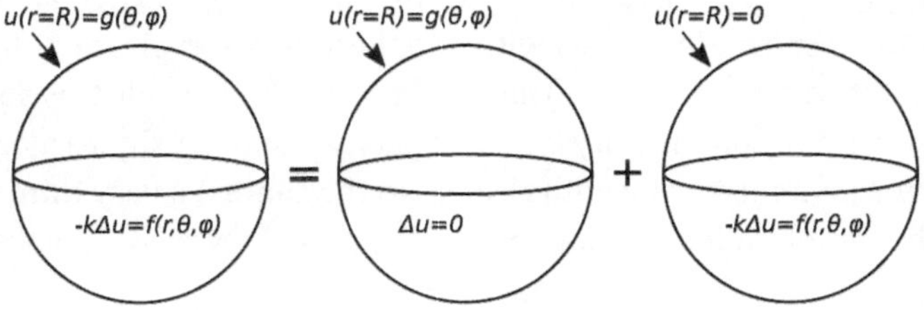

Figure 7.8

$$\left\{\begin{array}{l} \Delta u_1(r,\theta,\varphi) = -\dfrac{f(r,\theta,\varphi)}{k} \\ u_1(R,\theta,\varphi) = 0 \end{array}\right\} \tag{7.147}$$

Problem (2)

$$\left\{\begin{array}{l}\Delta u_2(r,\theta,\varphi)=0\\ \\ u_2(R,\theta,\varphi)=g(\theta,\varphi)\end{array}\right\} \tag{7.148}$$

General solution

Problem (2) is solved by expanding the solution in sums of spherical harmonics:

$$\begin{aligned}u_2(r,\theta,\varphi)&=\sum_{nm}A_{nm}\left(\frac{r}{R}\right)^n P_n^{(m)}(\cos(\theta))e^{im\varphi}\\ &=\sum_{nm}A_{nm}\left(\frac{r}{R}\right)^n Y_{n,m}(\theta,\varphi)\end{aligned} \tag{7.149}$$

Using the orthogonality of the spherical harmonics we find the A_{nm} coefficients.

$$g(\theta,\varphi)=\sum_{nm}A_{nm}Y_{n,m}(\theta,\varphi) \tag{7.150}$$

$$A_{nm}=\frac{\int\limits_0^{2\pi}\int\limits_0^{\pi}g(\theta,\varphi)Y_{n,m}(\theta,\varphi)\sin(\theta)d\theta d\varphi}{\left\|Y_{n,m}(\theta,\varphi)\right\|^2} \tag{7.151}$$

Problem (1) is solved by expanding the solution in terms of orthogonal eigenfunctions that solve the Sturm–Liouville problem.

$$\left\{\begin{array}{l}\Delta v(r,\theta,\varphi)+\lambda v=0\\ (0<r<R);(0<\theta<\pi);(0<\varphi<2\pi)\\ v(R,\theta,\varphi)=0\end{array}\right\} \tag{7.152}$$

The form of the corresponding eigenfunctions is:

$$v_{nmk}(r,\theta,\varphi)=j_n(\sqrt{\lambda_{nk}}r)P_n^{(m)}(\cos(\theta))e^{im\varphi}=j_n(\sqrt{\lambda_{nk}}r)Y_{n,m}(\theta,\varphi) \tag{7.153}$$

being $j_n(\sqrt{\lambda_{nk}}r)$ spherical Bessel functions.

$$\left\{j_l(z)=\sqrt{\frac{\pi}{2z}}J_{l+\frac{1}{2}}(z)\right\} \tag{7.154}$$

General solution

Then, expanding the solution in terms of the summation of $v_{nmk}(r, \theta, \varphi)$

$$u_1(r, \theta, \varphi) = \sum_{nmk} B_{nmk} \times j_n(\sqrt{\lambda_{nk}}r) P_n^{(m)}(\cos\theta) e^{im\varphi}$$
$$= \sum_{nmk} B_{nmk} \times j_n(\sqrt{\lambda_{nk}}r) Y_{n,m}(\theta, \varphi) \quad (7.155)$$

Final solution

Using the orthogonality we can solve the general problem:

$$-\sum_{nmk} \lambda_{nk} B_{nmk} \times j_n(\sqrt{\lambda_{nk}}r) Y_{n,m}(\theta, \varphi) = -\frac{f(r, \theta, \varphi)}{k} \quad (7.156)$$

Using the orthogonality of the orthogonal functions $v_{nmk}(r, \theta, \varphi)$, we find the B_{nmk} coefficients:

$$B_{nmk} = \frac{\int_0^R \int_0^{2\pi} \int_0^{\pi} f(r, \theta, \varphi) \times j_n(\sqrt{\lambda_{nk}}r) Y_{n,m}(\theta, \varphi) r^2 \sin(\theta) dr d\theta d\varphi}{k\lambda_{nk} \left\| v_{nmk}(r, \theta, \varphi) \right\|^2} \quad (7.157)$$

To solve the proposed case we need to replace in the general formulas the following functions for the given boundary conditions and the density of heat sources:

$$g(\theta, \varphi) = \begin{Bmatrix} T_1 \ (0 < \varphi < \pi) \\ T_2 \ (\pi < \varphi < 2\pi) \end{Bmatrix} \quad (7.158)$$

$$f(r, \theta) = \frac{C}{r^2 \sin(\theta)} \delta\left(r - \frac{R}{2}\right) \delta\left(\theta - \frac{\pi}{2}\right) \quad (7.159)$$

(due to the azimuthal symmetry of the heater). Normalization to find the C constant:

$$\int_0^R \int_0^{2\pi} \int_0^{\pi} f(r, \theta) r^2 \sin(\theta) d\theta d\varphi = C \cdot 2\pi = W \quad (7.160)$$

Then $C = \frac{W}{2\pi}$. We use the relations of the general solution previously found:

$$B_{nmk} = \frac{\int\limits_0^R\int\limits_0^{2\pi}\int\limits_0^{\pi} \frac{W}{2\pi r^2 \sin(\theta)}\delta(r-\frac{R}{2})\delta(\theta-\frac{\pi}{2})\, j_n(\sqrt{\lambda_{nk}}r)Y_{n,m}(\theta,\varphi)r^2\sin(\theta)drd\theta d\varphi}{k\lambda_{nk}\,\|v_{nmk}(r,\theta,\varphi)\|^2}$$

$$= W\frac{j_n(\sqrt{\lambda_{nk}}\frac{R}{2})P_n^0(0)}{k\lambda_{nk}\,\|v_{nmk}(r,\theta,\varphi)\|^2} \tag{7.161}$$

Only the term with index $m = 0$ remains due to the azimuthal symmetry. Furthermore, only the terms with even values of n ($n = 2l, l = 1, 2, \ldots$) persists, since $P_{2l+1}^0(0) = 0$

$$B_{nmk} = W\frac{j_n(\sqrt{\lambda_{nk}}\frac{R}{2})P_n(0)}{k\lambda_{nk}\,\|v_{nmk}(r,\theta,\varphi)\|^2} \tag{7.162}$$

Particular solution for the part of the solution described with the Laplace equation:

$$A_{nm} = \frac{\int\limits_0^{2\pi}\int\limits_0^{\pi} g(\theta,\varphi)Y_{n,m}(\theta,\varphi)\sin(\theta)d\theta d\varphi}{\left\|Y_{n,m}(\theta,\varphi)\right\|^2}$$

$$= \frac{\int\limits_0^{\pi} P_n^m(\theta)\sin(\theta)d\theta\{\int\limits_0^{\pi} T_1 e^{(im\varphi)}d\varphi + \int\limits_{\pi}^{2\pi} T_2 e^{(im\varphi)}d\varphi\}}{\left\|Y_{n,m}(\theta,\varphi)\right\|^2} =$$

$$= \frac{[T_1\frac{e^{(im\pi)}-1}{im} + T_2\frac{1-e^{(im\pi)}}{im}]}{\left\|Y_{n,m}(\theta,\varphi)\right\|^2}\int\limits_0^{\pi} P_n^m(\theta)\sin(\theta)d\theta =$$

$$= \frac{1-e^{(im\pi)}}{im}\frac{[T_1 - T_2]}{\left\|Y_{n,m}(\theta,\varphi)\right\|^2}\int\limits_{-1}^{1} P_n^m(x)dx \tag{7.163}$$

Note 1: In this case the A_{nm} coefficients will be complex numbers since we have used the expansion of the solution with complex azimuthal functions.

Note 2: Furthermore, due to the properties of Legendre polynomials only the coefficients with even values of $n + m$ remain, since for odd values of $n + m$ the function $P_n^m(x)$ is asymmetric.

7.8 Gas in Two Semi-Spheres

A gas is enclosed in: a **(a)** a spheric; **(b)** semi-spheric container. This container is divided in two equal halves and the densities of the gas in each of them are $\rho_0+\rho_1$ and $\rho_0-\rho_1$ ($\rho_1 << \rho_0$). At a given time the division disappears. Find the variation in the gas density afterwards considering that the radius is R and the speed of sound is $c = 1$.

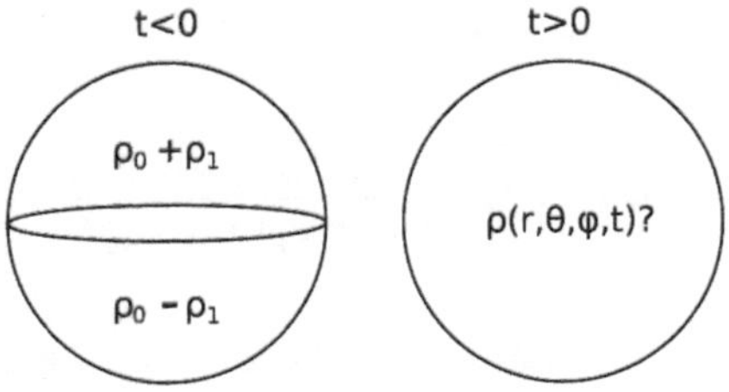

Figure 7.9

Part (a) Mathematical formulation

We will suppose that the plane separating both gases is at $\theta = \frac{\pi}{2}$. The problem will be solved for the $u = \rho - \rho_0$ variable.

General solution

The general solution for the variations from the equilibrium pressure (ρ_0) is:

$$u(r, \theta, \varphi, t) = \sum [A_{klm} \sin(\sqrt{\lambda_{kl}}t) + B_{klm} \cos(\sqrt{\lambda_{kl}}t)] j_l(\sqrt{\lambda_{kl}}r) P_l^{(m)}(\cos(\theta))e^{im\varphi} \tag{7.164}$$

where $j_l(\sqrt{\lambda_{kl}}r)$– are the spherical Bessel functions with boundary conditions $\left(\frac{\partial u}{\partial \vec{\mathbf{n}}} = 0\right)$ (being $\vec{\mathbf{n}}$ the direction normal to the surface).

Final solution

We impose the initial conditions. From the null initial velocity we get $A_{klm} = 0$. From the initial displacement:

$$u(r,\theta,\varphi,0)=\begin{Bmatrix}\rho_1 & \left(\theta<\frac{\pi}{2}\right)\\ -\rho_1 & \left(\theta>\frac{\pi}{2}\right)\end{Bmatrix}$$
$$=\sum B_{klm}j_l(\sqrt{\lambda_{kl}}r)P_l^{(m)}(\cos(\theta))e^{im\varphi} \quad (7.165)$$

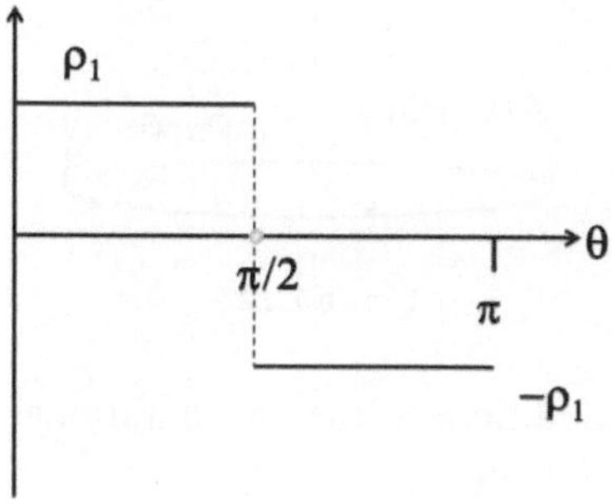

Figure 7.10

Due to the symmetry of the problem in the azimuthal angle it is clear that only exist terms with $m = 0$ in the sum. We use the orthogonality of the spherical Bessel functions and the Legendre polynomials to find the B_{kl} coefficients:

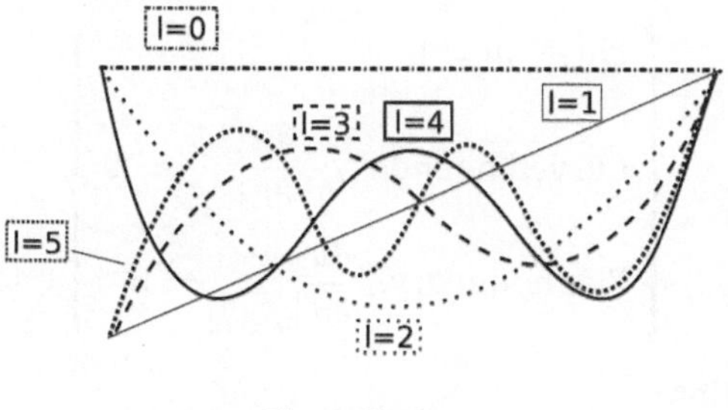

Figure 7.11

$$B_{kl}=\rho_1\int_0^R j_l(\sqrt{\lambda_{kl}}r)r^2dr$$
$$\times\frac{\int_0^{\pi/2}P_l(\cos(\theta))\sin(\theta)d\theta-\int_{\pi/2}^{\pi}P_l(\cos(\theta))\sin(\theta)d\theta}{\left\|j_l(\sqrt{\lambda_{kl}}r)\right\|^2\|P_l(\cos(\theta))\|^2} \quad (7.166)$$

The subtraction of the integrals is finite only for odd values of l. The physical reason is the anti-symmetry of the initial conditions with respect to the plane $\theta = \frac{\pi}{2}$.

Part (b)

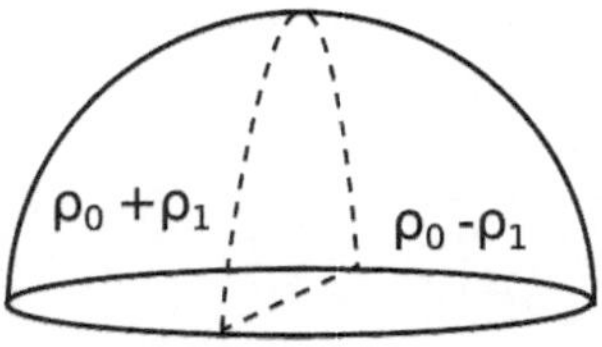

Figure 7.12

Now the initial conditions create an additional asymmetry with respect to the $\varphi = 0$ plane:

$$u(r, \theta, \varphi, 0) = \left\{ \begin{array}{ll} \rho_1 & (0 < \varphi < \pi) \\ -\rho_1 & (\pi < \varphi < 2\pi) \end{array} \right\} \tag{7.167}$$

Sturm–Liouville problem

We expand the solution in eigenfunctions of the Sturm–Liouville problem with boundary conditions where the gas oscillates:

$$\left\{ \begin{array}{l} \Delta u + \lambda u = 0 \\ \text{Curved boundary: } \left.\dfrac{\partial u}{\partial r}\right|_{r=R} = 0 \\ \text{Flat boundary: } \left.\dfrac{\partial u}{\partial \theta}\right|_{\theta=\pi/2} = 0 \end{array} \right\} \tag{7.168}$$

General solution

The solution for the variations from the equilibrium pressure (ρ_0) is:

$$\begin{aligned} u(r, \theta, \varphi, t) = \sum [&A_{klm} \sin(\sqrt{\lambda_{kl}} t) \\ &+ B_{klm} \cos(\sqrt{\lambda_{kl}} t)] j_l(\sqrt{\lambda_{kl}} r) P_l^{(m)}(\cos(\theta)) e^{im\varphi} \end{aligned} \tag{7.169}$$

Here $j_l(\sqrt{\lambda_{kl}} r)$ are spherical Bessel functions with the same boundary conditions ($\frac{\partial u}{\partial \vec{\mathbf{n}}} = 0$, being $\vec{\mathbf{n}}$ the direction normal to the surface). Furthermore, the second boundary condition converts into the following condition for the Legendre polynomials which are admitted in the solution.

$$\left.\frac{dP_l^{(m)}(x)}{dx}\right|_{x=0} = 0 \tag{7.170}$$

From the relation that allows to generate $P_l^{(m)}(x)$ by differentiating the Legendre polynomials m times, we deduce that the solution only admits those indices m, l for which $m + l$ is even, since:

$$P_l^{(m)}(x) = (1 - x^2)^{\frac{m}{2}} \frac{d}{dx^m}[P_l(x)] \tag{7.171}$$

and only $P_l(x)$ with even l satisfy the condition for which the derivative of the Legendre polynomial is zero at $x = 0$: $\frac{d}{dx}[P_l^{(m)}(x)](x = 0) = 0$

Final solution

We now impose the initial conditions. Since the initial velocity is zero: $A_{klm} = 0$. The initial displacement is:

$$u(r, \theta, \varphi, 0) = \left\{ \begin{array}{l} \rho_1 \quad (0 < \varphi < \pi) \\ \\ -\rho_1 \ (\pi < \varphi < 2\pi) \end{array} \right\}$$

$$= \sum B_{klm} j_l(\sqrt{\lambda_{kl}} r) P_l^{(m)}(\cos(\theta)) e^{im\varphi} \tag{7.172}$$

We use the orthogonality of the spherical Bessel functions, of the Legendre polynomials and of the azimuthal eigenfunctions to find the B_{klm} coefficients. Due to the symmetry of the problem (antisymmetric initial conditions with respect to $\varphi = 0$) in the azimuthal angle it is clear that $e^{im\varphi} \rightarrow \sin(m\varphi)$ (with $m = 1, 2, 3, \ldots$). We determine the coefficients as the solution of the equation:

$$B_{klm} \left\| j_l(\sqrt{\lambda_{kl}} r) \right\|^2 \| P_l(\cos(\theta)) \|^2 \| \sin(m\varphi) \|^2 =$$

$$= \rho_1 \int_0^R j_l(\sqrt{\lambda_{kl}} r) r^2 dr \int_0^{\pi/2} P_l^{(m)}(\cos(\theta)) d\cos(\theta)$$

$$\times \left[\int_0^{\pi} \sin(m\varphi) d\varphi - \int_{\pi}^{2\pi} \sin(m\varphi) d\varphi \right] \tag{7.173}$$

Note: in the final expansion of the solution only the coefficients with odd m will remain, and those for which $(m + l)$ is even. See the previous discussion about the condition: $\left.\frac{dP_l^{(m)}(x)}{dx}\right|_{x=0} = 0$

7.9 Case Study: Forced Oscillations in a Semi-Sphere

A gas is enclosed in a semi-spherical container. The curved surface is divided in two equal halves. One of them moves radially so that $r = R + \delta \sin(\omega t)$ is the radius of the recipient in the corresponding zone. The other curved half is open to the outer medium. Find the stationary oscillations of the gas density inside the recipient considering that $\delta << R$ and the equilibrium pressure is P_0. Consider the case (a) of slow (isothermal) oscillations and (b) adiabatic oscillations with constants $\left(\frac{\partial P}{\partial \rho}\right)_{i,a} = k$. In the (b) case consider that the open curved boundary from (a) is now closed.

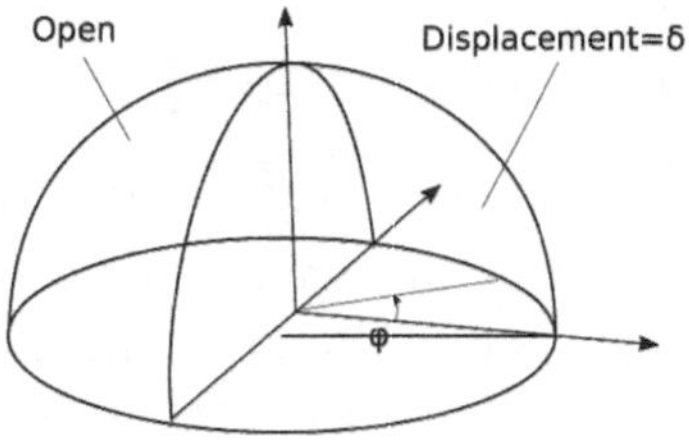

Figure 7.13

Part (a)

We first study the limit case of slow variations ($\omega \to 0$)). We estimate that the relative variation of the volume is:

$$\frac{\Delta V}{V} = \frac{\pi R^2 \delta \sin(\omega t)}{\frac{\pi R^3}{3}} = \frac{3\delta \sin(\omega t)}{R} \tag{7.174}$$

We can evaluate the relative variation of the pressure (in isothermal conditions) as: $\frac{\Delta P}{P_0} = \frac{\Delta V}{V} = \frac{3\delta \sin(\omega t)}{R}$. Recalling that $\Delta P =$

$\left(\frac{\partial P}{\partial \rho}\right)_i \Delta\rho = k\Delta\rho$

$$\Delta\rho = \frac{\Delta P}{k} = \frac{3P_0\delta\sin(\omega t)}{kR} = A\sin(\omega t) \quad (7.175)$$

Mathematical formulation for $\Delta\rho = u(r, \theta, \varphi, t)$

$$\left\{\begin{array}{l} \frac{\partial^2 u}{\partial t^2} = a^2 \Delta u \\ \left.\frac{\partial u}{\partial \theta}\right|_{\theta=\frac{\pi}{2}} = 0 \qquad \text{(flat closed base)} \\ u(R, \theta, \varphi, t) = f(\theta, \varphi)\sin(\omega t) = \left\{\begin{array}{ll} 0 & \left(\varphi < -\frac{\pi}{2};\ \varphi > \frac{\pi}{2}\right);\ \left(0 < \theta < \frac{\pi}{2}\right) \\ A\sin(\omega t) & \left(-\frac{\pi}{2} < \varphi < \frac{\pi}{2}\right);\ \left(0 < \theta < \frac{\pi}{2}\right) \end{array}\right\} \end{array}\right\} \quad (7.176)$$

We seek the stationary solution, which should be periodical and proportional to $\sin(\omega t)$

$$u(r, \theta, \varphi, t) = V(r, \theta, \varphi) \cdot \sin(\omega t) \quad (7.177)$$

Replacing in the wave equation and obtaining the following equation

$$\left\{\begin{array}{l} \Delta V + \frac{\omega^2}{a^2} V = 0 \\ \left.\frac{\partial V}{\partial \theta}\right|_{\theta=\frac{\pi}{2}} = 0 \\ V(R, \theta, \varphi) = f(\theta, \varphi) \end{array}\right\} \quad (7.178)$$

Sturm–Liouville problem

The previous is a Poisson problem that we can solve by expanding $V(r, \theta, \varphi)$ in a sum of orthogonal functions in the angular directions (spherical harmonics $Y(\theta, \varphi)$).

$$V(r, \theta, \varphi) = \mathcal{R}(r) \cdot Y(\theta, \varphi) \quad (7.179)$$

We know how to solve the Sturm–Liouville problem which appears in the angular part:

$$\Delta_{\theta,\varphi} Y + \mu Y = 0 \quad (7.180)$$

$$Y(\theta, \varphi) = P_l^m(\cos\theta)[A\cos(m\varphi) + B\sin(m\varphi)] \quad (m = 0, 1, 2, 3, \ldots) \quad (7.181)$$

Eigenvalues: $\mu_{lm} = l(l+1); l \geq |m|$

Applying the condition $\left.\frac{\partial V}{\partial \theta}\right|_{\theta=\frac{\pi}{2}} = 0$

$$\left.\frac{dP_l^m(\cos(\theta))}{d\theta}\right|_{\theta=\frac{\pi}{2}} = \left.\frac{dP_l^m(x)}{dx}\right|_{x=0} = 0 \tag{7.182}$$

is only satisfied if $(l+m)$ = even. This condition is obtained from the relation which allows to generate $P_l^{(m)}(x)$ differentiating m times Legendre polynomials in the following manner:

$$P_l^{(m)}(x) = (1-x^2)^{\frac{m}{2}} \frac{d^m P_l(x)}{dx^m} \tag{7.183}$$

Since for odd values of l, $P_l(x)$ is an asymmetric function and only the $P_l(x)$ with l = even are symmetric functions which satisfy the condition: $\left.\frac{dP_n(x)}{dx}\right|_{x=0} = 0$, only differentiating m = even times this function we keep its symmetry, from which we deduce that the solution only admits the m, l indices for which $m + l$ is even. Replacing $Y(\theta, \varphi)$ in the equation for $V(r, \theta, \varphi)$ we obtain an equation for $\mathcal{R}(r)$ (which is not a Sturm–Liouville problem):

$$r^2\left(\frac{d^2\mathcal{R}}{dr^2}\right) + r\left(\frac{d\mathcal{R}}{dr}\right) + \left[\frac{\omega^2}{a^2} - \frac{\mu_{lm}}{r^2}\right]\mathcal{R} = 0 \tag{7.184}$$

Its solutions are Bessel functions of semi even order:

$$\mathcal{R}_l(r) = C\frac{J_{l+1/2}(\frac{\omega}{a}r)}{\sqrt{r}} \tag{7.185}$$

General solution

With the previous, the general solution results:

$$u(r,\theta,\varphi,t) = \sum_{l,m} \frac{J_{l+1/2}(\frac{\omega}{a}r)}{\sqrt{r}} P_l^m(\cos(\theta))[A_{lm}\cos(m\varphi) + B_{lm}\sin(m\varphi)] \cdot \sin(\omega t) \tag{7.186}$$

Final solution

To find the coefficients we use the boundary condition for $r = R$ and the orthogonality for the eigenfunctions:

$$u(R, \theta, \varphi, t) = \sum_{l,m} \frac{J_{l+1/2}(\frac{\omega}{a}R)}{\sqrt{R}} P_l^m(\cos\theta)[A_{lm}\cos(m\varphi)$$
$$+B_{lm}\sin(m\varphi)] \cdot \sin(\omega t) =$$
$$= f(\theta, \varphi)\sin(\omega t) \quad (7.187)$$

The symmetry of the function $f(\theta, \varphi)$ with respect to $\varphi = 0$ imposes $B_{lm} = 0$

$$\sum_{l,m} A_{lm} \frac{J_{l+1/2}(\frac{\omega}{a}R)}{\sqrt{R}} P_l^m(\cos(\theta))\cos(m\varphi) = f(\theta, \varphi) \quad (7.188)$$

We find:

$$A_{lm} = \frac{A \int_0^{\frac{\pi}{2}} \int_{-\frac{\pi}{2}}^{\frac{\pi}{2}} P_l^m(\cos(\theta))\cos(m\varphi)\sin(\theta)d\theta d\varphi}{\frac{J_{l+1/2}(\frac{\omega}{a}R)}{\sqrt{R}} \int_0^{\frac{\pi}{2}} \int_{-\pi}^{\pi} [P_l^m(\cos(\theta))]^2[\cos(m\varphi)]^2\sin(\theta)d\theta d\varphi} \quad (7.189)$$

Simplification:

$$\int_{-\frac{\pi}{2}}^{\frac{\pi}{2}} \cos(m\varphi)d\varphi = \left|\frac{1}{m}\sin(m\varphi)\right|_{-\frac{\pi}{2}}^{+\frac{\pi}{2}} = \begin{pmatrix} \frac{2}{m} & (\text{m odd}) \\ 0 & (\text{m even}) \end{pmatrix} \quad (7.190)$$

Part (b)

Mathematical formulation

In the case of mathematical formulation for fast displacements of the border the boundary condition for the gas near it can be obtained in a more precise manner. The starting point is that the normal component of the velocity of the molecules of the gas must be equal to the normal velocity of the mobile boundary. We first estimate the relative variation of volume:

$$v_n = \delta\omega\cos(\omega t) \quad (7.191)$$

The following formula relates the velocity of the molecules with the s condensation parameter, which describes the relative variation in density: $s = \frac{\rho - \rho_0}{\rho_0}$.

$$v = v_0 - k \int_0^t \nabla s dt' \tag{7.192}$$

differentiating this relation with respect to time and considering the normal component to the mobile border we get:

$$\frac{\partial}{\partial t} v_n = -k \frac{\partial}{\partial n} s = -\frac{k}{\rho_0} \frac{\partial \rho}{\partial n} \tag{7.193}$$

Then the boundary condition for the mobile border will be:

$$\left. \frac{\partial \rho}{\partial n} \right|_{r=R} = -\frac{\rho_0}{k} \frac{\partial}{\partial t} v_n = \frac{\rho_0}{k} \delta \omega^2 \sin(\omega t) \tag{7.194}$$

The mathematical formulation for this case, in terms of the gas density, will be:

$$\begin{cases} \dfrac{\partial^2 u}{\partial t^2} = a^2 \Delta u \\ \left. \dfrac{\partial u}{\partial \theta} \right|_{\theta = \frac{\pi}{2}} = 0 \text{ (flat closed base)} \\ \left. \dfrac{\partial u}{\partial r} \right|_{r=R} (R, \theta, \varphi, t) = 0 \quad \left(\varphi < -\dfrac{\pi}{2}; \varphi > \dfrac{\pi}{2} \right); \left(0 < \theta < \dfrac{\pi}{2} \right) \\ \left. \dfrac{\partial u}{\partial r} \right|_{r=R} = \dfrac{\rho_0}{k} \delta \omega^2 \sin(\omega t) \left(-\dfrac{\pi}{2} < \varphi < \dfrac{\pi}{2} \right); \left(0 < \theta < \dfrac{\pi}{2} \right) \end{cases} \tag{7.195}$$

Final solution

This problem can have analytic solution using similar methods to those previously considered.

7.10 Heat Transfer in an Eight of a Sphere

An eight of a sphere has radius R, thermal conductivity k, heat capacity C and density ρ. Two of its flat surfaces (vertical, see the figure) are thermally insulated. Initially the base, as well as the curved surface, stay at a temperature T_0. Until $t = 0$ the body was at thermal equilibrium. Find the variation of temperature as a function of time if starting at $t = 0$, one half of its curved surface is put into contact with a thermal reservoir at a temperature T_1 while the other is put at T_2.

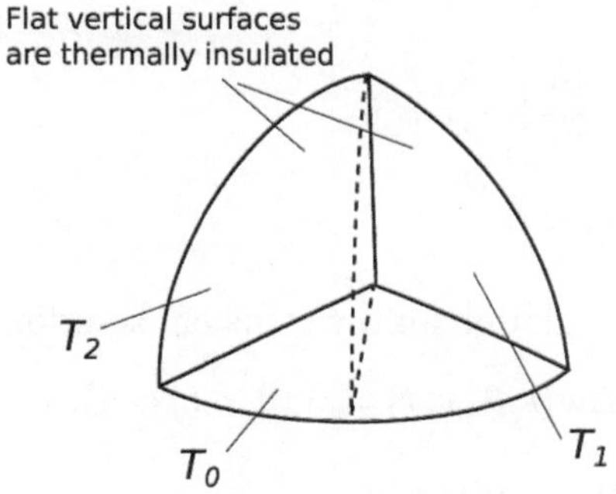

Figure 7.14

Mathematical formulation

We first formulate the problem by subtracting from the solution the constant temperature T_0:

$$\left\{\begin{array}{l} C\rho\dfrac{\partial u}{\partial t} - k\Delta u(r,\theta,\varphi,t) = 0 \\ u(r,\theta,\varphi,0) = 0 \\ u(R,\theta,\varphi,t>0) = f(\theta,\varphi) = \left\{\begin{array}{l} T_2 - T_0 \ (0 < \varphi < \pi/4) \\ T_1 - T_0 \ (\pi/4 < \varphi < \pi/2) \end{array}\right\} \\ \left.\dfrac{\partial u}{\partial \varphi}\right|_{\varphi=0} = \left.\dfrac{\partial u}{\partial \varphi}\right|_{\varphi=\frac{\pi}{2}} = 0 \\ u(\theta = \pi/2) = 0 \end{array}\right\} \tag{7.196}$$

We seek the solution by splitting the problem in two: one is stationary (the solution of Laplace's equation in the limit $t \to \infty$) and transient one:

$$u(r,\theta,\varphi,t) = v(r,\theta,\varphi) + w(r,\theta,\varphi,t) \tag{7.197}$$

Mathematical formulation for the stationary solution $v(r, \theta, \varphi)$

$$\left\{\begin{array}{l} \Delta v(r, \theta, \varphi) = 0) \\ \dfrac{1}{r^2}\dfrac{\partial}{\partial r}\left[r^2\left(\dfrac{\partial v}{\partial r}\right)\right] + \dfrac{1}{r^2}\Delta_{\theta,\varphi} v = 0 \\ v(r = R) = f(\theta, \varphi) = \left\{\begin{array}{l} T_2 - T_0 \quad (0 < \varphi < \pi/4) \\ T_1 - T_0 \ (\pi/4 < \varphi < \pi/2) \end{array}\right\} \\ \left.\dfrac{\partial v}{\partial \varphi}\right|_{\varphi=0} = \left.\dfrac{\partial v}{\partial \varphi}\right|_{\varphi=\frac{\pi}{2}} = 0 \\ v(\theta = \pi/2) = 0 \end{array}\right\} \tag{7.198}$$

Mathematical formulation for the transient solution $w(r, \theta, \varphi, t)$

$$\left\{\begin{array}{l} C\rho\dfrac{\partial w}{\partial t} - k\Delta w(r, \theta, \varphi, t) = 0 \ (0 < r < R);\ t > 0 \\ w(r, \theta, \varphi, 0) = -v(r, \theta, \varphi) \\ w(R, \theta, \varphi, t > 0) = 0 \\ \left.\dfrac{\partial w}{\partial \varphi}\right|_{\varphi=0} = \left.\dfrac{\partial w}{\partial \varphi}\right|_{\varphi=\frac{\pi}{2}} = 0 \\ w(r, \theta = \pi/2, \varphi, t) = 0 \end{array}\right\} \tag{7.199}$$

General solution

We first solve the stationary problem: The general solution of the Laplace's problem in spherical coordinates has previously been found in several Laplace's problems as, for example, the electric potential between two spherical shells. Without repeating the steps, we start directly from the result:

$$v(r, \theta, \varphi) = \sum_{n \geq m}^{\infty}\sum_{m=0}^{\infty}[r^n + D_{nm}r^{-n-1}]P_n^{(m)}(\cos(\theta)) \times [A_{nm}\sin(m\varphi) + B_{nm}\cos(m\varphi)] \tag{7.200}$$

As the solution must be finite for $r = 0 \rightarrow D_{nm} = 0$

$$v(r, \theta, \varphi) = \sum_{n \geq m}^{\infty}\sum_{m=0}^{\infty}[r^n]P_n^{(m)}(\cos(\theta)) \times [A_{nm}\sin(m\varphi) + B_{nm}\cos(m\varphi)] \tag{7.201}$$

Final solution

We now use the orthogonality of $\cos(m\varphi)$ and $\sin(m\varphi)$.

From the second boundary condition: $\left.\frac{\partial v}{\partial \varphi}\right|_{\varphi=0} = 0$ we get $A_{nm} = 0$.

From $\left.\frac{\partial v}{\partial \varphi}\right|_{\varphi=\pi/2} = 0$ we get:

$$\sin\left(m\frac{\pi}{2}\right) = 0 \qquad m\frac{\pi}{2} = k\pi \to m = 2k \qquad (k = 0, 1, 2, \ldots) \tag{7.202}$$

Then:

$$v(r, \theta, \varphi) = \sum_{n\geq 2k}^{\infty}\sum_{k=0}^{\infty} B_{nk}[r^n]P_n^{(2k)}(\cos(\theta))\cos(2k\varphi) \tag{7.203}$$

Imposing the first boundary condition: $v(r, \theta, \varphi)(\theta = \pi/2) = 0$. We arrive at the condition $P_n^{(2k)}(\cos\pi/2) = P_n^{(2k)}(0) = 0$

For this condition to be satisfied, it must happen that $n + 2k$ is odd. Finally we need to use the orthogonality of the angular functions (spherical harmonics) to find the B_{nk} coefficients. From the first boundary condition:

$$v(r = R) = f(\theta, \varphi) = \sum_{n\geq 2k}^{\infty}\sum_{k=0}^{\infty} B_{nk}[R^n]P_n^{(2k)}(\cos(\theta))\cos(2k\varphi)$$
$$(n + 2k = \text{odd}) \tag{7.204}$$

Using now the orthogonality of the $P_n^{(2k)}(\cos(\theta))$ and $\cos(2k\varphi)$ in the interval $(0 < \theta\pi/2)$ and $(0 < \varphi < \pi/2)$ we will get the B_{nk} coefficients. We multiply by $P_n^{(2k)}(\cos(\theta))$ and $\cos(2k\varphi)$ and integrate equation (7.204) between $\int_0^{\frac{\pi}{2}}\int_0^{\frac{\pi}{2}} \sin(\theta)d\theta d\varphi$

$$\int_0^{\frac{\pi}{2}}\int_0^{\frac{\pi}{2}} f(\theta, \varphi)P_n^{(2k)}(\cos(\theta))\sin(\theta)\cos(2k\varphi)d\theta d\varphi$$
$$= B_{nk}[R^n]\int_0^{\frac{\pi}{2}}\int_0^{\frac{\pi}{2}} [P_n^{(2k)}(\cos(\theta))]^2[\cos(2k\varphi)]^2\sin(\theta)d\theta d\varphi \tag{7.205}$$

$$B_{nk} = \frac{(T_2 - T_0)\int_0^{\frac{\pi}{2}}\int_0^{\frac{\pi}{4}} P_n^{(2k)}(\cos(\theta))\sin(\theta)\cos(2k\varphi)d\theta d\varphi + (T_1 - T_0)\int_0^{\frac{\pi}{2}}\int_{\frac{\pi}{4}}^{\frac{\pi}{2}} P_n^{(2k)}(\cos(\theta))\sin(\theta)\cos(2k\varphi)d\theta d\varphi}{[R^n]\int_0^{\frac{\pi}{2}}\int_0^{\frac{\pi}{2}}[P_n^{(2k)}(\cos(\theta))]^2[\cos(2k\varphi)]^2\sin(\theta)d\theta d\varphi} \tag{7.206}$$

Calculating the integrals:

$$\int_0^{\frac{\pi}{2}}[\cos(2k\varphi)]^2 d\varphi = \frac{\pi}{4} \tag{7.207}$$

$$\int_0^{\frac{\pi}{2}}[P_n^{(2k)}(\cos(\theta))]^2\sin(\theta)d\theta = \alpha_{kn} = \frac{1}{2}\int_0^{\pi}[P_n^{(2k)}(\cos\theta)]^2\sin(\theta)d\theta = \frac{1}{2}\frac{2(n+2k)!}{(2n+1)(n-2k)!} \tag{7.208}$$

$$\int_0^{\frac{\pi}{4}}\cos(2k\varphi)d\varphi = \frac{1}{2k}\left[\sin\left(\frac{\pi k}{2}\right) - 0\right] = \frac{1}{2k}\sin\left(\frac{\pi k}{2}\right) \tag{7.209}$$

$$\int_{\frac{\pi}{4}}^{\frac{\pi}{2}}\cos(2k\varphi)d\varphi = \frac{1}{2k}\left[\sin(\pi k) - \sin\left(\frac{\pi k}{2}\right)\right] = -\frac{1}{2k}\sin\left(\frac{\pi k}{2}\right) \tag{7.210}$$

Then

$$B_{nk} = \frac{(T_2 - T_0)\int_0^{\frac{\pi}{2}}\int_0^{\frac{\pi}{4}} P_n^{(2k)}(\cos(\theta))\sin(\theta)\cos(2k\varphi)d\theta d\varphi + (T_1 - T_0)\int_0^{\frac{\pi}{2}}\int_{\frac{\pi}{4}}^{\frac{\pi}{2}} P_n^{(2k)}(\cos(\theta))\sin(\theta)\cos(2k\varphi)d\theta d\varphi}{[R^n]\int_0^{\frac{\pi}{2}}\int_0^{\frac{\pi}{2}}[P_n^{(2k)}(\cos(\theta))]^2[\cos(2k\varphi)]^2\sin(\theta)d\theta d\varphi} \tag{7.211}$$

with

$$\sin\left(\frac{\pi k}{2}\right) = \left\{\begin{matrix} 0 & (k = 2l) \\ (-1)^l & (k = 2l+1) \end{matrix}\right\} \tag{7.212}$$

We now solve the transient problem by separating the solution in series of orthogonal functions.

Due to the presence of all homogeneous boundaries (the radial and the two angular, of the second type in the azimuthal angle) we seek the solution as:

$$w(r,\theta,\varphi,t)=\sum_{n,l,k}Q_{nlk}(t)V_{nlk}(r,\theta,\varphi)$$
$$=\sum_{n,l,k}Q_{lk}(t)\frac{J_{l+1/2}(\sqrt{\lambda_n}r)}{\sqrt{r}}P_l^{2k}(\cos(\theta))\cos(2k\varphi) \quad (7.213)$$

Here, in the same manner as was done in the Laplace problem we have already assumed the boundary conditions in the azimuthal direction.

Furthermore, we note that the values of λ_n are obtained when we apply the homogeneous boundary conditions of the first type for the radial part:

$$\frac{d}{dr}\left[\frac{J_{l+1/2}(\sqrt{\lambda_n}r)}{\sqrt{r}}\right]_{r=R}=\frac{d}{dr}\left[J_{l+1/2}(\sqrt{\lambda_n}r)\right]_{r=R}$$
$$-\frac{1}{2R}J_{l+1/2}(\sqrt{\lambda_n}R)=0 \quad (7.214)$$

Replacing the sum from equation (7.213) into the diffusion equation we need to solve, and recalling that the $V_{nlk}(r,\theta,\varphi)$ eigenfunctions are solutions of the Sturm–Liouville problem:

$$\left\{\begin{array}{l}\Delta V(r,\theta,\varphi)+\lambda V(r,\theta,\varphi)=0\ (0<r<R)\\ V(R,\theta,\varphi)=0\\ \left.\dfrac{\partial u}{\partial\varphi}\right|_{\varphi=0}=\left.\dfrac{\partial u}{\partial\varphi}\right|_{\varphi=\frac{\pi}{2}}\\ v(\theta=\pi/2)=0\end{array}\right\} \quad (7.215)$$

we arrive at an equation for the $Q_{nlk}(t)$ coefficients:

$$C\rho\frac{dQ_{nlk}(t)}{dt}+k\lambda_nQ_{nlk}(t)=0 \quad (7.216)$$

with exponential solutions:

$$Q_{nlk}(t)=D_{nlk}e^{-\frac{k\lambda_n}{C\rho}t} \quad (7.217)$$

Applying the initial condition and using the orthogonality of the eigenfunctions we get the D_{nlk} coefficients:

$$w(r, \theta, \varphi, 0) = \sum_{n,l,k} D_{nlk} \frac{J_{l+1/2}(\sqrt{\lambda_n} r)}{\sqrt{r}} P_l^{2k}(\cos(\theta)) \cos(2k\varphi)$$
$$= -v(r, \theta, \varphi) \qquad (7.218)$$

Just like in the Laplace's problem, in order to satisfy the third homogeneous boundary condition of the first kind for $\theta = \frac{\pi}{2}$ it's necessary that the members of the sum with odd $l + 2k$ remain, since only these Legendre polynomials are asymmetric (that is, satisfy the third boundary condition).

7.11 Case Study: Heated Quarter of a Sphere

The two flat surfaces of a region with the form of a quarter of a sphere with radius a are kept at a temperature T while the spherical surface is kept at zero temperature. Inside the region a thin metallic wire that is heated by a DC current I. Find the stationary distribution of temperature inside the region if the resistance of the wire is R.

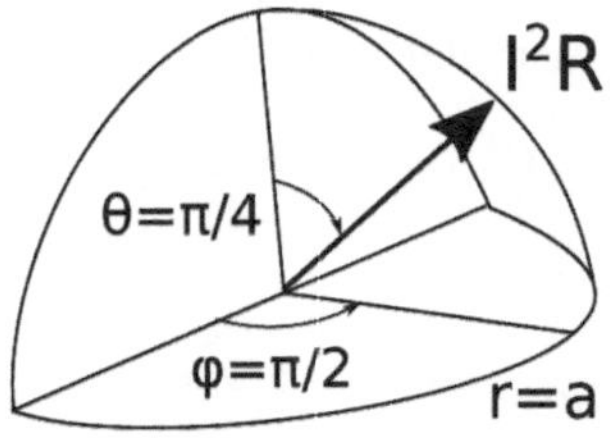

Figure 7.15

Mathematical formulation

Non-stationary case $T = u(r, \theta, \varphi, t)$:

$$\left\{ \rho \frac{\partial u}{\partial t} = k\Delta u + f(r, \theta, \varphi) \right\} \qquad (7.219)$$

Formulation of the stationary case:

$$\frac{\partial u}{\partial t} = 0 \tag{7.220}$$

We give a description of the problem considering the normalization:

$$f(r, \theta, \varphi) = \frac{I^2 R}{a} \frac{\delta(\theta - \frac{\pi}{4}) \times \delta(\varphi - \frac{\pi}{2})}{r^2 \sin(\theta)} \tag{7.221}$$

Note: an alternative to describe the heater could be to present it as a tube with infinitesimal but constant radius. In this case the dissipated local power density as a function of distance from the central point would be constant and would lead us to remove therm r^2 in the denominator with the corresponding difficulties of solving the problem in spherical coordinates.

Formulation with boundary conditions:

$$\left\{\begin{array}{l} -k\Delta u(r, \theta, \varphi) = f(r, \theta, \varphi) \\ u(r, \theta, 0) = u(r, \theta, \pi) = T \\ u(r, \frac{\pi}{2}, \varphi) = T \\ u(a, \theta, \varphi) = 0 \end{array}\right\} \tag{7.222}$$

Since we are capable of solving Laplace's equation with some homogeneous boundaries and Poisson's equation with some homogeneous boundaries, using the principle of superposition we seek the solution as the sum of two functions:

$$u(r, \theta, \varphi) = g(r, \theta, \varphi) + h(r, \theta, \varphi) \tag{7.223}$$

Poisson's problem (1)

$$\left\{\begin{array}{l} -k\Delta h = f(r, \theta, \varphi) \\ h(r, \theta, 0) = h(r, \theta, \pi) = 0 \\ h(r, \frac{\pi}{2}, \varphi) = 0 \\ h(a, \theta, \varphi) = 0 \end{array}\right\} \tag{7.224}$$

Laplace's problem (2)

$$\left\{\begin{array}{l} \Delta g = 0 \\ g(r,\theta,0) = g(r,\theta,\pi) = T \\ g(r,\frac{\pi}{2},\varphi) = T \\ g(a,\theta,\varphi) = 0 \end{array}\right\} \tag{7.225}$$

Sturm–Liouville problem

We first solve problem (1) and expand the solution as the sum of orthogonal eigenfunctions of the Sturm–Liouville problem:

$$\left\{\begin{array}{l} \Delta V + \lambda V = 0 \\ V(r,\theta,0) = V(r,\theta,\pi) = 0 \\ V(r,\frac{\pi}{2},\varphi) = 0 \\ V(a,\theta,\varphi) = 0 \end{array}\right\} \tag{7.226}$$

We separate variables as $V = \mathcal{R}(r) \cdot Y(\theta,\varphi)$

$$\frac{1}{r^2}\frac{\partial}{\partial r}\left[r^2\left(\frac{\partial V}{\partial r}\right)\right] + \frac{1}{r^2}\Delta_{\theta,\varphi}V + \lambda V = 0 \tag{7.227}$$

$$\frac{\frac{d}{dr}[r^2(\frac{d\mathcal{R}}{dr})] + \lambda r^2\mathcal{R}}{\mathcal{R}} = -\frac{\Delta_{\theta,\varphi}Y}{Y} = \mu > 0 \tag{7.228}$$

Solving the problem for the angular part:

$$\Delta_{\theta,\varphi}Y + \mu Y = 0 \tag{7.229}$$

Separating variables:

$$Y(\theta,\varphi) = \Theta(\theta) \cdot \Phi(\varphi) \tag{7.230}$$

$$\frac{\sin(\theta)\frac{d}{d\theta}[\sin(\theta)(\frac{d\Theta}{d\theta})] + \mu\sin^2(\theta)\cdot\Theta(\theta)}{\Theta(\theta)} = -\frac{1}{\Phi(\varphi)}\frac{d^2\Phi}{d\varphi^2} = \nu > 0 \tag{7.231}$$

We choose the positive sign for μ, ν since we expect to get periodic solutions for the angular variables.

The result of the problem for φ is:

$$\left\{\begin{array}{l}\dfrac{d^2\Phi}{d\varphi^2}+\nu\Phi(\varphi)=0\\ h(r,\theta,0)\rightarrow\Phi(0)=0\\ h(r,\theta,\pi)\rightarrow\Phi(\pi)=0\end{array}\right\} \tag{7.232}$$

$$\Phi(\varphi)=A\cos(m\varphi)+B\sin(m\varphi);\ \ \nu=m^2 \tag{7.233}$$

$\Phi(0)=0\rightarrow A=0\rightarrow$

$$\Phi(\varphi)=\sin(m\varphi) \tag{7.234}$$

Now that we have $\nu=m^2$, we can find the solution of the $\Theta(\theta)$ function:

$$\sin(\theta)\frac{d}{d\theta}\left[\sin(\theta)\left(\frac{d\Theta}{d\theta}\right)\right]+[\mu\sin^2(\theta)-m^2]\Theta(\theta)=0 \tag{7.235}$$

Its solution are Legendre polynomials: $\Theta(\theta)=P_n^{(m)}(\cos(\theta))$

Eigenvalues: $\mu=n(n+1)$

Imposing the boundary condition:

$$\Theta\left(\frac{\pi}{2}\right)=P_n^{(m)}(0)=0 \tag{7.236}$$

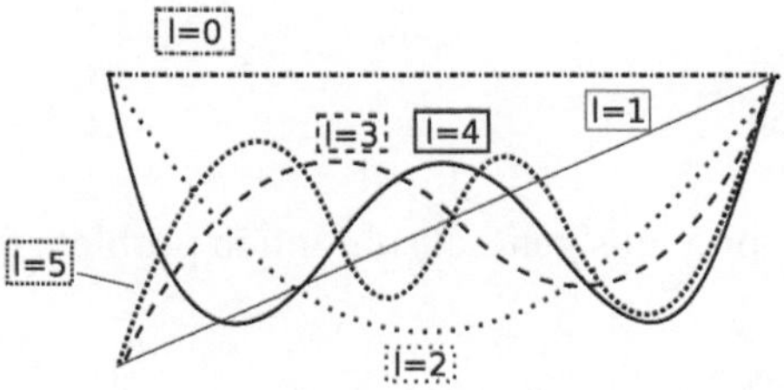

Figure 7.16

From the properties of the Legendre polynomials, this condition is only satisfied when:

$$n-m=\text{odd}=2l+1;\ m=n-2l+1;\ n=1,2,3\ldots;$$
$$l=0,1,2,3\ldots\ \text{with}\ m<n \tag{7.237}$$

In this way: $\Theta(\theta)=P_n^{[n-2l+1]}(\cos(\theta));\ \mu=n(n+1)$

We can finally solve the equation for $\mathcal{R}(r)$:

$$r^2\left(\frac{d^2\mathcal{R}}{dr^2}\right)+r\left(\frac{d\mathcal{R}}{dr}\right)+\left[\lambda r^2-\left(n+\frac{1}{2}\right)^2\right]\mathcal{R}=0 \tag{7.238}$$

Its solution is the Bessel function of half-integer order:

$$\mathcal{R}_n(r)=C_1\frac{J_{n+1/2}(\sqrt{\lambda_k^{(n)}}r)}{\sqrt{r}}+C_2\frac{N_{n+1/2}(\sqrt{\lambda_k^{(n)}}r)}{\sqrt{r}} \tag{7.239}$$

As the solution must be finite at $r = 0$ we have that $C_2 = 0$. The eigenvalues of the problem are obtained from the solution of the equation:

$$J_{n+1/2}\left(\sqrt{\lambda_k^{(n)}}a\right)=0 \tag{7.240}$$

So that the eigenfunctions of the Sturm–Liouville problem in a quarter of a sphere are:

$$V_{kmn}=\frac{J_{n+1/2}\left(\sqrt{\lambda_k^{(n)}}r\right)}{\sqrt{r}}P_n^{(m)}(\cos(\theta))\sin(m\varphi) \tag{7.241}$$

where

$$\left\{\begin{matrix} m=n-2l+1 \\ J_{n+1/2}(\sqrt{\lambda_k^{(n)}}a)=0 \end{matrix}\right\} \tag{7.242}$$

As these functions fulfill the boundary conditions for problem (1), we propose to seek its solutions as:

$$h(r,\theta,\varphi)=\sum_{km(l)n}C_{km(l)n}\times V_{kmn} \tag{7.243}$$

General solution

We replace the previous general solution in problem (1):

$$\left\{-k\Delta h=-k\Delta\sum_{km(l)n}C_{km(l)n}\cdot V_{kmn}=k\sum_{km(l)n}C_{km(l)n}\times\lambda_k^{(n)}V_{kmn}=f(r,\theta,\varphi)\right\} \tag{7.244}$$

Expanding more explicitly:

$$k\sum_{km(l)n}C_{km(l)n}\times\lambda_k^{(n)}\frac{J_{n+1/2}(\sqrt{\lambda_k^{(n)}}r)}{\sqrt{r}}P_n^{(m)}(\cos(\theta))\sin(m\varphi)$$

$$=\frac{I^2R}{a}\frac{\delta(\theta-\frac{\pi}{4})\times\delta(\varphi-\frac{\pi}{2})}{r^2\sin(\theta)} \tag{7.245}$$

Final solution

To find the coefficients we will use the orthogonality of the eigenfunctions:

We multiply by $\sin(m'\varphi)$ and integrate $\int_0^{\pi} d\varphi$

$$k \sum_{km(l)n} C_{km(l)n} \times \lambda_k^{(n)} \frac{J_{n+1/2}\left(\sqrt{\lambda_k^{(n)}} r\right)}{\sqrt{r}} P_n^{(m)}(\cos(\theta)) \int_0^{\pi} \sin(m'\varphi) \sin(m\varphi) d\varphi$$
$$= \frac{I^2 R}{a} \frac{\delta\left(\theta - \frac{\pi}{4}\right)}{r^2 \sin(\theta)} \int_0^{\pi} \sin(m'\varphi) \times \delta\left(\varphi - \frac{\pi}{2}\right) d\varphi \rightarrow \quad (7.246)$$

$$k \sum_{kn} C_{kmn} \times \lambda_k^{(n)} \frac{J_{n+1/2}\left(\sqrt{\lambda_k^{(n)}} r\right)}{\sqrt{r}} P_n^{(m)}(\cos(\theta)) \times \frac{\pi}{2}$$
$$= \frac{I^2 R}{a} \frac{\delta(\theta - \frac{\pi}{4}) \sin(m\frac{\pi}{2})}{r^2 \sin(\theta)} \quad (7.247)$$

We multiply now by $P_{n'}^{(m)}(\cos(\theta))$ and integrate $\int_0^{\pi/2} \sin(\theta) d\theta$

$$k \sum_{kn} C_{kmn} \times \lambda_k^{(n)} \frac{J_{n+1/2}\left(\sqrt{\lambda_k^{(n)}} r\right)}{\sqrt{r}} \frac{\pi}{2} \int_0^{\pi/2} P_{n'}^{(m)}(\cos(\theta)) P_n^{(m)}(\cos(\theta)) \sin(\theta) d\theta$$
$$= \frac{I^2 R}{a} \sin\left(m\frac{\pi}{2}\right) \int_0^{\pi/2} \frac{\delta(\theta - \frac{\pi}{4})}{r^2 \sin(\theta)} P_n^{(m')}(\cos(\theta)) \sin(\theta) d\theta \quad (7.248)$$

The expression becomes:

$$k \sum_{k} C_{kmn} \times \lambda_k^{(n)} \frac{J_{n+1/2}(\sqrt{\lambda_k^{(n)}} r)}{\sqrt{r}} \frac{\pi}{2} \frac{\left\| P_n^{(m)}(\cos(\theta)) \right\|^2}{2}$$
$$= \frac{I^2 R}{a} \frac{1}{r^2} P_n^{(m)}\left(\cos\left(\frac{\pi}{4}\right)\right) \sin\left(m\frac{\pi}{2}\right) \quad (7.249)$$

Finally we need to multiply this expression by $\frac{J_{n+1/2}\left(\sqrt{\lambda_{k'}^{(n)}}r\right)}{\sqrt{r}}$ and integrate $\int_0^a r^2 dr$. We get:

$$k\sum_k C_{kmn} \times \lambda_k^{(n)} \frac{\pi}{2} \frac{\left\| P_n^{(m)}(\cos(\theta)) \right\|^2}{2} \int_0^a \frac{J_{n+1/2}\left(\sqrt{\lambda_{k'}^{(n)}}r\right)}{\sqrt{r}} \times \frac{J_{n+1/2}\left(\sqrt{\lambda_k^{(n)}}r\right)}{\sqrt{r}} r^2 dr =$$

$$= \frac{I^2 R}{a} P_n^{(m)}\left(\cos\left(\frac{\pi}{4}\right)\right) \sin\left(m\frac{\pi}{2}\right) \int_0^a \left(\frac{1}{r^2}\right) \frac{J_{n+1/2}\left(\sqrt{\lambda_{k'}^{(n)}}r\right)}{\sqrt{r}} r^2 dr \tag{7.250}$$

$$\left\{ \begin{array}{l} \sin\left(m\frac{\pi}{2}\right) = \left\{ \begin{array}{ll} 0 & m = \text{even} \\ (-1)^{m+1} & m = \text{odd} \end{array} \right\} \\ \text{furthermore } m = n - 2l + 1 \end{array} \right\} \tag{7.251}$$

Then, the form of the coefficients for the expansion of $h(r, \theta, \varphi)$ is:

$$C_{lkn} = \frac{\frac{I^2 R}{a} P_n^{[n-2l+1]}\left(\cos\left(\frac{\pi}{4}\right)\right) \sin\left([n-2l+1]\frac{\pi}{2}\right) \int_0^a \frac{J_{n+1/2}\left(\sqrt{\lambda_{k'}^{(n)}}r\right)}{\sqrt{r}} dr}{k\lambda_k^{(n)} \frac{\pi}{2} \frac{\left\| P_n^{(m)}(\cos(\theta)) \right\|^2}{2} \left\| J_{n+1/2}\left(\sqrt{\lambda_{k'}^{(n)}}r\right) \right\|^2} \tag{7.252}$$

Mathematical formulation of problem (2)

$$\left\{ \begin{array}{l} \Delta g = 0 \\ g(r, \theta, 0) = g(r, \theta, \pi) = T \\ g(r, \frac{\pi}{2}, \varphi) = T \\ g(a, \theta, \varphi) = 0 \end{array} \right\} \tag{7.253}$$

It's more convenient to reformulate the problem to expand in angular eigenfunctions. We use a new function $g = g' + T_0$

The problem in Laplace's equation is:

$$\begin{cases} \Delta g' = 0 \\ g'(r,\theta,0) = g'(r,\theta,\pi) = 0 \\ g'(r,\frac{\pi}{2},\varphi) = 0 \\ g'(a,\theta,\varphi) = -T \end{cases} \tag{7.254}$$

Sturm–Liouville problem

We separate variables to seek the solution as a sum of angular orthogonal eigenfunctions $g' = \mathcal{R}(r) \cdot V(\theta,\varphi)$. We write the Laplacian as:

$$\frac{\frac{d}{dr}[r^2(\frac{d\mathcal{R}}{dr})]}{\mathcal{R}} = -\frac{\Delta_{\theta,\varphi}V}{V} = \lambda > 0 \tag{7.255}$$

Solving the problem for the angular part:

$$\Delta_{\theta,\varphi}V + \lambda V = 0 \tag{7.256}$$

Separating variables:

$$V(\theta,\varphi) = \Theta(\theta) \cdot \Phi(\varphi) \tag{7.257}$$

$$\frac{\sin(\theta)\frac{d}{d\theta}[\sin(\theta)(\frac{d\Theta}{d\theta})] + \lambda\sin^2(\theta)\cdot\Theta(\theta)}{\Theta(\theta)} = -\frac{1}{\Phi(\varphi)}\frac{d^2\Phi}{d\varphi^2} = \nu > 0 \tag{7.258}$$

We choose the positive sign for λ, ν because we expect to get periodic solutions (eigenfunctions) for the angular variables. The result of the problem for φ is:

$$\begin{cases} \frac{d^2\Phi}{d\varphi^2} + \nu\Phi(\varphi) = 0 \\ g'(r,\theta,0) \to \Phi(0) = 0 \\ g'(r,\theta,\pi) \to \Phi(\pi) = 0 \end{cases} \tag{7.259}$$

$$\Phi(\varphi) = A\cos(m\varphi) + B\sin(m\varphi); \quad \nu = m^2 \tag{7.260}$$

$$\Phi(0) = 0 \to A = 0 \to \tag{7.261}$$

$$\Phi(\varphi) = \sin(m\varphi) \tag{7.262}$$

Now that we have $\nu = m^2$, we can find the solution for the $\Theta(\theta)$ function:

$$\sin(\theta)\frac{d}{d\theta}\left[\sin(\theta)\left(\frac{d\Theta}{d\theta}\right)\right] + [\lambda\sin^2(\theta) - m^2]\Theta(\theta) = 0 \tag{7.263}$$

Its solution are Legendre polynomials:

$$\Theta(\theta) = P_n^{(m)}(\cos(\theta)) \tag{7.264}$$

Eigenvalues: $\lambda_n = n(n+1)$

Imposing the boundary condition in the plane: $\Theta(\frac{\pi}{2}) = P_n^{(m)}(0) = 0$

Just like for the previous function h, from the properties of Legendre polynomials we deduce that this condition is only satisfied when:

$$n - m = \text{odd} = 2l + 1 \qquad m = n - 2l + 1 \quad n = 1, 2, 3\ldots; \\ l = 0, 1, 2, 3, \ldots \text{ with } m < n \tag{7.265}$$

In this manner:

$$\Theta(\theta) = P_n^{[n-2l+1]}(\cos(\theta)); \ \lambda_n = n(n+1) \tag{7.266}$$

We can finally solve the equation for $\mathcal{R}(r)$:

$$\frac{\frac{d}{dr}[r^2(\frac{d\mathcal{R}}{dr})]}{\mathcal{R}} = \lambda_n \tag{7.267}$$

$$r^2\left(\frac{d^2\mathcal{R}}{dr^2}\right) + r\left(\frac{d\mathcal{R}}{dr}\right) - n(n+1)\mathcal{R} = 0 \tag{7.268}$$

Its solutions are a linear combination of: $\mathcal{R}_n(r) = C_1 r^n + C_2 r^{-(n+1)}$

Since the solution must be finite for $r = 0$ we have $C_2 = 0$.

General solution

With the previous we have the general solution:

$$g' = \sum_{n,m} C_{nm} r^n P_n^{(m)}(\cos(\theta))\sin(m\varphi) \tag{7.269}$$

Final solution

We impose the boundary conditions and use the orthogonality to find the C_{nm} coefficients:

$$\sum_{n,m} C_{nm} a^n P_n^{(m)}(\cos(\theta)) \sin(m\varphi) = -T \tag{7.270}$$

We multiply by $\sin(m'\varphi)$ and integrate $\int_0^{\pi} d\varphi$:

$$\sum_{n,m} C_{nm} a^n P_n^{(m)}(\cos(\theta)) \int_0^{\pi} \sin(m'\varphi)\sin(m\varphi) d\varphi = -T \int_0^{\pi} \sin(m'\varphi) d\varphi \tag{7.271}$$

$$\sum_{n} C_{nm} a^n P_n^{(m)}(\cos(\theta)) \times \frac{\pi}{2} = -T\frac{1}{m}[(-1)^m - 1] \tag{7.272}$$

We now multiplicate by $P_{n'}^{(m)}(\cos(\theta))$ and integrate $\int_0^{\pi/2} \sin(\theta) d\theta$:

$$\sum_{n} C_{nm} a^n \times \frac{\pi}{2} \int_0^{\pi/2} P_{n'}^{(m)}(\cos(\theta)) P_n^{(m)}(\cos(\theta)) \sin(\theta) d\theta$$

$$= -T\frac{1}{m}[(-1)^m - 1] \int_0^{\pi/2} P_{n'}^{(m)}(\cos(\theta)) \sin(\theta) d\theta \tag{7.273}$$

$$C_{nm} a^n \times \frac{\pi}{2} \frac{\left\| P_n^{(m)}(\cos(\theta)) \right\|^2}{2}$$

$$= -T \frac{[(-1)^m - 1]}{m} \int_0^{\pi/2} P_n^{(m)}(\cos(\theta)) \sin(\theta) d\theta \tag{7.274}$$

$$C_{nm} = -\frac{4T}{\pi} \frac{[(-1)^m - 1]}{\left\| P_n^{(m)}(\cos\theta) \right\|^2 m a^n} \int_0^{\pi/2} P_n^{(m)}(\cos(\theta)) \sin(\theta) d\theta \tag{7.275}$$

with $m = n - 2l + 1$; $n = 1, 2, 3 \ldots$; $l = 0, 1, 2, 3, \ldots$ and with $m < n$

So the final solution is:

$$u(r, \theta, \varphi) = g'(r, \theta, \varphi) + h(r, \theta, \varphi) + T \tag{7.276}$$

7.12 Case Study: Two Concentric Semi-Spheres

Find the stationary distribution of the displacement $u(r, \theta, \varphi, t)$ inside a semi-sphere composed of two semi-spheres (radii R_1, R_2) with different densities (ρ_1, ρ_2) and Young moduli E_1, E_2. The common base of the semi-spheres is fixed, while the curved part is subject to a force in the form of an acoustic plane pressure wave $P_0 \sin(\omega t)$ from $t = 0$ and directed along $\theta = 0$.

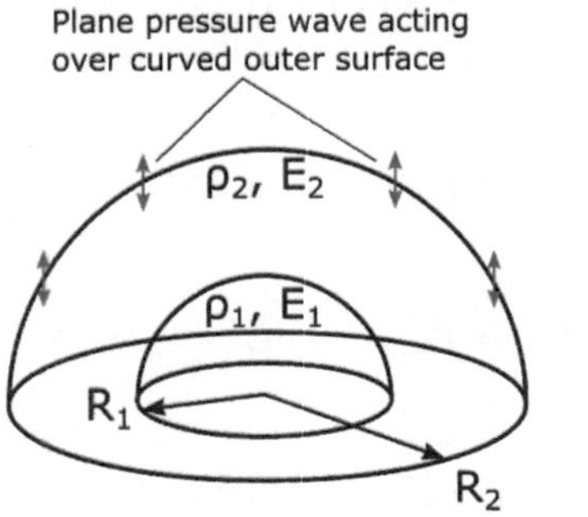

Figure 7.17

Mathematical formulation

$$\left\{\begin{array}{l} \dfrac{\partial^2 u}{\partial t^2} - \dfrac{1}{\rho(r)} \nabla[E(r)\nabla u(r, \theta, \varphi, t)] = 0 \\ |u(r = 0, \theta, \varphi, t)| < \infty \\ E_2 \left.\dfrac{\partial u(r, \theta, \varphi, t)}{\partial r}\right|_{r=R_2} = P_0 \sin(\omega t) \cos(\theta) \\ u(r, \theta = \pi/2, \varphi, t) = 0 \end{array}\right\} \tag{7.277}$$

Where

$$\rho(r) = \begin{cases} \rho_1 \; if \; 0 < r < R_1 \\ \rho_2 \; if \; R_1 < r < R_2 \end{cases}$$

$$E(r) = \begin{cases} E_1 \; if \; 0 < r < R_1 \\ E_2 \; if \; R_1 < r < R_2 \end{cases}$$

We seek the stationary solution ($t \to \infty$). Because the excitation is of the form $\sin(\omega t)$, at long times the displacement will change periodically as $\sin(\omega t)$ (can be seen in the boundary condition).

Then,

$$u(r, \theta, \varphi, t) = U(r, \theta, \varphi) \sin(\omega t).$$

Equation (7.277), dividing both sides by $\sin(\omega t)$, is of the form:

$$\left\{ \begin{array}{l} -\omega^2 U(r, \theta, \varphi) - \dfrac{1}{\rho(r)} \nabla[E(r) \nabla U(r, \theta, \varphi)] = 0 \\ |U(r = 0, \theta, \varphi, t)| < \infty \\ E_2 \left. \dfrac{\partial U(r, \theta, \varphi)}{\partial r} \right|_{r=R_2} = P_0 \cos(\theta) \\ U(r, \theta = \pi/2, \varphi) = 0 \end{array} \right\} \tag{7.278}$$

Sturm–Liouville problem

We expand in orthogonal eigenfunctions in the angular directions (θ, φ), since they have homogeneous boundary conditions. Separating variables:

$$U(r, \theta, \varphi) = R(r) V(\theta, \varphi)$$

$$\frac{1}{\rho(r)} \cdot \frac{1}{r^2} \left[\frac{\frac{d}{dr}\left(r^2 E(r) \frac{dR(r)}{dr}\right)}{R(r)} + \frac{\frac{1}{\sin\theta} \frac{\partial}{\partial\theta} \cdot \left(\sin\theta \frac{\partial V}{\partial\theta}\right) + \frac{1}{\sin^2\theta} \frac{\partial^2 V}{\partial\varphi^2}}{V(\theta, \varphi)} \right] = -\omega^2 \tag{7.279}$$

Sturm–Liouville problem for the angular part:

$$\Delta_{\theta,\varphi} V(\theta, \varphi) + \lambda V(\theta, \varphi) = 0$$

$$\frac{\frac{1}{\sin\theta} \frac{\partial}{\partial\theta} \left(\sin\theta \frac{\partial V}{\partial\theta}\right) + \frac{1}{\sin^2\theta} \frac{\partial^2 V}{\partial\varphi^2}}{V(\theta, \varphi)} = -\lambda \tag{7.280}$$

We separate variables and formulate an auxiliar Sturm–Liouville problem for φ.

$$V(\theta, \varphi) = \Theta(\theta) \Phi(\varphi)$$

$$\frac{\sin\theta \frac{\partial}{\partial\theta} \left(\sin\theta \frac{\partial\Theta(\theta)}{\partial\theta}\right)}{\Theta(\theta)} + \lambda \sin^2\theta = -\frac{\frac{d^2\Phi(\varphi)}{d\varphi^2}}{\Phi(\varphi)} = \mu \tag{7.281}$$

Azimuthal part (φ)

$$\begin{Bmatrix} \Phi''(\varphi) + \mu \cdot \Phi(\varphi) = 0 \\ \Phi(\varphi) = \Phi(\varphi) + 2\pi \end{Bmatrix}$$

$$\Phi(\varphi) = A_m \cos(m\varphi) + B_m \sin(m\varphi)$$

$$\mu = m^2$$

Polar part (θ)

$$\frac{1}{\sin\theta}\frac{\partial}{\partial\theta}\left(\sin\theta\frac{\partial\Theta(\theta)}{\partial\theta}\right) + \left(\lambda - \frac{m^2}{\sin^2\theta}\right)\Theta(\theta) = 0$$

With the change $x = \cos\theta$, becomes the associated Legendre equation, whose solution are the associated Legendre polynomials.

$$\Theta_n^m(\theta) = P_n^m(\cos\theta)$$

$$\lambda = n(n+1)$$

Applying the boundary condition:

$$U(r, \theta = \pi/2, \varphi) = 0 \Rightarrow \Theta(\theta = \pi/2) = 0 \Rightarrow P_n^m(0) = 0$$

Due to the properties of Legendre polynomials $n + m$ must be odd.

General solution

$$u(r, \theta, \varphi, t) = \sum_n \sum_m R_{nm}(r) P_n^m(\cos\theta) e^{im\varphi} \sin(\omega t) \tag{7.282}$$

By substituting equation (7.280) into (7.279), we get to the general solution (7.282) and arrive to an equation for the radial part.

$$\frac{\frac{1}{r^2}\frac{d}{dr}\left(r^2 E(r)\frac{dR(r)}{dr}\right)}{R(r)} - \frac{\lambda}{r^2} = -\omega^2\rho(r)$$

We divide the problem in regions $0 < r < R_1$ and $R_1 < r < R_2$, where ρ and E are constants in each region, resulting equations of the form:

$$\begin{cases} \dfrac{1}{\tilde{R}_1(r)} E_1 \dfrac{1}{r^2}\dfrac{d}{dr}\left(r^2\dfrac{d\tilde{R}_1(r)}{dr}\right) - \dfrac{\lambda}{r^2} = -\omega^2\rho_1 \; 0 < r < R_1 \\ \dfrac{1}{\tilde{R}_2(r)} E_2 \dfrac{1}{r^2}\dfrac{d}{dr}\left(r^2\dfrac{d\tilde{R}_2(r)}{dr}\right) - \dfrac{\lambda}{r^2} = -\omega^2\rho_2 \; R_1 < r < R_2 \end{cases} \tag{7.283}$$

$$\begin{cases} \dfrac{d^2\tilde{R}_1}{dr^2} + \dfrac{2}{r}\dfrac{d\tilde{R}_1}{dr} + \left(\dfrac{\omega^2}{a_1} - \dfrac{n(n+1)}{r^2}\right)\tilde{R}_1 = 0 \; 0 < r < R_1 \\ \dfrac{d^2\tilde{R}_2}{dr^2} + \dfrac{2}{r}\dfrac{d\tilde{R}_2}{dr} + \left(\dfrac{\omega^2}{a_2} - \dfrac{n(n+1)}{r^2}\right)\tilde{R}_2 = 0 \; R_1 < r < R_2 \end{cases} \tag{7.284}$$

With $a_i = E_i/\rho_i$, $a_1 \neq a_2$ since we cannot assume that the speed of sound is equal in both materials.

The solution of $\tilde{R}_i$ is:

$$\begin{cases} \tilde{R}_1(r) = \dfrac{A_{n,m}}{\sqrt{r}} J_{n+\frac{1}{2}}\left(\sqrt{\dfrac{\omega^2}{a_1}}r\right) \\ \tilde{R}_2(r) = \dfrac{B_{n,m}}{\sqrt{r}} J_{n+\frac{1}{2}}\left(\sqrt{\dfrac{\omega^2}{a_2}}r\right) + \dfrac{C_{n,m}}{\sqrt{r}} N_{n+\frac{1}{2}}\left(\sqrt{\dfrac{\omega^2}{a_2}}r\right) \end{cases} \tag{7.285}$$

Where we have used that $|\tilde{R}_1(0)| < \infty$.

The boundary conditions are:

- Continuity of the function:

$$\tilde{R}_1(R_1) = \tilde{R}_2(R_1) \tag{7.286}$$

- Continuity of the derivatives:

$$E_1 \left.\frac{d\tilde{R}_1(r)}{dr}\right|_{r=R_1} = E_2 \left.\frac{d\tilde{R}_2(r)}{dr}\right|_{r=R_1} \tag{7.287}$$

The third boundary condition is obtained by imposing

$$E_2 \left.\frac{\partial U(r,\theta,\varphi)}{\partial r}\right|_{r=R_2} = P_0(\cos\theta)) \tag{7.288}$$

Due to the symmetry of the problem, we see that the only index that remains is $m = 0$ (there is no azimuthal dependence). We will equate the $\cos\theta$ term to the first Legendre polynomial ($P_1(\cos\theta) = \cos\theta$).

$$\sum_n \frac{d\tilde{R}_2}{dr} P_n(\cos\theta) = \frac{P_0}{E_2} P_1(\cos\theta) \tag{7.289}$$

Multiplying both sides by $P_n(\cos\theta)\sin\theta$, and integrating in θ between 0 and $\pi/2$

$$\frac{d\tilde{R}_2}{dr}|P_n(\cos\theta)|^2 = \frac{P_0}{E_2}\int_{\theta=0}^{\theta=\pi/2} P_1(\cos\theta)P_n(\cos\theta)\sin\theta d\theta \quad (7.290)$$

$$\int_{\theta=0}^{\theta=\pi/2} P_1(\cos\theta)P_n(\cos\theta)\sin\theta d\theta = \int_{x=0}^{x=1} P_1(x)P_n(x)dx \quad (7.291)$$

Next we will evaluate this integral, considering that $P_n(x)$ is odd for odd values of n, and even when n is even.

Case with odd n:

$$\int_0^1 P_1(x)P_n(x)dx = \frac{1}{2}\int_{-1}^1 P_1(x)P_n(x)dx = \frac{1}{2}\frac{2}{2n+1}\delta_{1,n} \quad (7.292)$$

$$\int_0^1 P_1(x)P_n(x)dx = \frac{1}{2}\int_{-1}^1 P_1(x)P_n(x)dx = \begin{cases} 0 & if\, n \neq 1 \\ \frac{1}{3} & if\, n = 1 \end{cases} \quad (7.293)$$

Since we are considering the integral of an even function (product of two odd functions), that is symmetric with respect to 0. We have applied the orthogonality condition of the Legendre polynomials.

Case of even n:

$$\int_0^1 P_1(x)P_n(x)dx \neq 0 \quad (7.294)$$

However, from the boundary condition $U(r, \theta = \pi/2, \varphi) = 0$, we had concluded that n should be odd, because of which the only possibility is $n = 1$.

The third boundary condition has the form:

$$E_2\frac{d\tilde{R}_2(r)}{dr}\bigg|_{r=R_2} = P_0 \quad (7.295)$$

From 7.286, 7.287 and 7.295 we calculate A_1, B_1 and C_1 $(m = 0)$.

$$\begin{cases} \tilde{R}_1(r) = \frac{A_1}{\sqrt{r}}J_{3/2}\left(\sqrt{\frac{\omega^2}{a_1}}r\right) \\ \tilde{R}_2(r) = \frac{B_1}{\sqrt{r}}J_{3/2}\left(\sqrt{\frac{\omega^2}{a_2}}r\right) + \frac{C_1}{\sqrt{r}}N_{3/2}\left(\sqrt{\frac{\omega^2}{a_2}}r\right) \end{cases} \quad (7.296)$$

Using 7.286:

$$\frac{A_1}{\sqrt{R_1}}J_{3/2}\left(\sqrt{\frac{\omega^2}{a_1}}R_1\right) = \frac{B_1}{\sqrt{R_1}}J_{3/2}\left(\sqrt{\frac{\omega^2}{a_2}}R_1\right) + \frac{C_1}{\sqrt{R_1}}N_{3/2}\left(\sqrt{\frac{\omega^2}{a_2}}R_1\right) \quad (7.297)$$

Using 7.287:

$$E_1\frac{d}{dr}\left[\frac{A_1}{\sqrt{r}}J_{3/2}\left(\sqrt{\frac{\omega^2}{a_1}}r\right)\right]_{r=R_1} = E_2\frac{d}{dr}\left[\frac{B_1}{\sqrt{r}}J_{3/2}\left(\sqrt{\frac{\omega^2}{a_2}}r\right) + \frac{C_1}{\sqrt{r}}N_{3/2}\left(\sqrt{\frac{\omega^2}{a_2}}r\right)\right]_{r=R_1} \quad (7.298)$$

Using 7.295:

$$E_2\frac{d}{dr}\left[\frac{B_1}{\sqrt{r}}J_{3/2}\left(\sqrt{\frac{\omega^2}{a_2}}r\right) + \frac{C_1}{\sqrt{r}}N_{3/2}\left(\sqrt{\frac{\omega^2}{a_2}}r\right)\right]_{r=R_2} = P_0 \quad (7.299)$$

Equations (7.297), (7.298) and (7.299) are a system of three equations with three unknowns (A_1, B_1, C_1).

For simplicity we rename:

$$\frac{1}{\sqrt{r}}J_{3/2}\left(\sqrt{\frac{\omega^2}{a_i}}r\right) = j(a_i, r) \quad (7.300)$$

$$\frac{1}{\sqrt{r}}N_{3/2}\left(\sqrt{\frac{\omega^2}{a_i}}r\right) = n(a_i, r) \quad (7.301)$$

$$\frac{d}{dr}\left[\frac{1}{\sqrt{r}}J_{3/2}\left(\sqrt{\frac{\omega^2}{a_i}}r\right)\right]_{r=R_i} = j'(a_i, R_i) \quad (7.302)$$

$$\frac{d}{dr}\left[\frac{1}{\sqrt{r}}N_{3/2}\left(\sqrt{\frac{\omega^2}{a_i}}r\right)\right]_{r=R_i} = n'(a_i, R_i) \quad (7.303)$$

From 7.299 we get:

$$B_1 = \frac{-n'(a_2, R_2)C_1 + \frac{P_0}{E_2}}{j'(a_2, R_2)} \tag{7.304}$$

Replacing in 7.297

$$A_1 = \frac{n(a_2, R_1) - \frac{n'(a_2, R_2)}{j'(a_2, R_2)} j(a_2, R_1)}{j(a_1, R_1)} C_1 + \frac{\frac{P_0}{E_2} j(a_2, R_1)}{j'(a_2, R_2) j(a_1, R_1)} \tag{7.305}$$

Replacing now in 7.298

$$E_1 \left(\frac{n(a_2, R_1) - \frac{n'(a_2, R_2)}{j'(a_2, R_2)} j(a_2, R_1)}{j(a_1, R_1)} C_1 + \frac{\frac{P_0}{E_2} j(a_2, R_1)}{j'(a_2, R_2) j(a_1, R_1)} \right) j'(a_1, R_1) =$$

$$= E_2 \frac{-n'(a_2, R_2)C_1 + \frac{P_0}{E_2}}{j'(a_2, R_2)} j'(a_2, R_1) + E_2 C_1 n'(a_2, R_1) \tag{7.306}$$

We now isolate C_1

$$-\frac{\frac{P_0}{E_2} j(a_2, R_1)}{j'(a_2, R_2) j(a_1, R_1)} E_1 j'(a_1, R_1) + \frac{P_0 j'(a_2, R_1)}{j'(a_2, R_2)} =$$

$$= C_1 \left[E_1 \left(-\frac{n'(a_2, R_2) j(a_2, R_1)}{j'(a_2, R_2) j(a_1, R_1)} + \frac{n(a_2, R_1)}{j(a_1, R_1)} \right) j'(a_1, R_1) \right.$$
$$\left. + E_2 \left(\frac{n'(a_2, R_2)}{j'(a_2, R_2)} j'(a_2, R_1) - n'(a_2, R_1) \right) \right]$$

C_1 is:

$$C_1 = \frac{-P_0 \frac{E_1}{E_2} j(a_2, R_1) j'(a_1, R_1) + P_0 j'(a_2, R_1) j(a_1, R_1)}{E_1 \chi_1 j'(a_1, R_1) + E_2 \chi_2 j(a_1, R_1)} \tag{7.307}$$

$$\begin{cases} \chi_1 = -n'(a_2, R_2) j(a_2, R_1) + n(a_2, R_1) j'(a_2, R_2) \\ \chi_2 = n'(a_2, R_2) j'(a_2, R_1) - n'(a_2, R_1) j'(a_2, R_2) \end{cases}$$

The constants A_1 and B_1 are obtained in a similar fashion.

Final solution

$$u(r, \theta, \varphi, t) = \tilde{R}(r) \cos\theta \sin(\omega t) \tag{7.308}$$

$$\tilde{R}(r) = \begin{cases} \frac{A_1}{\sqrt{r}} J_{3/2}\left(\sqrt{\frac{\omega}{a_1}} r\right) & 0 < r < R_1 \\ \frac{B_1}{\sqrt{r}} J_{3/2}\left(\sqrt{\frac{\omega}{a_2}} r\right) + \frac{C_1}{\sqrt{r}} N_{3/2}\left(\sqrt{\frac{\omega}{a_2}} r\right) & R_1 < r < R_2 \end{cases}$$

7.13 Case Study: Variation of Temperature in a Semisphere

Find the distribution of temperature $u(r, \theta, \varphi, t)$ in a semisphere of radius $r = R$ with $\varphi \in (0, \pi) rad$, centered in (0,0,0). For $r = R$ the temperature is $u(r = R, \theta, \varphi, t) = T_0$. For $\varphi = 0$ and $\varphi = \pi$ the temperature is zero. Situated at a polar angle $\theta = \theta_0$ half of a ring (of radius $r_0 < R$) emits heat with a heat flux $J \cdot e^{-t} \cdot \varphi$, with $\varphi \in (0, \pi)$. At the initial instant $t = 0$ the temperature is $u(r, \theta, \varphi, t = 0) = 0$. C is the heat capacity, k is thermal conductivity and ρ_0 the density of mass.

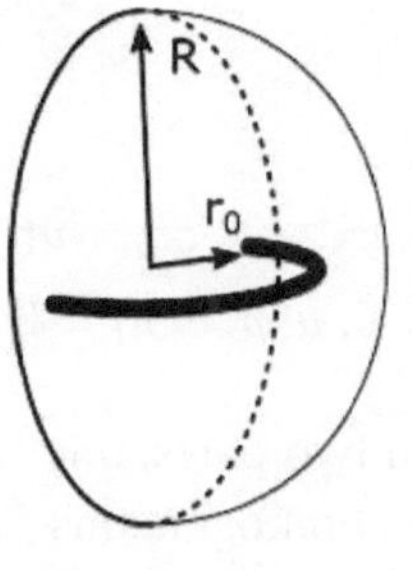

Figure 7.18

Mathematical formulation

$$\frac{\partial u}{\partial t} - k/(C\rho_0)\Delta u = f(r, \theta, \varphi, t) \quad (7.309)$$

With $\kappa = k/(C\rho_0)$ being the thermal diffusivity.

With $f = f(r, \theta, \varphi, t)$:

$$f = f_c \cdot f_r(r) \cdot f_\theta(\theta) \cdot f_\varphi(\varphi) \cdot f_t(t)$$
$$= -J/(C \cdot \rho_0 \cdot \pi \cdot r_0) \cdot \delta(r - r_0) \cdot \delta(\theta - \theta_0) \cdot e^{-t} \cdot \varphi/(\pi r^2 \sin(\theta)) \quad (7.310)$$

The term $J \cdot e^{-t} \cdot \varphi/(\pi \cdot r_0)$ represents the heat flux of the ring, being $(\pi \cdot r_0)$ the length of the half ring.

$$f_c \equiv \text{const} = -J/(C\rho_0\pi^2 r_0), \quad f_r(r) = \frac{\delta(r - r_0)}{r^2},$$
$$f_\theta(\theta) = \frac{\delta(\theta - \theta_0)}{\sin(\theta)}, \quad f_\varphi = \varphi, \quad f_t(t) = e^{-t} \quad (7.311)$$

With the Laplacian Δu:

$$\Delta u = \frac{1}{r^2}\frac{\partial\left(\frac{\partial u}{\partial r} \cdot r^2\right)}{\partial r} + \frac{1}{r^2}\Delta_{\theta,\varphi} u \quad (7.312)$$

$$\Delta_{\theta,\varphi} u = \frac{1}{\sin(\theta)} \cdot \frac{\partial\left(\frac{\partial u}{\partial \theta} \cdot \sin(\theta)\right)}{\partial \theta} + \frac{1}{\sin^2(\theta)} \cdot \frac{\partial^2 u}{\partial \varphi^2} \quad (7.313)$$

Initial conditions:

$$u(r, \theta, \varphi, t = 0) = 0 \quad (7.314)$$

Boundary conditions:

$$|u(r = 0, \theta, \varphi)| < +\infty, \quad u(r = R) = T_0,$$
$$u(\varphi = 0) = 0, \; u(\varphi = \pi) = 0 \quad (7.315)$$

We divide the problem in two parts, one is stationary (1) $w(r, \theta, \varphi)$ that we choose to be homogeneous with non homogeneous boundary conditions and the other is transient (2) $v(r, \theta, \varphi, t)$ and is non homogeneous but with homogeneous boundary conditions:

$$u(r, \theta, \varphi, t) = w(r, \theta, \varphi) + v(r, \theta, \varphi, t) \quad (7.316)$$

Problem 1

$$\Delta w(r, \theta, \varphi) = 0 \quad (7.317)$$

$$|w(r = 0, \theta, \varphi)| < +\infty, \quad w(r = R) = T_0,$$
$$w(\varphi = 0) = 0, \quad w(\varphi = \pi) = 0 \quad (7.318)$$

Sturm–Liouville problem

We separate variables: $w(r, \theta, \varphi) = \sum \sum C \cdot R(r) \cdot V_{\theta,\varphi}$.

$$\frac{\partial\left(r^2 \cdot \frac{\partial R}{\partial r}\right)}{\partial r}/R = -\frac{\Delta_{\theta,\varphi} V}{V} = \lambda > 0 \qquad (7.319)$$

And so, we have the value of V:

$$\frac{1}{\sin(\theta)} \cdot \frac{\partial\left(\frac{\partial V}{\partial \theta} \cdot \sin(\theta)\right)}{\partial \theta} + \frac{1}{\sin^2(\theta)} \cdot \frac{\partial^2 V}{\partial \varphi^2} + \lambda \cdot V = 0 \qquad (7.320)$$

At the same time we solve the Sturm–Liouville problem for $\Phi(\varphi)$, $\Theta(\theta)$, with $V(\theta, \varphi) = \Theta(\theta) \cdot \Phi(\varphi)$

$$\frac{\partial^2 \Phi(\varphi)}{\partial \varphi^2} + \lambda_\varphi \cdot \Phi = 0 \qquad (7.321)$$

Boundary conditions of $\Phi(\varphi)$

$$\Phi(0) = 0,\ \Phi(\pi) = 0 \qquad (7.322)$$

General solution

The general solution for $\Phi(\varphi)$ (7.321) is:

$$\Phi(\varphi) = C_\varphi \cdot \cos(\sqrt{\lambda_\varphi} \cdot \varphi) + C'_\varphi \cdot \sin(\sqrt{\lambda_\varphi} \cdot \varphi) \qquad (7.323)$$

Using the boundary conditions for $\Phi(\varphi)$ at $\varphi = 0, \pi$ we impose the temperature to be zero.

$$\Phi(\varphi) = C'_\varphi \cdot \sin(m \cdot \varphi);\ \sqrt{\lambda_\varphi} = m \qquad (7.324)$$

The polar component, which for which we need $\lambda = n \cdot (n + 1)$, $l \geq |m|$, so that the solution doesn't diverge:

$$1/\sin(\theta) \cdot \frac{d\left(\frac{\partial \Theta(\theta)}{\partial \theta} \cdot \sin(\theta)\right)}{\partial \theta} + \left(\frac{1}{\sin^2(\theta)} \cdot (-m^2) + n \cdot (n+1)\right) \cdot \Theta = 0 \qquad (7.325)$$

The solutions are Legendre polynomials:

$$\Theta(\theta) = P_n^{(m)}\cos(\Theta) \tag{7.326}$$

The radial part $R(r)$ doesn't satisfy a Sturm–Liouville problem, since its boundaries are not homogeneous ($u(r = R) = T_0 \neq 0$).

$$\rho^2 \cdot d^2R(r)/dr^2 + 2 \cdot \rho \cdot dR/dr - n(n+1)R = 0 \tag{7.327}$$

Has as a general solution:

$$R(r) = C_r \cdot r^n + C_r' \cdot r^{-(n+1)} \tag{7.328}$$

Imposing that $u(r = 0)$ must not diverge we get that the term $C_r' \cdot r^{-(n+1)}$ is zero:

$$R(r) = C_r \cdot r^n \tag{7.329}$$

The boundary condition $w(r = R) = T_0$ is now obtained by multiplying both sides of the equation by $\Theta(\theta) \cdot \Phi(\varphi) \cdot \sin(\theta)$ and integrating θ between 0 and π and φ between 0 and π:

$$C_{nm} = \frac{\int_0^\pi \int_0^\pi T_0 \cdot P_n^{(m)}(\cos(\theta)) \cdot \sin(\theta) \cdot \sin(m \cdot \varphi)\, d\theta\, d\varphi}{R^n \int_0^\pi (P_n^{(m)}(\cos(\theta)))^2 \cdot \sin(\theta)\, d\theta \int_0^\pi \sin^2(\varphi)\, d\varphi} \tag{7.330}$$

$$C_{nm} = \frac{T_0 \cdot \int_0^\pi P_n^{(m)}(\cos(\theta)) \cdot \sin(\theta)\, d\theta[(-1)^{(m+1)} + 1]/m}{R^n \int_0^\pi (P_n^{(m)}(\cos(\theta)))^2 \cdot \sin(\theta)\, d\theta \cdot \pi/2} \tag{7.331}$$

Considering $m = 2 \cdot l + 1$, with $l \in \mathbb{Z}$ (integer numbers):

$$C_{nl} = \frac{T_0 \cdot \int_0^\pi \cdot P_n^{(2 \cdot l+1)}(\cos(\theta)) \cdot \sin(\theta) \cdot\, d\theta \cdot 2/(2 \cdot l + 1)}{R^n \int_0^\pi (P_n^{(2 \cdot l+1)}(\cos(\theta)))^2 \cdot \sin(\theta)\, d\theta \cdot \pi/2} \tag{7.332}$$

And the general solution for the stationary problem is:

$$w(r, \Theta, \varphi) = \sum_{n,l} C_{nl} \cdot r^n \cdot P_n^{(m)}(\cos(\theta)) \cdot \sin(m \cdot \varphi) \tag{7.333}$$

Problem 2

$$\frac{\partial^2 v}{\partial t^2} - \Delta v(r, \theta, \varphi, t) = -\frac{J}{C\rho_0 \cdot 2\pi r_0} \cdot \delta(r - r_0) \cdot \delta(\theta - \theta_0) \cdot e^{-t} \times \varphi \frac{1}{2\pi r^2 \sin(\theta)} \tag{7.334}$$

$$|v(r=0,\theta,\varphi)| < +\infty,\ v(r=R)=0,\ v(\varphi=0)=0,\ v(\varphi=\pi)=0 \tag{7.335}$$

$$v(t=0) = -w(r,\theta,\varphi) \tag{7.336}$$

Sturm–Liouville problem

Separating variables $v = \sum T(t)\cdot N(r,\theta,\varphi)$:

For $N = N(r,\theta,\varphi)$:

$$\Delta N(r,\theta,\varphi) + \lambda' \cdot N = 0 \tag{7.337}$$

Next we separate variables once again: $N(r,\theta,\varphi) = R'(r)\cdot V'(\theta,\varphi)$.

And also for the θ and φ coordinates: $V' = v'(\theta)\cdot \Phi'(\varphi)$

$$\Delta V'(\theta,\varphi) + \lambda_{\theta'}\cdot V' = 0 \tag{7.338}$$

For the transient part v, we use m' as the index for the azimuthal part $\Phi'(\varphi)$ and n' for $\Theta'(\theta)$.

For $\Theta'(\theta)$ (for v) the equation and boundary conditions are the same as for $\Theta(\theta)$ (from the stationary problem w) (7.325), because of which its solution are Legendre polynomials:

$$\frac{1}{\sin(\theta)}\cdot\frac{\partial\left(\frac{\partial\Theta'(\theta)}{\partial\theta}\cdot\sin(\theta)\right)}{\partial\theta} + \left(\frac{1}{\sin^2(\theta)}\cdot(-m^2)+\lambda_{\theta'}\right)\cdot\Theta' = 0;$$
$$\lambda_{\theta'} = n'\cdot(n'+1) \tag{7.339}$$

$$\Theta(\theta) = P_{n'}^{m'}(\cos(\theta)) \tag{7.340}$$

In the same manner the index associated to θ, $\lambda_{\theta'}$ in this case, $\lambda_{\theta'} = n'\cdot(n'+1)$ so that $\Theta'(\theta)$ doesn't diverge.

Since the boundary conditions and the equation for φ are equal than for w, $\Phi'(\varphi)$ has the same solution: $\Phi'(\varphi) = C'_\varphi\cdot\sin(m'\cdot\varphi)$.

In the case of $R'(r)$:

$$\frac{1}{r^2}\cdot\frac{d(r^2\cdot dR'/dr)}{dr} + (\lambda' - \lambda_{\theta'}/r^2)R' = 0 \tag{7.341}$$

$$R'(r=0) < +\infty;\ R'(r=R) = 0 \tag{7.342}$$

We use that $\lambda_{\theta'} = n' \cdot (n'+1)$ so that $\Theta' = P_{n'}^{m'}$ doesn't diverge and using the change of variable: $b(r) = \sqrt{r} \cdot R'$:

$$\frac{d^2 b(r)}{dr^2} + \left(\lambda - \frac{(n'+1/2)^2}{r^2}\right) b(r) = 0 \tag{7.343}$$

With solution:

$$b_{n'}(r) = C_{r1} \cdot J_{n'+1/2}(r) + C_{r2} \cdot N_{n'+1/2}(r) \tag{7.344}$$

$$R'_{n'}(r) = C_{r1} \frac{J_{n'+1/2}(\lambda' \cdot r)}{\sqrt{r}} + C_{r2} \frac{N_{n'+1/2}(\lambda' \cdot r)}{\sqrt{r}} \tag{7.345}$$

Using $u(r=0) < +\infty$ we get that the Neumann term (divergent at $r=0$) must be zero ($C_{r2} = 0$):

$$R'_{n'}(r) = C_{r1} \cdot \frac{J_{n'+1/2}(\lambda' \cdot r)}{\sqrt{r}} \tag{7.346}$$

Imposing $R(r=R) = 0$ we get the equation of the eigenvalues λ':

$$\frac{J_{n'+1/2}(\lambda' \cdot R)}{\sqrt{R}} = 0 \tag{7.347}$$

Using (7.337) and $v = \sum T \cdot N(r, \theta, \varphi)$:

$$\left(\frac{\partial T}{\partial t} + \kappa \cdot \lambda' \cdot T\right) \cdot N = f(r, \theta, \varphi, t) \tag{7.348}$$

General solution

Using the orthogonality of $N = R'(r) \cdot V'(\theta, \varphi)$:

$$\left(\frac{\partial T}{\partial t} + \kappa \cdot \lambda' \cdot T\right) \cdot N = f(r, \theta, \varphi, t) \tag{7.349}$$

With $\varphi \in (0, \pi)$ rad (limits of the semisphere):

$$\frac{\partial T}{\partial t} + \kappa \cdot T = \frac{f_c \cdot f_t(t)}{|N|^2} \int_0^{\pi} \int_0^{\pi} \int_0^{R} f_r(r) f_\theta(\theta) R'(r) \times \Theta'(\theta) \cdot \Phi'(\varphi) r^2 \cdot \sin(\theta)\, dr\, d\theta\, d\varphi \tag{7.350}$$

On the other hand we indicate the explicit value of f_c, f_r, f_θ with (7.311):

$$\frac{\partial T}{\partial t} + \kappa \cdot T = -\frac{J}{C \cdot \rho_0 \cdot \pi^2 \cdot r_0 |N|^2} \cdot e^{-t} \cdot \pi \int_0^{R} \delta(r - r_0) \times dr \int_0^{\pi} \delta(\theta - \theta_0)\, d\theta \int_0^{\pi} \varphi d\varphi \tag{7.351}$$

Using the next property of the Dirac's delta function:
$\int_a^b f(x)\delta(x-x_0)dx = f(x_0)$, with $a < x_0 < b \in \mathbb{R}$ (real numbers), we get at:

$$\frac{\partial T}{\partial t} + \kappa \cdot \lambda' \cdot T = \frac{2/m' \cdot (-1)^{m'+1} \cdot f_c \cdot R'(r_0) \cdot \Theta'(\theta_0)}{\pi/2|N|^2 \int_0^R R'(r)^2 \cdot r^2\, dr \cdot \int_0^\pi \Theta'(\theta)^2 \sin(\theta)\, d\theta} \cdot e^{-t} \tag{7.352}$$

The term that multiplies e^{-t} is renamed for simplicity:

$$F_t = \frac{2/m' \cdot (-1)^{m'+1} \cdot f_c \cdot R'(r_0) \cdot \Theta'(\theta_0)}{\pi/2|N|^2 \int_0^R R'(r)^2 \cdot r^2\, dr \cdot \int_0^\pi \Theta'(\theta)^2 \sin(\theta)\, d\theta}$$

To find $T(t)$ (general solution) we find the sum of the solution of the homogeneous equation $T_h(t)$ and the particular solution $T_p(t)$. We first solve the homogeneous equation for $T_h(t)$:

$$\frac{\partial T_h}{\partial t} + \kappa \cdot \lambda' \cdot T_h = 0 \tag{7.353}$$

Whose solution is exponential:

$$T_h = C \cdot e^{-\kappa\lambda' t} \tag{7.354}$$

Final solution

To find the particular solution we use the method of the unknown coefficients, introducing a particular solution $C' \cdot e^{-t}$, suposing that $\kappa \cdot \lambda' - 1 \neq 0$.

$$T_p = F_t \cdot e^{-t}/(\kappa \cdot \lambda' - 1) \tag{7.355}$$

Then $T(t)$, with $\kappa \cdot \lambda' - 1 \neq 0$ is:

$$T(t) = T_h + Tp = C \cdot e^{-\lambda' \kappa t} + F_t \cdot e^{-t}/(\kappa \cdot \lambda' - 1) \tag{7.356}$$

If $\kappa \cdot \lambda' - 1 = 0$:

$$T(t) = C \cdot e^{-t} + F_t \cdot t \cdot e^{-t} \tag{7.357}$$

We use the initial condition of v (7.336) ($v(t=0) = -w(r, \theta, \varphi)$):

$$-w(r, \theta, \varphi) = \sum T(0) \cdot R'(r) \cdot \Theta'(\theta) \cdot \Phi'(\varphi) \tag{7.358}$$

To calculate the C constant of $T(t)$ we use the orthogonality of $R'(r)$, $\Theta'(\theta)$, $\Phi'(\varphi)$, and integrate:

$$\frac{\int_0^\pi \int_0^\pi \int_0^R -w(r,\theta,\varphi)\cdot R'(r)\cdot\Theta'(\theta)\cdot\Phi'(\varphi)\cdot r^2\cdot\sin(\theta)drd\theta d\varphi}{\int_0^\pi \int_0^\pi \int_0^R \cdot R'(r)^2\cdot\Theta'(\theta)^2\cdot\Phi'(\varphi)^2\cdot r^2\cdot\sin(\theta)drd\theta d\varphi} = T(0) \tag{7.359}$$

In the case of $\kappa\cdot\lambda'-1\neq 0$ (eq. 7.356):

$$C = e^{\lambda'\cdot\kappa t}\cdot\left(\frac{\int_0^\pi \int_0^\pi \int_0^R -w(r,\theta,\varphi)\cdot R'(r)\cdot\Theta'(\theta)\cdot\Phi'(\varphi)\cdot r^2\cdot\sin(\theta)drd\theta d\varphi}{\int_0^\pi \int_0^\pi \int_0^R \cdot R'(r)^2\cdot\Theta'(\theta)^2\cdot\Phi'(\varphi)^2\cdot r^2\cdot\sin(\theta)drd\theta d\varphi}\right.$$
$$\left. - F_t\cdot e^{-t}/(\kappa\cdot\lambda'-1)\right) \tag{7.360}$$

If $\kappa\cdot\lambda'-1=0$ (7.357):

$$C = e^{t}\cdot\left(\frac{\int_0^\pi \int_0^\pi \int_0^R -w(r,\theta,\varphi)\cdot R'(r)\cdot\Theta'(\theta)\cdot\Phi'(\varphi)\cdot r^2\cdot\sin(\theta)drd\theta d\varphi}{\int_0^\pi \int_0^\pi \int_0^R \cdot R'(r)^2\cdot\Theta'(\theta)^2\cdot\Phi'(\varphi)^2\cdot r^2\cdot\sin(\theta)drd\theta d\varphi}\right) - F_t\cdot t \tag{7.361}$$

Then the solution for $v = v(r,\theta,\varphi,t)$, with $\kappa\cdot\lambda'-1\neq 0$ is:

$$v(r,\theta,\varphi,t) = (C\cdot e^{-\lambda'\kappa t} + F_t\cdot e^{-t}/(\kappa\cdot\lambda'-1))$$
$$\times\sum\frac{J_{n'+1/2}(\sqrt{\lambda'}\cdot r)}{\sqrt{r}}\cdot P_{n'}^{m'}(\cos(\theta))\cdot\sin(m'\cdot\varphi) \tag{7.362}$$

The solution for $v = v(r,\theta,\varphi,t)$, with $\kappa\cdot\lambda'-1=0$:

$$v(r,\theta,\varphi,t) = (C\cdot e^{-t} + F_t\cdot t\cdot e^{-t})$$
$$\times\sum\frac{J_{n'+1/2}(\sqrt{\lambda'}\cdot r)}{\sqrt{r}}\cdot P_{n'}^{m'}(\cos(\theta))\cdot\sin(m'\cdot\varphi) \tag{7.363}$$

In this manner, the final solution is $u = w + v$ (stationary + transient solution):

$$u(r,\theta,\varphi) = \sum C\cdot r^n\cdot P_n^{(m)}(\cos(\theta))\cdot\sin(m\cdot\varphi) + T(t)$$
$$\times\sum\frac{J_{n'+1/2}(\sqrt{\lambda'}\cdot r)}{\sqrt{r}}\cdot P_{n'}^{m'}(\cos(\theta))\cdot\sin(m'\cdot\varphi) \tag{7.364}$$

7.14 Case Study: Oscillating Sphere Filled With Gas

Find the variations of pressure as a function of time and position when a sphere of radius a_0 filled with gas of density ρ_0 and with a speed of sound c oscillates periodically with a frequency ω and a maximum velocity v_0 $(v(t) = v_0 \cdot \sin(\omega t))$ along the z direction.

Consider that the periodic movement of the sphere has been going on for long enough, so that transient vibrations of the gas have damped out and there are no resonances.

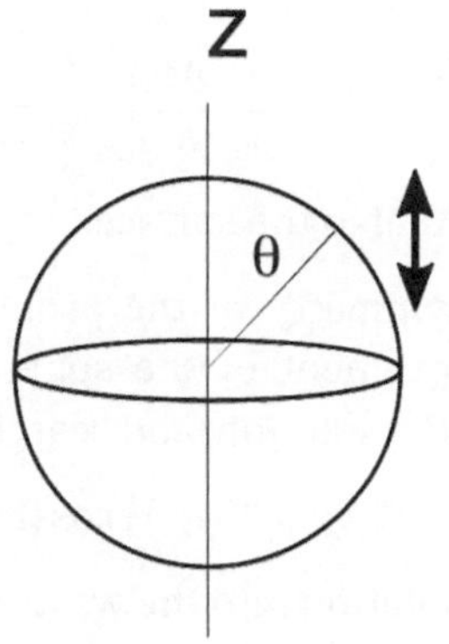

Figure 7.19

Mathematical formulation

This stationary and periodic process (stationary doesn't necessarily mean time independent) can be described with the following homogeneous equation:

$$\left\{ \begin{array}{l} \dfrac{\partial^2 P}{\partial t^2} - c^2 \Delta P = 0 \\ + \text{boundary condition?} \end{array} \right\} \tag{7.365}$$

We get the boundary conditions for the pressure from our knowledge of the variation of the velocity of the molecules.

Solution

The oscillations of the molecules inside the sphere are described by its velocity along z

$$\vec{v(t)} = v_0 \cdot \sin(\omega t)\hat{u}_z \tag{7.366}$$

where $\hat{u}_z$ is a unitary vector along z.

The boundary condition for the pressure is inhomogeneous since, in general, it depends on the angle. To find it we will use the ratio between the radial derivative of the pressure (i.e., the normal component of the pressure gradient near the surface) and the normal component of the velocity of the molecules near the sphere surface.

$$\left.\frac{\partial(\vec{v}\cdot\vec{n})}{\partial t} + \frac{1}{\rho_0}\frac{\partial P}{\partial r}\right|_{r=a_0} = 0 \tag{7.367}$$

where $\vec{n}$ is a vector normal to the surface.

Given the azimuthal symmetry of the problem and the temporal dependency of the movement of the sphere we can neglect the azimuthal variation of the solution, and search the solution as:

$$P(r,\theta,t) = \mathcal{P}(r,\theta)\cos(\omega t) \tag{7.368}$$

When we replace this solution into the wave equation we get:

$$-\mathcal{P}(r,\theta)\omega^2\cos(\omega t) - c^2\cos(\omega t)\Delta\mathcal{P}(r,\theta) = 0 \tag{7.369}$$

$$-\Delta\mathcal{P}(r,\theta) = \left(\frac{\omega}{c}\right)^2\mathcal{P}(r,\theta) \tag{7.370}$$

To solve the equation we will use the method of separation of variables as: $\mathcal{P}(r,\theta) = \mathcal{R}(r)\Theta(\theta)$,

Which yields the equation:

$$-\frac{1}{\mathcal{R}}\frac{d}{dr}\left(r^2\frac{d\mathcal{R}}{dr}\right) - \frac{1}{\Theta\sin\theta}\frac{d}{d\theta}\left(\sin\theta\frac{d\Theta}{d\theta}\right) = \left(\frac{\omega}{c}\right)^2 r^2 \tag{7.371}$$

Applying the angular and radial separation of variables with a constant of separation ν:

$$-\frac{1}{\Theta\sin\theta}\frac{d}{d\theta}\left(\sin\theta\frac{d\Theta}{d\theta}\right) = \frac{1}{\mathcal{R}}\frac{d}{dr}\left(r^2\frac{d\mathcal{R}}{dr}\right) + \left(\frac{\omega}{c}\right)^2 r^2 = \nu \tag{7.372}$$

With the change of variable $x = \cos\theta \rightarrow \frac{d}{d\theta} = -\sin\theta \frac{d}{dx}$ we can reformulate the angular equation Θ as:

$$\frac{1}{\Theta}\frac{d}{dx}\left(\sin^2\theta \frac{d\Theta}{dx}\right) + \nu = 0 \rightarrow \frac{d}{dx}\left((1-x^2)\frac{d\Theta}{dx}\right) + \nu\Theta = 0 \tag{7.373}$$

Considering the periodicity of the solution in the azimuthal and polar angles this equation has solutions in the form of Legendre polynomials:

$$\Theta_l(\theta) = P_l(x) = P_l(\cos\theta) \text{ where } \nu = l(l+1) \tag{7.374}$$

Taking into account that $\nu = l(l+1)$ we get an equation for the radial part:

$$\frac{1}{r^2}\frac{d}{dr}\left(r^2\frac{d\mathcal{R}}{dr}\right) + \left(\left(\frac{\omega}{c}\right)^2 - \frac{l(l+1)}{r^2}\right)\mathcal{R} = 0 \tag{7.375}$$

Its solution is: $\mathcal{R}_l(r) = j_l\left(\frac{\omega}{c}r\right)$

And the general solution will be:

$$\mathcal{P}(r,\theta) = \sum_{l=0}^{\infty} P_l(\cos\theta)\, j_l\left(\frac{\omega}{c}r\right) A_l \tag{7.376}$$

Applying the boundary condition for the derivative of the pressure in the radial variable we get:

$$-\rho_0\frac{\partial(\vec{v}_e\cdot\vec{n})}{\partial t} = v_0\rho_0\omega\cos(\omega t)\cos(\theta) = \left.\frac{\partial P}{\partial r}\right|_{r=a_0}$$
$$= -\left(\sum_{l=0}^{\infty} P_l(\cos\theta)\, j_l'\left(\frac{\omega}{c}a_0\right)\left(\frac{\omega}{c}\right) A_l\right)\cos(\omega t) \tag{7.377}$$

$$c v_0\rho_0\cos\theta = -\sum_{l=0}^{\infty} P_l(\cos\theta)\, j_l'\left(\frac{\omega}{c}a_0\right) A_l \tag{7.378}$$

where $j_l'()$ represents the derivative of the spherical Bessel function.

Using the orthogonality of the Legendre polynomials to find the coefficients:

$$A_l = \frac{-v_0\rho_0 c \int_0^{\pi} P_l(\cos\theta)\cos\theta\sin\theta d\theta}{j_l'\left(\frac{\omega}{c}a_0\right)\|P_l(\cos\theta)\|} \tag{7.379}$$

$$\left\{ \begin{array}{ll} A_l = 0 & l \neq 1 \\ A_1 = -\dfrac{v_0 \rho_0 c}{j_1'\left(\dfrac{\omega}{c} a_0\right)} & \end{array} \right\} \tag{7.380}$$

To conclude, we have the solution:

$$P(r, \theta, t) = -\frac{v_0 \rho_0 c}{j_1'\left(\frac{\omega}{c} a_0\right)} j_1\left(\frac{\omega}{c} r\right) \cos(\theta) \cos(\omega t) \tag{7.381}$$

7.15 Case Study: Stationary Distribution of Temperature in a Planet Close to a Star

Find the stationary temperature distribution of a spherical planet (neglect rotation) composed of two materials with different thermal conductivity coefficients (k_1 from the center of the planet to R_1; k_2 between R_1 and R_2). The planet absorbs a heat flux with density μ from the radiation of a star, far enough so that we can consider it as plane waves in the direction Θ=0. The whole surface of the planet radiates heat according to Newton's law with coefficient h, towards the outer space with temperature $T = 0$.

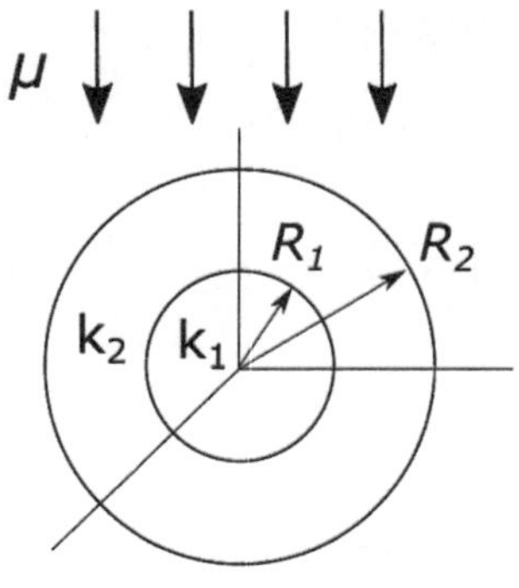

Figure 7.20

Mathematical formulation

We will solve the heat equation in spherical coordinates. Since we look for the stationary case, the heat equation becomes Laplace's equation.

$$\nabla \cdot (\kappa(r)\vec{\nabla} T) = 0 \tag{7.382}$$

$$\kappa(r) = \begin{cases} \kappa_1 & if \quad r \leq R_1 \\ \kappa_2 & if \quad R_1 \leq r \leq R_2 \end{cases} \tag{7.383}$$

In the surface of the sphere there are two heat fluxes simultaneously, one due to the radiation from the star, the other from the heat losses of the planet. The heat flux coming from the star is proportional to its power density μ, and will depend on the incidence on the planet. Then, we have:

$$f_{en} = \vec{f}_e \cdot \hat{n} = \mu \cdot (-\hat{z}) \cdot \hat{n} = -\mu \cdot \cos(\theta) \tag{7.384}$$

The radiation emitted by the planet is taken as normal to its surface:

$$f_{pn} = h \cdot T(r = R_2) \tag{7.385}$$

where $T(r = R_2)$ is the temperature distribution in the surface and h is the Newton's law constant.

From Fourier's law we get the boundary condition at the planet surface.

$$\vec{f} \cdot \hat{n} = -\kappa \frac{\partial T}{\partial \hat{n}} \tag{7.386}$$

$$\begin{cases} \kappa_2 \dfrac{\partial T}{\partial r}|_{r=R_2} = \mu \cdot \cos(\theta) - h \cdot T(r = R_2) & if \quad \theta \in \left(0, \dfrac{\pi}{2}\right) \\ \kappa_2 \dfrac{\partial T}{\partial r}|_{r=R_2} = -h \cdot T(r = R_2) & if \quad \theta \in \left(\dfrac{\pi}{2}, \pi\right) \end{cases} \tag{7.387}$$

Due to the discontinuity at κ we will assume that the solution is of the form:

$$T(r, \theta, \phi) = \begin{cases} T_1(r, \theta, \phi) & if \quad r \leq R_1 \\ T_2(r, \theta, \phi) & if \quad R_1 \leq r \leq R_2 \end{cases} \tag{7.388}$$

where the functions T_1 and T_2 satisfy Laplace's equation in their respective regions.

Since the solutions of the differential equation must be continuous, we have:

$$T_1(r = R_1, \theta, \phi) = T_2(r = R_1, \theta, \phi) \tag{7.389}$$

The second condition that relates T_1 and T_2 is obtained in the following way:

$$\begin{aligned}
&\nabla\cdot(\kappa(r)\vec{\nabla}T)=0\\
&\Rightarrow \lim_{\epsilon\to 0}\int_{\phi=0}^{\phi=2\pi}\int_{\theta=0}^{\theta=\pi}\int_{R-\epsilon}^{R+\epsilon}\nabla\cdot(\kappa(r)\vec{\nabla}T)dV\\
&\quad=\int_S \kappa_1\vec{\nabla}T_1 d\vec{S}-\int_S \kappa_2\vec{\nabla}T_2 d\vec{S}=0\\
&\Rightarrow \kappa_1\vec{\nabla}T_1(R_1)\cdot\hat{r}=\kappa_2\vec{\nabla}T_2(R_1)\cdot\hat{r}\\
&\Rightarrow \kappa_1\frac{\partial T_1}{\partial r}|_{r=R_1}=\kappa_2\frac{\partial T_2}{\partial r}|_{r=R_1}
\end{aligned}\tag{7.390}$$

This implies that the heat flux from region 1 to region 2 (and vice versa) compensate each other in the stationary regime.

To summarize:

$$T(r,\theta,\phi)=\begin{cases} T_1(r,\theta,\phi) & if \quad r\le R_1\\ T_2(r,\theta,\phi) & if \quad R_1\le r\le R_2\end{cases}\tag{7.391}$$

$$\nabla T_1=0\left\{\begin{array}{l} |T_1(r,\theta=0,\pi,\phi)|<\infty\\ T_1(r,\theta,\phi)=T_1(r,\theta,\phi+2\pi)\\ |T_1(r=0,\theta,\phi)|<\infty\end{array}\right\}\tag{7.392}$$

$$\nabla T_2=0\left\{\begin{array}{l} |T_2(r,\theta=0,\pi,\phi)|<\infty\\ T_2(r,\theta,\phi)=T_2(r,\theta,\phi+2\pi)\\ \kappa_2\frac{\partial T_2}{\partial r}|_{r=R_2}=\mu\cdot\cos(\theta)-h\cdot T_2(r=R_2)\quad if\ \theta\in\left(0,\frac{\pi}{2}\right)\\ \kappa_2\frac{\partial T_2}{\partial r}|_{r=R_2}=-h\cdot T_2(r=R_2)\quad if\ \theta\in\left(\frac{\pi}{2},\pi\right)\end{array}\right\}\tag{7.393}$$

Relation between T_1 and T_2

$$\left\{\begin{array}{l} T_1(r=R_1,\theta,\phi)=T_2(r=R_1,\theta,\phi)\\ \kappa_1\left.\frac{\partial T_1}{\partial r}\right|_{r=R_1}=\kappa_2\left.\frac{\partial T_2}{\partial r}\right|_{r=R_1}\end{array}\right\}\tag{7.394}$$

Solution of the problem

Due to the azimuthal symmetry of the problem, we can determine that both T_1 and T_2 will be independent functions of ϕ. Then, separating variables, we have:

$$T_{1,2}(r,\theta) = L_{1,2}(r)\Theta_{1,2}(\theta) \tag{7.395}$$

Replacing in the differential equation:

$$\frac{\nabla T_{1,2}}{T_{1,2}} = \frac{\frac{1}{r^2}\frac{\partial}{\partial r}(r^2\frac{\partial L_{1,2}}{\partial r})}{L_{1,2}} + \frac{\frac{1}{r^2\sin\theta}\frac{\partial}{\partial\theta}(\sin\theta\frac{\partial\Theta_{1,2}}{\partial\theta})}{\Theta_{1,2}} = 0 \tag{7.396}$$

Sturm–Liouville in the angular variable:

$$\frac{\frac{1}{\sin\theta}\frac{\partial}{\partial\theta}(\sin\theta\frac{\partial\Theta_{1,2}}{\partial\theta})}{\Theta_{1,2}} = -\frac{\frac{\partial}{\partial r}(r^2\frac{\partial L_{1,2}}{\partial r})}{L_{1,2}} = -\nu \tag{7.397}$$

The solutions of this equation for θ with boundary conditions $|\Theta_{1,2}(\theta = 0, \pi)| < \infty$ are Legendre polynomials.

$$\Theta_{1,2}(\theta) = P_n(\cos\theta) \quad \text{where} \quad \nu = n(n+1) \tag{7.398}$$

Then we are left with this equation for the radial variable:

$$r^2 L''_{1,2} + 2rL'_{1,2} - \nu L_{1,2} = 0 \tag{7.399}$$

$$\Rightarrow L_{1,2;n}(r) = c_1 r^n + c_2 r^{-(n+1)} \tag{7.400}$$

Since $r = 0$ is described by T_1 but not by T_2, the former must satisfy $|T_1(r = 0)| < \infty$, so that $L_1(r)$ and $L_2(r)$ get the following form:

$$L_{1;n}(r) = A_n r^n \qquad L_{2;n}(r) = B_n r^n + C_n r^{-(n+1)} \tag{7.401}$$

So finally we have:

$$T_1(r,\theta) = \sum_{n=0}^{\infty} A_n r^n P_n(\cos\theta) \quad T_2(r,\theta) = \sum_{n=0}^{\infty}[B_n r^n + C_n r^{-(n+1)}]P_n(\cos\theta) \tag{7.402}$$

To get the value of the constants we first apply the conditions relating T_1 to T_2.

$$T_1(r = R_1, \theta, \phi) = T_2(r = R_1, \theta, \phi) \Rightarrow A_n = B_n + C_n R_1^{-(2n+1)} \tag{7.403}$$

$$\kappa_1\frac{\partial T_1}{\partial r}|_{r=R_1} = \kappa_2\frac{\partial T_2}{\partial r}|_{r=R_1} \Rightarrow \kappa_1 A_n = \kappa_2 B_n + \frac{n+1}{n}\kappa_2 C_n R_1^{-(2n+1)} \tag{7.404}$$

Simplifying these equations we arrive at the following ratios of the coefficients:

$$A_n = B_n(1+\alpha_n) \quad ; \quad C_n = B_n \frac{\alpha_n}{R_1^{-(2n+1)}} \tag{7.405}$$

Being α_n parameters such that:

$$\alpha_n = \frac{1 - \frac{\kappa_1}{\kappa_2}}{\frac{\kappa_1}{\kappa_2} + \frac{n+1}{n}} \tag{7.406}$$

Since α_n diverges for $n = 0$ we cannot consider it for the solution, therefore $n = 1, 2, 3 \ldots$

Now that we have found the ratios between the coefficients, we will determine B_n from the inhomogeneous boundary condition at $r = R_2$.

$$\kappa_2 \frac{\partial T_2}{\partial r}|_{r=R_2} = \kappa_2 \sum_{n=1}^{\infty} B_n [n R_2^{n-1} - \frac{\alpha_n}{R_1^{-(2n+1)}}(n+1) R_2^{-(n+2)}] P_n(\cos\theta)$$
$$= f_e(\theta) - h \cdot T(r = R_2)$$

$$\text{where } f_e(\theta) = \begin{cases} \mu \cdot \cos(\theta) & if \quad \theta \in \left(0, \frac{\pi}{2}\right) \\ 0 \quad if \quad \theta \in \left(\frac{\pi}{2}, \pi\right) \end{cases}$$

$$\Rightarrow \kappa_2 \sum_{n=1}^{\infty} B_n [n R_2^{n-1} - \frac{\alpha_n}{R_1^{-(2n+1)}}(n+1) r^{-(n+2)}] P_n(\cos\theta)$$
$$= f_e(\theta) - h \sum_{n=1}^{\infty} B_n [R_2^n + \frac{\alpha_n}{R_1^{-(2n+1)}} R_2^{-(n+1)}] P_n(\cos\theta)$$

$$\Rightarrow \sum_{n=1}^{\infty} B_n [(h R_2^n + \kappa_2 n R_2^{n-1}) + \frac{\alpha_n}{R_1^{-(2n+1)}}(h R_2^{-(n+1)} - \kappa_2 (n+1) R_2^{-(n+2)})] P_n(\cos\theta)$$
$$= f_e(\theta) \tag{7.407}$$

Using the orthogonality of the Legendre polynomials and knowing that $||P_n(\cos\theta)||^2 = \frac{n}{2n+1}$, we get the value of the B_n coefficients.

$$B_n = \frac{2n+1}{n} \frac{\mu \int_0^{\frac{\pi}{2}} P_n(\cos\theta) \cos\theta \sin\theta d\theta}{(h R_2^n + \kappa_2 n R_2^{n-1}) + \frac{\alpha_n}{R_1^{-(2n+1)}}(h R_2^{-(n+1)} - \kappa_2 (n+1) R_2^{-(n+2)})} \tag{7.408}$$

Then the final solution is:

$$T(r,\theta) = \begin{cases} T_1 = \sum_{n=1}^{\infty} B_n (1+\alpha_n) r^n P_n(\cos\theta) \quad if \quad r \leq R_1 \\ T_2 = \sum_{n=1}^{\infty} B_n [r^n - \alpha_n R_1^{(2n+1)} r^{-(n+1)}] P_n(\cos\theta) \quad if \quad R_1 \leq r \leq R_2 \end{cases} \tag{7.409}$$

Note: Black body radiation

To be more realistic the problem should consider that both the star and the planet emit heat as black bodies. In the case of the star, the total power will be given by Stefan-Boltzmann's law:

$$P_{Total,e} = \epsilon_e A\sigma T^4_{surf,e} = 4\pi R_e^2 \epsilon_e \sigma T^4_{surf,e} \tag{7.410}$$

Where $T_{surf,e}$ is the temperature of the star surface, R_e is the radius of the star, ϵ_e the emissivity and σ the Stefan-Boltzmann constant.

Then, to find the power density μ, we will divide the total power between the surface of a sphere of radius D, being D the distance from the star to the planet, since the star emits radiation in an isotropic manner.

$$\mu = \frac{P_{total,e}}{4\pi D^2} = \frac{R_e^2}{D^2}\epsilon_e \sigma T^4_{surf,e} \tag{7.411}$$

The planet also emits heat as a black body and we have the equation:

$$\begin{aligned}
&\kappa_2 \sum_{n=1}^{\infty} B_n \left[n R_2^{n-1} - \frac{\alpha_n}{R_1^{-(2n+1)}}(n+1) R_2^{-(n+2)} \right] P_n(\cos\theta) \\
&\quad = f_e(\theta) - \epsilon_p \sigma T^4(r = R_2) \\
&\Rightarrow \kappa_2 \sum_{n=1}^{\infty} B_n \left[n R_2^{n-1} - \frac{\alpha_n}{R_1^{-(2n+1)}}(n+1) r^{-(n+2)} \right] P_n(\cos\theta) \\
&\quad = f_e(\theta) - \epsilon_p \sigma \left[\sum_{n=1}^{\infty} B_n \left[R_2^n + \frac{\alpha_n}{R_1^{-(2n+1)}} R_2^{-(n+1)} \right] P_n(\cos\theta) \right]^4
\end{aligned} \tag{7.412}$$

from which it's complicated to determine the B_n coefficients. A linear approximation can be done to simplify the calculations.

7.16 Pre-Heated Quarter of a Sphere

Find the distribution of temperature of a quarter of a sphere of radius R. One of its flat surfaces is kept at $T = 0$ and the other two surfaces are thermally insulated. The curved surface is pre heated from $t = -\infty$ to $t = 0$ by a heat source. The source is located at $\theta = \frac{\pi}{3}$, from $\phi = \frac{\pi}{4}$ to $\phi = \frac{3\pi}{4}$ and radiates heat with a power W. The thermal conductivity is k and the thermal diffusion coefficient is χ.

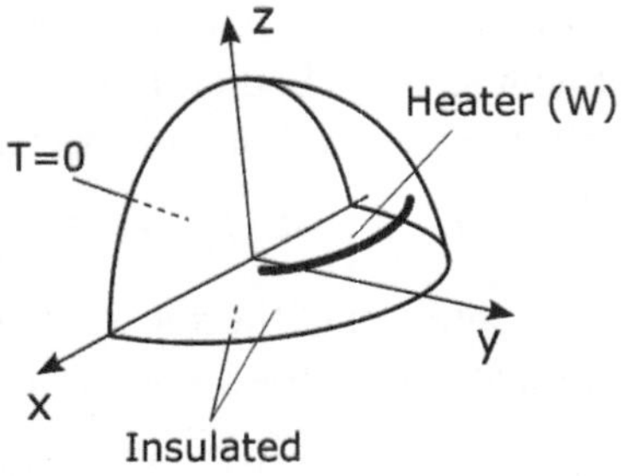

Figure 7.21

Solution:

Mathematical formulation

$$\frac{\partial T}{\partial t} - \chi \Delta T = 0 \tag{7.413}$$

$$T(r, \theta, 0, t) = T(r, \theta, \pi, t) = 0 \tag{7.414}$$

$$\left.\frac{\partial T}{\partial \theta}\right|_{\theta=\frac{\pi}{2}} = 0; \quad -k\left.\frac{\partial T}{\partial r}\right|_{r=R} = f(\theta, \phi) \tag{7.415}$$

$$|T(0, \theta, \phi, t)| < \infty \tag{7.416}$$

With:

$$f(\theta, \phi) = \left\{ \begin{array}{ll} 0 & \text{if } \phi < \frac{\pi}{4}, \\ \frac{2W}{\pi R^2 \sin(\theta)}\left[\delta\left(\theta - \frac{\pi}{3}\right)\right] & \text{if } \frac{\pi}{4} \le \phi \le \frac{3\pi}{4}, \\ 0 & \text{if } \phi > \frac{3\pi}{4}. \end{array} \right\} \tag{7.417}$$

We must solve two problems, a stationary one, $u(r, \theta, \phi)$ to find the initial conditions that we will have at $t = 0$ and a transient one $w(r, \theta, \phi, t)$, after the heat source has been turned off and we have homogeneous boundary conditions.

a) Stationary problem:

$$\left\{ \begin{array}{l} \Delta u = 0 \\ u(r, \theta, 0) = u(r, \theta, \pi) = 0 \\ \left.\frac{\partial u}{\partial \theta}\right|_{\theta=\frac{\pi}{2}} = 0; \quad -k\left.\frac{\partial u}{\partial r}\right|_{r=R} = f(\theta, \phi) |u(0, \theta, \phi)| < \infty \end{array} \right\} \tag{7.418}$$

Sturm–Liouville problem

Since we have homogeneous boundary conditions for θ and ϕ, we separate variables such that $u(r, \theta, \phi) = R(r)v(\theta, \phi)$ and we have the Sturm–Liouville problem for $v(\theta, \phi)$:

$$\Delta_{\theta,\phi}v + \mu v = 0 \tag{7.419}$$

$$v(\theta, 0) = v(\theta, \pi) = 0 \tag{7.420}$$

$$\left.\frac{\partial v}{\partial \theta}\right|_{\theta=\frac{\pi}{2}} = 0$$

Separating variables once again: $v(\theta, \phi) = \Theta(\theta)\Phi(\phi)$

$$\frac{1}{\Theta(\theta)\sin(\theta)}\frac{\partial}{\partial\theta}\left(\sin(\theta)\frac{\partial\Theta(\theta)}{\partial\theta}\right) + \frac{1}{\sin^2(\theta)\Phi(\phi)}\frac{\partial^2\Phi(\phi)}{\partial\phi^2} + \mu = 0 \tag{7.421}$$

Sturm–Liouville problem for $\Phi(\phi)$

$$\left\{\begin{array}{l} \dfrac{\partial^2\Phi(\phi)}{\partial\phi^2} + \nu\Phi(\phi) = 0 \\ \Phi(0) = \Phi(\pi) = 0 \end{array}\right\} \tag{7.422}$$

We have the eigenfunctions $\Phi(\phi) = \sin(m\phi)$, with $\nu = m^2$, so we have:

$$\sin(\theta)\frac{\partial}{\partial\theta}\left(\sin(\theta)\frac{\partial\Theta(\theta)}{\partial\theta}\right) + (\mu\sin^2(\theta) - m^2)\Theta(\theta) = 0 \tag{7.423}$$

The solutions are Legendre polynomials $\Theta(\theta) = P_n^{(m)}(\cos(\theta))$ and the eigenvalues are $\mu = n(n+1)$. Due to the boundary condition for $\theta = \frac{\pi}{2}$, the only valid solutions of $\Theta(\theta)$ will be those for which $m + n$ is an even number.

General solution

Replacing in the heat equation:

$$\Delta u = \frac{1}{r^2}\frac{\partial}{\partial r}\left(r^2\frac{\partial u(r, \theta, \phi)}{\partial r}\right) + \frac{1}{r^2}\Delta_{\theta,\phi}u(r, \theta, \phi) = 0 \tag{7.424}$$

We have:

$$\frac{1}{r^2}\frac{\partial}{\partial r}\left(r^2\frac{\partial R(r)}{\partial r}\right) - \frac{\mu}{r^2}R(r) = 0 \tag{7.425}$$

Whose solutions are $R(r) = Ar^n + Br^{-(n+1)}$. Due to the boundary condition for $r = 0$, $R(r) = Ar^n$.

The solution for u is:

$$u(r,\theta,\phi)=\sum_{n\geq m}^{\infty}\sum_{m=0}^{\infty}A_{nm}r^{n}P_{n}^{(m)}(\cos(\theta))\sin(m\phi) \tag{7.426}$$

Final solution

We will find the coefficients by using the boundary condition at $r = R$ and the orthogonality of the eigenfunctions.

$$-k\left.\frac{\partial u}{\partial r}\right|_{r=R}=f(\theta,\phi) \tag{7.427}$$

$$\sum_{n\geq m}^{\infty}-kA_{nm}nR^{(n-1)}P_{n}^{(m)}(\cos(\theta))\frac{\pi}{2}=\frac{2W}{\pi R^{2}\sin(\theta)}\left[\delta\left(\theta-\frac{\pi}{3}\right)\right]\int_{\frac{\pi}{4}}^{\frac{3\pi}{4}}\sin(m\phi)d\phi \tag{7.428}$$

$$\sum_{n\geq m}^{\infty}kA_{nm}nR^{(n-1)}P_{n}^{(m)}(\cos(\theta))\frac{\pi}{2}=\frac{2W}{m\pi R^{2}\sin(\theta)}\left[\delta\left(\theta-\frac{\pi}{3}\right)\right]\times\left[\cos\left(\frac{3\pi m}{4}\right)-\cos\left(\frac{\pi m}{4}\right)\right] \tag{7.429}$$

$$kA_{nm}nR^{(n-1)}\frac{2(n+m)!}{(2n+1)(n-m)!}\frac{\pi}{2}=\\ =\frac{2W}{m\pi R^{2}}\left[\cos\left(\frac{3\pi m}{4}\right)-\cos\left(\frac{\pi m}{4}\right)\right]\int_{0}^{\frac{\pi}{2}}P_{n}^{(m)}(\cos(\theta))\delta\left(\theta-\frac{\pi}{3}\right)d\theta \tag{7.430}$$

Finally we find the A_{nm} coefficients:

$$A_{nm}=\frac{2W}{\pi^{2}R^{2}k}\frac{(2n+1)(n-m)!}{mn(n+m)!}P_{n}^{(m)}\left(\cos\left(\frac{\pi}{3}\right)\right)\times\left[\cos\left(\frac{3\pi m}{4}\right)-\cos\left(\frac{\pi m}{4}\right)\right] \tag{7.431}$$

It can be observed that if m is even A_{nm} will be zero. Therefore only the terms with odd values of n and m remain, since $n+m$ must be even due to the boundary conditions of the θ variable.

We already have the initial condition from the transient problem $w(r, \theta, \phi, t)$, which can be formulated as:

$$\left\{\begin{array}{l} \dfrac{\partial w}{\partial t} - \chi \Delta w = 0 \\ w(r, \theta, 0, t) = w(r, \theta, \pi, t) = 0 \\ \left.\dfrac{\partial w}{\partial \theta}\right|_{\theta=\frac{\pi}{2}} = \left.\dfrac{\partial w}{\partial r}\right|_{r=R} = 0 \\ |w(0, \theta, \phi, t)| < \infty \\ w(r, \theta, \phi, 0) = u(r, \theta, \phi) \end{array}\right\} \tag{7.432}$$

Since we have homogeneous boundary conditions for r, θ and ϕ we separate variables so that $w(r, \theta, \phi, t) = Q(t)g(r, \theta, \phi)$ and we formulate a Sturm–Liouville problem for $g(r, \theta, \phi)$:

$$\left\{\begin{array}{l} \Delta g + \lambda g = 0 \\ g(r, \theta, 0) = g(r, \theta, \pi) = 0 \\ \left.\dfrac{\partial g}{\partial \theta}\right|_{\theta=\frac{\pi}{2}} = \left.\dfrac{\partial g}{\partial r}\right|_{r=R} = 0 \\ |g(0, \theta, \phi)| < \infty \end{array}\right\} \tag{7.433}$$

We once again separate variables: $g(r, \theta, \phi) = R(r)v(\theta, \phi)$.

We already have the eigenfunctions of the angular variables and their boundary conditions remain the same, therefore $v(\theta, \phi) = P_n^{(m)}(\cos(\theta))\sin(m\phi)$. Replacing in the Sturm–Liouville problem, we get an equation for $R(r)$

$$\frac{1}{r^2}\frac{\partial}{\partial r}\left(r^2\frac{\partial R(r)}{\partial r}\right) + \left(\lambda - \frac{\mu}{r^2}\right)R(r) = 0 \tag{7.434}$$

We had previously found the values of μ to be $\mu = n(n+1)$. The solutions for $R(r)$, due to the condition at $r = 0$, are Bessel functions of order $n + \frac{1}{2}$; $R_n(r) = \frac{1}{\sqrt{r}} J_{n+\frac{1}{2}}(\sqrt{\lambda} r)$.

The values of λ can be found using the boundary condition for $r = R$. There exists an infinite set of eigenvalues for every n. If we label $z'_{k,n}$ the k-fold root of the equation $\frac{dj_n(z)}{dz}$, where $j_n(z) = \sqrt{\pi/(2z)} J_{n+\frac{1}{2}}(z)$ are spherical Bessel functions. The eigenvalues can be written as $\lambda_{k,n} = \left(\frac{z'_{k,n}}{R}\right)^2$.

With this we have an equation for the temporal part:

$$\frac{\partial Q}{\partial t} + \chi \lambda_{k,n} Q = 0 \tag{7.435}$$

Its solutions are proportional to $e^{-\chi \lambda_{k,n} t}$.

The solutions of w will be:

$$w(r, \theta, \phi, t) = \sum_{n \geq m}^{\infty} \sum_{m=0}^{\infty} B_{nmk} e^{-\chi \lambda_{k,n} t} \frac{J_{n+\frac{1}{2}}(\sqrt{\lambda_{k,n}} r)}{\sqrt{r}} P_n^{(m)}(\cos(\theta)) \sin(m\phi) \tag{7.436}$$

Where $m + n$ must be even. To find the B_{nm} constants we will use the initial condition and the orthogonality of the eigenfunctions.

$$\sum_{n \geq m}^{\infty} \sum_{m=0}^{\infty} B_{nmk} \frac{J_{n+\frac{1}{2}}(\sqrt{\lambda_{k,n}} r)}{\sqrt{r}} P_n^{(m)}(\cos(\theta)) \sin(m\phi) =$$

$$= \sum_{n' \geq m'}^{\infty} \sum_{m'=0}^{\infty} A_{n'm'} r^{n'} P_{n'}^{(m')}(\cos(\theta)) \sin(m'\phi) \tag{7.437}$$

The orthogonality of the angular eigenfunctions lets us know that $n = n'$, $m = m'$. Therefore in w only the odd values of n and m will not be zero. For the orthogonality of the Bessel functions, we have:

$$B_{nmk} = A_{nm} \frac{\int_0^R r^n \frac{J_{n+\frac{1}{2}}(\sqrt{\lambda_{k,n}} r)}{\sqrt{r}} r^2 dr}{\left|\left| J_{n+\frac{1}{2}}(\sqrt{\lambda_{k,n}} r) \right|\right|^2} \tag{7.438}$$

The final solution is:

$$T(r, \theta, \phi, t) = \begin{cases} u(r, \theta, \phi) & \text{if} \quad t \leq 0, \\ w(r, \theta, \phi, t) & \text{if} \quad t > 0 \end{cases}$$

Chapter 8

Fourier Transform and Its Applications

To solve the problems described with PDEs and defined in infinite or semi-infinite spaces we will apply the integral Fourier transform (FT). The properties of the FT are detailed in Chapter 9 of [1], whereas the details of the method to solve PDEs using the FT can be found in Chapter 7 of [5] both of the main texts in the bibliography.

In this section, the following nomenclature will be used:

- We will use $\mathfrak{F}$ to refer to the Fourier transform operator and $\mathfrak{F}^{-1}$ to refer to the inverse Fourier transform.
- The functions to which the Fourier transform is applied will be written between square brackets [and]. In this way, $\mathfrak{F}[f(x)]$ will be the operation consisting in applying the Fourier transform to the $f(x)$ function, which returns a $g(\omega)$ function, being ω the variable in reciprocal space (if x is the variable in real space).
- The parentheses (and) will be used as usual, to indicate the variable on which a function depends, which includes the function consisting in applying the Fourier transform to another function. So sometimes we will see: $\mathfrak{F}[f](\omega)$, which means that the Fourier transform is applied to the f function (which

Mathematical Methods for Physics: Problems and Solutions
Farkhad G. Aliev and Antonio Lara

ISBN 978-981-4968-71-3 (Hardcover), 978-1-003-41088-1 (eBook)
www.jennystanford.com

depends on the x variable, although it is omitted for simplicity) and the result is a function of the ω variable (this implicitly tells us that f is a function of x).

8.1 Reciprocity of the Fourier Transform

Using the symmetric version of the Fourier transform show that the following relation hods in real space (x)

$$\mathfrak{F}[f](x) = \mathfrak{F}^{-1}[f](-x) \tag{8.1}$$

In other words: the Fourier transform of a function $f(\omega)$, evaluated at x is equal to the inverse transform of the same function $f(\omega)$ evaluated at $(-x)$. Based on the relation (8.1) deduce the rules of the multiple application (several consecutive times) of the Fourier transform.

Using the symmetric version of the Fourier transform:

$$F(x) = \mathfrak{F}[f](x) = \frac{1}{\sqrt{2\pi}} \int_{-\infty}^{+\infty} e^{-i\omega x} f(\omega) d\omega \tag{8.2}$$

Also:

$$\mathcal{F}(-x) = \mathfrak{F}^{-1}[f](-x) = \frac{1}{\sqrt{2\pi}} \int_{-\infty}^{+\infty} e^{i\omega(-x)} f(\omega) d\omega$$

$$= \{\text{we use the kernel: } e^{+i\omega x}\} \tag{8.3}$$

$$\mathfrak{F}[f](x) = \frac{1}{\sqrt{2\pi}} \int_{-\infty}^{+\infty} e^{-i\omega x} f(\omega) d\omega \tag{8.4}$$

Then $\mathcal{F}(-x) = \mathfrak{F}[f](x)$. Applying $\mathfrak{F}^{-1}$ we have $\mathfrak{F}^{-1}[\mathcal{F}(-x)] = \mathfrak{F}^{-1}\mathfrak{F}[f](x) = f(\omega)$

From here we deduce the property of reciprocity with respect to multiple applications of the Fourier transform. We start with the double application:

$$\mathfrak{F}^2[F(x)] = \mathfrak{F}\mathfrak{F}[F(x)] \tag{8.5}$$

Replacing $\mathfrak{F}[F(x)]$ by $f(\omega) = \mathfrak{F}^{-1}F(-x)$ and changing the order of application of $\mathfrak{F}$ with $\mathfrak{F}^{-1}$

$$\mathfrak{F}\mathfrak{F}[F(x)] = \mathfrak{F}^{-1}\mathfrak{F}[F(-x)] = F(-x) \tag{8.6}$$

Then: $F(x)$ is even only if $\mathfrak{F}^2[F(x)] = F(x)$. On the other hand, $F(x)$ is odd only if $\mathfrak{F}^2[F(x)] = -F(x)$. In any case the relation: $\mathfrak{F}^4[F(x)] = F(x)$ holds.

Graphical representation of the consecutive action of the $\mathfrak{F}$ operator on a function defined in real space:

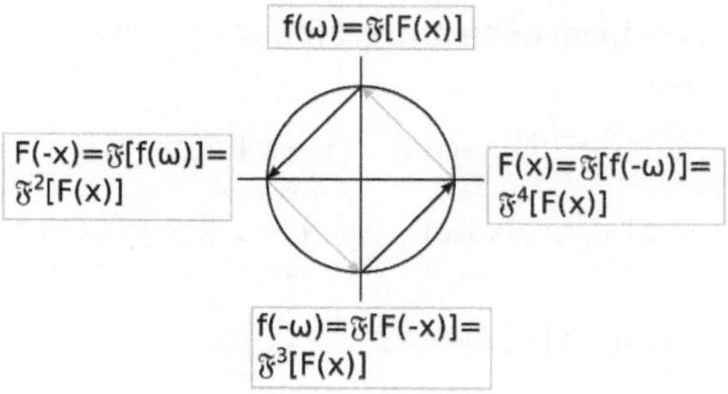

Figure 8.1

8.2 Fourier Transform of a Bidirectional Pulse

Using the symmetric version of the Fourier transform find the spectrum of a bidirectional pulse:

$$f(x) = \begin{Bmatrix} -1 \ (-1 \leq x \leq 0) \\ +1 \ (0 \leq x \leq 1) \\ 0 \quad (x < -1; x > 1) \end{Bmatrix} \tag{8.7}$$

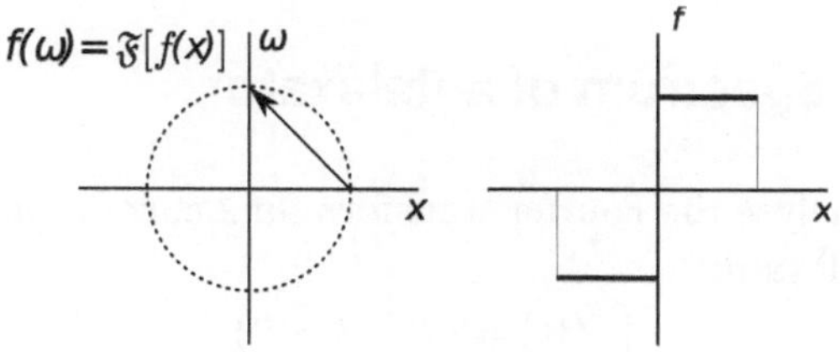

Figure 8.2

$$\mathfrak{F}[f](\omega) = \frac{1}{\sqrt{2\pi}} \int_{-\infty}^{+\infty} e^{-i\omega x} f(x)dx = \frac{1}{\sqrt{2\pi}} \left[-\int_{-1}^{0} e^{-i\omega x} dx + \int_{0}^{1} e^{-i\omega x} dx \right] =$$

$$= \frac{1}{\sqrt{2\pi}} \left[-\frac{(1 - e^{i\omega})}{-i\omega} + \frac{(e^{-i\omega} - 1)}{-i\omega} \right] = \frac{2\cos(\omega) - 2}{-i\omega\sqrt{2\pi}}$$

$$= i\sqrt{\frac{2}{\pi}} \frac{\cos(\omega) - 1}{\omega} \tag{8.8}$$

The Fourier transform of an antisymmetric function is an imaginary function because:

$$e^{-i\omega x} = \cos(\omega x) - i \sin(\omega x) \tag{8.9}$$

Then, the integral cancels out $\frac{1}{\sqrt{2\pi}} \int_{-\infty}^{+\infty} \cos(\omega x) f(x)dx$ and we are left with: $-i \int_{-\infty}^{+\infty} \sin(\omega x) f(x)dx$

Graphical representation (schematic only for positive frequencies) of the square of the modulus of the Fourier transform (spectral power):

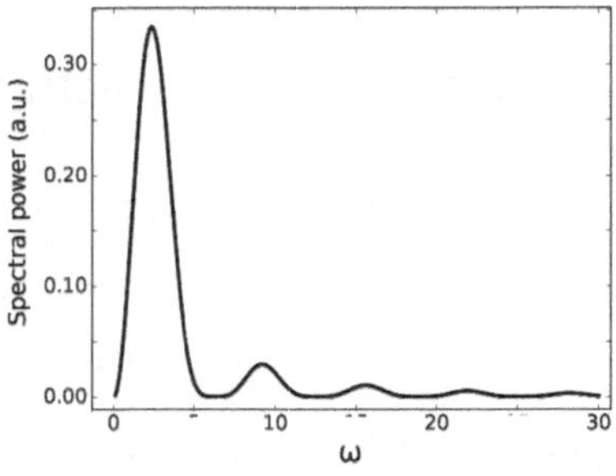

Figure 8.3

8.3 Loss Spectrum of a Relaxator

Find and analyze the Fourier transform of a current pulse with the form of a relaxator:

$$\left\{ \begin{array}{ll} f(t) = e^{-\alpha t} & (t > 0) \\ f(t) = 0 & (t < 0) \end{array} \right\} \tag{8.10}$$

The obtained imaginary part represents the energy losses. Show a graph of the losses of the relaxator during a period of the excitation as a function of frequency.

We use the symmetric form of the Fourier transform:

$$\mathfrak{F}[f](\omega) = \frac{1}{\sqrt{2\pi}} \int_{-\infty}^{+\infty} e^{-i\omega t} f(t)dt = \frac{1}{\sqrt{2\pi}} \int_{0}^{+\infty} e^{-\alpha t} e^{-i\omega t} dt$$

$$= \frac{1}{\sqrt{2\pi}} \int_{0}^{+\infty} e^{-(\alpha+i\omega)t} dt \qquad (8.11)$$

The integral is solved:

$$\mathfrak{F}[f](\omega) = -\frac{1}{\alpha+i\omega} e^{-(\alpha+i\omega)t}\Big|_{0}^{+\infty} = \frac{1}{\alpha+i\omega} = \frac{1}{\alpha+i\omega}\frac{\alpha-i\omega}{\alpha-i\omega}$$

$$= \frac{\alpha}{\alpha^2+\omega^2} - i\frac{\omega}{\alpha^2+\omega^2} \qquad (8.12)$$

The value of the imaginary part during a period (T) is:

$$\frac{\omega}{\alpha^2+\omega^2} \cdot \frac{1}{T} = \frac{1}{2\pi}\frac{\omega^2}{\alpha^2+\omega^2} \qquad (8.13)$$

We have the corresponding graph (only positive frequencies are shown): $\frac{1}{2\pi}\frac{\omega^2}{\alpha^2+\omega^2}$ vs. ω

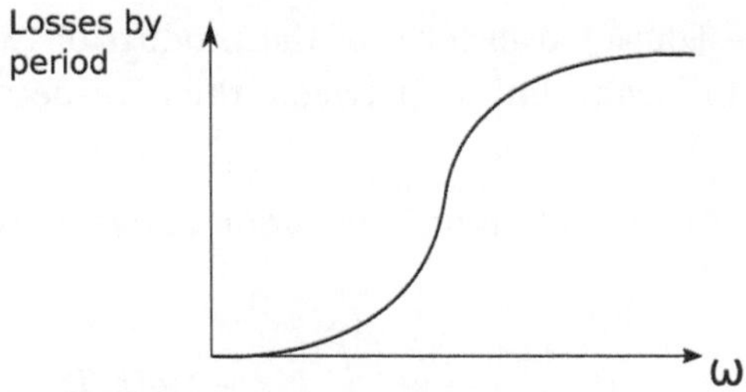

Figure 8.4

8.4 Inverse Fourier Transform of a Function

Find the inverse Fourier transform $\mathfrak{F}^{-1}[f(\omega)]$ of the following function which is defined in the frequency domain.

$$f(\omega) = \frac{1}{\omega^2+4\omega+13} \qquad (8.14)$$

We first note that:

$$\omega^2 + 4\omega + 13 = (\omega + 2)^2 + 9 \tag{8.15}$$

Using Fourier transform tables we know:

$$\mathfrak{F}^{-1}\left[\frac{2a}{\omega^2 + a^2}\right] = e^{-a|t|} \tag{8.16}$$

Then:

$$\mathfrak{F}^{-1}\left[\frac{1}{\omega^2 + 3^2}\right] = \frac{1}{6}e^{-3|t|} \tag{8.17}$$

Here we have also used the rule: $\mathfrak{F}^{-1}[aF(\omega)] = a\mathfrak{F}^{-1}[F(\omega)]$ and we have multiplied and divided the expression by a factor 6 to be able to use the expression of the transform from the tables with $a = 3$. Applying the displacement in reciprocal space to $\omega_0 = 2$, (using the rule $\mathfrak{F}^{-1}[F(\omega - \omega_0)] = f(t)e^{i\omega_0 t}$) we obtain the answer for our case:

$$\mathfrak{F}^{-1}\left[\frac{1}{\omega^2 + 4\omega + 9}\right] = \frac{1}{6}e^{-2it}e^{-3|t|} \tag{8.18}$$

8.5 Fourier Transform of the Product of Two Functions

Show that the Fourier transform of the product of two functions is equal to the convolution between their respective Fourier transforms.

We first recall the definition of convolution between two functions $f * g$:

$$[f * g](x) = \frac{1}{\sqrt{2\pi}} \int_{-\infty}^{+\infty} f(x - t)g(t)dt \tag{8.19}$$

Furthermore, the convolution has the following reciprocity property:

$$[f * g](x) = \frac{1}{\sqrt{2\pi}} \int_{-\infty}^{+\infty} f(\xi)g(x - \xi)d\xi = [g * f](x) \tag{8.20}$$

We will use the symmetric form of the Fourier transform:

$$\mathfrak{F}[f](k) = \frac{1}{\sqrt{2\pi}} \int_{-\infty}^{+\infty} e^{-ikx} f(x)dx \tag{8.21}$$

Applying the Fourier transform to the product of two functions $f \cdot g$:

$$\mathfrak{F}[f \cdot g] = \frac{1}{\sqrt{2\pi}} \int_{-\infty}^{+\infty} e^{-ikx} f(x) g(x) dx = \{\text{presenting g(x) with } \mathfrak{F}^{-1} of \text{ G(k)}\}$$

$$= \mathfrak{F}[f \cdot g] = \frac{1}{\sqrt{2\pi}} \int_{-\infty}^{+\infty} dx e^{-ikx} f(x) \left[\frac{1}{\sqrt{2\pi}} \int_{-\infty}^{+\infty} e^{ik'x} G(k') dk' \right]$$

$$= \{\text{Changing the order of integration}\}$$

$$= \frac{1}{\sqrt{2\pi}} \int_{-\infty}^{+\infty} G(k') dk' \left[\frac{1}{\sqrt{2\pi}} \int_{-\infty}^{+\infty} e^{-ikx} f(x) e^{ik'x} dx \right]$$

$$= \frac{1}{\sqrt{2\pi}} \int_{-\infty}^{+\infty} G(k') dk' \left[\frac{1}{\sqrt{2\pi}} \int_{-\infty}^{+\infty} f(x) e^{-i(k-k')x} dx \right]$$

$$= \frac{1}{\sqrt{2\pi}} \int_{-\infty}^{+\infty} G(k') F(k-k') dk' = [G * F](k) \tag{8.22}$$

In the end we get:

$$\mathfrak{F}[f \cdot g] = [G(k) * F(k)] \tag{8.23}$$

In a similar manner it can be shown that:

$$\mathfrak{F}[f * g] = F(k) \cdot G(k) \tag{8.24}$$

Applying $\mathfrak{F}^{-1}$ to the relation (8.23) we get:

$$f \cdot g = \mathfrak{F}^{-1}[G(k) * F(k)] \tag{8.25}$$

Finally, applying $\mathfrak{F}^{-1}$ to the relation (8.24) we get:

$$[f * g] = \mathfrak{F}^{-1}[F(k) \cdot G(k)] \tag{8.26}$$

8.6 Example of the Calculation of the Fourier Transform of a Product of Two Functions from the Convolution Operation

From the relation of the Fourier transform of the product of two functions and the convolution operation, find the Fourier transform of a rectangular pulse of amplitude 1 in the temporal range $(-1 < t < 1)$ with a periodic modulation $\cos(\omega_0 t)$.

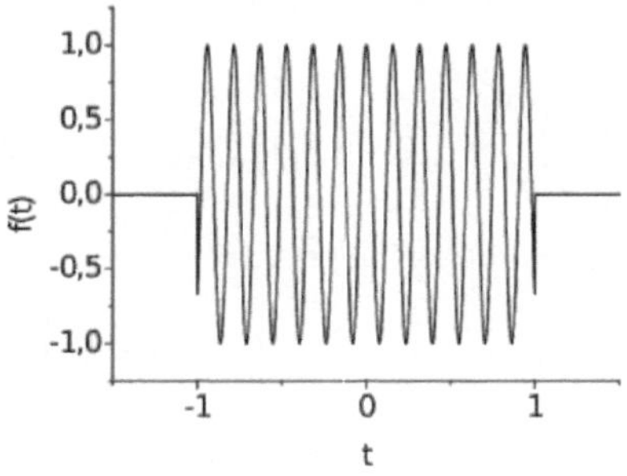

Figure 8.5

We first recall the definition of the convolution between two functions $f * g$:

$$[f * g](x) = \int_{-\infty}^{+\infty} f(x-t)g(t)dt \tag{8.27}$$

We have previously obtained the following relation (the numeric factor depends on how the convolution is defined):

$$\mathfrak{F}[f(x) \cdot g(x)] = \frac{1}{\sqrt{2\pi}}[F * G](k) \tag{8.28}$$

Rewriting this expression in terms of the Fourier transform to pass from the time domain to the frequency domain:

$$\mathfrak{F}[f(t) \cdot g(t)] = \frac{1}{\sqrt{2\pi}}[F * G](\omega) = \frac{1}{\sqrt{2\pi}} \int_{-\infty}^{+\infty} F(\omega')G(\omega - \omega')d\omega' \tag{8.29}$$

In our case:

$$f(t) = \left\{ \begin{array}{l} 0 \ (t < -1) \\ 1 \ (-1 < t < +1) \\ 0 \ t > 1) \end{array} \right\} \tag{8.30}$$

$$g(t) = \{\cos(\omega_0 t) \ (-\infty < t < +\infty)\} \tag{8.31}$$

$$\mathfrak{F}[\cos\omega_0 t] = \frac{1}{\sqrt{2\pi}}\int_{-\infty}^{+\infty}[\cos\omega_0 t]e^{-i\omega t}dt = \frac{1}{\sqrt{2\pi}}\frac{1}{2}\int_{-\infty}^{+\infty}[e^{-i\omega_0 t}+e^{+i\omega_0 t}]e^{-i\omega t}dt$$

$$= \frac{1}{2\sqrt{2\pi}}\int_{-\infty}^{+\infty}e^{+i(-\omega_0)t}e^{-i\omega t}dt + \frac{1}{2\sqrt{2\pi}}\int_{-\infty}^{+\infty}e^{i\omega_0 t}e^{-i\omega t}dt$$

$$= \frac{1}{2\sqrt{2\pi}}[\delta(\omega+\omega_0)+\delta(\omega-\omega_0)] \quad (8.32)$$

On the other hand, the Fourier transform of the rectangular pulse $f(t)$ is well known:

$$\mathfrak{F}[f(t)] = \frac{1}{\sqrt{2\pi}}\int_{-1}^{+1}e^{-i\omega t}dt = \frac{1}{(-i\omega)\sqrt{2\pi}}[e^{-i\omega}-e^{+i\omega}] = \frac{2}{\sqrt{2\pi}}\frac{\sin(\omega)}{\omega} \quad (8.33)$$

We need to use the definition of the Fourier transform of the product of functions, finding the convolution:

$$\mathfrak{F}[\cos(\omega_0 t)\cdot f(t)] = \frac{1}{\sqrt{2\pi}}\frac{1}{2\sqrt{2\pi}}\frac{2}{\sqrt{2\pi}}\int_{-\infty}^{+\infty}\frac{\sin(\omega')}{\omega'}[\delta([\omega+\omega_0]-\omega')$$

$$+\delta([\omega-\omega_0]-\omega')]d\omega' =$$

$$= \frac{1}{2\pi\sqrt{2\pi}}\left[\frac{\sin(\omega+\omega_0)}{\omega+\omega_0}+\frac{\sin(\omega-\omega_0)}{\omega-\omega_0}\right] \quad (8.34)$$

8.7 Parseval Theorem Formulated for Two Different Functions

Show that the integral of the product of a function $f(x)$ and a different conjugated function $g^*(x)$ is equal to the integral of the product between the corresponding Fourier transforms (Parseval theorem applied to two different functions):

$$\int_{-\infty}^{+\infty} f(x)g^*(x)dx = \int_{-\infty}^{+\infty} F(k)G^*(k)dk \quad (8.35)$$

where $F(k); G(k)$ are the Fourier transforms of the functions $f(x); g(x)$.

We will use the asymmetric definition of the Fourier transform to calculate:

$$g^*(x) = \int_{-\infty}^{+\infty} G^*(k)e^{ikx}dk \tag{8.36}$$

Then:

$$\int_{-\infty}^{+\infty} f(x)g^*(x)dx = \int_{-\infty}^{+\infty} f(x)\int_{-\infty}^{+\infty} G^*(k)e^{ikx}dkdx = \int_{-\infty}^{+\infty}\int_{-\infty}^{+\infty} G^*(k)f(x)e^{ikx}dxdk =$$

$$= \int_{-\infty}^{+\infty} dkG^*(k)\int_{-\infty}^{+\infty} f(x)e^{ikx}dx = \int_{-\infty}^{+\infty} G^*(k)F(k)dk \tag{8.37}$$

In the particular case $g^*(x) = f^*(x)$ we arrive at the well-known Parseval identity:

$$\int_{-\infty}^{+\infty} f(x)f^*(x)dx = \int_{-\infty}^{+\infty} |f(x)|^2 dx = \int_{-\infty}^{+\infty} |F(k)|^2 dk \tag{8.38}$$

8.8 Wiener–Khinchin (WK) Theorem

Show that the spectral density of a signal (in power) is the Fourier transform of the autocorrelation of this signal.

Before explaining the solution, spectral power of a signal will be defined. We know that the spectral power of an electromagnetic, sound, etc., signal is proportional to the square of the modulus of its amplitude.

Parseval identity applied to Fourier series indicates that the total power of a signal is proportional to the sum of the squares of the amplitudes of all its harmonics.

If the accumulated power in the dω interval is $|g(\omega)|^2 d\omega$, then $|g(\omega)|^2$ will be the spectral density of a function $f(t)$.

This problem considers the method to calculate the spectral power from the autocorrelation of a function.

Formal solution

We will use the symmetric version of the Fourier transform:

$$\frac{1}{2\pi}\int_{-\infty}^{+\infty} |g(\omega)|^2 e^{-i\omega t} d\omega = \frac{1}{2\pi}\int_{-\infty}^{+\infty} g(\omega) g^*(\omega) e^{-i\omega t} d\omega \tag{8.39}$$

Note: as can be seen, in this case we define the direct transform as the transition from frequency the domain to the time domain. We do this to remark the freedom that there exists in the definition of the Fourier transform. Depending on this definition the form of the theorems of convolution and correlation can change.

Then:

$$\frac{1}{2\pi}\int_{-\infty}^{+\infty} g^*(\omega) g(\omega) e^{-i\omega t} d\omega = \tag{8.40}$$

$\{$g(ω) is written as a Fourier transform$\}$

$$= \int_{-\infty}^{+\infty} \frac{d\omega}{2\pi} g^*(\omega) e^{-i\omega t} \int_{-\infty}^{+\infty} f(t') e^{i\omega t'} dt' = \tag{8.41}$$

$\{$Exchanging the order of integration:$\}$

$$\int_{-\infty}^{+\infty} dt' f(t') \left[\int_{-\infty}^{+\infty} \frac{d\omega}{2\pi} g(\omega) e^{-i\omega(t'-t)} \right]^* = \int_{-\infty}^{+\infty} dt' f(t') [f(t'-t)]^* \tag{8.42}$$

Then:

$$\frac{1}{2\pi}\int_{-\infty}^{+\infty} |g(\omega)|^2 e^{-i\omega t} d\omega = \int_{-\infty}^{+\infty} dt' f(t') [f(t'-t)]^* \tag{8.43}$$

Is the autocorrelation of the $f(t)$ function. Generalizing this result to the cross correlation between two different functions $f(t)$ and $g(t)$ we can deduce the analogous relation to the theorem of convolution—specifically that the cross correlation between two functions equals the Fourier transform of the product of their corresponding Fourier transforms.

$$[f \star g] = \mathfrak{F}[F^*(\omega) \cdot G(\omega)] \tag{8.44}$$

Applying $\mathfrak{F}^{-1}$ to both sides of the relation of autocorrelation we get the Wiener–Khinchin theorem.

$$\mathfrak{F}^{-1}\left[\frac{1}{2\pi}\int_{-\infty}^{+\infty}|g(\omega)|^2 e^{-i\omega t}d\omega\right] = |g(\omega)|^2 = g(\omega)g^*(\omega)$$

$$= \mathfrak{F}^{-1}\left[\int_{-\infty}^{+\infty} dt' f(t')[f(t'-t)]^*\right] \quad (8.45)$$

This is an important relation which is employed in signal processing in many electronic devices, as those which measure the spectral power of signals such as noise, vibrations, etc. In the case of real signals, the W-K theorem becomes:

$$\mathfrak{F}^{-1}\left[\int_{-\infty}^{+\infty} dt' f(t')[f(t'-t)]^*\right] = \mathfrak{F}^{-1}\left[\int_{-\infty}^{+\infty} dt' f(t')[f(t'-t)]\right] =$$

$$= |g(\omega)|^2 = g(\omega)g^*(\omega) = g(\omega)g(-\omega) \quad (8.46)$$

The change of sign is due to the antisymmetric character of the imaginary part of the Fourier transform of a real signal.

8.9 Fourier Transform of an Oscillation Modulated by a Gaussian Pulse

Find the Fourier transform $g(\omega)$ of the following function $f(t) = \cos(\omega_0 t) \cdot e^{-(\frac{t}{\tau})^2}$.

Present the results graphically.

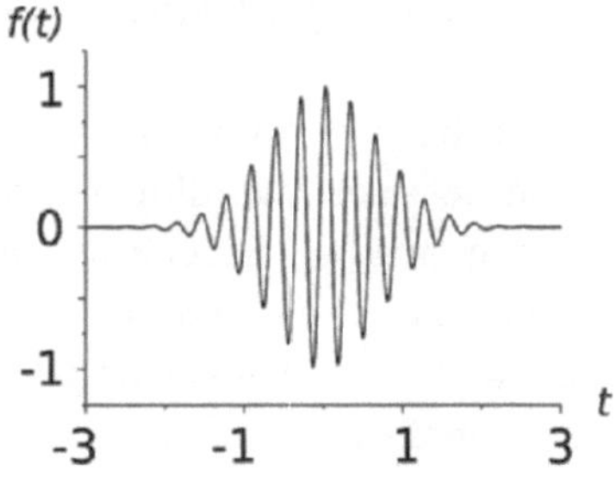

Figure 8.6

We will use the symmetric form of the Fourier transform:

$$g(\omega) = \frac{1}{\sqrt{2\pi}} \int_{-\infty}^{+\infty} e^{-i\omega t} f(t)dt = \frac{1}{\sqrt{2\pi}} \int_{-\infty}^{+\infty} e^{-i\omega t} \cos(\omega_0 t) \cdot e^{-(\frac{t}{\tau})^2} dt =$$

$$= \frac{1}{2\sqrt{2\pi}} \int_{-\infty}^{+\infty} e^{-i\omega t} e^{-i\omega_0 t} \cdot e^{-(\frac{t}{\tau})^2} dt + \frac{1}{2\sqrt{2\pi}} \int_{-\infty}^{+\infty} e^{-i\omega t} e^{i\omega_0 t} \cdot e^{-(\frac{t}{\tau})^2} dt =$$

$$= \frac{1}{2\sqrt{2\pi}} \int_{-\infty}^{+\infty} e^{-i(\omega+\omega_0)t} \cdot e^{-(\frac{t}{\tau})^2} dt + \frac{1}{2\sqrt{2\pi}} \int_{-\infty}^{+\infty} e^{-i(\omega-\omega_0)t} \cdot e^{-(\frac{t}{\tau})^2} dt =$$

$$= \frac{1}{2}\mathfrak{F}[f](\omega+\omega_0) + \frac{1}{2}\mathfrak{F}[f](\omega-\omega_0) \tag{8.47}$$

From integrals tables we have:

$$\mathfrak{F}[f](\omega) = \frac{1}{\sqrt{2\pi}} \int_{-\infty}^{+\infty} e^{-i\omega t} \cdot e^{-(\frac{t}{\tau})^2} dt = \frac{\tau}{\sqrt{2}} e^{-(\frac{\omega\tau}{2})^2} \tag{8.48}$$

Then

$$g(\omega) = \frac{\tau}{\sqrt{2}} \left\{ \frac{e^{-\left(\frac{(\omega+\omega_0)\tau}{2}\right)^2} + e^{-\left(\frac{(\omega-\omega_0)\tau}{2}\right)^2}}{2} \right\} \tag{8.49}$$

Schematic graphic representation of the result:

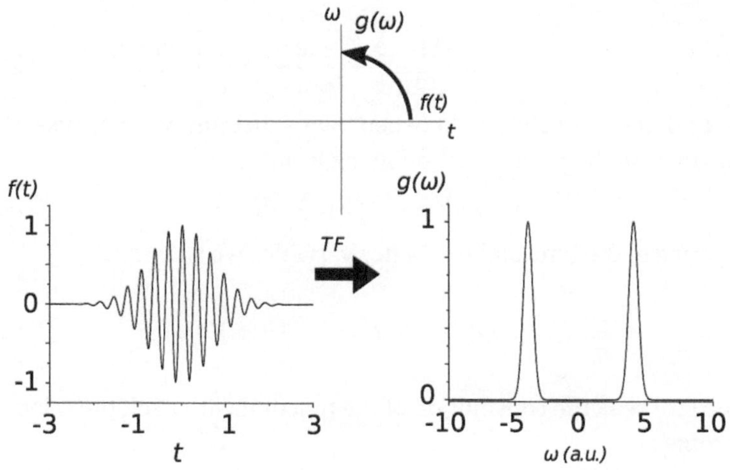

Figure 8.7

8.10 Autoconvolution of a Rectangular Pulse

Find the convolution of the following rectangular pulse with itself. Use Fourier transforms from tables to facilitate calculations. Present the result graphically.

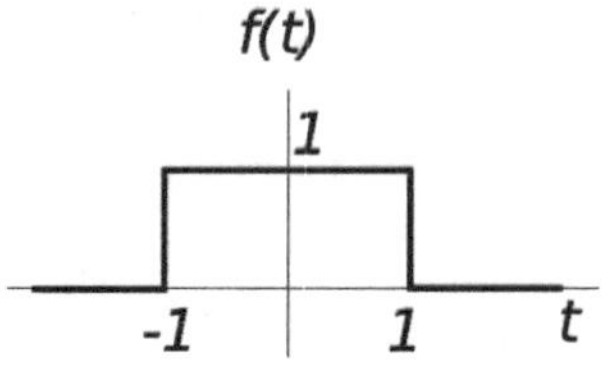

Figure 8.8

Mathematical description of the pulse:

$$f(t) = \left\{ \begin{matrix} 1 & |t| \leq 1 \\ 0 & |t| > 1 \end{matrix} \right\} \tag{8.50}$$

The Fourier transform of this pulse is:

$$F(\omega) = \mathfrak{F}[f](\omega) = [\mathfrak{F}\text{ symmetrical}] = \frac{1}{\sqrt{2\pi}} \int_{-1}^{+1} e^{-i\omega t} dt$$

$$= \frac{1}{\sqrt{2\pi}} \frac{e^{-i\omega} - e^{i\omega}}{i\omega} = \sqrt{\frac{2}{\pi}} \frac{\sin(\omega)}{\omega} \tag{8.51}$$

To find the convolution between two functions we will use the convolution theorem in the following form:

$$[f * g] = \mathfrak{F}^{-1}[F(\omega) \cdot G(\omega)] \tag{8.52}$$

In terms of the time and frequency variables we can write:

$$\frac{1}{\sqrt{2\pi}} \int_{-\infty}^{+\infty} f(t')g(t - t')dt' = \mathfrak{F}^{-1}[F(\omega)G(\omega)] \tag{8.53}$$

Since we seek the convolution of the function with itself the relation becomes:

$$\frac{1}{\sqrt{2\pi}} \int_{-\infty}^{+\infty} f(t')f(t - t')dt' = \mathfrak{F}^{-1}[F(\omega)F(\omega)] \tag{8.54}$$

We know that $F(\omega) = \sqrt{\frac{2}{\pi}}\frac{\sin(\omega)}{\omega}$. Then:

$$\frac{1}{\sqrt{2\pi}}\int_{-\infty}^{+\infty} f(t')f(t-t')dt' = \mathfrak{F}^{-1}\left[(F(\omega))^2\right] = \frac{1}{\sqrt{2\pi}}\int_{-\infty}^{+\infty}\frac{2}{\pi}\left[\frac{\sin(\omega)}{\omega}\right]^2 e^{+i\omega t}d\omega = \tag{8.55}$$

From integrals tables:

$$\begin{Bmatrix} \sqrt{\frac{2}{\pi}}\left(1-\frac{|t|}{2}\right) & |t| \leq 2 \\ 0 & |t| > 2 \end{Bmatrix} \tag{8.56}$$

Graphic representation of the result:

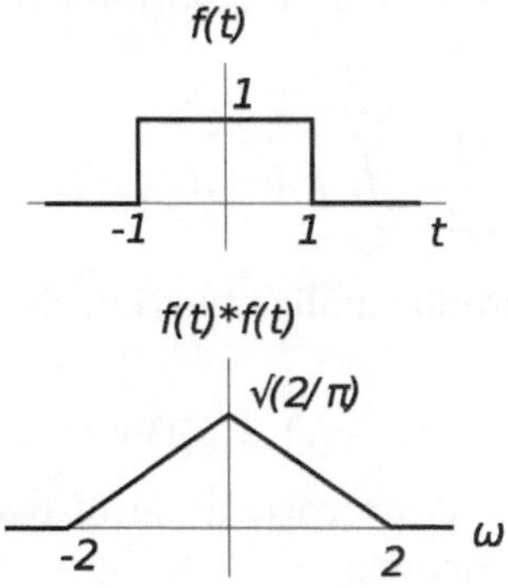

Figure 8.9

Note: the operation of convolution between two functions $f(t)*g(t)$ must not be mixed up with the cross correlation (CC) $f(t) \star g(t)$. In a way the correlation is conceptually simpler than the convolution since the two functions in a correlation are not conceptually different as can be those in a convolution. In the correlation the functions are in general different or they represent data sets that can contain similarities. We investigate their "correlation" by comparing them directly, by superposing them, with one of them shifted either left or right.

The definition of crossed correlation between the $f(t)$ and $g(t)$ functions is:

$$f(t) \star g(t) = \frac{1}{\sqrt{2\pi}}\int_{-\infty}^{+\infty} f^*(t')g(t+t')dt' \tag{8.57}$$

Where $f^*(t)$ is the conjugated of $f(t)$. The relation between CC and convolution is as follows:

$$f(t) \star g(t) = f^*(-t) * g(t) \tag{8.58}$$

Precisely:

$$f^*(-t) * g(t) = \frac{1}{\sqrt{2\pi}} \int_{-\infty}^{+\infty} f^*(-t')g(t-t')dt' \tag{8.59}$$

Changing variables: $(-t' \rightarrow \tau)$ and exchanging the limits of integration we get the result we were looking for:

$$f^*(-t) * g(t) = -\frac{1}{\sqrt{2\pi}} \int_{+\infty}^{-\infty} f^*(\tau)g(t+\tau)d\tau$$

$$= \frac{1}{\sqrt{2\pi}} \int_{-\infty}^{+\infty} f^*(\tau)g(t+\tau)d\tau = f(t) \star g(t) \tag{8.60}$$

Note that the CC operation, unlike the convolution, is not commutative:

$$f(t) \star g(t) \neq g(t) \star f(t) \tag{8.61}$$

Note: the case of the cross correlation of two complex functions is considered in the following website: http://mathworld.wolfram.com/Cross-CorrelationTheorem.html

8.11 Fourier Transform of a Bipolar Triangular Pulse

Find the Fourier transform of the pulse shown in the figure. How is the Fourier transform of the convolution of the signal with itself?

Solution:

We describe mathematically the function:

$$f(x) = \begin{Bmatrix} 1-x & (0 < x \leq 1) \\ -1-x & (-1 \leq x < 0) \\ 0 & (|x| > 1) \end{Bmatrix} \tag{8.62}$$

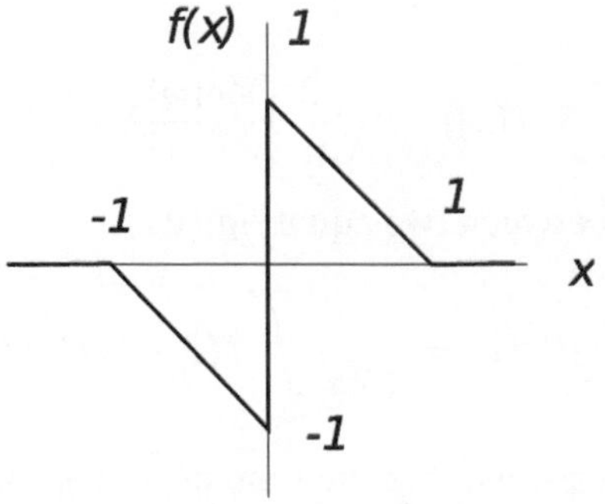

Figure 8.10

Alternatively:

$$f(x) = \begin{Bmatrix} \text{sign}(x) - x \ (|x| < 1) \\ 0 \qquad\qquad (|x| > 1) \end{Bmatrix} \tag{8.63}$$

$$\mathfrak{F}[f(x)] = \frac{1}{\sqrt{2\pi}} \int_{-\infty}^{+\infty} f(x) e^{(-i\omega x)} dx = \frac{1}{\sqrt{2\pi}} \int_{-\infty}^{+\infty} f(x)[\cos(\omega x) - i\sin(\omega x)] \tag{8.64}$$

Since $f(x)$ is an odd function and we integrate between symmetric limits:

$$-\frac{i}{\sqrt{2\pi}} \int_{-\infty}^{+\infty} f(x)\sin(\omega x)dx$$

$$= -\frac{i}{\sqrt{2\pi}} \left\{ \int_{-1}^{0} (-1 - x)\sin(\omega x)dx + \int_{0}^{1} (1 - x)\sin(\omega x)dx \right\} \tag{8.65}$$

$$\int x\sin(\omega x)dx = -\frac{x\cos(\omega x)}{\omega} + \frac{\sin(\omega x)}{\omega^2} \tag{8.66}$$

Then:

$$\int_{-1}^{0} (-1 - x)\sin(\omega x)dx = \left[\frac{\cos(\omega x)}{\omega} + \frac{x\cos(\omega x)}{\omega} - \frac{\sin(\omega x)}{\omega^2} \right]_{-1}^{0} = \frac{1}{\omega} - \frac{\sin(\omega)}{\omega^2} \tag{8.67}$$

On the other hand:

$$\int_{0}^{1} (1 - x)\sin(\omega x)dx = \frac{1}{\omega} - \frac{\sin(\omega)}{\omega^2} \tag{8.68}$$

Then:

$$\mathfrak{F}[f(x)] = i\sqrt{\frac{2}{\pi}}\left[\frac{\sin(\omega)-\omega}{\omega^2}\right] \tag{8.69}$$

The convolution between two equal signals is:

$$[f(t)*f(t)](\tau) = \frac{1}{\sqrt{2\pi}}\int_{-\infty}^{+\infty} f(\tau-t)f(t)dt \tag{8.70}$$

We will use the version of the theorem of convolution which relates the Fourier transform of the convolution of two signals to the product of their Fourier transforms. This relation can be rewritten in terms of the spectral (ω) and temporal (τ) variables as:

$$\mathfrak{F}[[f(t)*f(t)](\tau)] = F(\omega)\cdot F(\omega) \tag{8.71}$$

where $F(\omega)$ is the Fourier transform of $[f(t)]$. In our particular case:

$$\mathfrak{F}[[f(t)*f(t)](\tau)] = F(\omega)\cdot F(\omega) = -\frac{2}{\pi}\left[\frac{\sin(\omega)-\omega}{\omega^2}\right]^2 \tag{8.72}$$

8.12 Fourier Transform of a Rectangular Pulse

Using the symmetric version of the Fourier transform find the spectrum of a triangular pulse:

$$f(x) = \left\{\begin{matrix} 1-|x| \; (-1\le x\le 1) \\ 0 \quad\quad (x<-1;\; x>1)\end{matrix}\right\} \tag{8.73}$$

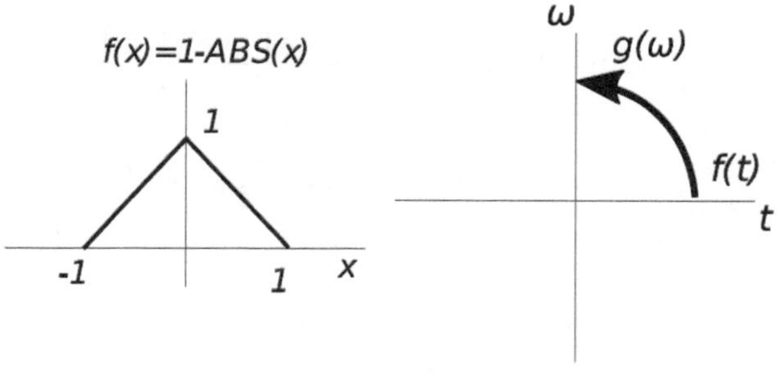

Figure 8.11

Using the symmetric Fourier transform:

$$\mathfrak{F}[f](\omega) = \frac{1}{\sqrt{2\pi}}\int_{-\infty}^{+\infty} e^{-i\omega x} f(x)dx = \frac{1}{\sqrt{2\pi}}\int_{-1}^{+1}(1-|x|)e^{-i\omega x}dx \tag{8.74}$$

We will use the fact that the transform is applied on a symmetric function. Since $e^{-i\omega x} = \cos(\omega x) - i\sin(\omega x)$

$$\frac{1}{\sqrt{2\pi}}\int_{-1}^{+1}(1-|x|)[\cos(\omega x) - i\sin(\omega x)]dx \tag{8.75}$$

$$\frac{i}{\sqrt{2\pi}}\int_{-1}^{+1}(1-|x|)\sin(\omega x)dx = 0 \tag{8.76}$$

The integral is composed of the sum of a symmetric function and an antisymmetric one. The limits are symmetric.

$$\frac{1}{\sqrt{2\pi}}\int_{-1}^{+1}(1-|x|)\cos(\omega x)dx = \frac{2}{\sqrt{2\pi}}\int_{0}^{1}(1-|x|)\cos(\omega x)dx \tag{8.77}$$

Is the integral of a symmetric function between symmetric limits. This integral can be solved by parts with the change of variables:

$$\{u = (1-|x|);\ dv = d[\sin(\omega x)]\} \tag{8.78}$$

And the integral is expressed as:

$$\left|\frac{2}{\omega\sqrt{2\pi}}(1-|x|)\times[\sin(\omega x)]\right|_0^1 + \frac{2}{\omega\sqrt{2\pi}}\int_0^1 \sin(\omega x)dx =$$

$$= \frac{2}{\omega\sqrt{2\pi}}\int_0^1 \sin(\omega x)dx = -\sqrt{\frac{2}{\pi}}\left|\frac{\cos(\omega x)}{\omega^2}\right|_0^1$$

$$= -\sqrt{\frac{2}{\pi}}\frac{\cos(\omega)-1}{\omega^2} = \sqrt{\frac{2}{\pi}}\frac{1-\cos(\omega)}{\omega^2} \tag{8.79}$$

The graphic representation of the result is shown next (only for positive frequencies).

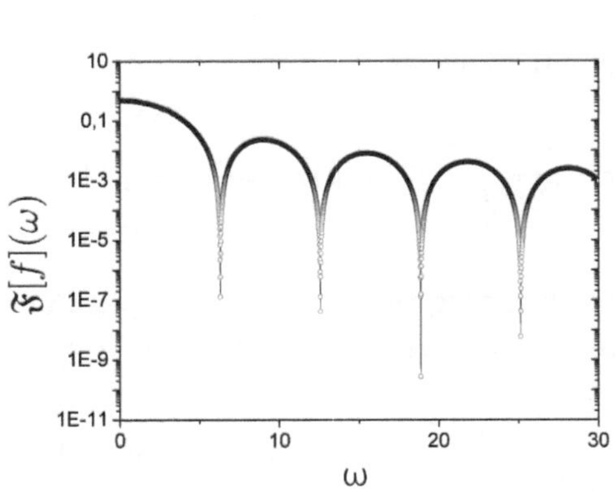

Figure 8.12

8.13 Fourier Transform of the Convolution of a Triangular Pulse with Itself

Find the Fourier transform of the convolution of the triangular pulse shown in the figure below with itself.

$$f(x) = \begin{Bmatrix} 1 - |x| \ (-1 \leq x \leq 1) \\ 0 \qquad (x < -1;\ x > 1) \end{Bmatrix} \tag{8.80}$$

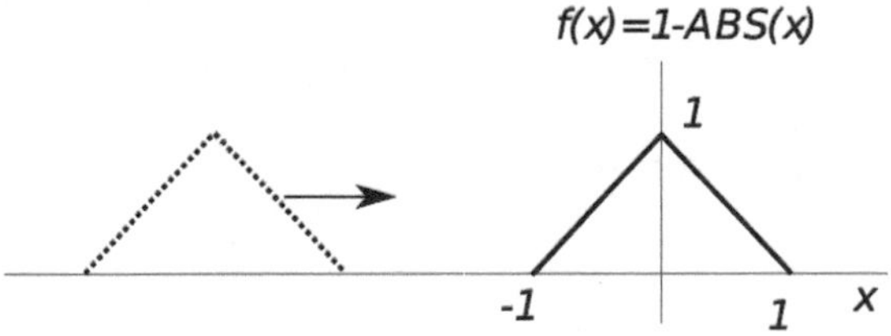

Figure 8.13

We know that, by definition, the convolution between two functions is the inverse Fourier transform of the product of their Fourier transforms, which without normalization is:

$$f * g[x] = \int_{-\infty}^{+\infty} f(x-t)g(t)dt = \mathfrak{F}^{-1}[F(k)G(k)] \tag{8.81}$$

Appling the Fourier transform to this relation an supposing that $f = g$, the Fourier transform of the auto-convolution is:

$$\mathfrak{F}[f * f[x]] = \mathfrak{F}\left[\int_{-\infty}^{+\infty} f(x-t)f(t)dt\right] = \mathfrak{F}\mathfrak{F}^{-1}[F(k)F(k)] = [F(k)]^2 \tag{8.82}$$

The Fourier transform of the triangular pulse is:

$$F(k) = \sqrt{\frac{2}{\pi}}\frac{1-\cos(k)}{k^2} \tag{8.83}$$

Furthermore we can simplify it using the relation:

$$1 - \cos(k) = 2\sin^2\left(\frac{k}{2}\right) \tag{8.84}$$

$$F(k) = \sqrt{\frac{8}{\pi}}\frac{\sin^2\left(\frac{k}{2}\right)}{k^2} \tag{8.85}$$

The solution is:

$$\mathfrak{F}\left[\int_{-\infty}^{+\infty} f(x-t)f(t)dt\right] = \frac{8}{\pi}\left[\frac{\sin^2\left(\frac{k}{2}\right)}{k^2}\right]^2 \tag{8.86}$$

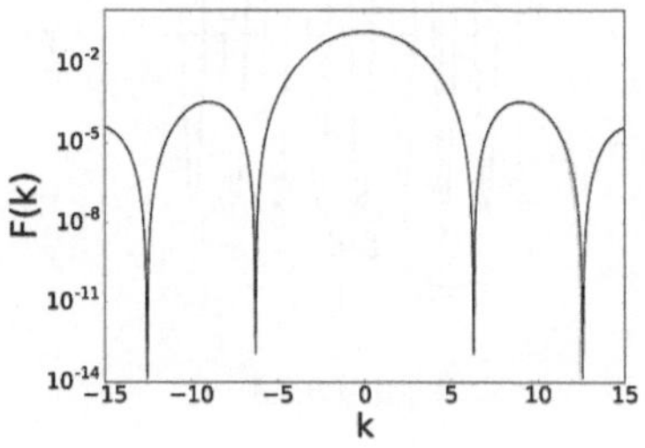

Figure 8.14

Note: the next figure indicates the schematic form of the convolution of a triangular pulse with itself (this is, proportional to inverse Fourier transform of $\sin^4(x/2)]/x^4$):

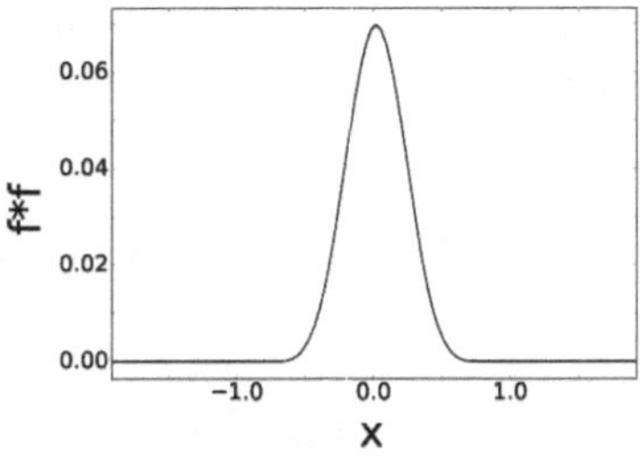

Figure 8.15

8.14 Fourier Transform of a Shifted Rectangular Pulse with a Sine Modulation

Find the Fourier transform of a modulated rectangular pulse, shifted in time to t_0. Before the shift the pulse was defined in a temporal range ($-\tau < t < \tau$), with an amplitude $I/2\tau$ (the surface of the total pulse is I) and with a periodic modulation ($\sin\omega_0 t$). Consider the case of the shift of the pulse without shifting the modulation.

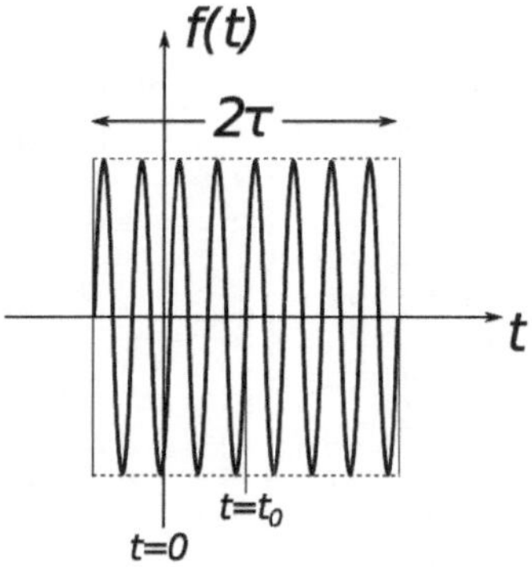

Figure 8.16

Mathematical description of the pulse:

$$f_m(t) = \left\{ \begin{array}{ll} \dfrac{I}{2\tau} & (t_0 - \tau < t < t_0 + \tau \\ 0 & (t > t_0 + \tau) \\ 0 & (t < t_0 - \tau) \end{array} \right\} \times \sin(\omega_0 t) = f(t) \times \sin(\omega_0 t) \quad (8.87)$$

We first use the Fourier transform of a shifted rectangular pulse without modulation.

$$\mathfrak{F}[f(t-t_0)] = \frac{1}{\sqrt{2\pi}} \int_{-\infty}^{+\infty} f(t-t_0)e^{-i\omega t}dt = \{t' = t - t_0\}$$

$$= \frac{1}{\sqrt{2\pi}} \int_{-\infty}^{+\infty} f(t')e^{-i\omega(t'+t_0)}dt' =$$

$$e^{-i\omega t_0} F(\omega) = \frac{I}{\sqrt{2\pi}} e^{-i\omega t_0} \frac{\sin(\omega\tau)}{\omega\tau} \tag{8.88}$$

The product of $f(t)$ by $\sin(\omega_0 t)$ is equivalent to:

a) Shifting the Fourier transform to $(+\omega_0)$
b) Shifting the Fourier transform to $(-\omega_0)$
c) Subtract these shifted Fourier transforms and divide by $2i$.

Using this rule we find:

$$\mathfrak{F}[f_m(t)] = \frac{I}{i\sqrt{2\pi}} \left[e^{-i(\omega+\omega_0)t_0} \frac{\sin[(\omega+\omega_0)\tau]}{(\omega+\omega_0)\tau} - e^{-i(\omega-\omega_0)t_0} \frac{\sin[(\omega-\omega_0)\tau]}{(\omega-\omega_0)\tau} \right] \tag{8.89}$$

The strictly mathematical method consists in using the theorem of convolution, seeking the Fourier transform of $f_m(t) = f(t) \times \sin(\omega_0 t)$ as the transform of the product of two functions from the operation of convolution between the respective Fourier transforms.

$$\mathfrak{F}[f(t-t_0) \cdot g(t)] = F(\omega) * G(\omega) = \int_{-\infty}^{+\infty} F(\omega-\eta)G(\eta)d\eta \tag{8.90}$$

with $f(t)$ previously defined and $g(t) = \sin(\omega_0 t)$. Using for $F(\omega)$ the previously obtained result (8.88), as well as $G(\omega) = i\sqrt{\frac{\pi}{2}}[\delta(\omega + \omega_0) - \delta(\omega - \omega_0)]$, we arrive at the previous result (8.89).

8.15 Case Study: Solution of a PDE Using the Fourier Transform: Case 1

Using the Fourier method, solve the following partial derivative equation with initial condition:

$$\frac{\partial u}{\partial t} - t^2 \frac{\partial u}{\partial x} = 0 \tag{8.91}$$

$$u(x, 0) = f(x) = \cos(x) \tag{8.92}$$

The x variable is defined in the range $(-\infty < x < +\infty)$

First the Fourier transform is applied to the equation to be solved. We will use the symmetric form of the transform.

We define a new function $v(k, t) = \mathfrak{F}[u] = \frac{1}{\sqrt{2\pi}} \int_{-\infty}^{+\infty} u(x, t)e^{-ikx}dx$

Applying the property of the Fourier transform of the derivative of a function:

$$\frac{\partial}{\partial t}v(k, t) - t^2(ik)v(k, t) = 0 \tag{8.93}$$

and with the initial condition:

$$v(k, t = 0) = \mathfrak{F}[\cos(x)] = F(k) \tag{8.94}$$

We arrive at an ordinary differential equation:

$$\frac{dv(k, t)}{v(k, t)} = (ik)t^2dt = d\left[\frac{ikt^3}{3}\right] \tag{8.95}$$

Then:

$$v(k, t) = C(k)e^{\frac{ikt^3}{3}} \tag{8.96}$$

being $C(k)$ an unknown function which is determined by the initial conditions $(t = 0)$.

Since $v(k, 0) = F(k)$ we get the solution:

$$v(k, t) = F(k)e^{\frac{ikt^3}{3}} \tag{8.97}$$

To find the final solution we must apply the inverse transform to $v(k, t)$:

$$u(x, t) = \frac{1}{\sqrt{2\pi}}\int_{-\infty}^{+\infty} v(k, t)e^{ikx}dk = \frac{1}{\sqrt{2\pi}}\int_{-\infty}^{+\infty} F(k)e^{\frac{ikt^3}{3}}e^{ikx}dk =$$

$$= \frac{1}{\sqrt{2\pi}}\int_{-\infty}^{+\infty} F(k)e^{ik\left(x+\frac{t^3}{3}\right)}dk = \cos\left(x + \frac{t^3}{3}\right) \tag{8.98}$$

since

$$u(x, 0) = \frac{1}{\sqrt{2\pi}}\int_{-\infty}^{+\infty} F(k)e^{ikx}dk = \cos(x) \tag{8.99}$$

Note: **An alternative method** to find the inverse transform is to use the particular value of the function $F(k)$.

From Fourier transform tables we know:

$$F(k)=\mathfrak{F}[\cos(x)]=\frac{1}{\sqrt{2\pi}}\int_{-\infty}^{+\infty}\cos(x)e^{-ikx}dx=\sqrt{\frac{\pi}{2}}[\delta(k-1)+\delta(k+1)] \tag{8.100}$$

Then:

$$u(x,t)=\frac{1}{\sqrt{2\pi}}\int_{-\infty}^{+\infty}F(k)e^{\frac{ikt^3}{3}}e^{ikx}dk=$$

$$=\frac{1}{2}\int_{-\infty}^{+\infty}[\delta(k-1)+\delta(k+1)]e^{ik(x+\frac{t^3}{3})}dk$$

$$=\frac{e^{-i(x+\frac{t^3}{3})}+e^{+i(x+\frac{t^3}{3})}}{2}=\cos\left(x+\frac{t^3}{3}\right) \tag{8.101}$$

8.16 Case Study: Solution of a PDE Using the Fourier Transform: Case 2

Solve the wave equation with known initial conditions using the Fourier method. Show the solution as an inverse Fourier transform.

$$\left\{\begin{array}{l}\frac{\partial^2 u}{\partial t^2}=c^2\frac{\partial^2 u}{\partial x^2}\\ u(x,0)=\sqrt{\frac{2}{\pi}}\frac{\sin(x)}{x}\\ \left.\frac{\partial u}{\partial t}\right|_{t=0}=0\\ -\infty<x<\infty;\ t>0\end{array}\right\} \tag{8.102}$$

We keep (t) constant and find the Fourier transform of the wave equation with respect to x.

$$\frac{\partial^2}{\partial t^2}\mathfrak{F}[u](\omega,t)=-c^2\omega^2\mathfrak{F}[u](\omega,t) \tag{8.103}$$

From the first initial condition:

$$\mathfrak{F}[u](\omega,0) = \mathfrak{F}\left[\sqrt{\frac{2}{\pi}}\frac{\sin(x)}{x}\right](\omega) = f(\omega) = \begin{Bmatrix} 1 & |\omega| < 1 \\ 0 & |\omega| > 1 \end{Bmatrix} \quad (8.104)$$

Furthermore, from the second initial condition $\frac{d}{dt}\mathfrak{F}[u](\omega, 0) = 0$. The general solution of (8.103) is that of a harmonic oscillator. $\mathfrak{F}[u](\omega, t) = A(\omega)\sin(c\omega t) + B(\omega)\cos(c\omega t)$. From the second initial condition $A(\omega) = 0$ and from the first one, $B(\omega) = f(\omega)$. Then $\mathfrak{F}[u](\omega, t) = f(\omega)\cos(c\omega t)$. In this way, using the inverse Fourier transform the solution will be:

$$u(x,t) = \mathfrak{F}^{-1}[f(\omega)\cos(c\omega t)] = \frac{1}{\sqrt{2\pi}}\int_{-\infty}^{+\infty}[f(\omega)\cos(c\omega t)]e^{(ix\omega)}d\omega =$$

$$= \frac{1}{\sqrt{2\pi}}\int_{1}^{+1}\cos(c\omega t)e^{(ix\omega)}d\omega \quad (8.105)$$

8.17 Case Study: Solution of a PDE Using the Fourier Transform: Case 3

Find the solution to:

$$\frac{d^2X(t)}{dt^2} - a^2X(t) = f(t) \quad (8.106)$$

applying the Fourier transform.

Notes:

a) We need to impose the condition that the solution satisfies $f(t = \pm\infty) \to 0$ to ensure the existence of the Fourier transform of the functions we'll consider.
b) The solution requires the use of the definition of the Fourier transform of the product of two functions, as well as the use of Fourier transform tables.

Equation to be solved:

$$\frac{d^2X(t)}{dt^2} - a^2X(t) = f(t) \quad (8.107)$$

To solve the problem we will suppose that: $\mathfrak{F}[X(t)] = G(\omega)$ and $\mathfrak{F}[f(t)] = F(\omega)$. Applying the Fourier transform operator to the equation:

$$\mathfrak{F}\left[\frac{d^2X(t)}{dt^2} - a^2X(t) = f(t)\right] \tag{8.108}$$

or

$$-\omega^2 G(\omega) - a^2 G(\omega) = F(\omega) \tag{8.109}$$

$$-[\omega^2 + a^2]G(\omega) = F(\omega) \tag{8.110}$$

$$G(\omega) = -\frac{F(\omega)}{[\omega^2 + a^2]} \tag{8.111}$$

We now use Fourier transform tables:

$$\mathfrak{F}\left[e^{(-a|t|)}\right] = \frac{2a}{[\omega^2 + a^2]} \tag{8.112}$$

Then, defining a new function $Y(\omega)$ we can write:

$$Y(\omega) = -\frac{1}{\left[\omega^2 + a^2\right]} = \mathfrak{F}[y(t)] = \mathfrak{F}\left[-\frac{1}{2a}e^{(-a|t|)}\right] \tag{8.113}$$

We will now use the convolution theorem, which relates the Fourier transform of the product of two functions to a convolution. In the present case we can write:

$$G(\omega) = Y(\omega)F(\omega) = \mathfrak{F}[y(t) * f(t)] \tag{8.114}$$

where an asterisk indicates the convolution operation between the functions $y(t)$ and $f(t)$.

Think to find the solution:

$$\begin{aligned} X(t) &= \mathfrak{F}^{-1}[G(\omega)] = \mathfrak{F}^{-1}[Y(\omega)F(\omega)] \\ &= \mathfrak{F}^{-1}\mathfrak{F}[y(t) * f(t)] = [y(t) * f(t)] \end{aligned} \tag{8.115}$$

(that is, the solution of the problem is the convolution between the functions $y(t)$ and $f(t)$). Using the definition of the convolution (not normalized) we get the solution for the displacement of the forced relaxator:

$$X(t) = -\frac{1}{2a}\int_{-\infty}^{+\infty} e^{(-a|t-\tau|)} f(\tau)d\tau \tag{8.116}$$

Note: the solution could describe the movement of an object in one dimension in a medium with a friction proportional to the displacement under an external random force.

8.18 Case Study: Solution of a PDE Using the Fourier Transform: Case 4

Using the Fourier method, solve the following PDE with initial condition:

$$\frac{\partial u}{\partial t} - t^3 \frac{\partial^2 u}{\partial x^2} = 0$$

$$u(x, 0) = f(x) = \sin(x)$$

The x variable is defined in the range $(-\infty < x < +\infty)$

We first apply the Fourier Transform to the equation to be solved:

We will use the symmetric version of the transform.

We define a new function $v(k, t) = \mathfrak{F}[u] = \frac{1}{\sqrt{2\pi}} \int_{-\infty}^{+\infty} u(x, t)e^{-ikx}dx$

Applying the property of the Fourier Transform of the derivative of a function:

$$\frac{\partial}{\partial t}v(k, t) + t^3k^2v(k, t) = 0 \tag{8.117}$$

and with the initial condition:

$$v(k, t = 0) = \mathfrak{F}[\sin(x)] = F(k) \tag{8.118}$$

We get to an ordinary differential equation:

$$\frac{dv(k, t)}{v(k, t)} = -k^2t^3dt = d\left[\frac{-k^2t^4}{4}\right] \tag{8.119}$$

Then:

$$v(k, t) = C(k)e^{\frac{-k^2t^4}{4}} \tag{8.120}$$

being $C(k)$ an unknown function that is determined by the initial conditions $(t = 0)$.

Since $v(k, 0) = F(k)$ we get the solution:

$$v(k, t) = F(k)e^{\frac{-k^2t^4}{4}} \tag{8.121}$$

From the tables of Fourier Transforms we know:

$$F(k) = \mathfrak{F}[\sin(x)] = \frac{1}{\sqrt{2\pi}} \int_{-\infty}^{+\infty} \sin(x)e^{-ikx}dx = i\sqrt{\frac{\pi}{2}}[\delta(k-1) - \delta(k+1)] \tag{8.122}$$

To find the final solution we must apply the inverse Fourier Transform to $v(k, t)$:

$$u(x,t) = \frac{1}{\sqrt{2\pi}} \int_{-\infty}^{+\infty} F(k) e^{\frac{-k^2 t^4}{4}} e^{ikx} dk =$$

$$= \frac{i}{2} \int_{-\infty}^{+\infty} [\delta(k-1) - \delta(k+1)] e^{\left(ikx - k^2 \frac{t^4}{4}\right)} dk = e^{\left(-\frac{t^4}{4}\right)} \sin(x)$$

(8.123)

8.19 Case Study: Solution of a PDE Using the Fourier Transform: Case 5

Using the Fourier method, solve the next PDE with initial condition:

$$\frac{\partial u}{\partial t} + \sin(t)\frac{\partial u}{\partial x} = 0$$

$$u(x, 0) = f(x) = \sin(x)$$

The x variable is defined in the range $(-\infty < x < +\infty)$

We first apply the Fourier Transform to the equation to be solved. We will use the symmetric version of the transform.

We define a new function $v(k,t) = \mathfrak{F}[u] = \frac{1}{\sqrt{2\pi}} \int_{-\infty}^{+\infty} u(x,t) e^{-ikx} dx$

Applying the property of the Fourier Transform of the derivative of a function:

$$\frac{\partial}{\partial t} v(k,t) + \sin(t)(ik) v(k,t) = 0 \tag{8.124}$$

and with the initial condition:

$$v(k, t = 0) = \mathfrak{F}[\sin(x)] = F(k) \tag{8.125}$$

We arrive at an ordinary differential equation:

$$\frac{dv(k,t)}{v(k,t)} = -(ik)\sin(t)dt = d\,[ik\cos(t)] \tag{8.126}$$

Then:

$$v(k,t) = C(k) e^{ik\cos(t)} \tag{8.127}$$

where $C(k)$ an unknown function that is determined thanks to the initial conditions $(t = 0)$. Since $v(k, 0) = F(k)$ we get the solution:

$$C(k) = \frac{F(k)}{e^{ik}} \tag{8.128}$$

$$v(k, t) = F(k)e^{ik(\cos(t)-1)} \tag{8.129}$$

From the Fourier Transforms tables, we know:

$$F(k)=\mathfrak{F}[\sin(x)]=\frac{1}{\sqrt{2\pi}}\int_{-\infty}^{+\infty}\sin(x)e^{-ikx}dx = i\sqrt{\frac{\pi}{2}}[\delta(k-1)-\delta(k+1)] \tag{8.130}$$

To find the final solution we must apply the inverse transform to $v(k, t)$:

$$\begin{aligned} u(x, t) &= \frac{1}{\sqrt{2\pi}}\int_{-\infty}^{+\infty} F(k)e^{ik(\cos(t)-1)}e^{ikx}dk = \\ &= \frac{i}{2}\int_{-\infty}^{+\infty}[\delta(k-1)-\delta(k+1)]e^{ik(x+\cos(t)-1)}dk \\ &= \sin(\cos(t)+x-1)) \end{aligned} \tag{8.131}$$

8.20 Case Study: Solution of a PDE Using the Fourier Transform: Case 6

Using the Fourier method, solve the next PDE with initial condition:

$$\frac{\partial u}{\partial t} + e^{-t}\frac{\partial^2 u}{\partial x^2} = 0$$

$$u(x, 0) = f(x) = e^{-|x|}$$

The x variable is defined in the range $(-\infty < x < +\infty)$

We first apply the Fourier Transform to the equation to be solved:

We will use the symmetric version of the transform.

We define a new function $v(k, t) = \mathfrak{F}[u] = \frac{1}{\sqrt{2\pi}}\int_{-\infty}^{+\infty} u(x, t)e^{-ikx}dx$

Applying the property of the Fourier Transform of the derivative of a function:

$$\frac{\partial}{\partial t}v(k,t) + e^{-t}(-k^2)v(k,t) = 0 \tag{8.132}$$

and with the initial condition we get from The Fourier Transforms tables:

$$v(k,t=0) = \mathfrak{F}[f(x)] = F(k) = \sqrt{\frac{2}{\pi}}\frac{1}{1+k^2} \tag{8.133}$$

We arrive at an ordinary differential equation:

$$\frac{dv(k,t)}{v(k,t)} = -(k^2)e^{-t}dt \tag{8.134}$$

Then:

$$v(k,t) = C(k)e^{(-k^2e^{-t})} \tag{8.135}$$

where $C(k)$ an unknown function that is determined thanks to the initial condition $(t = 0)$.

Since $v(k, 0) = F(k)$ we get the solution:

$$v(k,0) = C(k)e^{(-k^2)} = \sqrt{\frac{2}{\pi}}\frac{1}{1+k^2} \tag{8.136}$$

$$v(k,t) = \frac{1}{e^{(-k^2)}}\sqrt{\frac{2}{\pi}}\frac{1}{1+k^2}e^{(-k^2e^{-t})} \tag{8.137}$$

To find the final solution we must apply the inverse transform to $v(k,t)$:

$$u(x,t) = \frac{1}{\sqrt{2\pi}}\int_{-\infty}^{+\infty} C(k)e^{(-k^2e^{-t})}e^{ikx}dk \tag{8.138}$$

8.21 Case Study: Solution of the Diffusion Equation in an Infinite String with Convection Using the Fourier Transform

Solve the diffusion with convection problem for a 1-D infinite system (the additional term is due to the mass transfer) using the Fourier method.

(i) Find the variation of the concentration with time $u(x, t)$, considering that the distribution of the concentration at $t = 0$ is $u(x, 0) = f(x)$.
(ii) Consider the case of an initial point like distribution: $u(x, 0) = \delta(x)$.

Note: the equation of convection-diffusion is basically the diffusion equation, modified with an inhomogeneous term, proportional to the velocity $c = \text{Const}$, multiplied by the gradient of the concentration, i.e., $c\frac{\partial u}{\partial x}$.

We formulate the problem:

$$\left\{\begin{array}{l} \dfrac{\partial u(x,t)}{\partial t} = k\dfrac{\partial^2 u(x,t)}{\partial x^2} + c\dfrac{\partial u}{\partial x} \ (-\infty < x < \infty);\ k, c > 0 u(x,0) = f(x) \\ f(\pm\infty) = 0 \end{array}\right\} \tag{8.139}$$

We apply the Fourier transform in order to pass the diffusion equation and the initial condition to reciprocal space $(x \to \omega)$.

$$\mathfrak{F}\left[\frac{\partial^2 u(x,t)}{\partial x^2}\right] = -\omega^2 U(\omega, t) \tag{8.140}$$

$$\mathfrak{F}\left[\frac{\partial u(x,t)}{\partial t}\right] = \frac{d}{dt}U(\omega, t) \tag{8.141}$$

$$\mathfrak{F}\left[\frac{\partial u(x,t)}{\partial x}\right] = i\omega U(\omega, t) \tag{8.142}$$

$$\mathfrak{F}[f(x)] = F(\omega) \tag{8.143}$$

$$\left\{\begin{array}{l} \dfrac{dU(\omega,t)}{dt} - ic\omega U(\omega,t) + k\omega^2 U(\omega,t) = 0 \\ U(\omega,0) = F(\omega) \end{array}\right\} \tag{8.144}$$

$$\left\{\begin{array}{l} \dfrac{dU(\omega,t)}{dt} + [k\omega^2 - ic\omega]U(\omega,t) = 0 \\ U(\omega,0) = F(\omega) \end{array}\right\} \tag{8.145}$$

We seek the solution in the form $U(\omega, t) = A(\omega)e^{\alpha t}$

$$\frac{dU(\omega, t)}{dt} = A(\omega)\alpha e^{\alpha t} \tag{8.146}$$

The equation to be solved is:

$$A(\omega)\alpha e^{\alpha t} + [k\omega^2 - ic\omega]A(\omega)e^{\alpha t} = 0 \tag{8.147}$$

$$\alpha = -(k\omega^2 - ic\omega) \tag{8.148}$$

$$U(\omega, t) = A(\omega)e^{(-[k\omega^2 - ic\omega]t)} \tag{8.149}$$

Applying the initial condition:

$$U(\omega, t) = F(\omega)e^{(-[k\omega^2 - ic\omega]t)} \tag{8.150}$$

Final solution (inverse symmetric Fourier transform):

$$u(x, t) = \frac{1}{\sqrt{2\pi}} \int_{-\infty}^{+\infty} F(\omega)e^{(-[k\omega^2 - ic\omega]t)}e^{(i\omega x)}d\omega \tag{8.151}$$

We will consider the case of a particular point like distribution: $u(x, 0) = \delta(x)$. In this case:

$$F(\omega) = \frac{1}{\sqrt{2\pi}} = \text{const} \tag{8.152}$$

$$u(x, t) = \frac{1}{\sqrt{2\pi}} \int_{-\infty}^{+\infty} \frac{1}{\sqrt{2\pi}} e^{(-[k\omega^2 - ic\omega]t)}e^{(i\omega x)}d\omega$$

$$= \frac{1}{2\pi} \int_{-\infty}^{+\infty} e^{(-k\omega^2 t)}e^{[i\omega(x+ct)]}d\omega \tag{8.153}$$

Since $\mathfrak{F}^{-1}[e^{(-a\omega^2)}](\tau) = \frac{e^{\left(-\frac{\tau^2}{4a}\right)}}{\sqrt{2a}}$

Inserting the new variables:

$$x + ct = \tau; kt = a \tag{8.154}$$

$$u(x, t) = \frac{1}{\sqrt{2\pi}} \frac{e^{(-\frac{\tau^2}{4a})}}{\sqrt{2a}} = \frac{e^{[-\frac{(x+ct)^2}{4kt}]}}{\sqrt{4\pi kt}} \tag{8.155}$$

From this solution we deduce the Green's function of the diffusion equation with convection, replacing x by $x - \xi$, being $x = \xi$ the point of application of the initial pulse. In a similar fashion, the Green's function in the case of the heat equation in 1D ($c = 0$) and with $u(x, 0) = f(x - \xi)$ will be:

$$u(x, \xi, t) = \frac{e^{[-\frac{(x-\xi)^2}{4kt}]}}{\sqrt{4\pi kt}} \tag{8.156}$$

8.22 Application of the Fourier Transform to Find the Displacements of a String Attached to an Elastic Fabric

Using the Fourier transform method, find the solution for a wave which propagates along an infinite thread connected to an elastic fabric if all initial conditions (displacement and velocity) are known.

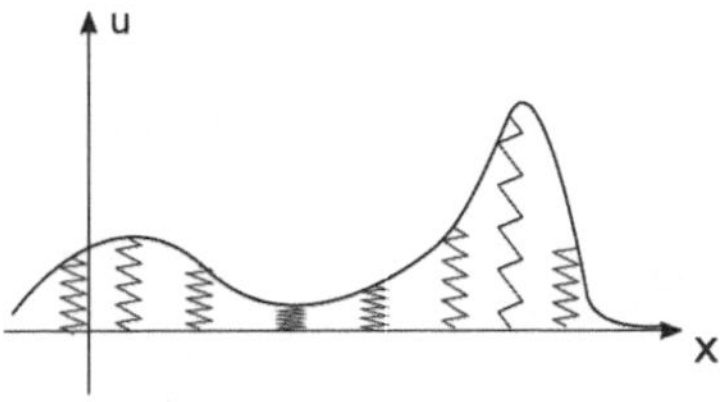

Figure 8.17

Formulation of the problem:

$$\left\{\begin{array}{l} \dfrac{\partial^2 u(x,t)}{dt^2} - c^2 \dfrac{\partial^2 u(x,t)}{dx^2} = -ku(x,t) \\ u(x,0) = f(x) \\ \left.\dfrac{\partial u}{\partial t}\right|_{t=0} = \psi(x) \\ f(\pm\infty); \psi(\pm\infty) = 0 \\ c, k > 0 \end{array}\right\} \tag{8.157}$$

We apply the Fourier transform to pass to reciprocal space both the equation and the initial conditions.

$$\mathfrak{F}[u(x,y)] = U(\omega,t) \tag{8.158}$$

$$\mathfrak{F}\left[\frac{\partial^2 u(x,y)}{\partial x^2}\right] = -\omega^2 U(\omega,t) \tag{8.159}$$

$$\mathfrak{F}\left[\frac{\partial^2 u(x,y)}{\partial t^2}\right] = \frac{\partial^2 U(\omega,t)}{\partial t^2} \tag{8.160}$$

$$\mathfrak{F}[f(x)] = F(\omega) \tag{8.161}$$

$$\mathfrak{F}[\psi(x)] = \Psi(\omega) \tag{8.162}$$

$$\left\{\begin{array}{l} \dfrac{\partial^2 U(\omega, t)}{\partial t^2} + c^2\omega^2 U(\omega, t) + kU(\omega, t) = 0 \\ \dfrac{\partial^2 U(\omega, t)}{\partial t^2} - C^2 U(\omega, t) = 0 \\ C^2 = c^2\omega^2 + k \\ U(\omega, 0) = F(x) \\ \left.\dfrac{\partial U}{\partial t}\right|_{t=0} = \Psi(x) \end{array}\right\} \tag{8.163}$$

The general solution is:

$$U(\omega, t) = A(\omega)\cos(C\omega t) + B(\omega)\sin(C\omega t) \tag{8.164}$$

$$U(\omega, 0) = F(x) = A(\omega) \tag{8.165}$$

$$\left.\frac{\partial U}{\partial t}\right|_{t=0} = \Psi(\omega) = C\omega B(\omega) \tag{8.166}$$

$$B(\omega) = \frac{\Psi(\omega)}{C\omega} \tag{8.167}$$

$$U(\omega, t) = F(\omega)\cos(C\omega t) + \frac{\Psi(\omega)}{C\omega}\sin(C\omega t) \tag{8.168}$$

Final result:

$$u(x, t) = \frac{1}{\sqrt{2\pi}}\int_{-\infty}^{+\infty}\left[F(\omega)\cos(C\omega t) + \frac{\Psi(\omega)}{C\omega}\sin(C\omega t)\right]e^{(i\omega x)}d\omega \tag{8.169}$$

8.23 Case Study: Oscillations in an Infinite String with Friction

Solve the oscillations in an infinite string with a friction proportional to the vertical component of the velocity (with a proportionality coefficient K) using the Fourier method. Consider that the initial displacement is $f(x)$ and the initial velocity is $\psi(x)$. The speed of sound is equal to c.

Figure 8.18

Mathematical formulation:

$$\left\{\begin{array}{l} \dfrac{\partial^2 u}{\partial t^2} - \dfrac{T}{\rho}\dfrac{\partial^2 u}{\partial x^2} = -\dfrac{K}{\rho}\dfrac{\partial u}{\partial t} \ (-\infty < x < \infty) \\ \dfrac{\partial^2 u}{\partial t^2} - c^2\dfrac{\partial^2 u}{\partial x^2} = -k\dfrac{\partial u}{\partial t} \quad (-\infty < x < \infty) \\ u(x, 0) = f(x) \\ \left.\dfrac{\partial u}{\partial t}\right|_{t=0} = \psi(x) \\ f(\pm\infty); \psi(\pm\infty) = 0 \\ c, k > 0 \end{array}\right\} \tag{8.170}$$

We will apply the Fourier transform to pass to reciprocal space the equation and the initial conditions. We use ω to refer to the wave vector.

$$\mathfrak{F}[u(x, t)] = U(\omega, t) \tag{8.171}$$

$$\mathfrak{F}\left[\frac{\partial^2 u(x, t)}{\partial x^2}\right] = -\omega^2 U(\omega, t) \tag{8.172}$$

$$\mathfrak{F}\left[\frac{\partial^2 u(x, t)}{\partial t^2}\right] = \frac{\partial^2 U(\omega, t)}{\partial t^2} \tag{8.173}$$

$$\mathfrak{F}[f(x)] = F(\omega) \tag{8.174}$$

$$\mathfrak{F}[\psi(x)] = \Psi(\omega) \tag{8.175}$$

$$\left\{\begin{array}{l} \dfrac{d^2 U(\omega, t)}{dt^2} + k\dfrac{dU(\omega, t)}{dt} + c^2\omega^2 U(\omega, t) = 0 \\ U(\omega, 0) = F(\omega) \\ \left.\dfrac{\partial U}{\partial t}\right|_{t=0} = \Psi(\omega) \end{array}\right\} \tag{8.176}$$

We seek the solution in the form $U(\omega, t) = A(\omega)e^{\alpha t}$

$$\frac{d^2U(\omega, t)}{dt^2} = A(\omega)\alpha^2 e^{\alpha t} \tag{8.177}$$

$$\frac{dU(\omega, t)}{dt} = A(\omega)\alpha e^{\alpha t} \tag{8.178}$$

Equation to be solved:

$$A(\omega)\alpha^2 e^{\alpha t} + kA(\omega)\alpha e^{\alpha t} + c^2\omega^2 A(\omega)e^{\alpha t} = 0 \tag{8.179}$$

$$\alpha^2 + k\alpha + c^2\omega^2 = 0 \tag{8.180}$$

$$\alpha_{1,2} = -\frac{k}{2} \pm \frac{\sqrt{k^2 - 4c^2\omega^2}}{2} = -\frac{k}{2} \pm \beta \tag{8.181}$$

We will have real values for frequencies that satisfy:

$$k^2 - 4c^2\omega^2 > 0; |\omega| \leq \frac{k}{2c} \tag{8.182}$$

There will be two linearly independent solutions:

$$\begin{Bmatrix} U_1(\omega, t) = A(\omega)e^{\alpha_1 t} \\ U_2(\omega, t) = A(\omega)e^{\alpha_2 t} \end{Bmatrix} \tag{8.183}$$

We can simplify the solution creating two linearly independent combinations:

$$V_1(\omega, t) = \frac{U_1(\omega, t) + U_2(\omega, t)}{2} = A(\omega)e^{(-\frac{k}{2}t)}\frac{e^{(+\beta t)} + e^{(-\beta t)}}{2} = \\ = A(\omega)e^{(-\frac{k}{2}t)}\cosh(\beta t) \tag{8.184}$$

$$V_2(\omega, t) = \frac{U_1(\omega, t) - U_2(\omega, t)}{2} = A(\omega)e^{(-\frac{k}{2}t)}\frac{e^{(+\beta t)} - e^{(-\beta t)}}{2} = \\ = A(\omega)e^{(-\frac{k}{2}t)}\sinh(\beta t) \tag{8.185}$$

The solution of equation (8.176) for the frequency range $|\omega| < \frac{k}{2c}$ must be presented in the form of a combination of two independent solutions , in general with different weights:

$$U(\omega, t) = e^{(-\frac{k}{2}t)}[A(\omega)\cosh(\beta t) + B(\omega)\sinh(\beta t)] \tag{8.186}$$

Finally for frequencies $|\omega| > \frac{k}{2c} \rightarrow \alpha_{1,2} =$, complex values:

$$\alpha_{1,2} = -\frac{k}{2} \pm \frac{\sqrt{k^2 - 4c^2\omega^2}}{2} = -\frac{k}{2} \pm i\gamma \tag{8.187}$$

Analogously, we get:

$$U(\omega, t) = e^{-\left(\frac{k}{2}t\right)}[A(\omega)\cos(\gamma t) + B(\omega)\sin(\gamma t)] \qquad (8.188)$$

Applying the Fourier transform to the initial conditions we get $A(\omega)$; $B(\omega)$:

$$\left\{\begin{array}{l} U(\omega, 0) = F(\omega) = A(\omega) \\ \left.\dfrac{\partial U}{\partial t}\right|_{t=0} = \Psi(\omega) = -\dfrac{k}{2}A(\omega) + \beta B(\omega) \ \left(|\omega| \le \dfrac{k}{2c}\right) \\ \left.\dfrac{\partial U}{\partial t}\right|_{t=0} = \Psi(\omega) = -\dfrac{k}{2}A(\omega) + \gamma B(\omega) \ \left(|\omega| > \dfrac{k}{2c}\right) \end{array}\right\} \qquad (8.189)$$

Final solution:

$$u(x, t) = \frac{1}{\sqrt{2\pi}} \int_{-\infty}^{+\infty} U(\omega, t) e^{(i\omega x)} d\omega \qquad (8.190)$$

8.24 Case Study: Fourier Transform to Find the Distribution of Temperature in a Semi-Infinite Bar

Using the method of the Fourier transform find the stationary distribution of temperature $u(x, y)$ in a flat, semi-infinite bar (2-D) if the variation of temperature in the inner boundary is known: $u(x, 0) = f(x)$, and the rest of the boundaries are in contact with a thermal reservoir at zero temperature.

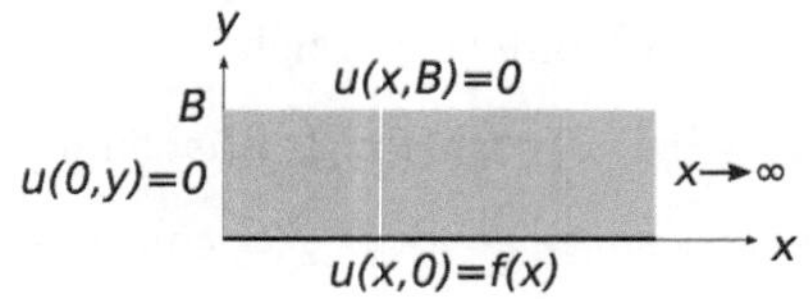

Figure 8.19

Formulation of the problem:

$$\left\{\begin{array}{l} \dfrac{\partial^2}{dx^2}u(x,y)+\dfrac{\partial^2}{dy^2}u(x,y)=0 \\ (0\le x<+\infty)\,;(0\le y\le b) \\ u(x,0)=f(x) \\ u(x,B)=0 \\ f(x\to\infty)\to 0 \\ u(x,y)<N\text{ (i.e., the solution is finite)} \end{array}\right\} \qquad (8.191)$$

One of the possible methods is to create a virtual image of the system for $x < 0$ and extend the boundary conditions anti-symmetrically there for the boundary at $y = 0$, this is, reflect anti-symmetrically the function $u(x,0) = f(x)$.

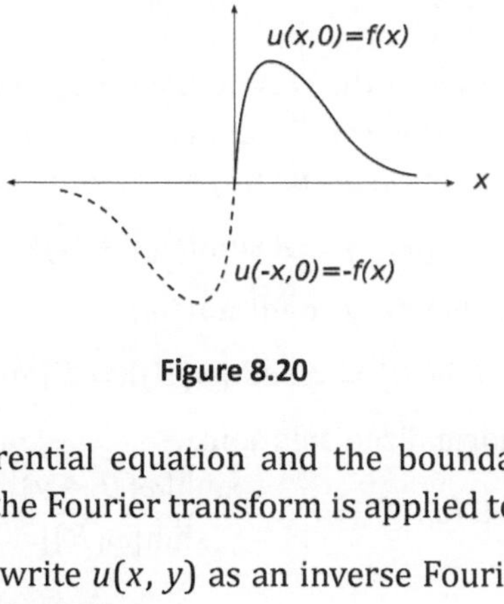

Figure 8.20

To find the differential equation and the boundary conditions in reciprocal space the Fourier transform is applied to convert $x \to \omega$.

Alternatively, we write $u(x,y)$ as an inverse Fourier transform and replace it into the equation:

$$u(x,y)=\frac{1}{\sqrt{2\pi}}\int_{-\infty}^{+\infty}U(\omega,y)e^{(i\omega x)}d\omega \qquad (8.192)$$

$$\frac{\partial^2 u}{\partial x^2}=\frac{1}{\sqrt{2\pi}}\int_{-\infty}^{+\infty}-\omega^2U(\omega,y)e^{(i\omega x)}d\omega \qquad (8.193)$$

$$\frac{\partial^2 u}{\partial y^2} = \frac{1}{\sqrt{2\pi}} \int_{-\infty}^{+\infty} \frac{\partial^2 U(\omega, y)}{dy^2} e^{(i\omega x)} d\omega \tag{8.194}$$

$$\frac{\partial^2 u}{dx^2} + \frac{\partial^2 u}{dy^2} = \frac{1}{\sqrt{2\pi}} \int_{-\infty}^{+\infty} \left[-\omega^2 U(\omega, y) + \frac{\partial^2 U(\omega, y)}{dy^2} \right] e^{(i\omega x)} d\omega = 0 \tag{8.195}$$

The equation to be solved in reciprocal space is:

$$\left\{ \begin{array}{l} \dfrac{\partial^2 U(\omega, y)}{dy^2} = \omega^2 U(\omega, y) \\ U(\omega, 0) = F(\omega) \\ U(\omega, B) = 0 \end{array} \right\} \tag{8.196}$$

From the first boundary condition we obtain its transform $F(\omega)$:

$$\frac{1}{\sqrt{2\pi}} \int_{-\infty}^{+\infty} U(\omega, 0) e^{(i\omega x)} d\omega = \frac{1}{\sqrt{2\pi}} \int_{-\infty}^{+\infty} F(\omega) e^{(i\omega x)} d\omega \tag{8.197}$$

Analogously we arrive at the second boundary condition $u(x, B) = 0 \rightarrow U(\omega, B) = 0$. This condition in reciprocal space allows us to write the general solution of (8.196) more transparently:

$$U(\omega, y) = A \sinh[\omega(B - y)] \tag{8.198}$$

Applying the first boundary condition (ω):

$$U(\omega, 0) = A \sinh[\omega(B)] = F(\omega) \tag{8.199}$$

we arrive at the normalized solution:

$$U(\omega, y) = F(\omega) \frac{\sinh[\omega(B - y)]}{\sinh[\omega(B)]} \tag{8.200}$$

Final solution:

$$u(x, y) = \frac{1}{\sqrt{2\pi}} \int_{-\infty}^{+\infty} F(\omega) \frac{\sinh[\omega(B - y)]}{\sinh[\omega(B)]} e^{(i\omega x)} d\omega =$$

$$= \frac{1}{2\pi} \int_{-\infty}^{+\infty} \left[\int_{-\infty}^{+\infty} f(z) e^{(-i\omega z)} dz \right] \frac{\sinh[\omega(B - y)]}{\sinh[\omega(B)]} e^{(i\omega x)} d\omega \tag{8.201}$$

8.25 Case Study: Application of the Fourier Transform to Find the Distribution of Temperature in a Semi-Plane

Using the Fourier Transform method find the stationary distribution of temperature $u(x, y)$ in a semi-plane if we know the variation of temperature at its boundary: $u(x, 0) = f(x)$.

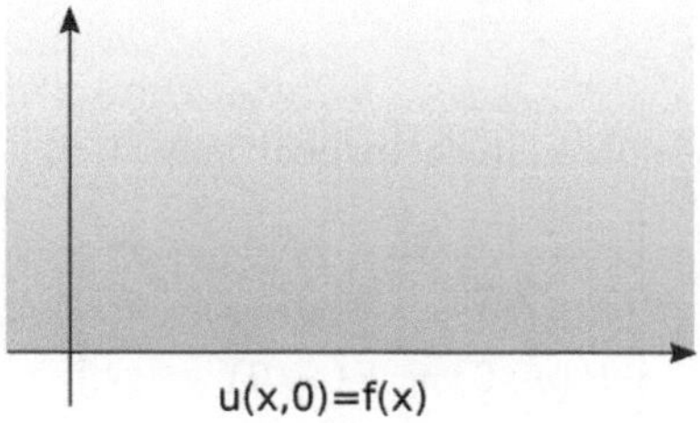

Figure 8.21

Mathematical formulation

$$\left\{\begin{array}{ll} \frac{\partial^2}{dx^2}u(x,y) + \frac{\partial^2}{dy^2}u(x,y) = 0 & (-\infty < x < +\infty); y > 0 \\ \text{Boundary condition:} & u(x,0) = f(x) \\ f(|x| \to \infty) \to 0 & \\ u(y \to \infty) \to 0; u(|x| \to \infty) \to 0 & \\ u(x,y) < N & \text{(The solution is finite)} \end{array}\right\} \quad (8.202)$$

We can apply the Fourier Transform to the x variable, leaving y as a parameter $(x \to \omega)$ and solve the problem in reciprocal space. This is possible since x is defined in the range $[-\infty, +\infty]$.

The alternative mode (with some more calculations, but equivalent) is to use $u(x, y)$ from the inverse Fourier Transform and replace the derivatives in the equation.

$$u(x,y) = \frac{1}{\sqrt{2\pi}} \int_{-\infty}^{+\infty} U(\omega, y) e^{(i\omega x)} d\omega \quad (8.203)$$

$$\frac{\partial^2 u(x,y)}{\partial x^2} = \frac{1}{\sqrt{2\pi}} \int_{-\infty}^{+\infty} -\omega^2 U(\omega, y) e^{(i\omega x)} d\omega \tag{8.204}$$

$$\frac{\partial^2 u(x,y)}{\partial y^2} = \frac{1}{\sqrt{2\pi}} \int_{-\infty}^{+\infty} \frac{\partial^2 U(\omega, y)}{dy^2} e^{(i\omega x)} d\omega \tag{8.205}$$

$$\frac{\partial^2 u(x,y)}{dx^2} + \frac{\partial^2 u(x,y)}{dy^2} = \frac{1}{\sqrt{2\pi}} \int_{-\infty}^{+\infty} \left[-\omega^2 U(\omega, y) + \frac{\partial^2 U(\omega, y)}{dy^2} \right] e^{(i\omega x)} d\omega = 0 \tag{8.206}$$

The equation to be solved in reciprocal space is:

$$\left\{ \begin{array}{l} \dfrac{\partial^2 U(\omega, y)}{dy^2} = \omega^2 U(\omega, y) \\ \\ U(\omega, 0) = F(\omega, 0) \end{array} \right\} \tag{8.207}$$

The boundary condition in reciprocal space (ω) has been obtained by transforming the boundary conditions in (x,0)

$$u(x, 0) = f(x) \tag{8.208}$$

Writing this relation using the inverse Fourier Transform:

$$\frac{1}{\sqrt{2\pi}} \int_{-\infty}^{+\infty} U(\omega, 0) e^{(i\omega x)} d\omega = \frac{1}{\sqrt{2\pi}} \int_{-\infty}^{+\infty} F(\omega, 0) e^{(i\omega x)} d\omega \tag{8.209}$$

The general solution is $U(\omega, y) = A(\omega)e^{(\omega y)} + B(\omega)e^{(-\omega y)}$

Due to the need that the solution be convergent for $y \to \infty$, we are going to impose $A(\omega) = 0$.

In this case we must use the modulus of the ω variable in the exponential variation of the solution to guarantee that the solution will not diverge, regardless of the sign of ω.

Using the other boundary condition (in reciprocal space) so that $y = 0$ we write the solution (in reciprocal space):

$$U(\omega, y) = F(\omega) e^{(-|\omega| y)} \tag{8.210}$$

Finally, the general form of the solution to the Laplace equation in the semi-plane (2-D) will be:

$$u(x, y) = \frac{1}{\sqrt{2\pi}} \int_{-\infty}^{+\infty} F(\omega) e^{(-|\omega| y)} e^{(i\omega x)} d\omega \tag{8.211}$$

To arrive at the more detailed form of the solution we must solve the integral by writing $F(\omega)$ as a Fourier Transform: $F(\omega) = \frac{1}{\sqrt{2\pi}} \int_{-\infty}^{+\infty} f(z)e^{(-i\omega z)}dz$

$$u(x, y) = \frac{1}{2\pi} \int_{-\infty}^{+\infty} \left[\int_{-\infty}^{+\infty} f(z)e^{(-i\omega z)}dz \right] e^{(-|\omega|y)}e^{(i\omega x)}d\omega \qquad (8.212)$$

Changing the order of integration:

$$u(x, y) = \frac{1}{2\pi} \int_{-\infty}^{+\infty} \left[\int_{-\infty}^{+\infty} e^{[-i\omega(x-z)]}e^{(-|\omega|y)}d\omega \right] f(z)dz \qquad (8.213)$$

The value of the integral is:

$$\int_{-\infty}^{+\infty} e^{[-i\omega(x-z)]}e^{(-|\omega|y)}d\omega = \frac{2y}{y^2 + (x-z)^2} \qquad (8.214)$$

So we arrive to so-called Poisson's integral formula:

$$u(x, y) = \frac{1}{\pi} \int_{-\infty}^{+\infty} \frac{y}{y^2 + (x-z)^2} f(z)dz \qquad (8.215)$$

Example of application:

We suppose the following distribution of temperature at the boundary:

$$f(x) = \left\{ \begin{matrix} T_0 \ (A < x < B) \\ 0 \ \ (x < A; x > B) \end{matrix} \right\} \qquad (8.216)$$

Then we will have:

$$u(x, y) = \frac{T_0}{\pi} \int_A^B \frac{y}{y^2 + (x-z)^2} dz = \frac{T_0}{\pi} \int_A^B \frac{y^2}{y^2 + (x-z)^2} d\left(\frac{z}{y}\right) =$$

$$= \frac{T_0}{\pi} \int_A^B \frac{1}{\left[1 + \frac{(z-x)^2}{y^2}\right]} d\left(\frac{z}{y}\right) = \left\{ \xi = \left(\frac{z-x}{y}\right) \right\}$$

$$= \frac{T_0}{\pi} \int_{\frac{A-x}{y}}^{\frac{B-x}{y}} \frac{1}{[1+\xi^2]} d\xi \left\{ \int \frac{1}{[1+\xi^2]} d\xi = \tan^{-1}(\xi) + \text{const} \right\}$$

$$= \frac{T_0}{\pi} \left[\tan^{-1}\left(\frac{B-x}{y}\right) - \tan^{-1}\left(\frac{A-x}{y}\right) \right] = \frac{T_0}{\pi}[\varphi_B - \varphi_A] \qquad (8.217)$$

Where $\varphi_{A,B}$ are angles of visibility of the points A and B from the point (x, y)

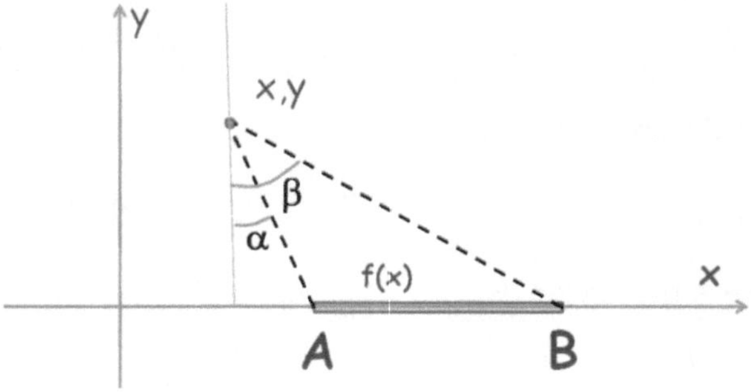

Figure 8.22

Note: The formulas of this type also exist in 3D coordinates to present the solutions of the Laplace equation in integral form.

8.26 Fourier Transform of the General Solution of Laplace's Problem in a Disk

Show that the Fourier Transform of the general solution of Laplace's problem in a disk has the form of an infinite sum of Dirac's Delta functions.

Solution:

The mathematical description of the solution of Laplace's problem in a disk is expanded in periodic angular functions in the azimuthal angle φ:

$$u(\rho, \varphi) = \sum_{n\geq 0}[A(n, \rho)\cos(n\varphi) + B(n, \rho)\sin(n\varphi)] \tag{8.218}$$

Applying the Fourier Transform to the solution we get:

$$\mathfrak{F}[u(\rho,\varphi)] = F(\rho,\omega) = \frac{1}{\sqrt{2\pi}}A(n,\rho)\sum_{n\geq 0}\left[\int_{-\infty}^{+\infty}\cos(n\varphi)e^{-i\omega\varphi}d\varphi\right] +$$

$$+\frac{1}{\sqrt{2\pi}}B(n,\rho)\sum_{n\geq 0}\left[\int_{-\infty}^{+\infty}\sin(n\varphi)e^{-i\omega\varphi}d\varphi\right] \quad (8.219)$$

Writing

$$\cos(n\varphi) = \frac{1}{2}[e^{+in\varphi} + e^{-in\varphi}] \quad (8.220)$$

$$\sin(n\varphi) = \frac{1}{2i}[e^{+in\varphi} - e^{-in\varphi}] \quad (8.221)$$

We get integrals of the type:

$$\int_{-\infty}^{+\infty} e^{+in\varphi}e^{-i\omega\varphi}d\varphi = \delta(n-\omega) \quad (8.222)$$

or

$$\int_{-\infty}^{-\infty} e^{-in\varphi}e^{-i\omega\varphi}d\varphi = \delta(n+\omega) \quad (8.223)$$

This implies that the Fourier Transform of the general solution of Laplace's problem in a disk is an infinite sum of Dirac's Delta type of functions. This is a consequence of applying the Fourier Transform to a function which is periodic every 2π.

Appendix

Dirac's Delta Function in Cylindrical and Spherical Coordinates

In spherical coordinates: $\delta(\vec{r}-\vec{r_0}) = \frac{1}{r^2 \sin(\theta)}\delta(r-r_0)\delta(\theta-\theta_0)\delta(\varphi-\varphi_0)$

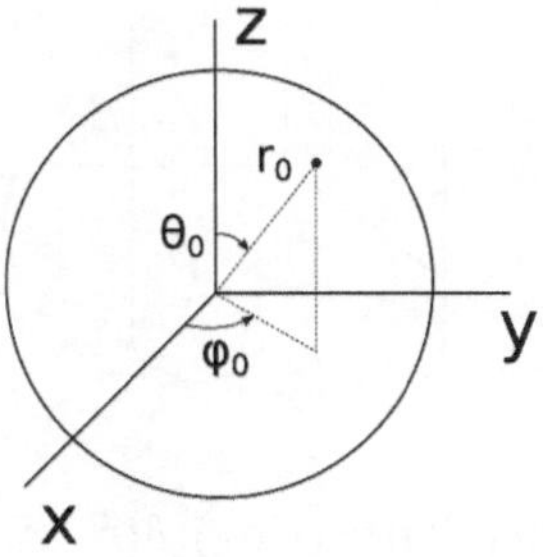

Without azimuthal dependence (ring): $\delta(\vec{r} - \vec{r_0}) = \frac{1}{2\pi r^2 \sin(\theta)}\delta(r - r_0)\delta(\theta - \theta_0)$

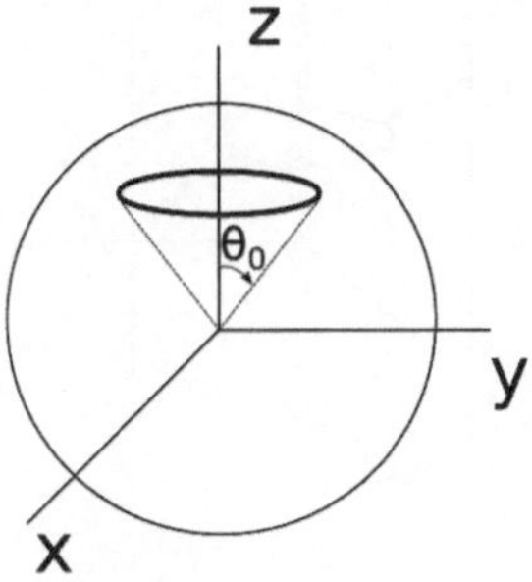

Without polar nor azimuthal dependence: $\delta(\vec{r} - \vec{r_0}) = \frac{1}{4\pi r^2}\delta(r - r_0)$

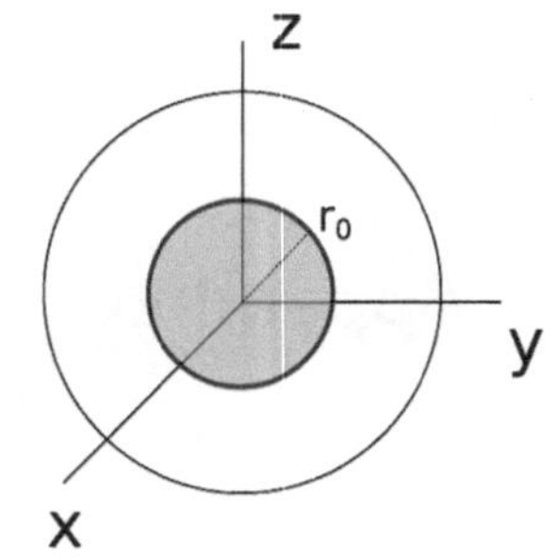

In cylindrical coordinates: $\delta(\vec{r}-\vec{r_0}) = \frac{1}{\rho}\delta(\rho-\rho_0)\delta(\varphi-\varphi_0)\delta(z-z_0)$

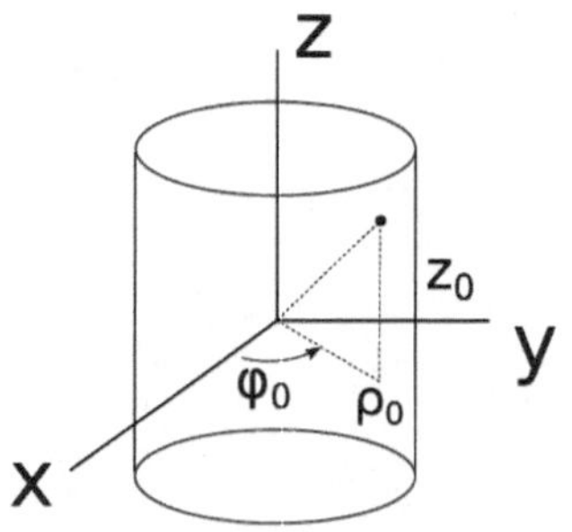

Without azimuthal dependence (ring): $\delta(\vec{r}-\vec{r_0}) = \frac{1}{2\pi\rho}\delta(\rho-\rho_0)\delta(z-z_0)$

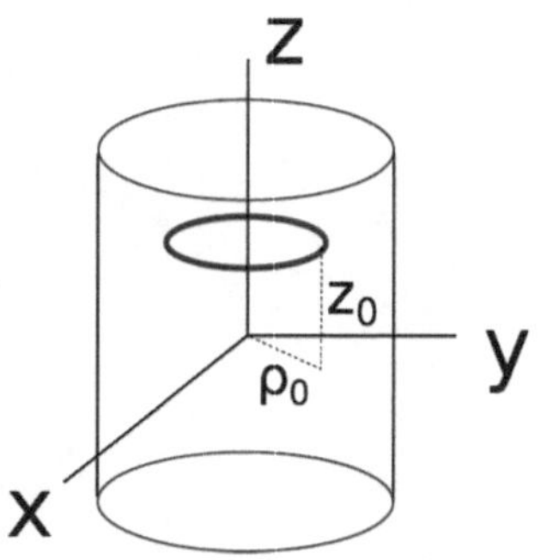

Without azimuthal nor polar dependence (very thin tube): $\delta(\vec{r}-\vec{r_0}) = \frac{1}{2\pi\rho}\delta(\rho-\rho_0)$

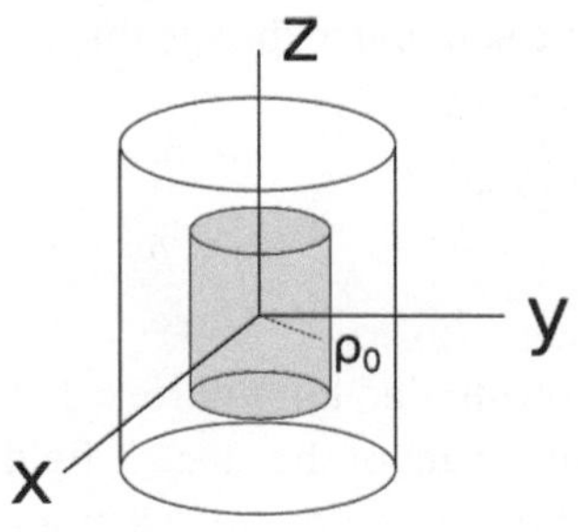

Physical Equations

- Transversal displacement

$$\frac{d^2u}{dt^2} - \frac{T}{\rho(x)}\frac{d^2u}{dx^2} = \frac{1}{\rho(x)}f(x,t)$$

Boundary conditions:

  - Fixed boundaries: $u(\Sigma, t) = 0$
  - Free boundaries: $\frac{du}{dx}(\Sigma, t) = 0$
  - Boundary linked to a spring: right: $T\frac{du}{dx}(L, t) + \beta u(L, t) = 0$; left: $T\frac{du}{dx}(0, t) - \beta u(0, t) = 0$

- Longitudinal oscillations along a rod

$$\frac{d^2u}{dt^2} - \frac{1}{\rho(x)}\frac{d}{dx}\left[E(x)\frac{du}{dx}\right] = \frac{1}{\rho(x)}f$$

Boundary conditions:

  - Fixed boundaries: $u(\Sigma, t) = 0$
  - Free boundaries or linked to a spring: section 5.1.2 of [1]

- Heat transport in three dimensions

$$C(\vec{n})\rho(\vec{n})\frac{dT}{dt} - \frac{d}{d\vec{n}}\left[\kappa(\vec{n})\frac{dT}{d\vec{n}}\right] = f(\vec{n}, t)$$

Boundary conditions:

  - Boundaries at zero temperature: $T(\Sigma, t) = 0$
  - Boundaries thermally insulated: $\frac{dT}{d\vec{n}}(\Sigma, t) = 0$
  - Thermal flux proportional to the temperature at the boundaries, with the temperature of the outer medium equal to zero: $\kappa(\Sigma)\frac{dT}{d\vec{n}}(\Sigma, t) + \alpha T(\Sigma, t) = 0$

- Note that the sign must change due to the change between the boundaries Σ

- Electrostatic potential

$$\Delta\phi = -\frac{\rho}{\epsilon\epsilon_0}$$

  - Electrically grounded boundaries: $\phi(\Sigma, t) = 0$
  - Normal component of the electric field when the boundary is a charged metal interfacing vacuum: $E_n = \frac{\rho}{\epsilon\epsilon_0}$
  - Some other conditions are described in [1]

- Gas pressure in three dimensions

$$\frac{d^2P}{dt^2} - a^2\Delta P = -a^2\nabla \cdot f$$

  - Open boundaries: $P(\Sigma, t) = P_{ext}$
  - Closed boundaries: $\frac{dP}{dn}(\Sigma, t) = 0$
  - For other types of boundary conditions see [1]

- Diffusion in one dimension

$$\frac{dn}{dt} - \frac{d}{dx}\left(D(x)\frac{dn}{dx}\right) = f(x, t)$$

Some Sturm–Liouville Problems for the Oscillations in a String

$$\frac{d^2X}{dx^2} + \lambda X = 0$$

- Both boundaries fixed:
  - Boundary conditions: $X(0) = X(L) = 0$
  - Eigenfunctions: $X_n(x) = \sin(\sqrt{\lambda_n}x)$
  - Eigenvalues: $\lambda_n = (\frac{n\pi}{L})^2$

- Both boundaries free:
  - Boundary conditions: $\left.\frac{dX}{dx}\right|_{x=0} = \left.\frac{dX}{dx}\right|_{x=L} = 0$
  - Eigenfunctions: $X_n(x) = \cos(\sqrt{\lambda_n}x)$
  - Eigenvalues: $\lambda_n = (\frac{n\pi}{L})^2$

- Free right boundary, fixed left boundary:
  - Boundary conditions: $X(0) = \left.\frac{dX}{dx}\right|_{x=L} = 0$
  - Eigenfunctions: $X_n(x) = \sin(\sqrt{\lambda_n}x)$
  - Eigenvalues: $\lambda_n = (\frac{2n+1}{2}\frac{\pi}{L})^2$
- Free left boundary, fixed right boundary:
  - Boundary conditions: $X(L) = \left.\frac{dX}{dx}\right|_{x=0} = 0$
  - Eigenfunctions: $X_n(x) = \cos(\sqrt{\lambda_n}x)$
  - Eigenvalues: $\lambda_n = (\frac{2n+1}{2}\frac{\pi}{L})^2$
- Right boundary linked to a spring, free left boundary:
  - Boundary conditions: $T\left.\frac{dX}{dx}\right|_{x=L} + \beta X(L) = 0; \left.\frac{dX}{dx}\right|_{x=0} = 0$
  - Eigenfunctions: $X_n(x) = \cos(\sqrt{\lambda_n}x)$
  - Eigenvalues: $\sqrt{\lambda_n} = \frac{1}{L}\left[\arctan\left(\frac{\beta}{T\sqrt{\lambda_n}}\right) + 2n\pi\right]$

Bibliography

Main texts

1. A. Levanyuk, A. Cano. *Métodos Matemáticos de la Física. Método de Fourier* (UAM, 2006).
2. G. B. Arfken, H. Weber, F. E. Harris. *Mathematical Methods for Physicists*, 7th ed. (Academic Press, 2013).
3. A. N. Tijonov, A. A. Samarski. *Ecuaciones de la Física Matemática* (Pueblo y Ciencia, 1975).
4. B. M. Budak, A. D. Samarski, A. N. Tijonov. *Problemas de la Física Matemática v. 1,2* (McGraw-Hill, Mir, 1984).
5. N. H. Asmar, *Partial Differential Equations with Fourier Series and Boundary Value Problems*, 2nd ed. (Pearson, 2005).

Complementary texts

6. V. P. Pikulin, S. I. Pohozaev. *Equations in Mathematical Physics. A Practical Course* (Springer, 2001).
7. C. H. Edwards, D. E. Penney, with the assistance of D. Calvis, *Elementary Differential Equations* (Pearson, 2015).
8. C. H. Edwards, D. E. Penney, *Ecuaciones diferenciales y problemas con valores en la frontera* (Pearson, 2009).
9. R. Haberman. *Applied Partial Differential Equations with Fourier Series and Boundary Value Problems* (Pearson, 2013).
10. *Handbook of Mathematical Functions with Formulas, Graphs and Mathematical Tables* (Editors: M. Abramowitz and I. A. Stegun, 1965).
11. M. R. Spiegel y L. Abellanas, *Fórmulas y tablas de matemática aplicada*, serie Schaum (McGraw-Hill, 1991).

Extra references

12. D. G. Zill, M. R. Cullen, *Matemáticas Avanzadas para la ingeniería, v.1: Ecuaciones Diferenciales, Tercera Edición* (McGraw-Hill Interamericana 2008).

13. W. E. Boyce, R. C. DiPrima, *Elementary Differential Equations and Boundary Value Problems* (Wiley, 2012).
14. L. C. Evans, *Partial Differential Equations: Second Edition, Graduate Studies in Mathematics, v. 19* (American Mathematical Society, 2010).
15. D. G. Zill, M. R. Cullen, *Differential Equations with Boundary-Value Problems*, 7th ed. (Brooks/Cole, Cengage Learning 2009).

Index